Structural Integrity

Volume 22

Series Editors

The *Structural Integrity* book series is a high level academic and professional series publishing research on all areas of Structural Integrity. It promotes and expedites the dissemination of new research results and tutorial views in the structural integrity field.

The Series publishes research monographs, professional books, handbooks, edited volumes and textbooks with worldwide distribution to engineers, researchers, educators, professionals and libraries.

Topics of interested include but are not limited to:

- Structural integrity
- Structural durability
- Degradation and conservation of materials and structures
- Dynamic and seismic structural analysis
- Fatigue and fracture of materials and structures
- Risk analysis and safety of materials and structural mechanics
- Fracture Mechanics
- Damage mechanics
- Analytical and numerical simulation of materials and structures
- Computational mechanics
- Structural design methodology
- Experimental methods applied to structural integrity
- Multiaxial fatigue and complex loading effects of materials and structures
- Fatigue corrosion analysis
- Scale effects in the fatigue analysis of materials and structures
- Fatigue structural integrity
- Structural integrity in railway and highway systems
- Sustainable structural design
- Structural loads characterization
- Structural health monitoring
- Adhesives connections integrity
- Rock and soil structural integrity.

**** Indexing: The books of this series are submitted to Web of Science, Scopus, Google Scholar and Springerlink ****

This series is managed by team members of the ESIS/TC12 technical committee.

Springer and the Series Editors welcome book ideas from authors. Potential authors who wish to submit a book proposal should contact Dr. Mayra Castro, Senior Editor, Springer (Heidelberg), e-mail: mayra.castro@springer.com

More information about this series at http://www.springer.com/series/15775

Joshua Pelleg

Cyclic Deformation in Oxides, Carbides and Nitrides

Alumina, Magnesia, Yttria, SiC, B_4C and Si_3N_4

Springer

Joshua Pelleg
Department of Materials Engineering
Ben-Gurion University of the Negev
Beer Sheva, Israel

ISSN 2522-560X ISSN 2522-5618 (electronic)
Structural Integrity
ISBN 978-3-030-86117-9 ISBN 978-3-030-86118-6 (eBook)
https://doi.org/10.1007/978-3-030-86118-6

This Springer imprint is published by the registered company Springer Nature Switzerland AG
The registered company address is: Gewerbestrasse 11, 6330 Cham, Switzerland

Ideas you do not spell out loud and clearly
may be lost in the misty abyss.

Joshua Pelleg

To my wife Ada and my children, Deenah and her late husband Gidon Barak, Ruth and Christer Kallevag, Shlomit and Asher Pelleg and to my grandchildren: Roy, Tal, Rotem and Noa Barak; Ella and Maya Kallevag; and Ofir and Ori Pelleg.

Preface—Cyclic Deformation in Oxides, Carbides and Nitrides

In the light of ceramics becoming more and more important technologically in various industrial applications and in the extensive research activity in both pre-industrial and post-research endeavors, it is felt that the time is ripe for collecting and summarizing current knowledge in a book concentrating on the subject. The ceramics of significant technical importance are the oxides, carbides and nitrides. Of these ceramic groups the following pure unalloyed or reinforced compounds were chosen: alumina (Al_2O_3), magnesia (MgO), yttria (Y_2O_3), zirconia (ZrO_2), SiC (possibly B_4C) and Si_3N_4. Of these only zirconia contains a stabilizing agent, the yttria, to avoid its decomposition and hence the term yttria stabilized zirconia (YSZ). For complete stabilization at least 8% yttria is essential.

The purpose of this book is to present and test the fatigue properties of these relatively new structural materials produced by various techniques currently available for the production of ceramics and explore their various applications.

The chapters considered in this book are outlined in the content. Accordingly, Chap. 1 considers the basic concepts of fatigue in the ceramic compounds outlined above. The S-N curve and the endurance limit characterizing these compounds is discussed in Chap. 2. Fatigue is also known as cyclic deformation, and the number of cycles stress related to the stress applied without failure is generally presented in S-N curves, where the letters S and N refer to stress and the number of cycles. Within the applied cycles, it is customary to consider three cases known as low cycle, high cycle and very high cycle all related to the concept "stress cycles" which is continued over to Chap. 3 because they are directly related to the S-N curves (Chap. 1) and the endurance limit (Chap. 2). To understand cyclic deformation, it is in particular important to be aware of the fatigue life of the constituents considered in this book, in particular because about 90% of all failures occur by fatigue. Chapter 4 is devoted to this subject. Clearly, deformation is directly associated with dislocations, but an overall discussion on dislocations related to fatigue is postponed to Chap. 8. However, work hardening (or softening), a strictly and directly related phenomenon to the effects of dislocations present in the material, not only in static deformation, but also in fatigue, is discussed in Chap. 5. Fatigue induces structural changes in material undergoing cyclic deformation. Outstanding feature observed is the hysteresis, which shows the variation of the stress–strain relation on the stress reversal. The hysteresis loop obtained by loading in the plastic region above the yield stress, unloading and reloading in the reverse direction results in a lower yield stress level than the original level on loading. This topic is covered in Chap. 6. During the process of loading and unloading and the application of the cyclic cycles, a cumulative damage is generally observed as outlined in Chap. 7. The origin of the damage may start from a micro-crack formation already during underloading growing to observable cracks at accelerated growth rate they coalescence, and the induced damage leads to failure by unexpected fracture of the specimen. Overloading can cause a retardation or arrest of crack growth, but the effect depends on if the applied stress is tensile or compressive. The delaying effect caused by overloading is related to (a) the residual stresses at the crack tip, (b) crack tip closure and (c) crack tip blunting. Fatigue life and the associated structural changes observed in fatigued specimens comprises Chaps. 8 and 9. Included in Chap. 9 are the important factors of stress, grain and temperature that strongly influences the work hardening. Significant structural changes are the progression markings (beach marks) and the striations and importantly those changes which are dislocations related (see Chap. 8). Fatigue fracture and the temperature effect on fracture for some of the ceramics considered in this book are discussed in Chap. 10. The notch effect is considered in Chap. 11. Fatigue usually starts at the surface, which generally is not smooth (never perfect, and machining results and preparation effects are always present), where micro-notches can act as stress-raisers which reduces the fatigue strength. Regardless of the origin of stress when cycling is applied, fatigue damage may result. Premature failure resulting from cyclic stresses due to temperature changes is related to the concept of thermal fatigue which might occur in high temperature components. Chapters 12 and 16 is devoted to thermal effects. Chap. 11 considers the important environmental effect on fatigue and discusses this problem in detail. Surface treatments such as shot pinning, sand blasting and laser treatment are dealt with in Chap. 13. One cannot emphasize sufficiently the importance of surface treatment, since as known fatigue failure is initiated at the surface. Modern technology applications pivot around nanomaterials, and as such, fatigue consideration is of utmost interest. It is discussed in Chap. 14. Often components fail by static fatigue, and it should be considered as indeed done in Chap. 15. Factors reducing resistance to failure have to be spelled out in design in pre-application of a component in technological use. Thus, design has to be carried out in such a way that the major factors leading to reduced fatigue resistance should be considered.

In the closing, I would like to express my gratitude to all publishers and authors for permission to use and reproduce some of their illustrations and microstructures.

Finally, without the tireless devotion, help, understanding and unlimited patience of my wife Ada, I could never have completed this book, despite my decades of teaching in this field; her encouragement was essential, and her helpful attitude was instrumental in inspiring to write this book.

Beer Sheva, Israel Joshua Pelleg

Contents

About the Author

Joshua Pelleg received his B.Sc. in Chemical Engineering at the Technion—Institute of Technology, Haifa, Israel, M.Sc. in Metallurgy at the Illinois Institute of Technology, Chicago, IL, and Ph.D. in Metallurgy at the University of Wisconsin, Madison, WI. He has been in the Ben-Gurion University of the Negev (BGU) Materials Engineering Department in Beer-Sheva, Israel, since 1970, and was among the founders of the department and served as its second chairman. He was the recipient of the Samuel Ayrton Chair in Metallurgy. He specializes in the mechanical properties of materials and the diffusion and defects in solids. He has chaired several university committees and served four terms as the chairman of Advanced Studies at Ben-Gurion University of the Negev. Prior to his work at BGU, he acted as an assistant professor and then associate professor in the Department of Materials and Metallurgy at the University of Kansas, Lawrence, KS. He was also a visiting professor: in the Department of Metallurgy at Iowa State University; at the Institute for Atomic Research, US Atomic Energy Commission, Ames, IA; at McGill University, Montreal, PQ; at the Tokyo Institute of Technology, Applied Electronics Department, Yokohama, Japan; and in Curtin University, Department of Physics, Perth, Australia. His non-academic research and industrial experience includes: Chief Metallurgist in Urdan Metallurgical Works Ltd., Netanya, Israel; Research Engineer in International Harvester Manufacturing Research, Chicago, IL; Associate Research Officer for the National Research Council of Canada, Structures and Materials, National Aeronautical Establishment, Ottawa, ON; Physics Senior Research Scientist, Nuclear

Research Center, Beer-Sheva, Israel; Materials Science Division, Argonne National Labs, Argonne, IL; Atomic Energy of Canada, Chalk River, ON; Visiting Scientist, CSIR, National Accelerator Centre, Van de Graaf Group Faure, South Africa; Bell Laboratories, Murray Hill, NJ; and GTE Laboratories, Waltham, MA. His current research interests are mechanical properties, diffusion in solids, thin film deposition and properties (mostly by sputtering) and the characterization of films, among them various silicides. His research activity is summarized in ten books published by Springer and Elsevier.

Chapter 1
Basic Concepts of Fatigue

Abstract Basic concepts of fatigue in oxides, carbides and nitrides are the subject of this chapter. Cyclic cycles are illustrated, and the fundamental equations are presented. S–N relations discussed in the section are important aspect of cyclic deformation. The most common oxides, carbides and nitrides discussed are alumina, magnesia, zirconia and silicon nitride.

Keywords S–N · Cycles · Partially stabilized zirconia · Stress range · Stress amplitude · Mean stress · Stress ratio

The basic concept pivots around the S–N relation and the various factors affecting it to be discussed in the coming chapters. S stands for the stress and N for the number of cycles which determine the fatigue lifetime at a specific stress to which a material (specimen) is exposed during service. The origin of the applied stress is of no consequence as long as it is cyclic, thus thermal fluctuations (thermal stress) or vibrations (aircraft wings for example) can induce fatigue failure in a component. The main issue is that the stress applied is acting alternately in reverse direction in a forward and reversed cycle. The cycle can be tension-tension, tension compression or compression-compression. Various wave forms of cyclic stress may be actually acting in practice or applied to a specimen for testing its fatigue resistance under a set of varying factors predetermined for the test. An example of stress cycles to which machine components might be exposed in industrial applications is illustrated in Fig. 1.1. As mentioned in the preface about 90% of all failures occur by fatigue fracture with a characteristic surface appearance. The fracture, regardless of if the material is ductile or brittle has a smooth cleavage-like surface, characterizing brittle fracture and failure occurs without any plastic deformation and usually without warning. On the surface progression markings known in the professional jargon as beach marks appear as seen in Fig. 1.2. Other features such as striations will be discussed in later chapters.The reason for fatigue fracture initiation on the surface is due to the local stress concentration as a consequence of the inherent small local discontinuities resulting from the fabrication process. Among the other surface effects which can induce fatigue failure one should note environmental influences such as corrosion and hydrogen-induced cracking, welding process defects, porosities which might cause lack of local fusion if the pores are of sufficient size and also

J. Pelleg, *Cyclic Deformation in Oxides, Carbides and Nitrides*, Structural Integrity 22,
https://doi.org/10.1007/978-3-030-86118-6_1

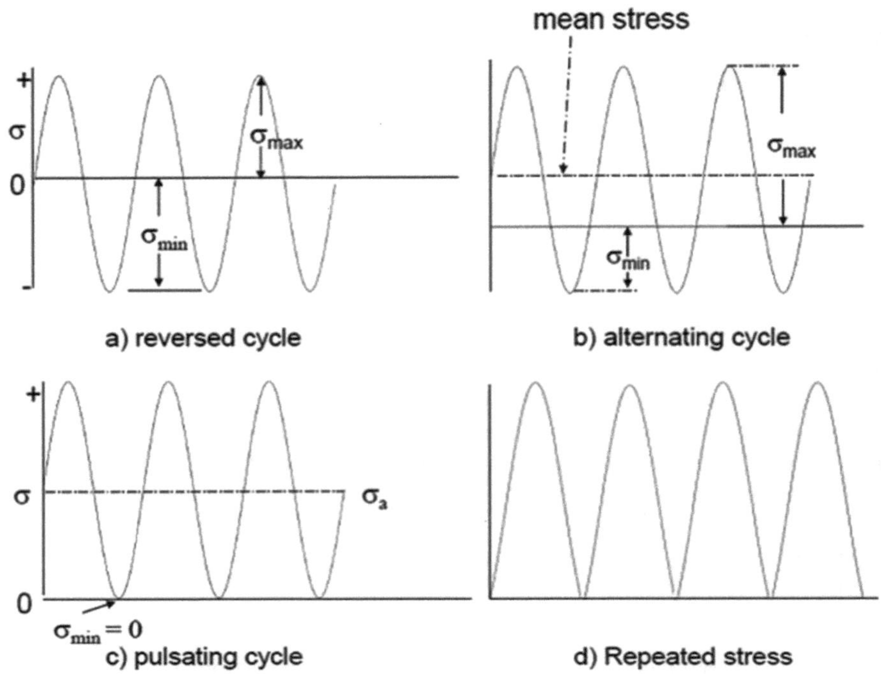

Fig. 1.1 Various forms of cyclic stresses. **a–c** Are sinusoidally varying cycles; **d** represents a repeating cycle

Fig. 1.2 SEM fractograph of a fractured specimen with fatigue beach marks and crack origin region, Bhaumik et al. [1]. With kind permission of Elsevier

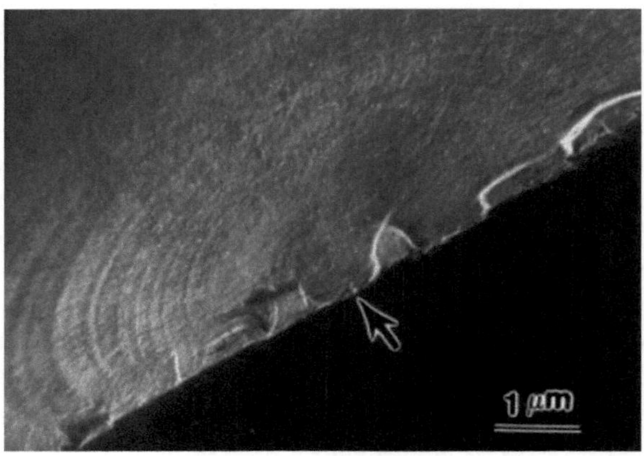

defects induced following welding. They directly are to be blamed for premature fracture of material at smaller number of cycles at a given applied stress or for a given cycle at lower applied stress since their presence determines the potential location for initiation of a fatigue crack. This is the reason why a material, and specifically

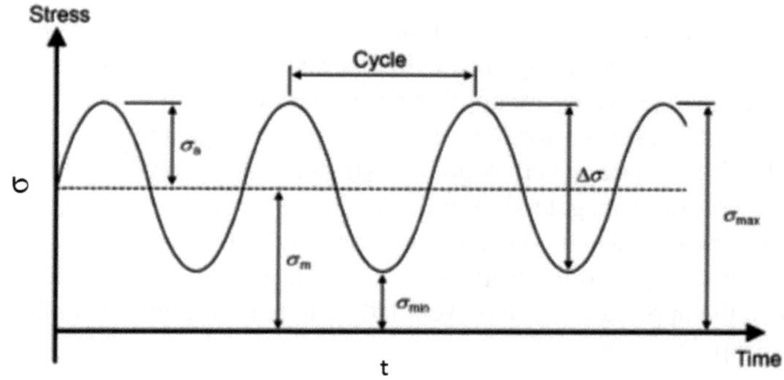

Fig. 1.3 Several cycles in a fatigue stress cycle

a test specimen preparation should be done with the utmost care to provide smooth surfaces, preferentially by electro-polishing.

Coming back to Fig. 1.1 the following relations can be defined [2] as illustrated for the stress variation with time in Fig. 1.3:

$$\text{stress range:} \quad \Delta\sigma = \sigma_{max} - \sigma_{min} \tag{1.1}$$

$$\text{stress amplitude:} \quad \sigma_a = \frac{\sigma_{max} - \sigma_{min}}{2} \tag{1.2}$$

$$\text{mean stress:} \quad \sigma_m = \frac{\sigma_{max} + \sigma_{min}}{2} \tag{1.3}$$

$$\text{stress ratio:} \quad R = \frac{\sigma_{min}}{\sigma_{max}} \tag{1.4}$$

$$\text{amplitude ratio:} \quad A = \frac{\sigma_a}{\sigma_m} = \frac{1 - R}{1 + R} \tag{1.5}$$

In Fig. 1.3 an oscillatory stress is superimposed on the mean stress (dashed line).

It is the belief of the author that a few general remarks which are a common knowledge will help the reader to be acquainted with the subject getting the taste of fatigue even before further reading of this book. Some of them are enumerated below:

1. Structures subjected to repeated cyclic loadings can undergo progressive damage which shows itself by the nucleation and propagation of cracks. This damage is called fatigue. It manifests itself by the loss of resistance to failure with time.
2. Two regions are commonly considered in fatigue: (a) low cycle fatigue up to about 10^4 cycles where plasticity is observed and fatigue life is related to the growth of a small fatigue crack under a general-yielding cyclic loading, and

(b) high cycle fatigue requiring more than 10^4 (10^4–10^8) cycles to failure; the stress is low and primarily elastic. In this case small elastic strains under a high number of load cycles are applied before failure occurs.

3. Fatigue failures, both for high and low cycle, follow the same basic steps of crack initiation, stage I crack growth, stage II crack growth, and finally failure.
4. These cracks propagate slowly during stage I crack growth along crystallographic planes, where shear stresses are highest. Once they reach a critical size quick propagation during stage II crack growth in a direction perpendicular to the applied force and eventually can lead to failure usually in a brittle catastrophic mode.
5. At large enough growth rate the crack becomes large enough, inducing striations formations observed on the fractured surface. Each striation represents the growth from one loading cycle.
6. Ferrous metals and Ti alloys show a definite fatigue limit exhibiting a horizontal line known as the endurance limit below which continued loading does not lead to fatigue failure, namely, the material can endure a large number of stress cycles.
7. Fast fracture sets in when the stress exceed a critical value known as the fracture toughness. The process is characterized at this stage by microvoid coalescence.
8. Fatigue performance is characterized—as mentioned above—by the S–N curve, also known as a Wöhler curve, which is a plot of the cyclic stress (S) against the number of cycles, N on a logarithmic scale.
9. The fatigue lifetime of a material can be determined by:

    (a) Stress-life method
    (b) Strain-life method
    (c) Crack-growth method

10. Because the large scatter in the results between tests, many specimens have to be tested for averaging the results.

The topics listed above will be considered below in details. One has to emphasize that the items listed have to be considered very carefully before or at designing specimens or rather components for good fatigue resistance and long cycle lifetime.

1.1 S–N Relations

The general presentation of the S–N curve (also known as Wöhler curve) is illustrated in Fig. 1.4. Two types of S–N curves are illustrated one with a horizontal line representing ferrous metals and Ti alloys and a continuously decreasing non-linear curve which represents most if not all the other materials. In the ferrous and Ti alloys a well-defined endurance limit is unambiguously observed by the horizontal line, below which no failure occurs by the applied stress for a very large number of cycles. The S–N curves are obtained from tests on samples (often called as coupons)

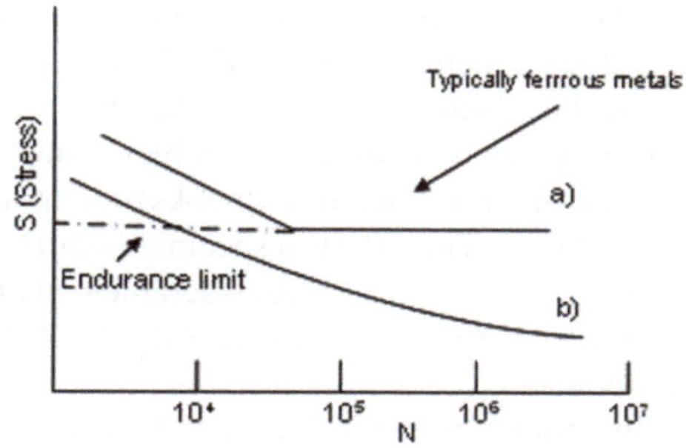

Fig. 1.4 S–N curves: **a** with a well-defined endurance limit, **b** without a definite fatigue limit

of the material to be characterized by applying a proper stress, say sinusoidal and counting the cycles to failure. Each coupon (sample) test provides a point on the plot resulting in the S–N curve. There is a run-out where the specimen does not fail after applying a large number of cycles, namely the time to failure exceeds the specified numbers of cycles, which is usually at least 10^7 cycles. Thus, a **runout** is a **test** which is interrupted before it fails, namely discontinued after 10^7–10^8 cycles. Fatigue run-out specimens (up to the mentioned load cycles of ~5×10^7) are used to determine the stress level named "fatigue limit".

S–N data are obtained either on smooth or notched specimens until failure. The usual technique is to test the first specimen at a high stress where failure is expected to occur within a relatively small number of cycles. Then the test stress is decreased for each succeeding specimen tested until one or two stress specimens do not fail at the specific cycle applied, usually 10^7 cycles. The highest stress at which non-failure (run-out) occurs is taken as the fatigue threshold or endurance limit. As mentioned above, non-ferrous alloys do not have an endurance limit, and for these materials tests are terminated after the arbitrarily chosen ~$10^8 - 5 \times 10^8$ cycles. An experimental illustration showing both cases the one with a fatigue threshold and one without it is shown in Fig. 1.5. As seen, it is about of the same form and shape as the schematic curves in Fig. 1.4.

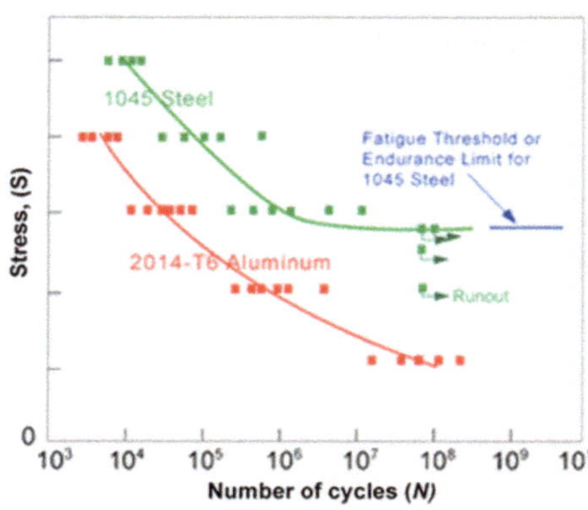

Fig. 1.5 S–N plot for 1045 steel and 2014-T6 aluminum. Center for Nondestructive Evaluation, Iowa State University, Ames, Iowa 50011, USA

There might be deviations from the S–N curves illustrated in Fig. 1.4 since the actual specimens are never flawless. To represent actual service conditions, one has to take into consideration the state of the surface; for example, pitting from corrosion, or porosity might influence the S–N relation and the mentioned curve (commonly illustrated in textbooks) does not represent precisely the actual in-service S–N relation. Thus, the test specimens will have significantly different fatigue performance. Further, there is a considerable scatter in fatigue data, therefore usually a large number of actual tests are performed to obtain reliable S–N curves for the design of components.

In addition to the indicated S–N relation, namely to the stress-controlled fatigue test, cyclic strain-controlled tests are also common in particular when the cycling is thermal (fluctuations in the operating temperature) when the component expands and contracts. In such test the strain amplitude is held constant during cycling. Stain controlled low-cycle fatigue tests on extruded $AZ_{31}B$ Mg alloy (with repeated tests at each strain level) is shown in Fig. 1.6. In the strain-life diagram of the extruded $AZ_{31}B$ Mg alloy, as the total strain amplitude decreases the number of cycles can be increased as seen in the figure, with run-out at about 0.1 $\Delta\varepsilon_t/2$ and no failure occurred at 10^7 cycles. The lifetime is usually predicted by the Coffin-Manson relationship which is expressed by Eq. (1.6)

$$\frac{\Delta\varepsilon_p}{2} = \varepsilon_f^{'}(2N_f)^c \tag{1.6}$$

Here c is the fatigue ductility exponent and $\varepsilon_f^{'}$ is the fatigue ductility coefficient. For high cycle fatigue tests Basquin's equation is used given as

$$\sigma_a = \frac{\Delta\varepsilon_e}{2}E = \sigma_f^{'}(2N_f)^b \tag{1.7}$$

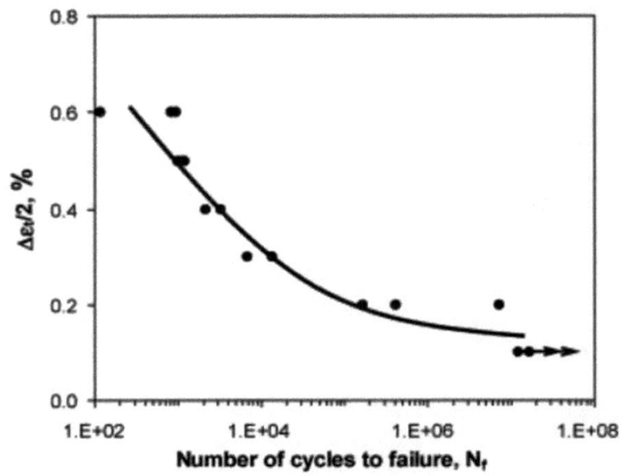

Fig. 1.6 Fatigue lifetime of the extruded $AZ_{31}B$ magnesium alloy tested at room temperature, Lin and Chen [3]. With kind permission of Elsevier

where b is the fatigue strength exponent and $\sigma_f^{'}$ is the fatigue strength coefficient. SEM images on the fatigue crack initiation site—all from the specimen surface—, and propagation zone of specimens tested at different strain amplitudes are illustrate in Fig. 1.7. The area of the crack increased with decreasing total strain amplitude propagation (dashed line on the images). A log scale is almost always used for N in the S–N curve. However, S–N test data can be displayed on a log–log plot, with the actual S–N line representing the mean of the data from several tests. Such a plot—idealized S–N curve—is shown in Fig. 1.8, where the stress amplitude is plotted against the number of cycles to failure. Since in most cases the line shown in Fig. 1.8 is linear, a power law can define the S–N relationship as

$$N_1 = N_2 \left(\frac{S_1}{S_2} \right)^{\frac{1}{b}} \tag{1.8}$$

where b is the slope of the line, and it is referred to as the Basquin slope, which is obtained from Eq. (1.8) as

$$b = \frac{-(\log S_1 - \log S_2)}{\log N_2 - \log N_1} \tag{1.9}$$

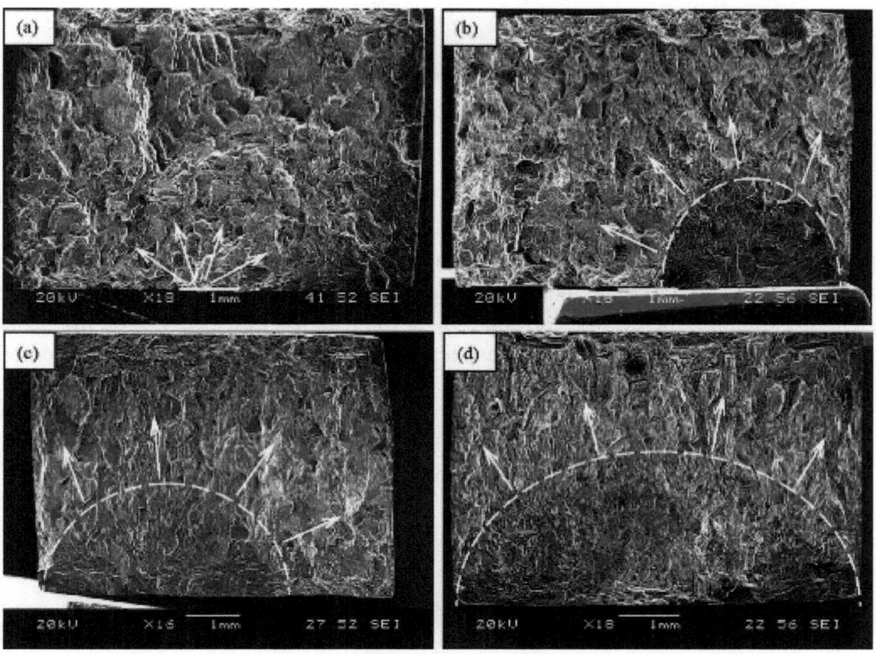

Fig. 1.7 Typical low-magnification SEM images showing the overall view of the fracture surfaces at the total strain amplitudes of **a** 0.6%, **b** 0.4%, **c** 0.3%, and **d** 0.2%, Lin and Chen [3]. With kind permission of Elsevier

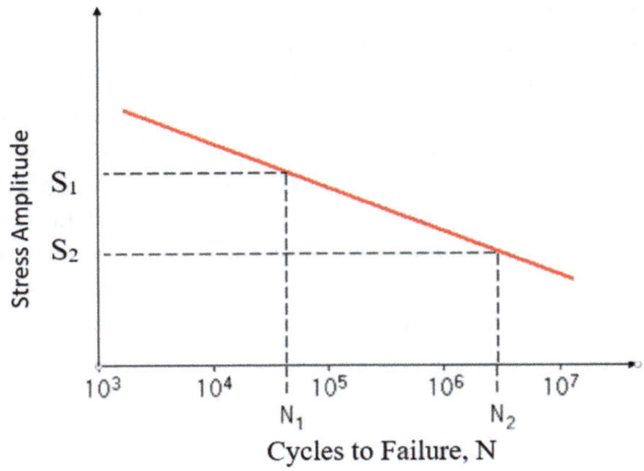

Fig. 1.8 The cycles to failure for specific stress amplitudes, S_1 and S_2

On using a logarithmic scale relative effects (change in %) are obtained, and a linear scale for obtaining absolute (change in units). On double logarithmic scale the mean value of the test results can be expressed in a range of cycles more generally than Eqs. (1.8)–(1.9) as

$$N = C\Delta\sigma^{-m} \tag{1.10}$$

where N is the number of cycles of stress range $\Delta\sigma$, C a constant (representing structural details), $\Delta\sigma$ constant amplitude stress range. Express Eq. (1.10) logarithmically to obtain

$$\log N = \log C - m\log(\Delta\sigma) \tag{1.11}$$

Several empirical relationships relate the alternating stress to the mean stress. The corresponding curves connect the endurance limit on the alternating axis to either the yield strength, S_y, the ultimate strength S_u or the true fracture stress σ_f on the mean stress axis as seen in Fig. 1.9. The respective equations of the curves are presented as

$$\text{Goodman} \quad \frac{S_a}{S_e} + \frac{S_m}{S_u} = 1 \tag{i}$$

$$\text{Gerber} \quad \frac{S_a}{S_e} + \left(\frac{S_m}{S_u}\right)^2 = 1 \tag{ii}$$

$$\text{Söderberg} \quad \frac{S_a}{S_e} + \frac{S_m}{S_y} = 1 \tag{iii}$$

$$\text{Morrow} \quad \frac{S_a}{S_e} + \frac{S_m}{\sigma_f} = 1 \tag{iv}$$

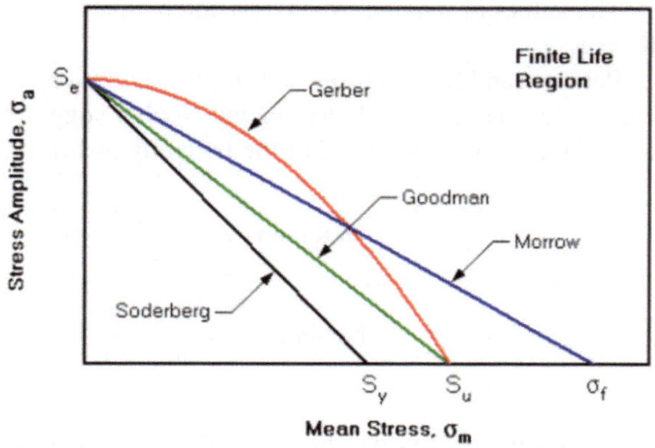

Fig. 1.9 Comparison of mean stress equations

In the above S_e, S_m, S_a, S_u, S_y and σ_f are the stress are the endurance level, mean stress, stress amplitude, ultimate strength, yield strength and the fracture stress, respectively.

It was indicated already that tests are also performed on notched specimens. The above relations of stress-life analysis are based on the assumption that specimens are smooth, namely unnotched. However, actual fatigue failures occur at flaws-whatever might be the origin-mostly at the surface acting as stress risers at these notch-like locations. The stress at these locations acting as stress risers, is much greater than the nominal stress, S. i.e., these are local stresses, let us refer to them as S_{max}. The ratio of these stresses, the theoretical stress concentration factor, K_T is given as

$$K_T = \frac{S_{\max}}{S} \tag{1.12}$$

The effect of notches is considered by the fatigue notch factor, K_f (also known as fatigue stress concentration factor) and is related to the unnotched fatigue strength (in ferrous alloys and Ti it is the endurance limit) as

$$K_f = \frac{S_e(unnotched)}{S_e(notched)} \tag{1.13}$$

In most cases the fatigue notch factor is less than the stress concentration factor, and is less than 1, thus,

$$1 \leq K_f \leq K_T \tag{1.14}$$

The notch factor, K_f is dependent on the type of material and notch size, and therefore a factor, the notch sensitivity factor takes these into account, which is given as

$$q = \frac{K_f - 1}{K_T - 1} \tag{1.15}$$

The values of q range from 0 (no notch effect, $K_f = 1$) to 1 (full theoretical effect, $K_f = K_T$).

Peterson and Neuber proposed empirical relationship to determine q by correlating with experimental data. Peterson and Neuber suggested, instead the peak stress, the stress at a given point ahead of the notch tip as Eqs. (1.16) and (1.17), respectively

$$\text{Peterson} \quad q = \frac{K_f - 1}{K_T - 1} = \frac{1}{1 + \frac{c_p}{\rho}} \tag{1.16}$$

$$\text{Neuber} \quad q = \frac{K_f - 1}{K_T - 1} = \frac{1}{1 + \sqrt{\frac{c_N}{\rho}}} \tag{1.17}$$

where ρ is the notch tip radius, c_p and c_N are material constants, respectively.

From Eq. 1.13 one obtains Eq. (1.18). The notch factor can be used to correct the fatigue strength (the S–N curve) for the notched specimen.

$$S_e = \frac{S_e'}{K_f} \tag{1.18}$$

A few basic relations, such as Paris' equation are presented below to be considered very carefully before or at designing specimens for good fatigue resistance and long cycle lifetime. The Paris equation is a crack growth relation, and since crack growth leads to failure if not healing occurs and, familiarization of the rate of growth of a fatigue crack is important for designing fatigue resistant components or parts. The basic formula of the Paris law takes into consideration the stress intensity factor, K which characterizes the load around a crack tip and is given as

$$\frac{\mathrm{d}a}{\mathrm{d}N} = C\Delta K^m \tag{1.19}$$

where a is the crack length, $\frac{\mathrm{d}a}{\mathrm{d}N}$ is the crack growth rate C and m are constants that depend on the material, environment and stress ratio. The range of the stress intensity factor during the fatigue cycle, ΔK is given by

$$\Delta K = K_{max} - K_{min} \tag{1.20}$$

1.1.1 S–N Relation in Alumina

Impact fatigue behavior of isostatically pressed, fully sintered, fine grained alumina was investigated for fatigue behavior. The endurance limit based on S–N relations and the fatigue resistance were estimated. Repeated impact tests using a special apparatus with a swinging pendulum was instrumented for repeated blows up to failure. The resulting number of impacts was registered thus obtaining N. The impact energy is expressed by

$$E = mgh(1 - \cos\theta) \tag{1.21}$$

where m is the mass of the hammer (hammer weight = 0.2538 kg), g is the gravitational constant, h is the height of the pendulum arm (pendulum length = 28.3 cm) and θ is the angle of release. Single-blow impact strength is obtained by impacting from a high angle, and for repeated impact tests the angle was set at decreasing intervals of 5° below the high angle required for a single impact breaking strength. Impact frequency was set at 8 impacts/min. The collected data (Table 3 of Maity et al.) were plotted in the usual way as an S–N curve shown in Fig. 1.10. The plot in Fig. 1.11 is presented on both axes in logarithmic scale. The S–N curve can be represented by a straight line up to about 10^4 cycles of impact and the corresponding stress is 270 MPa. The line can be expressed as

$$\sigma^n \mathrm{N} = \text{constant} \tag{1.22}$$

In this specials experiment, σ is impact stress and N is the number of impacts to failure. The resulting n which can be obtained from the slope is unusually low at a level of

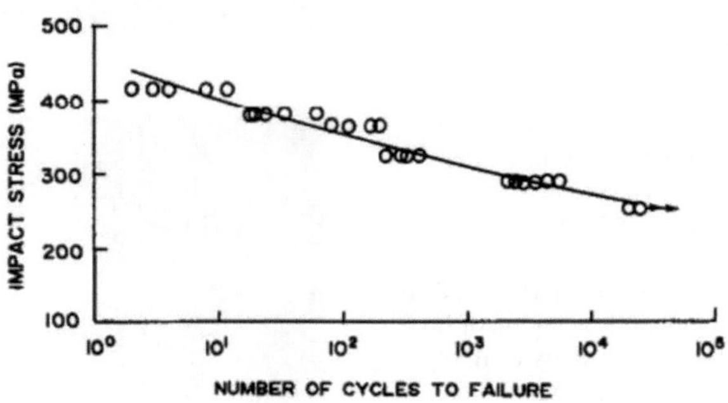

Fig. 1.10 Stress-endurance plot of sintered fine-grained alumina. Points with arrows indicate unbroken samples for which tests were terminated, Maity et al. [4]. With kind permission of Elsevier

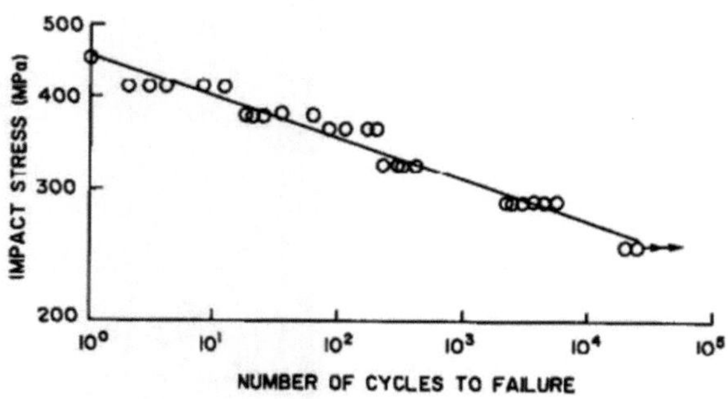

Fig. 1.11 S–N curve including single impact strength, Maity et al. [4]. With kind permission of Elsevier

17.12 compared which is different than the value reported in the literature (between 12.7 and 64). The microstructure of the fine-grained alumina before the fatigue test is shown in Fig. 1.12. The fractured surface after fatigue is seen in Fig. 1.13. Point (A) on the top of the specimen indicates the fracture origin. Smooth mirror region (B) radially centered around the fracture initiation characterizes the fatigue failure,

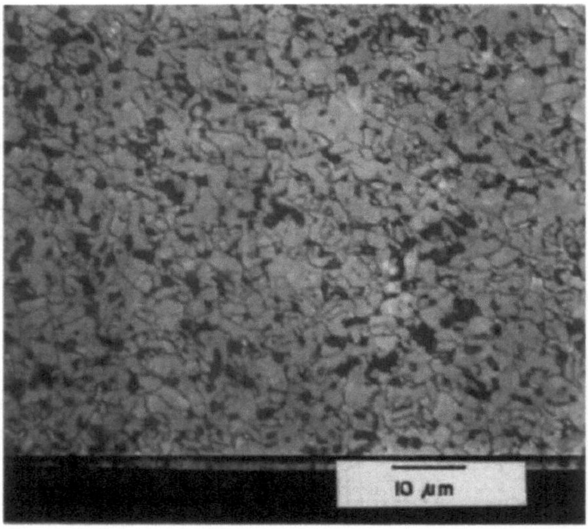

Fig. 1.12 Typical microstructure of fine-grained alumina used, Maity et al. [4]. With kind permission of Elsevier

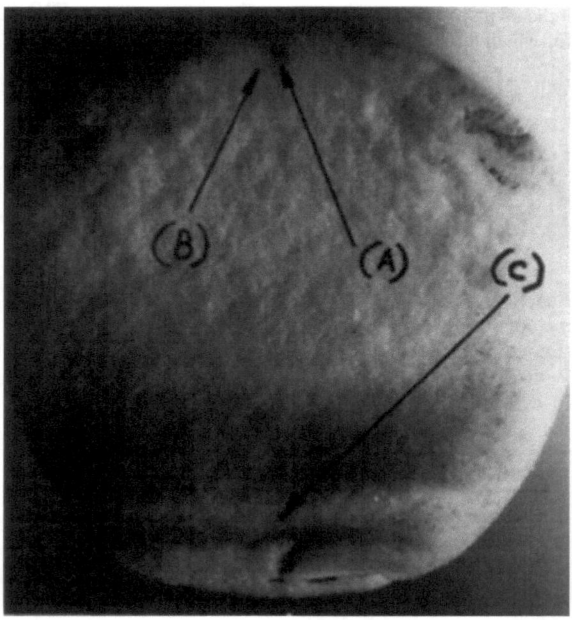

Fig. 1.13 Macrophotograph of fractured surface after fatigue test (σ = 288.65 MPa, N = 2864 impacts). (A) = point of crack initiation, (B) = mirror region, (C) = region of lip formation, Maity et al. [4]. With kind permission of Elsevier

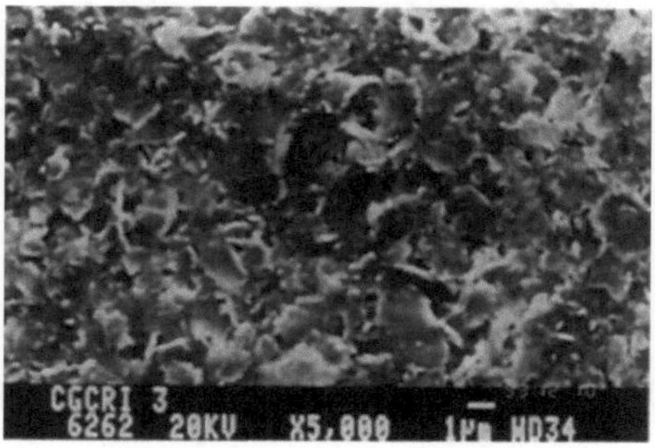

Fig. 1.14 Magnified view of the mirror region B (Fig. 1.13), indicating a severe transgranular fracture pattern, Maity et al. [4]. With kind permission of Elsevier

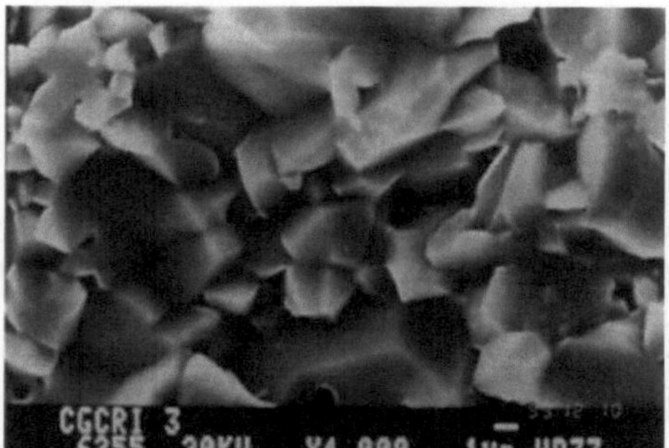

Fig. 1.15 Magnified view of region C (Fig. 1.13), showing mainly intergranular fatigue crack propagation path, Maity et al. [4]. With kind permission of Elsevier

and (C) the catastrophic failure where lip formation is observed. Magnified views of (B) and (C) of Fig. 1.13 are shown in Figs. 1.14 and 1.15, respectively. In summary, the fatigue limit of 270 MPa and the fatigue resistance of 17.12 are higher than the values found in coarse-grained alumina. It was also observed that in the fast crack growth region near the fatigue crack initiation point, the crack is transgranular, while in the slow crack growth region they are intergranular.

1.1.2 S–N Relation in Magnesia

Surprisingly no direct work on fatigue life of MgO is currently presented in the literature, despite the fact that ceramic materials are widely used in various fields of engineering because of their many excellent properties such as high hardness, high melting point, resistance against oxidation and chemical corrosion, and high-temperature stability. Nevertheless, indirect indication of the flexural strength against the number of cycles has been reported in the framework of the investigation of the susceptibility of ceramic materials to thermal shock, which limits their application. Such a plot is presented in Fig. 1.16 where among the many material MgO was also shown (in blue). The results indicate that the flexural strength of most of the ceramic materials declines markedly after single thermal cycle as shown in

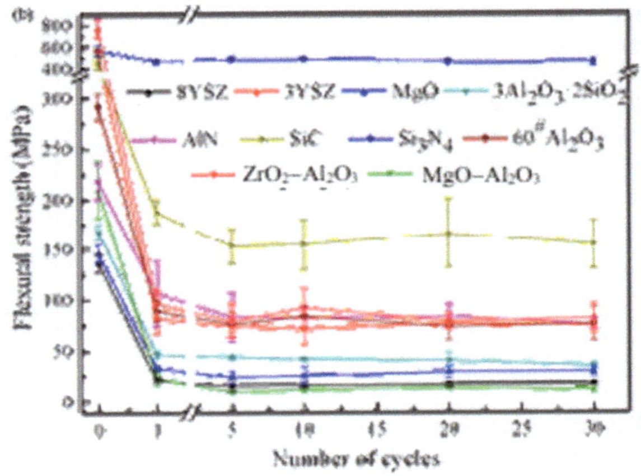

Fig. 1.16 Retained strength of investigated materials as a function of number of thermal cycles at temperature difference of 600 °C, Li et al. [5]. Open access

Fig. 1.16. However, after five thermal cycles the flexural strength does not degrade as the number of cycles increase (apparently their damage saturation level has been reached). One has to note that in MgO as in most ceramics, the coefficient of thermal expansion (CTE) is the most critical factor affecting the thermal shock and leading to failure. Clearly, other factors such as porosity and fracture toughness also affect thermal shock.

In another work Si implanted MgO is compared with unimplanted MgO in order to get information on the fatigue resistance on cycling, which is considered as the major cause for failure of ceramics. A low-cycle thermal fatigue at high cycle temperatures is of a particular problem in ceramics since high temperatures during cycling reduce the resistance to plastic deformation producing cycling plastic strain that can lead to the development of fatigue cracks.

Magnesia single crystals were used in the experiments. Samples of freshly cleaved (100) surface was irradiated with Si^- 90 keV ions with a dose of 5.0×10^{16} cm^{-2}, and unimplanted ones are tested under identical thermal cyclic conditions. A typical illustration of the temperature variation during and after thermal cycling is seen in Fig. 1.17. The fatigue life has been measured as a number of thermal cycles, made at the same peak cycle temperature, prior to an appearance of fatigue fracture. The measurements made for various peak cycle temperatures are presented in Fig. 1.18. As expected, the characteristic feature of S–N curves is seen. It is seen that with decreasing stress the number of cycles to fracture increases. Ion implantation of MgO crystals with the Si^- ions increases their fatigue life and fatigue limit. Also indicated in the curve, that for the ion implanted area 400 thermal cycles are insufficient to reach fracture at the cycle peak temperature of 700 °C which is apparently close to the fatigue limit of the ion implanted material. Contrary to this for the unimplanted area 300 cycles are sufficient to produce fatigue fracture at the peak cycle temperature of 675 °C. The number of cracks were used as the criterion for evaluating the number of thermal cycles prior to fracture. To eliminate nonuniform defect distribution four cracks appearance was taken as the criterion which proved to be more reliable for

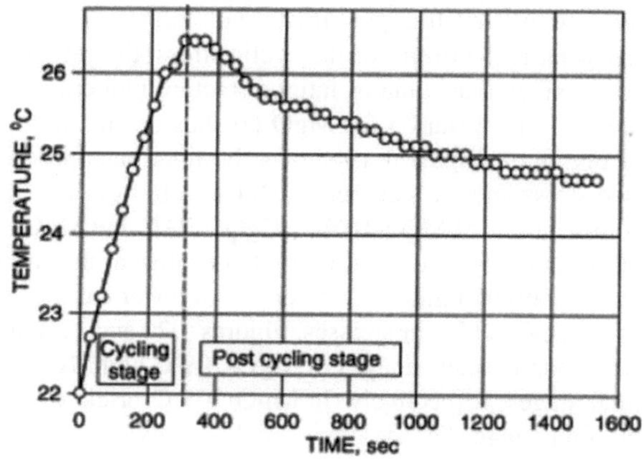

Fig. 1.17 Variation of sample temperature during thermal cycling and subsequent cooling, Gurarie et al. [6]. With kind permission of Elsevier

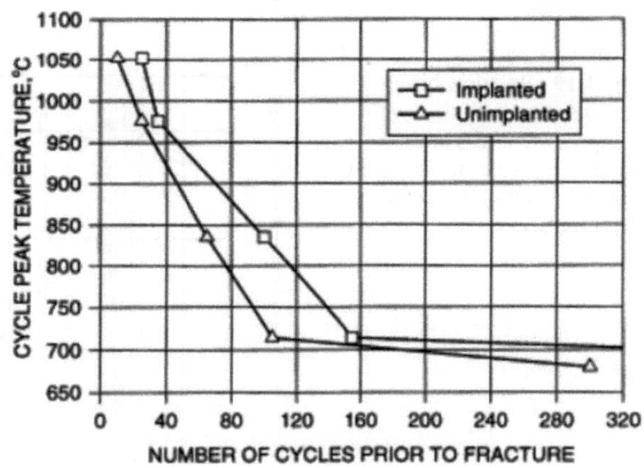

Fig. 1.18 Fatigue curves for implanted and unimplanted MgO crystals, Gurarie et al. [6]. With kind permission of Elsevier

the evaluation of the fatigue resistance than the one based on the appearance of a single crack. The higher resistance of the implanted crystals to cycling fatigue is related to the implantation-induced lattice damage. A high density of dislocations and other defects is produced which suppress the dislocation mobility. This (the decreased dislocation mobility) is associated with higher yield stress and hence the elastic strain per cycle and the reduction in plastic deformation strain.

The lack of sufficient direct experimental work on the fatigue life of MgO, necessitated to use other materials based on MgO. Such a material is MgO with C addition. The cycling fatigue is often considered as a major cause of failure of ceramics and presents one of the major concerns in the design of reliable ceramic components. This material is important in steel industry, where MgO bricks (+C) are used as furnace (converter) and ladle lining and damage can occur by cracks generated by cyclical changes of stress (thermal in nature). For example, the ladles are repeatedly filled with molten metal and emptied again and thus the lining is under repeated long term thermal cycling (and fracture of ceramics). As mentioned earlier, it is immaterial what is the origin of the cyclic stress, the fatigue failure by cyclic stress has to be considered as a major fatigue fracture inducing factor. A typical example of the temperature variation for MgO crystals during and after thermal cycling is shown in Fig. 1.19. Figure 1.19 shows the relationship between the maximum compressive stress and the number of cycles to failure (i.e., the S–N curve) for bricks with compositions of MgO-10%C, MgO-15%C and MgO-20%C, respectively. The slope of this curve becomes increasingly negative as the carbon content decreases. In other words, fatigue failure of bricks with a lower carbon content occurs after fewer load cycles, even at lower stresses. Figures 1.20 and 1.21 show X-ray CT scan (computed tomography scan) images of MgO-20%C and MgO-10%C bricks in the middle and final stages, respectively. In MgO-20%C brick, no large cracks are observed until the final stage. On the other hand, there were some observable cracks even in the middle stage in MgO-10%C brick, and the propagation and branching of relatively large cracks were also observed in the final stage.

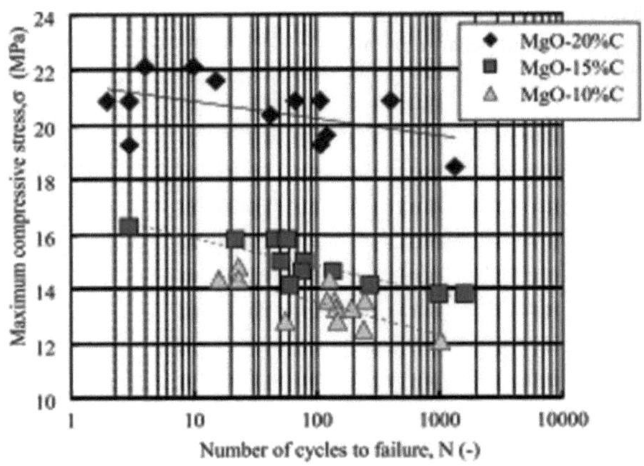

Fig. 1.19 Relationship between maximum compressive stress and number of cycles to failure in MgO–C bricks, Hino et al. [7]. Free access

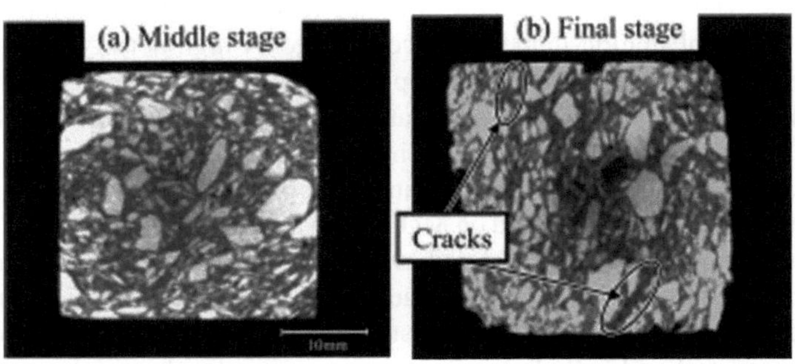

Fig. 1.20 X-ray CT images of MgO-20%C brick during fatigue failure test (**a** middle stage and **b** final stage), Hino et al. [7]. Free access

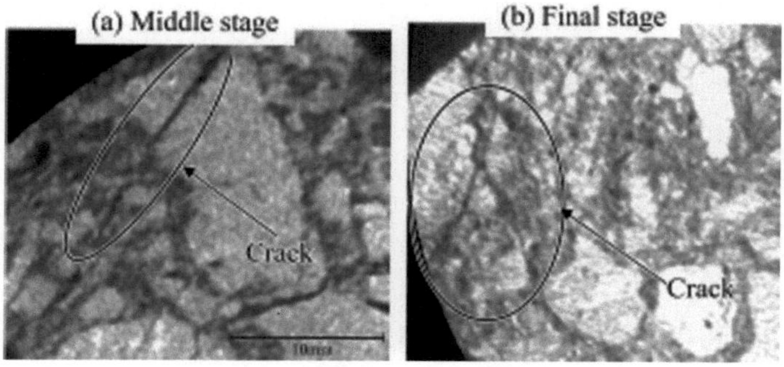

Fig. 1.21 X-ray CT images of MgO-10%C brick during fatigue failure test (**a** middle stage and **b** final stage), Hino et al. [7]. Free access

In summary, the fracture toughness of MgO–C bricks decreases as the carbon content decreases, and the fracture toughness of MgO–C brick is lower at high temperature of 1473 K than at room temperature. In inspecting X-ray CT scans no large cracks were observed until just before fracture in fatigue failure tests of higher C content MgO. On the other hand, in tests of bricks with a lower carbon content, large cracks were observed even in the middle of the tests. Thus, in particular, it was found that the fatigue failure of MgO–C bricks is unaffected by atmospheric temperature, and that the fatigue failure life under the same acting stress ratio decreases as the carbon content in the brick decreases.

1.1.3 S–N Relation in Zirconia

S–N relation of monolithic ZrO_2 and with ZrO_2/Nb ZrO_2/Ta is illustrated in Figs. 1.22 and 1.23. The stress represents flexural strength in MPa. In order to predict the fatigue life of the indented materials with equivalent initial flaw size, indentation load of 98 N and 294 N were chosen for monolithic ceramic and for ceramic/metal composites, respectively. The initial flaw size was about 335 μm for all studied compositions. The indentation fatigue life for disc specimens is presented in semi logarithmic form as peak stress versus cycles to failure in Fig. 1.24. The indented specimens were tested under the same cyclic conditions as the un-indented ones. Any flaw present in monolithic zirconia under very low subcritical loading can lead to catastrophic failure. This is shown by the black arrow in Fig. 1.24 indicating the number of cycles where failure occurs already at around 170 MPa when the crack is about 450 μm.

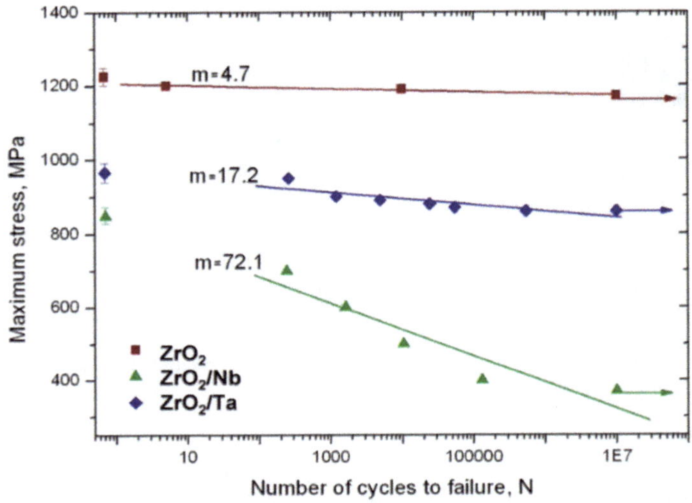

Fig. 1.22 Fatigue resistance S–N curves with values of slopes for un-indented polished specimens, Smirnov et al. [8]. Open access

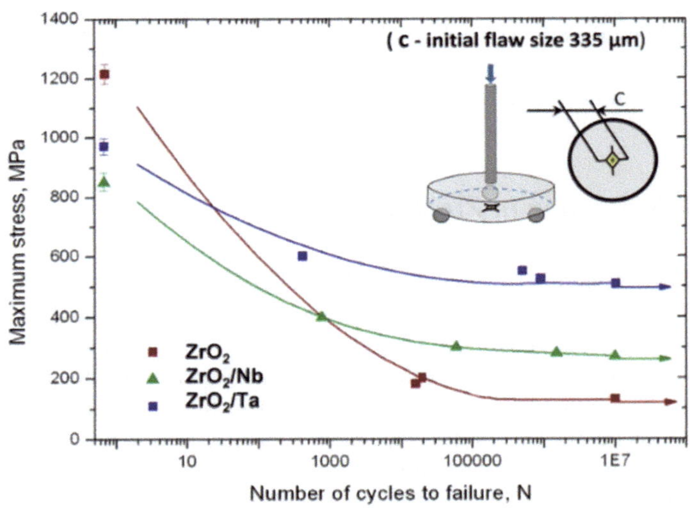

Fig. 1.23 S–N plot of the cycles to failure (N) of indented ZrO_2, ZrO_2/Ta and ZrO_2/Nb composites, Smirnov et al. [8]. Open access

1.1.3.1 Yttria Stabilized Zirconia

Zirconia is an unstable oxide, and it has to be stabilized. The most common stabilizer is yttria and for full stabilization at least 8 mol% is required. Partial stabilized

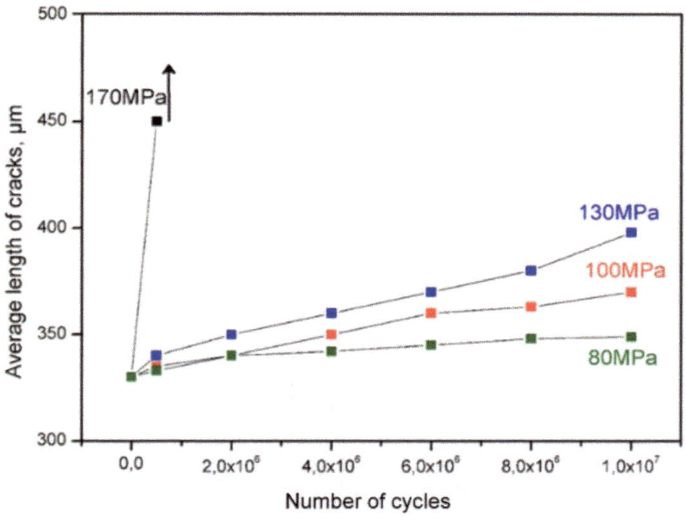

Fig. 1.24 Number of cycles and crack length as a function of the applied load for ZrO_2, Smirnov et al. [8]. Open access

zirconia (YPSZ) contains at least 3 mol% yttria having a tetragonal structure and the polycrystals are also known as TZP. Stabilization involves a transformation of tetragonal metastable remains cubic in a wide temperature range. (XRD) patterns of the ZrO_2 samples, before and after sintering at 1600 °C and of the fracture surface after bending testing indicate tetragonal phase and residual monoclinic phase (about 13 vol %), respectively. Figure 1.25 presents a micrograph of the polished and etched ZrO_2 surface and the fracture surface of a specimen grain growth of ZrO_2 grains has been observed in this material. In Fig. 1.25(b) a typical brittle fracture surface is

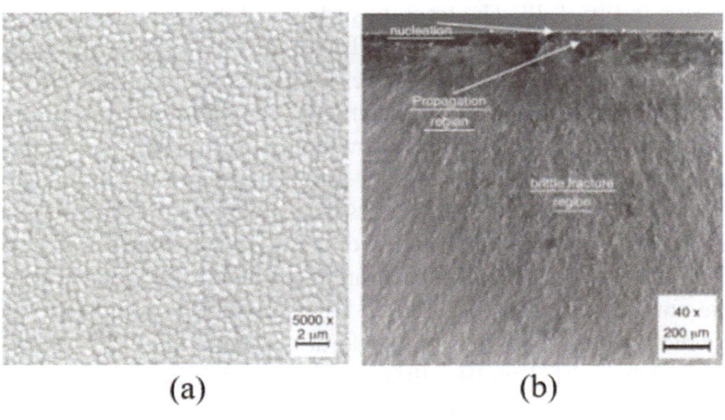

Fig. 1.25 Micrograph of the ZrO_2 samples: **a** polished surface, **b** fracture surface, Souza et al. [9]. With kind permission of Elsevier

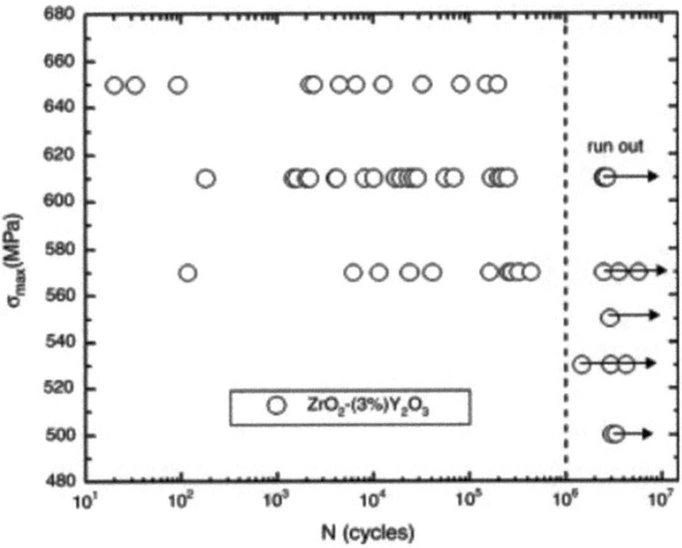

Fig. 1.26 Cyclic fatigue tests results of the ZrO_2 samples: $\sigma_{max} \times N$ curves, Souza et al. [9]. With kind permission of Elsevier

seen, which clearly shows that the initial crack nucleation and propagation region are located in the upper side of the illustration, corresponding to the region of maximum tensile stress in the bending test. The curves resulting from the cyclic fatigue tests are shown in Fig. 1.26 in terms of the stress-life curve which correlates to the maximum bending stress level, σ_{max} to the number of cycles to failure, N_f. The tests were interrupted at $N_f = 2 - 5 \times 10^6$ cycles, where it is estimated to provide infinite life. The specimens which did not fracture in the test are marked by an arrow symbol (run out). Polished surfaces of SEM micrographs of specimens that survived fatigue testing are seen in Fig. 1.27. The fracture surfaces of the sintered samples fractured during fatigue tests are shown in Fig. 1.28. The rupture in fatigue starts in the polished tensile surface and occurs in a brittle mode, which depends on the critical flaw size. For its propagation, this flaw must overcome the compression stresses generated by the tetragonal–monoclinic phase transformation.

In summary, cyclic fatigue tests by four-point bending tests resulted in the stress-life $\sigma \times N$ curve. The experimental results indicate that 3Y-TZP ceramics suffered cyclic fracture. The analysis (see Souza et al.) indicates that the material suffers tetragonal-monolithic phase transformation due to cyclic as well as static loading. The fatigue strength limit is ~550 MPa. High scattering of the fatigue data was observed, which is not surprising. In an additional work uniaxial tension–compression fatigue behavior of 3-mol%-yttria-stabilized tetragonal zirconia polycrystals was investigated. In Fig. 1.29 cumulative plastic strain is plotted against fatigue cycle at various Rs. Recall R was defined in Eq. 1.4 as the ration of σ_{min} to σ_{max}.. Note in this figure, that the strain accumulation slows down with increasing cycles.

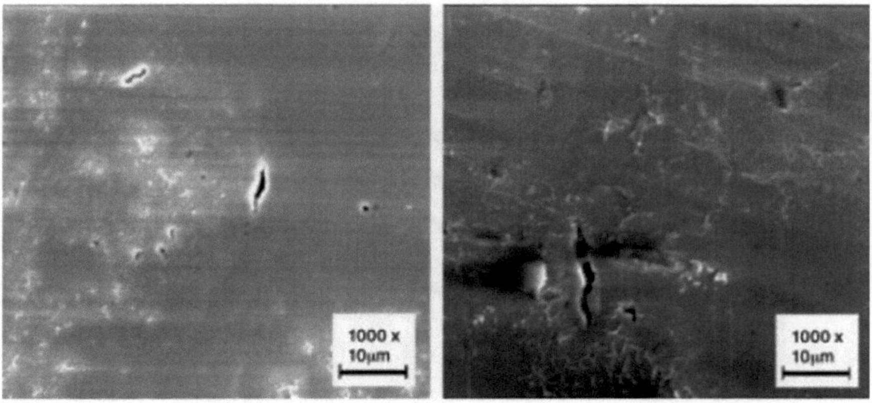

(a) (b)

Fig. 1.27 SEM micrographs of the polished surface in survived samples after fatigue bending testing: **a** 530 MPa; **b** 610 MPa, Souza et al. [9]. With kind permission of Elsevier

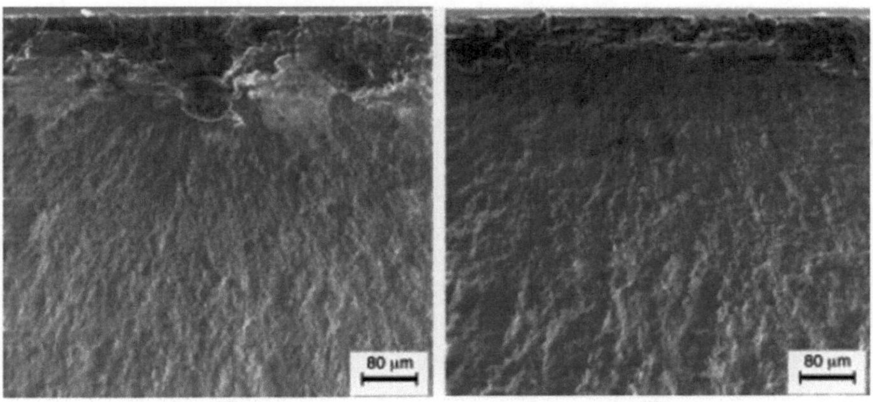

(a) (b)

Fig. 1.28 Fracture surface of the ZrO_2 specimens after fatigue testing: **a** $N = 1000$ cycles; **b** $N =$ 200,000 cycles (100×), Souza et al. [9]. With kind permission of Elsevier

The S–N curves plotted as cycle to failure, N_f versus the stress amplitude increasing cycles. The S–N curves plotted as cycle to failure, N_f versus the stress amplitude are shown in Fig. 1.30 on a log–log scale. This form is a modification of the conventional S–N curve, which is normally plotted on a semilog scale. These curves are actually straight lines with the same slope for all the R ratios. Thus, these data can be represented by the following expression:

$$\log N_f = -m \log \sigma_a + \log\left(\frac{Z}{A}\right) \tag{1.23}$$

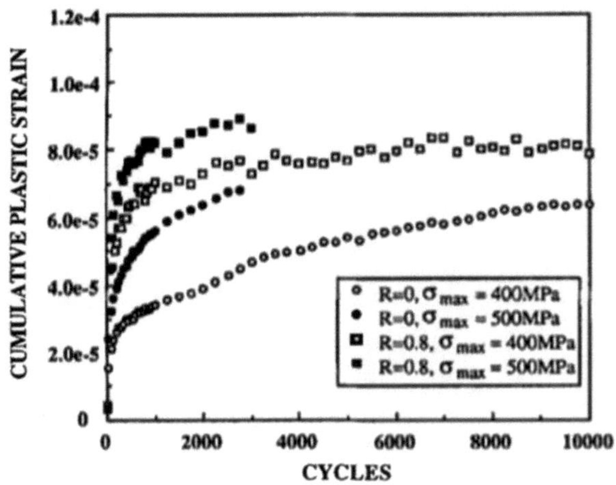

Fig. 1.29 Cumulative plastic strain versus cycles, Liu and Chen [10]. With kind permission of John Wiley and Sons

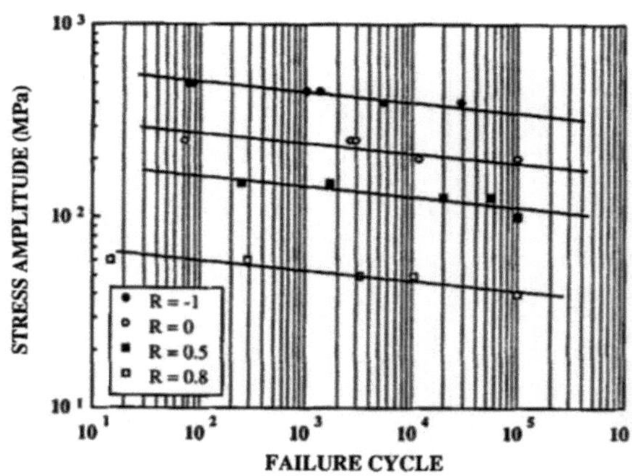

Fig. 1.30 Stress amplitude-failure cycle (S–N) curve of 3Y-TZP. Tests terminated at 10^5 cycles if failure did not occur, Liu and Chen [10]. With kind permission of John Wiley and Sons

where N_f is cycles to failure, m is the reciprocal slope of Fig. 1.30, and Z/A is a constant whose value depends on the R ratio. From Fig. 1.30 we find m to be ~21. The values of Z/A are listed in Table 1.1.

Several typical fatigue fracture surfaces of 3Y-TZP are shown in Fig. 1.31. At higher magnification, the fracture path was always intergranular. The fracture surface at Fig. 1.31a, b show the fracture origin. Further, crack origin are flaws at the near surface seen as closed pores in Fig. 1.31c, d and the ones connected to the free surface seen in Figs. 1.31e, f. All these have been identified as the crack origins. The appearance of these fracture origins suggests that they formed during or before sintering and not by machining. XRD indicates that ~30% monoclinic phase was present, but this phase fraction was not different from that of uniaxially overloaded fracture surface from tensile test specimens.

Table 1.1 Values of m, Z, and A in Eq. (1.23) for different R ratios, Liu and Chen [10]. With kind permission of John Wiley and Sons

R ratio	*m*	log (*Z*/*A*)	*Z*	*A**
−1	21	58.05	1.35×10^{35}	1.20×10^{-23}
0	21	52.52	1.35×10^{35}	4.06×10^{-18}
0.5	21	48.00	1.35×10^{35}	1.35×10^{-13}
0.8	21	38.87	1.35×10^{35}	1.82×10^{-4}

*Values for *A* given for ΔK in $MPa \cdot m^{1/2}$ and *a* in meters.

1.1.4 S–N Relation in Si_3N_4

Rotary bending fatigue test results are illustrated in Fig. 1.32. The S–N curve generally is within the 10^4–10^5 stress-cycle range, but in some cases the number of the cycles was extended to more than 10^8 to examine whether a knee (horizontal line defining endurance limit) exists. Arrows of some data indicate those specimens where testing was stopped. The plot is on a logarithmic scale, and the curve can be expressed by the formula

$$\sigma^n N = \text{constant} \tag{1.24}$$

The exponent n is 49.7 as determined from the slope of the straight line (up to 10^8 cycles). In Fig. 1.33 macroscopic observation illustrates the fractured surface of the fatigue test in comparison to static fatigue. The micrograph is of the fractured surface after rotary bending test. Further, high frequency cyclic fatigue test was performed on Si_3N_4 samples. Contrary to metals such as ferritic steels, where the fatigue limit is ~10^7 cycles, in ceramic materials the number of cycles to failure exceed 10^7 cycles and no fatigue limit appears at these cycles (Fig. 1.34).To accelerate the fatigue tests to level of 10^{10} cycles resonant bending method was used.

No heat was generated in ceramic specimens (Si_3N_4) under cyclic loading at high frequencies (unlike in metals or polymers). The S–N curve for sintered Si_3N_4 is illustrated in Fig. 1.35. It appears that the fatigue in this ceramic depends principally on the number of the cycles than on time, because the fatigue strength at various frequencies did not show a systematic variation in the number of cycles to failure (see Fig. 1.35).

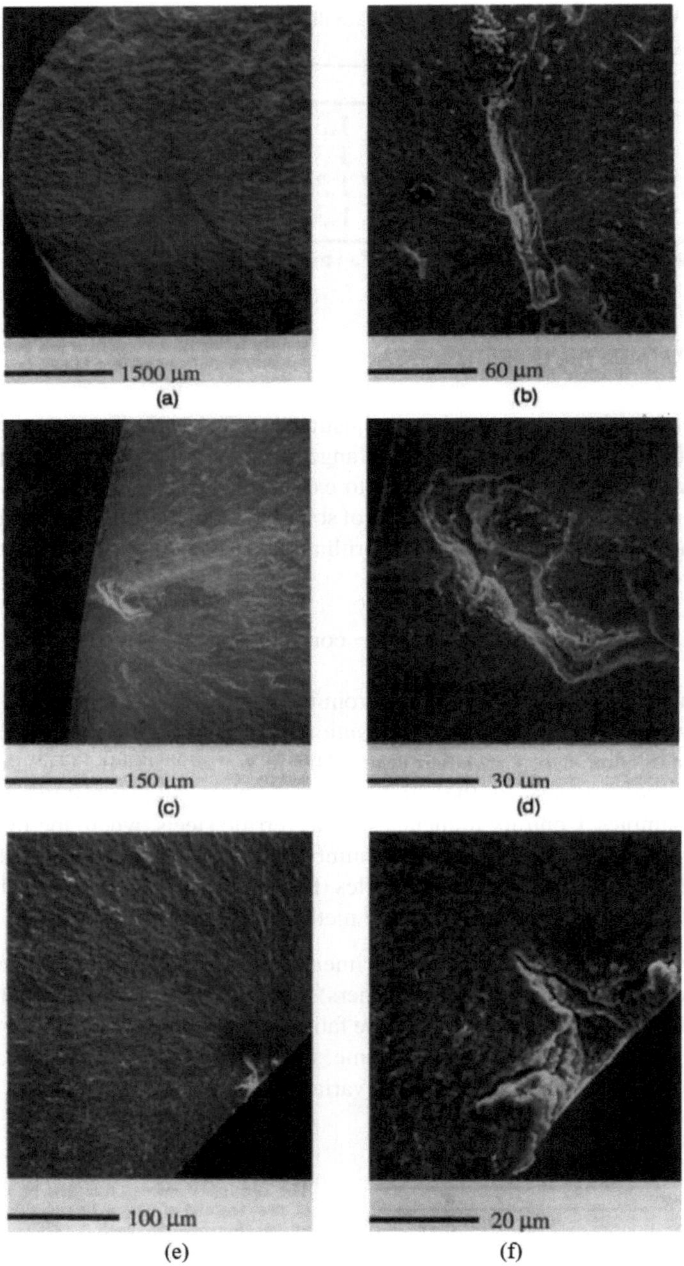

Fig. 1.31 Fracture origins of fatigue failure in 3Y-TZP: **a**, **b** a flaw near the center of the specimen; **c**, **d** a close pore near the surface; **e**, **f** a surface flaw, Liu and Chen [10]. With kind permission of John Wiley and Sons

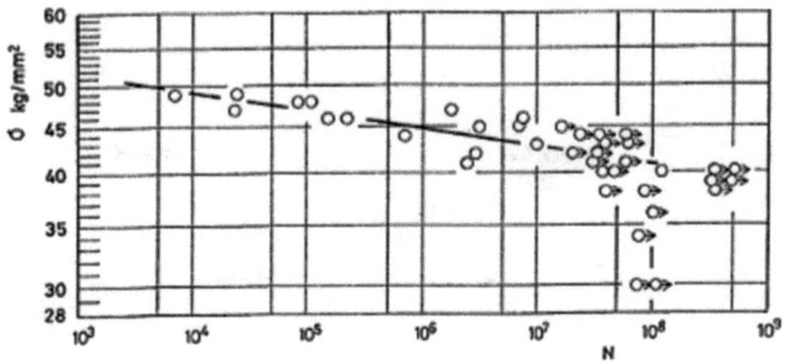

Fig. 1.32 S–N curve, Ko [11]. Free access

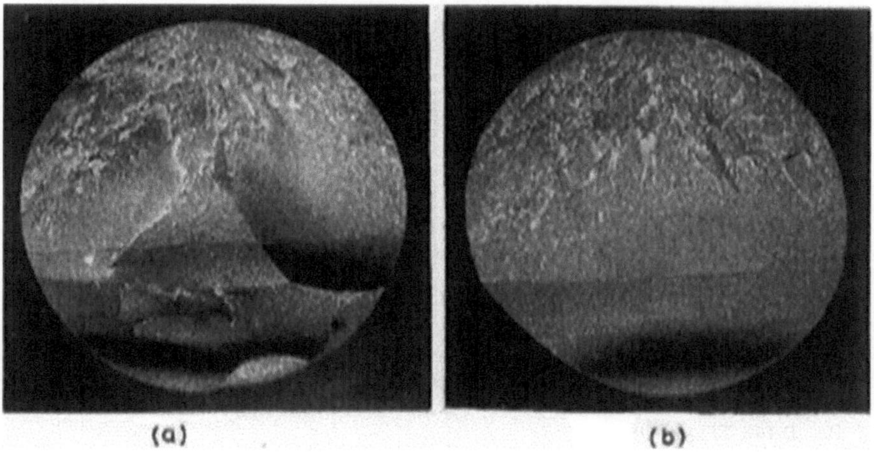

Fig. 1.33 Overall fractured surfaces after fatigue tests. **a** After rotary bending test ($\sigma = 42$ kg/mm^2, $N = 2.94 \times 10^6$), **b** after static fatigue test ($\sigma = 58$ kg/mm^2, $t = 2.49 \times 10^5$ s), Ko [11]. Free access

Fig. 1.34 Typical microphotographs of fractured surface after rotary bending test ($\sigma = 44$ kg/mm^2, $N = 6.96 \times 10^5$), **a** of mirror-like region, **b** of the area surrounding mirror-like region, Ko [11]. Free access

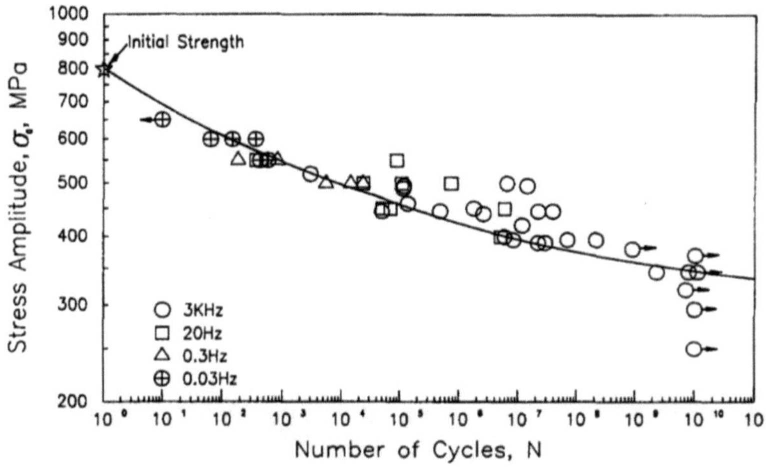

Fig. 1.35 S–N curve for a sintered silicon nitride, Soma et al. [12]. With kind permission of Elsevier

References

1. S.K. Bhaumik, R. Rangaraju, M.A. Venkataswamy, T.A. Bhaskaran, M.A. Parameswara, Eng. Fail. Anal. **9**, 255 (2002)
2. J. Pelleg, *Mechanical Properties of Materials* (Springer, Berlin, 2013)
3. X.Z. Lin, D.L. Chen, Mater. Sci. Eng. A **496**, 25 (2008)
4. S. Maity, D. Basuand, B.K. Sarkar, Mater. Lett. **2**(1), 63 (1994)
5. K. Li, D. Wang, H. Chen, L. Guo, J. Adv. Cer. **3**, 250 (2014)
6. V.N. Gurarie, D.N. Jamieson, A.V. Orlov, J.S. Williams, M. Conway, Nuc. Instr. Methods Phys. Res. B **148**, 773 (1999)
7. Y. Hino, K. Yoshida, Y. Kiyota, M. Kuwayama, ISIJ Int. **53**, 1392 (2013)
8. A. Smirnov, J.I. Beltrán, T. Rodriguez-Suarez, C. Pecharromán, M.C. Muñoz, J.S. Moya, J.F. Bartolomé, Sci. Rep. **7**, 44922 (2017)
9. R.C. Souza, C. dos Santos, M.J.R. Barboza, L.A. de Bicalho, C.A.R.P. Baptista, C.N. Elias, J. Mater. Res. Tech. **3**, 48 (2014)
10. S.-Y. Liu, I.-W. Chen, J. Am. Cerarn. Soc. **74**, 1197 (1991)
11. H.-N. Ko, J. Ceram. Soc. Jpn. **99**, 533 (1991)
12. T. Soma, M. Masuda, M. Matsui, I. Oda, Int. J. High Tech. Cer. **4**, 289 (1988)
13. L.F. Coffin, A note on low cycle fatigue laws. J. Mater. **6**, 388 (1971)
14. S.S. Manson, *Thermal Stress and Low Cycle Fatigue* (McGraw Hill, New York, 1966), pp. 125–192
15. J. Goodman, *Mechanics Applied to Engineering* (Longman, London, 1899)
16. H. Gerber, Bestimung der zulassigen spannungen in Eisen konstructionzen. Zeischrift Bayerischen Architeckten Ingenieur-vereins **6**, 101 (1874)
17. J. Morrow, *"Fatigue Properties of Metals," Section 3.2 of Fatigue Design Handbook, Pub. No. AE-4* (Society of Automotive Engineers, Warrendale, PA, 1968)
18. C.R. Söderberg, Trans. ASME **52**, 13 (1930)
19. R.E. Peterson, Notch sensitivity, in *Metal Fatigue*, ed. by G. Sines, J.L. Waisman (MacGraw-Hill, New York, 1959), p. 293
20. H. Neuber, *Kerbspannungslehre* (Springer-Verlag, Vienna, 1987)
21. A.A. Kozulin, A.S. Narikovich, S.N. Kulkov, V.N. Leitsin, S.S. Kulkov, AIP Conf. Proc. **1760**, 020036 (2016)

Chapter 2
Endurance Limit

Abstract The endurance limit of alumina, yttria stabilized zirconia, SiC and Si3N4 ceramics was evaluated. Their fatigue strength was determined. Relations of the fatigue strength and the number of cycles to failure is presented graphically and as mathematical equations. Connections relating the ultimate tensile strength to the fraction of the fatigue strength remaining after a specific number of cycles is shown. SEM illustrations of micrographs obtained after fatigue tests are illustrated. The effect of the mean stress on the cyclic fatigue was measured for sintered zirconia. Unlike in ferrous alloys in other materials a continuously decreasing S–N curve is exhibited.

Keywords · Endurance limit · Fatigue strength · UTS · Failure · Cracks · Pre-cracking · Smooth and notched specimens · Fiber · Nicalon · SiC/SiC · Crack bridgeing

Endurance limit is the stress level below which a specimen can withstand cyclic stress indefinitely without fatigue failure. This definition applies to ferrous metals and Ti alloys where a distinct endurance limit exists and recognized by the so-called knee, which basically is a horizontal line in the S–N plot. It is believed that in ferrous alloys the endurance limit is due to interstitials such as C or N which pin dislocations. This prevents the slip mechanism that might lead to the formation of microcracks and ultimate failure. Clearly overload unpins dislocations and should be avoided and when the temperature is not very high even periodic overloads can have the same effect (unpinning). Further if high temperatures are applied dislocations are mobilized. Both cases thus unpin dislocations.

However most nonferrous alloys (Al, Mg, Cu, etc.) have no endurance limit, and therefore usable cycles set arbitrarily is up to ~$1 \times 10^8 - 5 \times 10^8$ cycles. The materials display a continuously decreasing S–N curve. Up to this level of cycles it is safe to use components in practice under an appropriate stress applied.

Smooth specimens act differently than notched specimens. Notched specimens may have a completely different behavior due to residual stresses set up by overload and are considered as active stress raisers. Their use is for the sake of simulating minute flaws, (usually resulting from machining) on the surface of components.

J. Pelleg, *Cyclic Deformation in Oxides, Carbides and Nitrides*, Structural Integrity 22,
https://doi.org/10.1007/978-3-030-86118-6_2

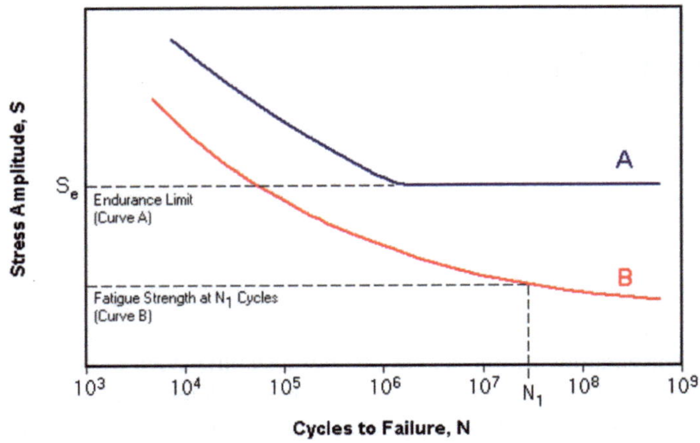

Fig. 2.1 Typical S–N curves

Figure 1.4 illustrates the two types of curves, the one with endurance limit and a second without it, which is reproduced here for convenience. Such curves are determined experimentally from testing material of interest. However sometimes all available of a specific material is its tensile strength only, and evaluation of the fatigue strength can be approximated by using a logarithmic relation between the number of cycles and the fatigue strength. This relation can be approximated as

$$S_f = aN^b \tag{2.1}$$

S_f is the fatigue strength, N is the number of cycles for life, a and b are constants which depend on the UTS, endurance limit and f the fraction of the fatigue strength remaining at 10^3 cycles (see Fig. 2.1). Equation (2.1) in logarithmic form is

$$\ln S_f = \ln a + b \ln N \tag{2.2}$$

When S_f is known N can be expressed as

$$N = \left(\frac{\sigma_a}{a}\right)^{\frac{1}{b}} \tag{2.3}$$

As mentioned, a and b are constants and can be evaluated at 10^3 cycles as follows: From (2.3) at $N = 10^3$ cycles

$$N = 10^3 = \left(\frac{\sigma_a}{a}\right)^{\frac{1}{b}} \tag{2.4}$$

$$\log 10^3 = \frac{1}{b}\left[\log\left(\frac{\sigma_a}{a}\right)\right] = \frac{1}{b}[\log \sigma_a - \log a] \tag{2.5}$$

$$3 = \frac{1}{b}[\log S_e - \log(f S_{UT}) \tag{2.6}$$

$$3 = -\frac{1}{b}\left[\log(f S_{UT}) - \log S_e\right] \tag{2.7}$$

$$\mathrm{b} = -1/3 \log \frac{f S_{UT}}{S_e} \tag{2.8}$$

In Eq. 2.5 the stress has been taken as S_e representing the endurance limit in Fig. 2.1. σ_a has been indicated to be proportional to the S_{UT} as indicated, namely

$$\sigma_a = fS_{UT} \tag{2.9}$$

with substitutions for a and σ_a in the respective equations at 10^3 cycles b has been evaluated as Eq. 2.8. To evaluate the other constant, constant a, one can follow the following steps:

From Eq. (2.3) one can write

$$N^b = \frac{\sigma_a}{a} \tag{2.10}$$

$$\mathrm{a} = \frac{\sigma_a}{N^b} \tag{2.11}$$

$$\log a = \log \sigma_a - b \log N \tag{2.12}$$

Substituting for b from Eq. (2.8) and for 10^3 cycles (the logarithm is 3), one obtains

$$\begin{aligned} \log a &= \log \sigma_a - \left[-1/3 \log\left(\frac{f S_{UT}}{S_e}\right) \times \log N\right] \\ &= \left[\log \sigma_a + \log\left(\frac{f S_{UT}}{S_e}\right)\right] \end{aligned} \tag{2.13}$$

With Eq. (2.9) for σ_a and since log 10^3 is 3 in we obtain for Eq. 2.13

$$\mathrm{a} = \frac{(fS_{UT})^2}{S_e} \tag{2.14}$$

Equations (2.8) and (2.14) are for 10^3 cycles. f has to be known or approximated. This exercise can be performed for another chosen number of cycles, say 10^5.

For say 10^3cycle f can be estimated as follows:

From Eq. (2.10)

$$\sigma_a = aN^b \tag{2.15}$$

substitute for (a) from Eq. (2.14) and obtain for σ_a

$$\sigma_a = \frac{(fS_{UT})^2}{Se} N^b \tag{2.16}$$

rearrange as

$$\frac{\sigma_a S_e}{N^b} = (fS_{UT})^2 \tag{2.17}$$

take logarithm as

$$\log \frac{\sigma_a S_e}{N^b} = 2 \log f + 2 \log S_{UT} \tag{2.18}$$

or

$$\log \sigma_a = 2 \log f + 2 \log S_{UT} + 2 \log N^b - 2 \log S_e \tag{2.19}$$

or from (2.19)

$$2 \log f = \log \sigma_a - 2 \log S_{UT} - \log N^b + \log S_e \tag{2.20}$$

or rewrite as

$$2 \log f = \frac{\log(\sigma_a S_e)}{2 \log(S_{UT}) N^b} \tag{2.21}$$

For $N = 10^3$

$$2 \log f = \frac{\log \sigma_a S_e}{\log S_{UT} (2 \times 10^3)^b} \tag{2.22}$$

$$f = \frac{\sigma_a S_e}{2S_{Ut}} (2 \times 10^3)^{-b} \tag{2.23}$$

simplify by calling $\sigma_a \times S_e/2$ as σ_f' to obtain for f at 10^3 cycles f as

$$f = \frac{\sigma_f'}{S_{UT}} (2 \times 10^3)^{-b} \tag{2.24}$$

Also b can be evaluated. Let us do it for the endurance cycles $N_e = 10^3$. From (2.24) we can write

$$f = \frac{\sigma_f^{'}}{S_{UT}}(2N_e)^{-b} \tag{2.25}$$

$$\log f = \log \frac{\sigma_f^{'}}{S_{UT}} - b \log 2(N_e) \tag{2.26}$$

$$\log(f S_{UT}) = \log \sigma_f^{'} - b \log(2N_e) \tag{2.27}$$

From Eq. (2.9) substituting for fs_{UT}

$$-b = \log \sigma_a - \log \sigma_f^{'} + \log(2N_e) \tag{2.28}$$

And finally for b we obtain

$$b = -\frac{\log(\sigma_f^{'}/\sigma_a)}{\log 2N_e} \tag{2.29}$$

As an example of using the above relations from knowing the UTS at a different N, say 10^5 one can evaluate a point a fatigue strength S_f on the S–N curve at this number of cycles. Assume that the endurance limit, say of steel is $S_e = 290$ MPa and the fraction f of the ultimate strength, S_{UT} at 10^3 cycles is $f = 0.77$. Also assume that the ultimate strength $S_{UT} = 655$ MPa.

The fatigue strength can be evaluated from Eqs. 2.1 and 2.14 as

$$\mathrm{a} = \frac{(f S_{UT})^2}{S_e} = \frac{(0.77 \times 6.55)^2}{290} = 877.1$$

$$\mathrm{a} = 877.13$$

For evaluating Eq. (2.1) b is needed and obtained from Eq. (2.18) as

$$b = -\frac{1}{3}\log \frac{f S_{UT}}{S_e} = -\frac{1}{3}\log \frac{0.77 \times 655}{290} = -0.08$$

$$\mathrm{b} = -0.08$$

Now using Eq. (2.1) to obtain the fatigue strength at 10^5 cycles

$$\mathrm{S_f} = \mathrm{aN^b} = 877.13 \times (10^5)^{-0.08} = 349.2$$

Thus, the fatigue strength is

$$\mathrm{S_f} = 349.2$$

One can also determine from Eq. 2.15 the life at a completely reversed load of 410 MPa as

$$N = \left(\frac{\sigma_a}{a}\right)^{\frac{1}{b}} = \left(\frac{410}{877.13}\right)^{\frac{1}{-0.08}} = 13{,}418\ cycles.$$

Thus, the life is only 13,418 cycles.

2.1 Fatigue Strength in Alumina

Fatigue strength of pure or monolithic alumina is not readily available—if at all—in the literature per se, despite its enormous importance in industry. Although no direct data on its behavior when cyclic stresses are applied or acting, extensive information exists on alumina acting as a strengthener or when Al_2O_3 by itself is reinforced by other constituents. As a common knowledge, when no endurance limit can be observed, which is the case of most metals, in such cases the fatigue strength S_f for a given number of cycles must be specified. As mentioned, an effective endurance limit for these materials is sometimes defined as the stress that causes failure at 1×10^8 or 5×10^8 loading cycles.

Nevertheless, the bending stress of alumina as a function of N, the number of cycles has been included in a publication characterizing static and fatigue strength of alumina ceramic joints (Tamai and Hirano). The S–N curve is linear since the stress and N in the plot are on logarithmic scale. The solid line represents monolithic alumina. It might be noted that the open squares are for Al_2O_3–Al_2O_3 joint (ticusil is a brazing alloy). The arrows indicate that no fracture occurs up to 10^7 cycles. The curve shows the relationship between maximum stress, σ_{max} and the cycles to failure N_f. It is worth mentioning that the joints between Al_2O_3/Al_2O_3 are almost the same as for the monolithic Al_2O_3 with the ticusil brazing material, but the stress in the S–N relation for the other cases shown in the Fig. 2.2 is lower.

2.2 In Yttria-Stabilized Zirconia (YSZ)

Fatigue behavior by uniaxial tension–compression of 3 mo% YSZ polycrystals is given in this section. It is determined that micro-cracking is the dominant mechanism of fatigue damage, and the fatigue lifetime is primarily controlled by crack propagation which is very sensitive to the acting maximum stress. Surface modification can improve fatigue strength. One of the commonly used method of modifying the surface to improve fatigue resistance is shot peening (SP). The fatigue strength of a 3 mol % yttria stabilized zirconia was improved by shot peening. Yttria stabilized zirconia exhibits superior fracture toughness compared to other ceramics—which as dentists

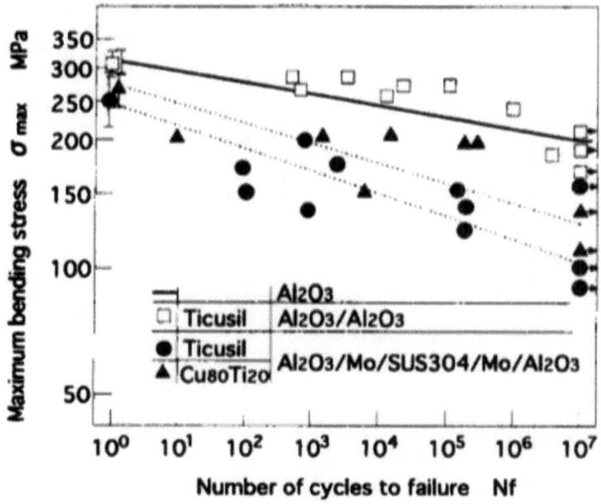

Fig. 2.2 S–N diagram, Tamai and Hirano [1]. Free access

know—makes it suitable for dental applications (as crowns) because it provides special aesthetic appearance in addition to mechanical properties. The mechanical properties are subject to repeated stress during their dental use. The dental prostheses are susceptible to fracture from repetitive occlusal loading [Occlusion is the relationship between the maxillary (upper) and mandibular (lower) teeth when they approach each other, as occurs during chewing or at rest], which determines its survival rate.

The stabilization of zirconia depends on the amounts of the stabilizing agent, and in the case of 3 mol% yttria the stabilization is partial (PSZ). Improving the surface by shot peening—as mentioned—is associated with introducing compressive residual stresses near the surface and used in other ceramics also, not only in zirconia. Large compressive stresses even up to 2000 MPa can be achieved; the static and cyclic load capacity is much increased. The SP specimens investigated were smooth and pre-cracked. In the Fig. 2.3 S–N curves show the relation between the maximum bending (flexural) stress, σ_{max} and the number of cycles to failure, N_f in the cyclic fatigue test. The bending strength for each of the specimens are also shown in the figure. Some specimens fractured outside the pre-crack area and are indicated by asterisks. The fatigue limit, σ_{f0} is defined as the value of σ_{max} at which the specimen endured 2×10^6 fatigue cycles and is indicated by the arrows. The fatigue limit values, σ_{f0} of smooth and pre-cracked specimen of various crack depth pre-cracks are compared in Fig. 2.4. The fatigue limit of the pre-cracked and SPed specimens increased compared with the non-SPed specimens; the fatigue limit of the smooth and SPed specimen also increased compared to the non-SPed ones. Thus, shot peening is effective in increasing the fatigue life. Clearly seen in the figure that the fatigue limit of the pre-cracked specimens decreased with increase in the depth of the pre-crack. SEM micrographs of the fractured surfaces after fatigue is seen in Fig. 2.5. Regardless of the pre-crack depth all non-shot peened specimens fractured at the pre-crack region. The origin of the fracture in the different specimens is seen in Fig. 2.5

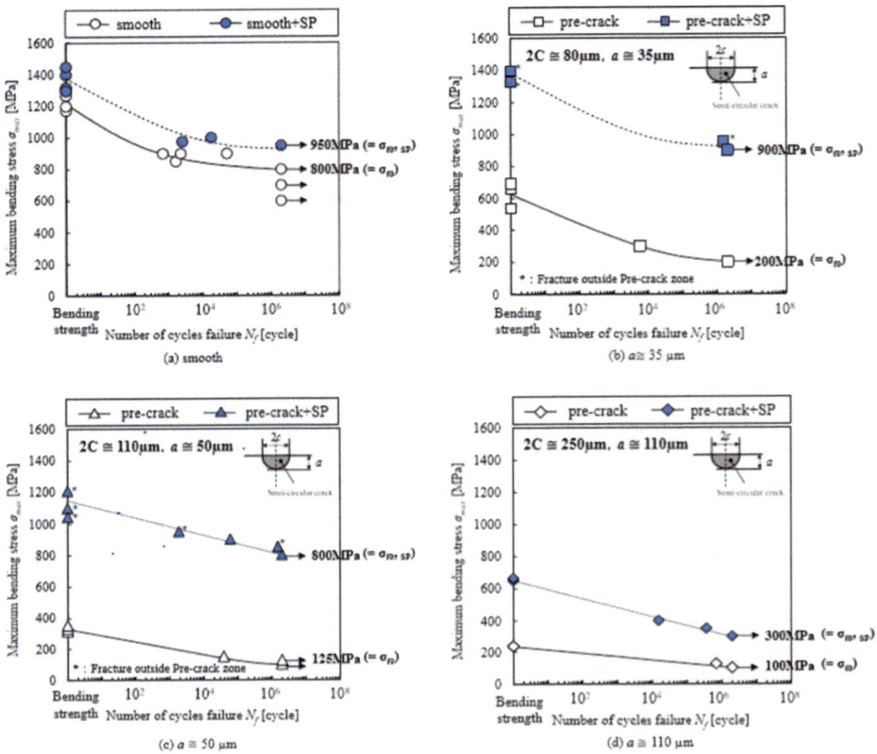

Fig. 2.3 S–N curves for the following specimens: **a** smooth, **b** pre-cracked with a ≅ 35 μm, **c** pre-cracked with a ≅ 50 μm, and **d** pre-cracked with a ≅ 110 μm. The open and solid symbols represent the non-SP and SP specimens, respectively, Takahashi et al. [2]. With kind permission of Taylor and Francis

Fig. 2.4 Comparison of the fatigue limits for the smooth specimens, and for the pre-cracked specimens with varying crack depth, a, Takahashi et al. [2]. With kind permission of Taylor and Francis

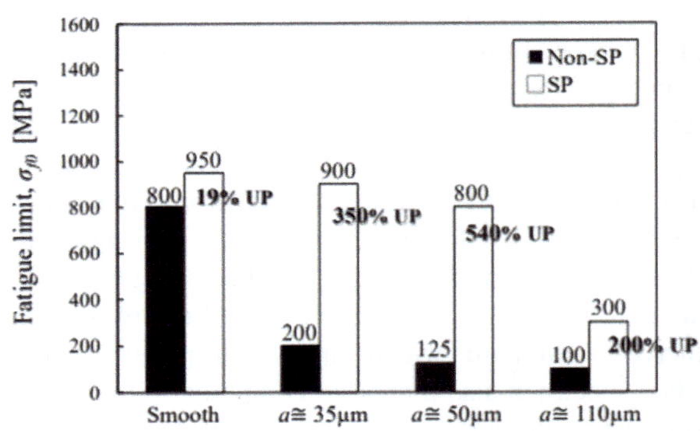

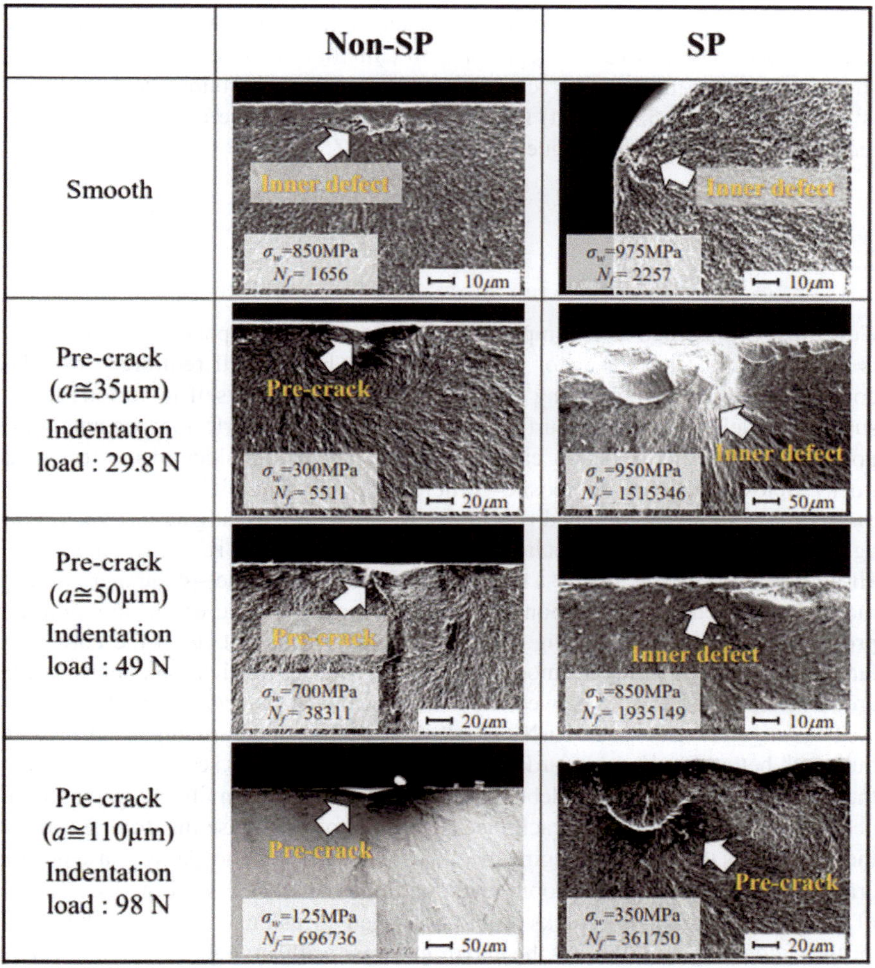

Fig. 2.5 SEM micrographs of the fractured surfaces obtained after the fatigue tests, showing the origin of fracture in the different specimens, Takahashi et al. [2]. With kind permission of Taylor and Francis

after the fatigue test as SEM micrographs of the fractured surfaces. The majority of the specimens in the pre-cracked + SPed specimens with a crack depth a ≅ 35 and 50 μm fractured outside the pre-crack zone.

In summary, the effect of shot peening on the fatigue strength of the PSZ and the pre-crack depth is: (a) The fatigue limits of the pre-cracked specimens after shot peening were significantly increased, by 200–540%, compared to those of the pre-cracked specimens without shot peening. The compressive residual stress induced by the local plastic deformation and the transformation of zirconia from the tetragonal to the monoclinic phase contributed to the improvement in the fatigue limits. (b)

The majority of pre-cracked + Sped specimens fractured outside the pre-crack zone. (c) The surface cracks with a depth of ≤50 μm could be considered harmless by shot peening. The maximum crack depth, a_{max} considered harmless by shot peening is calculated to be ~53 μm. (d) Finally, shot peening is a useful surface treatment method for improving the fatigue properties of PSZ.

2.3 In SiC

Generally, SiC is used as a composite rather than as an independent compound. It is either a reinforcing agent for other material or being itself reinforced by other components. Such a reinforcing component of SiC is SiC itself in various forms, such as particles, fibers etc. Further a fatigue test can be performed on smooth or notched specimens (the notch can be in the center or the sides of the test specimen). Below, a SiC/SiC composite—where the reinforcing SiC is in fiber form—is discussed. Smooth and notched specimens were used to investigate the crack propagation in various SiC/SiC specimens including Hi-Nicalon/SiC reinforcing agents. Hi-Nicalon is a ceramic fiber—a multi-filament silicon carbide—ceramic reinforcements (manufactured by Nippon Carbon Co., Japan). Cyclic crack initiation and propagation is illustrated in Fig. 2.6. The fatigue cracks initiate at the corners of large pores during loading in smooth specimens, while in notched specimens, cracks are formed at the interfaces between fibers and the matrix that connect with the notch.

The fatigue resistance of Hi-Nicalon/SiC is similar to that of enhanced SiC/SiC but much better than the standard SiC/SiC at 1300 °C in air as seen in Fig. 2.7. Here the cracks meet 0° fibers and debonding of interfaces between fibers and the matrix occurs. The 0° fibers bridge crack faces and therefore decrease the driving force at the crack tip as the general bridging mechanism. At 1300 °C the glassy phase fills the cracks and solidifies again at room temperature in the cracks (see Fig. 2.7). The fill of the glassy phases in the cracks prohibits oxygen diffusion along the crack path. In air the cyclic fatigue life versus the maximim stress is illustrated in Fig. 2.8. It can be seen that the cyclic fatigue life versus the maximum stress of Hi-Nicalon/SiC is about the same as that of enhanced SiC/SiC, but much smaller than the standard SiC/SiC at 1300 °C. Cyclic crack initiation is illustrated in Fig. 2.6. Regarding glass forming, it was achieved by adding boron-based particulate that reacts with oxygen to produce a sealant glass that inhibits oxidation. The modified SiC/SiC by this technology is the enhanced SiC/SiC. It might be of interest to show the variation of the normalized modulus with the number of cycles. It is shown in Fig. 2.9.

2.4 In Si_3N_4

The fatigue behavior of sintered Si_3N_4 under rotary bending and static fatigue is presented below. The character of the fractured surface is also considered. The rotary

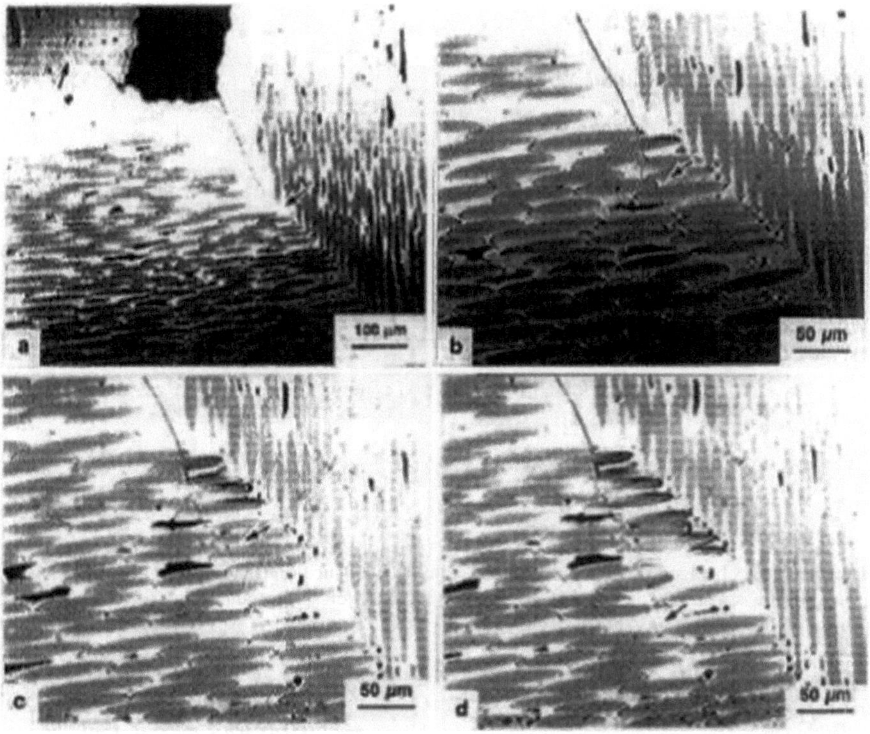

Fig. 2.6 Cyclic crack initiation and propagation process during cyclic fatigue in a smooth specimen. **a** After loading to 100 MPa; **b** cyclic fatigue at the maximum stress of 100 MPa for 100 cycles; **c** cyclic fatigue at amaximum stress of 100 MPa for 3500 cycles; **d** cyclic fatigue at the maximum stress of 100 MPa for 6300 cycles, Zhu and Kagawa [3]. With kind permission of Prof. Dr. S. Zhu

bending test performed in the range 10^4–10^8 stress cycles is illustrated in Fig. 2.10. It is worth noting that 10^8 stress-cycles is equivalent to about 20 days under the test conditions. A straight line by least square method was fitted to the point of the logarithmic graph and the line was approximated by Eq. (2.1.22) rewritten here for convenient reading as

$$\sigma^n N = \text{const} \tag{1.22}$$

The exponent n = 49.7 was obtained from the slope of the straight line. No knee is observed in the figure (characteristic for ferrous and Ti alloys), but it is claimed based on earlier investigation (Ko) that knee seems to exist beyond 10^8 cycles in sintered Si_3N_4. The macro-photographs are indicated in Fig. 2.11, where rotary bending test and static fatigue tests are compared. Macroscopic observation indicates that each fractured surface after both fatigue tests had a flat region and a rough region. The flat region was referred to as mirror-like region. The overall fractured surface after

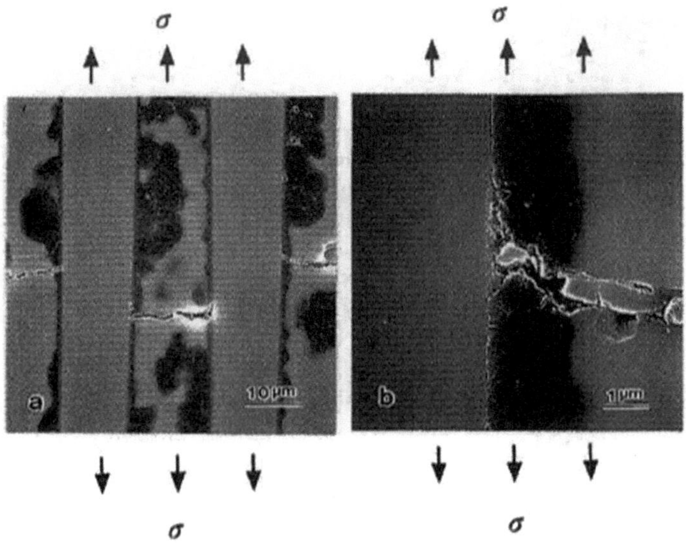

Fig. 2.7 Cracks bridged by 0° fibers in the specimen of HI-nicalon/SiC fatigued at 1300 °C under the maximum stress of 150 MPa in air for 2.5×10^4 cycles, Zhu and Kagawa [3]. With kind permission of Prof. Dr. S. Zhu

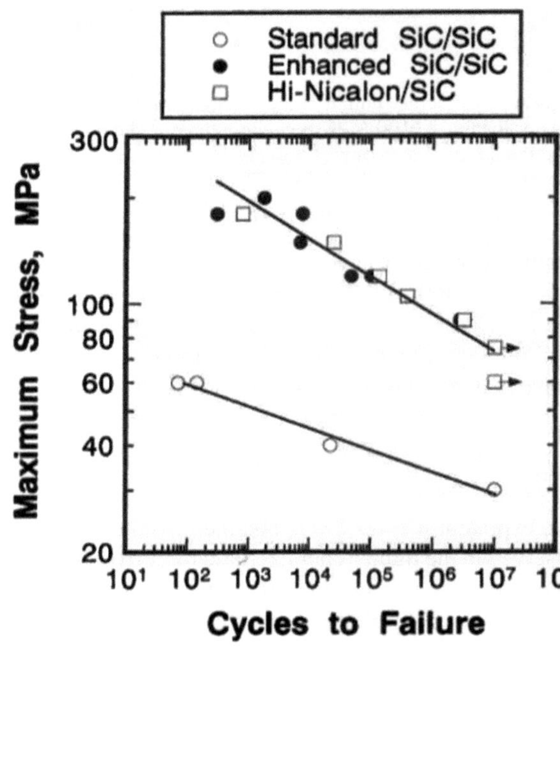

Fig. 2.8 The maximum stress versus cycles to failure for fatigue in Hi-Nicalon/SiC, enhanced SiC/SiC and standard SiC/SiC in air at 1300 °C, Zhu and Kagawa [3]. With kind permission of Prof. Dr. S. Zhu

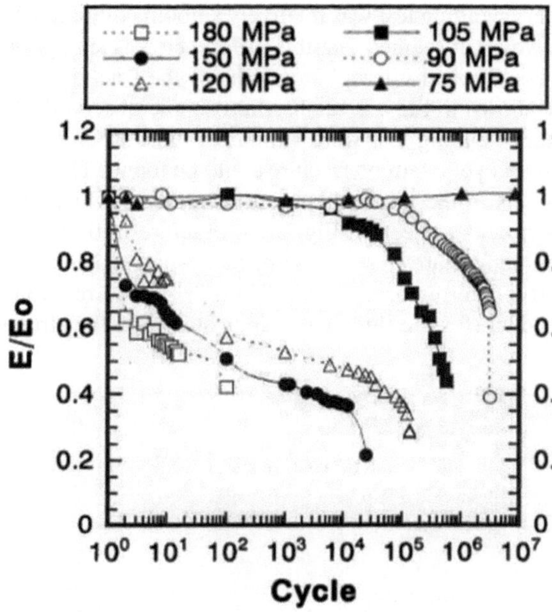

Fig. 2.9 The elastic modulus normalized by the value of the modulus under the first loading (E/E_o) versus cycles for fatigue for fatigue of Hi-Nicalon/SC in air at 1300 °C, Zhu and Kagawa [3]. With kind permission of Prof. Dr. S. Zhu

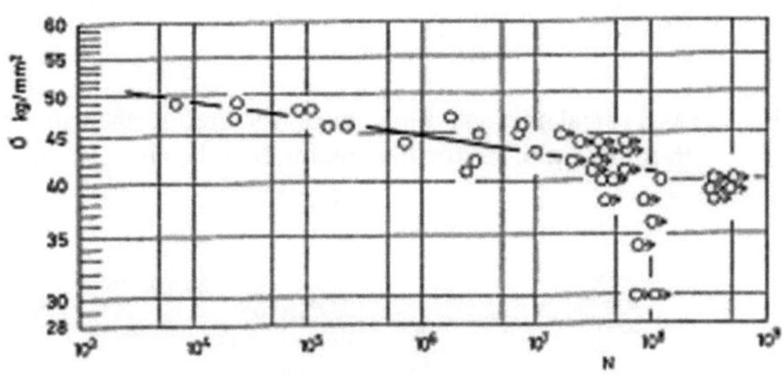

Fig. 2.10 S–N curve, Ko [4]. Free access

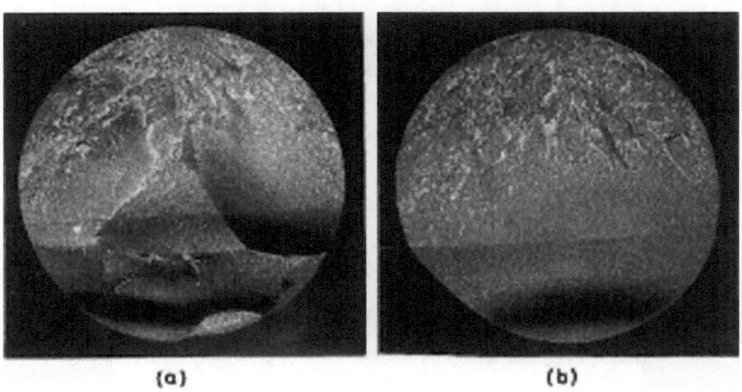

Fig. 2.11 Overall fractured surfaces after fatigue tests. **a** After rotary bending test ($\sigma = 42$ kg/mm^2, $N = 2.94 \times 10^6$), **b** after static fatigue test ($\sigma = 58$ kg/mm^2, $t = 2.49 \times 10^5$ s), Ko [4]. Free access

rotary bending test was relatively smooth compared with that after static fatigue test. However, it seemed that this difference was simply due to the magnitude of applied stress. Typical microphotographs of the fractured surface after rotary bending test are shown in Figs. 2.12a,b. Microscopic observations on the fractured surface after rotary bending test were similar to those after static fatigue test; no characteristic sign of cyclic fatigue fracture could be found. The fracture origin are pores or porous regions in the mirror-like region, some of which are located in the interior of the specimens as seen in Fig. 2.13. Fracture initiation was near the surface—as usual fatigue fracture starts at the surface—and propagated radially from the semicircular mirror-like region as seen in Fig. 2.14. The relation between the applied stress to the data (for the three cases indicated) and the straight line applies for all of them

Fig. 2.12 Typical microphotographs of fractured surface after rotary bending test ($\sigma = 44$ kg/mm^2, $N = 6.96 \times 10^5$), **a** of mirror-like region, **b** of the area surrounding mirror-like region, Ko [4]. Free access

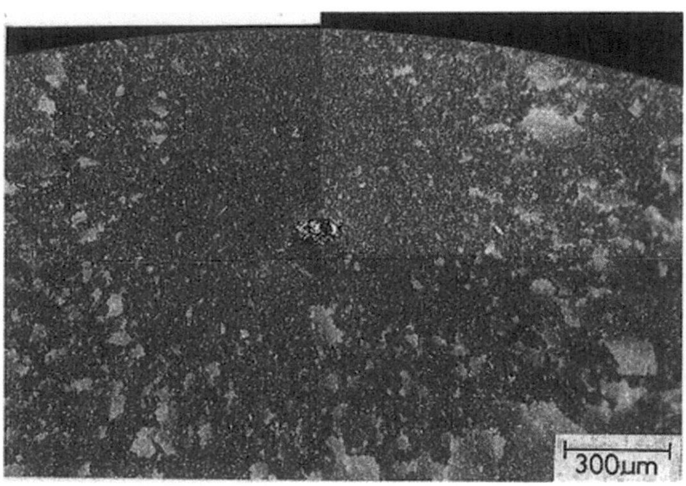

Fig. 2.13 Mirror-like region in the interior of fractured surface ($\sigma = 43$ kg/mm^2, $N = 1.02 \times 10^7$), Ko [4]. Free access

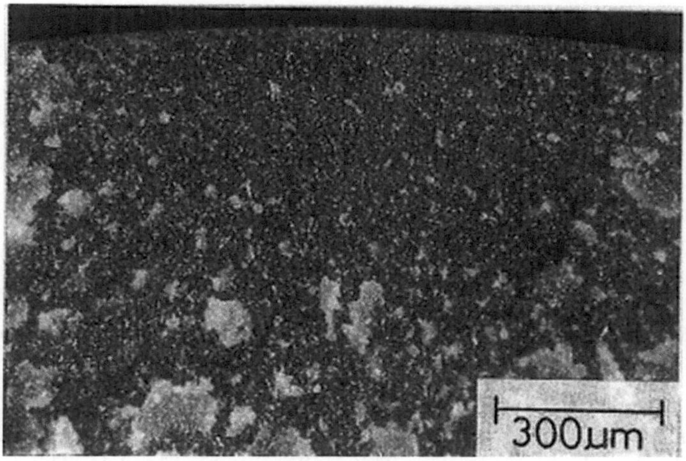

Fig. 2.14 Typical mirror-like region of fractured surface ($\sigma = 42$ kg/mm^2, N $= 2.94 \times 10^6$), Ko [4]. Free access

resulting in a similar relation to Eq. (2.1.22) given as

$$\sigma d^s = \text{const} \tag{2.30}$$

The exponent s = 0.59. It might be of interest to indicate the stress intensity factor K_I relating it to the bending stress, σ and the crack depth by Eq. (2.31) and the depth of the mirror-like region, d is shown in Fig. 2.15. Least squares fit was applied

$$K_I = F\sigma\sqrt{\pi b} \tag{2.31}$$

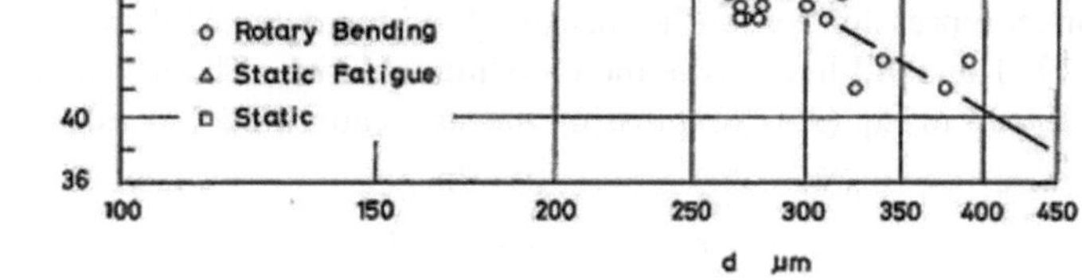

Fig. 2.15 Relation between depth of mirror-like region and applied stress, Ko [4]. Free access

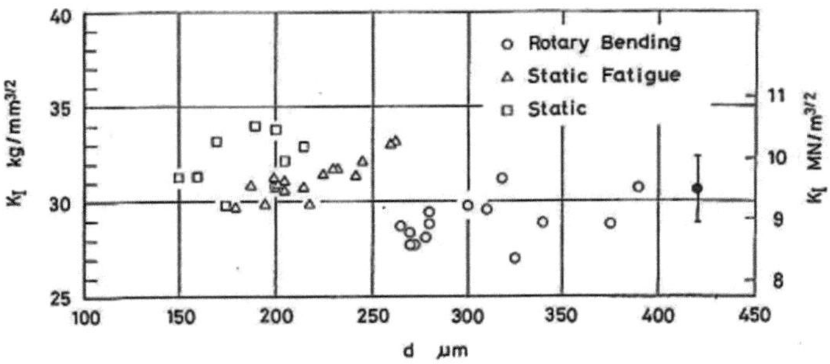

Fig. 2.16 Relation between depth of mirror-like region and stress intensity factor, Ko [4]. Free access

F is a dimensionless quantity dependent on the shape of the crack and the ratio of the crack depth to the specimen size. This is illustrated in Fig. 2.16. The mean value of K_I is $K_I = 9.5$ MN/m$^{3/2}$.

In summary, the fatigue life of sintered Si_3N_4 increased as the stress amplitude of the applied stress decreased. The S–N curve (also the S-t curve not shown) can be represented by a straight line up to about 10^8 stress-cycles. The stress exponent under rotary bending was evaluated. It is thought that the difference of fatigue behavior under rotary bending and the static fatigue is due to subcritical crack growth in the mirror-like region.

In an earlier work on cyclic fatigue testing of ceramic material the fatigue life of sintered Si_3N_4 by resonant bending fatigue test was evaluated. The fracture strengths of ceramics are usually evaluated by means of bend tests instead of tensile tests. The reason is that, owing to the brittleness of ceramics, the tensile test needs specifically designed jigs and specimens to achieve good alignment, together with special techniques and time for the sample preparation and measurement. However, in this work, the tensile system has been extended for cyclic fatigue testing at elevated temperatures under various cyclic loading conditions. Figure 2.17 shows the S–N curve derived for the sintered silicon nitride. Figure 2.18 shows an example of a stress–strain curve for a sintered silicon nitride tested under tension–compression cyclic load at 1200 °C. The effect of mean stress on the cyclic fatigue behavior in a sintered silicon nitride was measured by superimposing a steady stress on to the cyclic stress. The frequency was 20 Hz and the ratio, R, of the minimum stress (σ_{min}) to the maximum stress (σ_{max}) was between −1 and +1 in tension-tension and tension–compression tests. The measured fatigue data at 10^7 cycles are shown in Figs. 2.19. The solid line shows the Goodman relation. The modified Goodman relation is shown in Eq. (2.32) expressing the stress amplitude, σ_a as

$$\sigma_a = \sigma_e \left(1 - \frac{\sigma_m}{\sigma_t}\right) \tag{2.32}$$

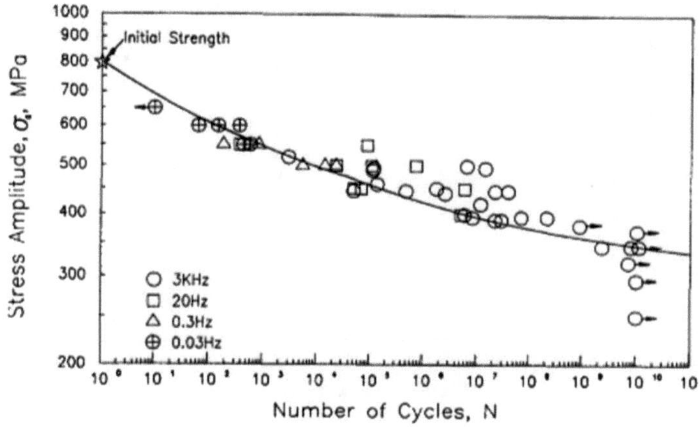

Fig. 2.17 S–N curve for a sintered silicon nitride, Soma et al. [5]. With kind permission of Elsevier

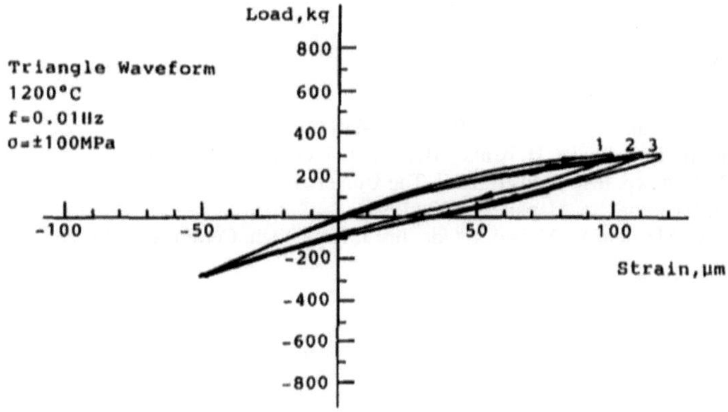

Fig. 2.18 Stress–strain curve in the cyclic tension–compression test at 1200 °C, Soma et al. [5]. With kind permission of Elsevier

Here σ_m is the mean stress, σ_t is the initial tensile strength and σ_e is the alternating fatigue strength at $R = -1$.

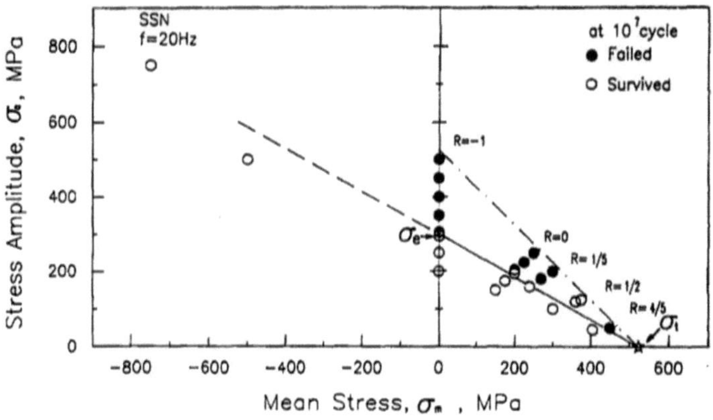

Fig. 2.19 Stress amplitude versus mean stress for fatigue failure in a sintered silicon nitride at 10^7 cycles in tension–compression and compression-compression tests. The modified Goodman relation is shown by the solid line, Soma et al. [5]. With kind permission of Elsevier

References

1. F. Tamai, K. Hirano, JSME Int. J. **39**(4), 613–619 (1996)
2. K. Takahashi, K. Iwanaka, H. Koike, Mater. Sci. Tech. **33**, 623 (2017)
3. S. Zhu, Y. Kagawa, Inst. Industrial Sci. The University of Tokyo
4. H.-N. Ko, J. Ceram. Soc. Japan **99**, 533 (1991)
5. T. Soma, M. Masuda, M. Matsui, I. Oda, Int. J. High Tech. Ceram. **4**, 289 (1988)

Chapter 3
Stress Cycles

Abstract Samples subjected to low cycle fatigue (LCF) under relatively high stress, fail at a low number of cycles. Below 10^5 cycles low cycle fatigue prevails. Thermal cycles induce low cycle fatigue in material and stresses arise from the difference in the thermal expansion coefficients and repeated shut down and reopen of the service temperature. Several empirical relations have been suggested for low cycle fatigue such as Morrow's and Coffin—Manson Concept. Alternating stresses in the elastic region are associated with high cycle fatigue. Stress life and strain life methods are present for fatigue life prediction. Very high cycle fatigue tests are also presented in this section.

Keywords Low cycle fatigue · Morrow's relation · Coffin-Manson concept · High cycle fatigue · Stress life · Strain life prediction · Very vigh cycle fatigue · Empirical relations · Gerber's equation · Goodman's equation

3.1 Introduction

In Chap. 1, Fig. 1.1 various forms of cyclic stress cycles were indicated, mainly as sinusoidally varying cycles. However, not irregularly (also known as random) varying cycles were shown. Before proceeding with the concept of cycles in fatigue it would be reasonable to show also fatigue cycles occurring as random cycles. They occur usually in aircraft wings which has to be considered for safe air carrier performance. An illustration is shown in Fig. 3.1, but the subject is outside the scope of this book.

Fatigue failures set in because of the application of fluctuating stresses, which are much lower than the stress required to failure during a single application of stress such as in tensile tests. Components subjected to repeated loads are likely to fail during mechanical service unexpectedly as indicated in Chap. 1, which makes their use critical if their design is not flawless and proper. Stress cycles is one of the three major factors necessary to induce fatigue failure, these are: (1) Maximum tensile stress of sufficiently high magnitude, (2) a sufficiently large fluctuation in the applied stress, and as stated (3) a sufficiently large number of cycles of the applied stress. There are many types of fluctuating stresses. Several of the more common

J. Pelleg, *Cyclic Deformation in Oxides, Carbides and Nitrides*, Structural Integrity 22,
https://doi.org/10.1007/978-3-030-86118-6_3

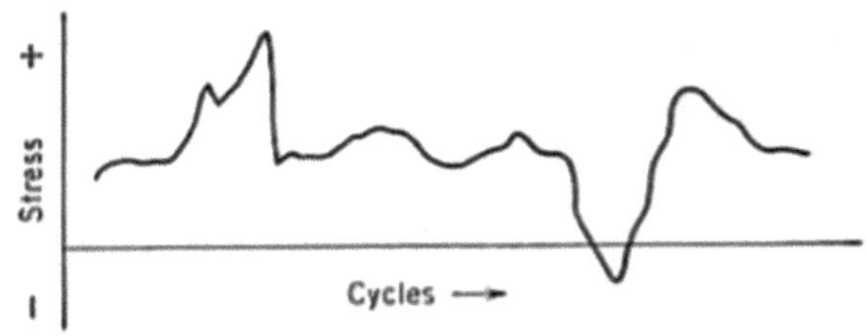

Fig. 3.1 Irregular (random) stress cycle

types encountered were shown in Fig. 1.1, and the one shown in Fig. 3.1 for random stress cycle.

Further division of the applied cycles-to be discussed below-are: low cycle fatigue, high cycle fatigue and very high cycle fatigue. Sometimes ultra-low cycles fatigue is indicated in the literature. Further it should be noted that fatigue tests are also performed by using strain-controlled tests.

3.2 Low Cycle Fatigue

Components subjected to relatively high stress are likely to fail at a low number of cycles, and they are subject to low cycle fatigue (LCF). Generally, the demarcation line between low cycle and high cycle fatigue is ~10^5 cycles, below which low cycle fatigue prevails and above it the fatigue is considered as high cycle fatigue (HCF). Thermal stresses arising from thermal expansion in the material might create LCF conditions when the repeated stresses are of thermal origin. Since thermal expansion is associated with strain changes, \the cyclic fatigue results from cyclic strain rather than cyclic stress. It should be noted that fatigue is a localized damage process of a component produced by cyclic loading. The various cycles are classified as shown in Fig. 3.2.

The following relations have been suggested for the LCF:

3.2.1 Morrow's Relation

The Morrow's relation is given by

$$\varepsilon_a = \varepsilon_a^{el} + \varepsilon_a^{pl} = \frac{\sigma_f^{'}}{E}(2N_f)^b + \varepsilon_f^{'}(2N_f)^c \tag{3.1}$$

where N_f is the number of load cycles, ε_a^{el} and ε_a^{pl} are the elastic and plastic strain amplitudes, respectively and $\sigma_f^{'}$ and $\varepsilon_f^{'}$ are the stress and stain to failure respectively and E is the elastic modulus. Equation (3.1) can be redefined according to Morrow,

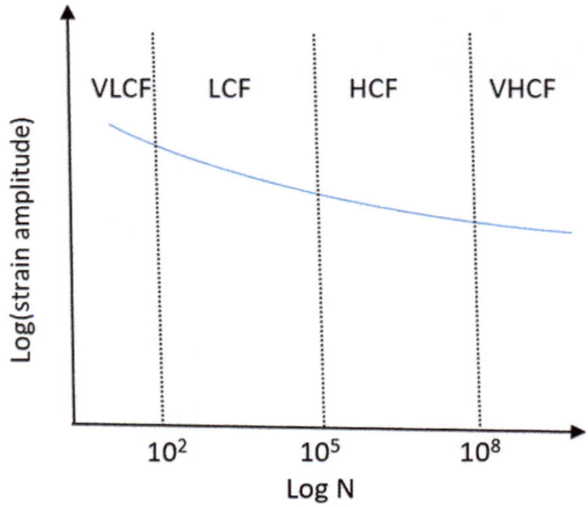

Fig. 3.2 Classification of the cycles according to the number of cycles. The schematic plot is log strain amplitude versus log number of cycles

taking into account the mean stress, σ_m as

$$\varepsilon_a = \varepsilon_a^{el} + \varepsilon_a^{pl} = \frac{\left(\sigma_f^{'} - \sigma_m\right)}{E}\left(2N_f\right)^b + \varepsilon_f^{'}\left(2N_f\right)^c \tag{3.2}$$

A schematic stress–strain relation showing the elastic strain and the plastic strain up to fracture is seen in Fig. 3.3. The logarithms of ε_a^{el} and ε_a^{pl} vs the logarithm of $2N_f$ presented in Fig. 3.4 for each in Eq. (3.3) and (3.4) give straight lines with intercepts of $\log \varepsilon_f^{'}$ and $\log\left(\frac{\sigma_f^{'}}{E}\right)$ respectively as seen below:

$$\log \varepsilon_a = \log \varepsilon_f^{'} + c \log 2N_f \tag{3.3}$$

$$\log \varepsilon_a = \log \frac{\sigma_f^{'}}{E} + b \log 2N_f \tag{3.4}$$

From the slopes the parameters c and b are evaluated. The lines are shown schematically.

with their sum according to Eq. (3.1).

3.2.2 *Coffin—Manson Concept: (The Strain-Life Equation)*

The Coffin-Manson concept that related strain and life is given by:

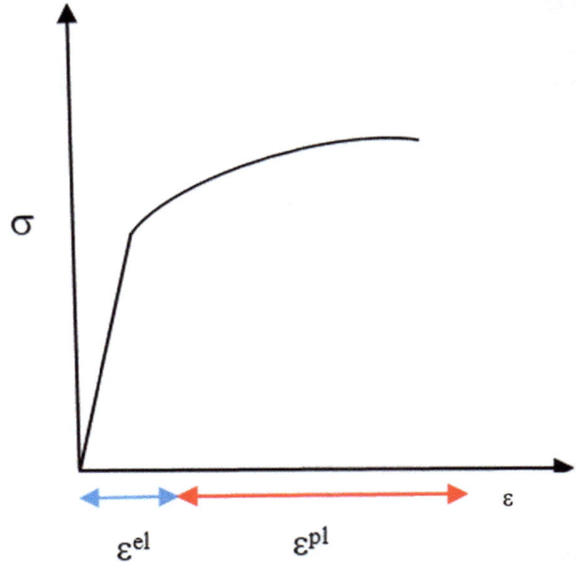

Fig. 3.3 Schematic stress–strain showing elastic and plastic strains

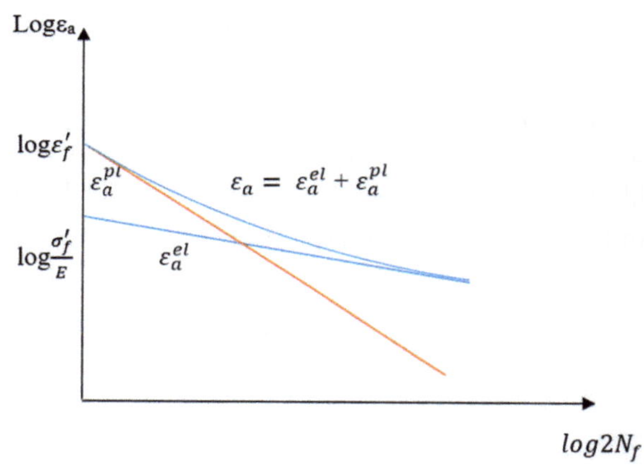

Fig. 3.4 Schematic illustrations of the lines related to Eqs. (3.1), (3.3) and (3.4)

$$\frac{\Delta\varepsilon_e}{2} = \frac{\sigma_f^{'}}{E}(N_i)^b + \varepsilon_f^{'}(N_i)^c \tag{3.5}$$

For obtaining this relation let us start with the uniaxial stress–strain relation; the linear elastic strain is

$$\varepsilon_{el} = \frac{\sigma}{E} \tag{3.6}$$

The plastic strain ε_{pl}

$$\varepsilon_{pl} = \left(\frac{\sigma}{K'}\right)^{1/n'} \tag{3.7}$$

Here K' is the strain hardening coefficient and n is the strain hardening exponent. The total strain, ε_t is the sum of Eqs. (3.6) and (3.7) given as

$$\varepsilon_t = \varepsilon_{el} + \varepsilon_{pl} = \frac{\sigma}{E} + \left(\frac{\sigma}{K'}\right)^{1/n'} \tag{3.8}$$

The strain life expression known as Basquin's equation composed of an elastic and plastic term. The elastic term is given as

$$\frac{\Delta\varepsilon_{el}}{2} = \frac{\sigma_f^{'}}{E}(N_i)^b \tag{3.9}$$

where $\sigma_f^{'}$ is the fatigue strength coefficient and b is the fatigue strength exponent. Clearly, E is the elastic modulus.

The plastic strain life term is known as the Coffin-Mason Equation and given as

$$\frac{\Delta\varepsilon_{pl}}{2} = \varepsilon_f^{'}(N_i)^c \tag{3.10}$$

where $\varepsilon_f^{'}$ is the strain ductility coefficient and c is the strain ductility exponent. The total strain life terms, the elastic and plastic are given as the sum of Eqs. (3.9) and (3.10), which is Eq. (3.5)

$$\frac{\Delta\varepsilon_t}{2} = \frac{\sigma_f^{'}}{E}(N_i)^b + \varepsilon_f^{'}(N_i)^c \tag{3.5}$$

Note that the strain life equation is given in terms of cycles (N). However, in most publications the strain life equation is shown as a function of the number of reversals to failure ($2N_f$) and here two reversals are one cycle. To eliminate confusion, the term (N) here is used to define N cycles.

(N_i) cycles relates to the initiation and (N_f) to the cycles to failure. More exactly N_i is the number of cycles until initiation of a crack is determined (percentage of load drop) during strain-controlled testing. And similarly, N_f is defined as the number of cycles until specimen failure occurs. Original publications of Basquin and Coffin-Mason were in terms of N_f the number of cycles to failure. More modern practice of the Basquin and Coffin-Mason equations (together they define the strain life equation), the strain life equation is defined in terms of N_i the cycles for initiation and N_f the cycles to failure.

In Eq. (3.5) four constants ($\sigma_f^{'}$, b, $\varepsilon_f^{'}$ and c) have to be evaluated in strain-life curve. Various approximations were suggested the most well-known is Manson's Four Point method. From the equations for Manson's Universal Slopes approximation method the derived constants are as follows:

$$\sigma_f^{'} = 3.5\sigma_{\mathrm{UTS}}$$
$$\mathrm{b} = -0.12$$
$$\mathrm{c} = -0.6$$
$$\varepsilon_f^{'} = \left\{ \ln\left[\frac{1}{1 - RA} \right] \right\}^{-0.6} \tag{3.11}$$

Note that the total strain range ($\Delta\varepsilon_t$) is defined as the sum of the elastic ($\Delta\varepsilon_{el}$) and plastic ($\Delta\varepsilon_{pl}$) strain ranges, as indicated in Eq. (3.8). The stress amplitude (σ_a) and strain amplitude (ε_a) are half of the stress range ($\Delta\sigma$) and strain range ($\Delta\varepsilon$) and are respectively defined as

$$\sigma_a = \frac{\Delta\sigma}{2} \tag{3.12}$$

$$\varepsilon_a = \frac{\Delta\varepsilon}{2} \tag{3.13}$$

and similarly, to the stress ratio of Eq. (1.4), the strain ratio is defined as

$$R_\varepsilon = \frac{\varepsilon_{\min}}{e_{\max}} \tag{3.14}$$

For fully reversed fatigue testing an $R_\varepsilon = -1$.
Equation 3.5 can be written in logarithmic form as

$$\log\frac{\Delta\varepsilon_t}{2} = \log\frac{\sigma_f^{'}}{E} + b\log N_i + \log\varepsilon_f^{'} + c\log N_i \tag{3.5}$$

With the parameters indicated in Eq. (3.11) and recalling that 2 reversals at failure, N_f is one cycle, Eq. 3.5 can be expressed after some manipulation as

$$\frac{\Delta\varepsilon_t}{2} = \frac{1}{2}\frac{3.5\sigma_{UTS}}{E}(N_f)^{-0.12} + \frac{1}{2}\left[\ln\left(\frac{1}{1 - RA}\right)\right]^{0.6}(N_f)^{-0.6} \tag{3.15}$$

In the plastic term the reduction of the area is given as

$$RA = \frac{A_f - A_0}{A_f} = 1 - \frac{A_0}{A_f} \tag{3.16}$$

and the term can be expressed after substitution as

$$\frac{1}{2}\left[\ln\left(\frac{1}{1 - 1 + \frac{A_0}{A_f}}\right)\right]^{0.6}(N_f)^{-0.6} = \frac{1}{2}\left[\ln\left(\frac{1}{\frac{A_0}{A_f}}\right)\right]^{0.6}(N_f)^{-0.6} \tag{3.17}$$

The reduction of area is proportional to the strain at fracture, ε_f

$$\ln\frac{A_0}{A_f} \approx \varepsilon_f \tag{3.18}$$

and thus, we can write for the plastic term

$$\frac{1}{2}\left[\ln\left(\frac{1}{\varepsilon_f}\right)\right]^{0.6}\left(N_f\right)^{-0.6} \tag{3.19}$$

Finally, Eq. 3.15) can be rewritten as

$$\varepsilon_a = \frac{\Delta\varepsilon_t}{2} = \frac{1.75\sigma_{UTS}}{E}\left(N_f\right)^{-0.12} + 0.5\left[\ln\left(\frac{1}{\varepsilon_f}\right)\right]^{0.6}\left(N_f\right)^{-o.6} \tag{3.20}$$

3.3 High Cycle Fatigue

High cycle fatigue is a type of fatigue caused by alternating stresses in the elastic range. Before failure small elastic strains under a high number of load cycles occurs. Fatigue cracks start after long periods of use such as hundreds of thousands or millions of cycles. It can occur in any moving part but is more common in rotating parts such as shafts, gears, disks, and turbine blades. The stress comes from a combination of mean and alternating stresses. It has lower stresses than LCF, and the stresses are also lower than the yield strength of the material. As to the question why cycle fatigue cause failure in critical components, one should be aware to the observation, that the component is first subjected to repeated loading and unloading stresses. Microscopic size cracks begin to form in the high stress locations. Eventually, the crack will reach a critical size, propagate suddenly, and the material will fracture. These life-limited components must undergo rigorous, expensive inspections and maintenance requirements to prevent a catastrophic failure. Following this short description of high cycle fatigue, presentation of the main approaches to its analysis is considered. Two approaches are discussed: 3.3.1 the Stress-Life (S–N) method and 3.3.2 the Strain-life (E-N) prediction of fatigue failure.

3.3.1 Stress-Life (S–N) Method

In Chap. 1 Fig. 1.3, reproduced here indicate all the important parameters of Eqs. 1.1–1.5, among them the applied and mean stresses. The lifetime assessment of Wöhler compares the calculated maximum alternate stress, σ_a with the experimental endurance stress, σ_e which serves to estimate the high cycle fatigue (HCF). To determine an influence of mean stress σ_m on the fatigue life, Wöhler's method was developed empirically further by Gerber (1874), Goodman (1899), Soderberg (1930) and Morrow (1960), which are expressed by the following relations:

Gerber's equation

$$\frac{\sigma_a}{\sigma_e} + \left(\frac{\sigma_m}{\sigma_u}\right) = 1 \quad (3.21)$$

Goodman's equation

$$\frac{\sigma_a}{\sigma_e} + \left(\frac{\sigma_m}{\sigma_u}\right) = 1 \quad (3.22)$$

Söderberg's equation

$$\frac{\sigma_a}{\sigma_e} + \left(\frac{\sigma_m}{\sigma_y}\right) = 1 \quad (3.23)$$

Morrow's equation

$$\frac{\sigma_a}{\sigma_e} + \left(\frac{\sigma_m}{\sigma_f}\right) = 1 \quad (3.24)$$

where σ_y and σ_f are the yield and the fracture stresses, respectively. However, these equations predict a region of infinite life by combining the endurance strength σ_e with either the yield strength σ_y, ultimate strength σ_u, or true fracture stress σ_f evaluated from monotonic uni-axial tensile tests. Therefore, these life equations ought to be applied only to tensile mean stresses. (Also, for small mean loadings ($R << 1$) these methods do not differ among each too much). In present design practice, Haigh's equation becomes the most popular in engineering analyses (developed from Goodman's relation). The Wöhler (Fig. 3.5) curve for the life time is given as

$$\sigma_a = a\left(N_f\right)^b \quad (3.25)$$

From goodman's equation and from the value of the endurance stress when $N_f = N_e$

$$\sigma_e = a(N_e)^b \quad (3.26)$$

one can write Haig's equation as

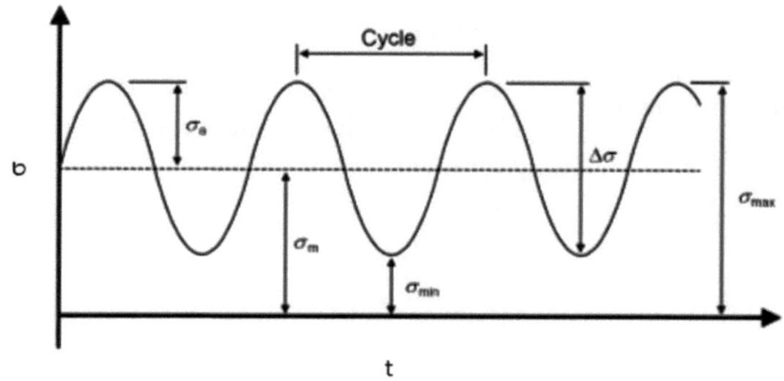

Fig. 3.5 Several cycles in a fatigue stress cycle including the various stress parameters

$$\sigma_a = a(N_e)^b\left(1 - \frac{\sigma_m}{\sigma_u}\right) \tag{3.27}$$

a and b are empirical constants estimated as

$$a = \frac{(0.90\sigma_u)^2}{\sigma_e} \tag{3.28}$$

$$b = \frac{1}{3}log\left(\frac{0.9\sigma_u}{\sigma_e}\right) \tag{3.29}$$

All stresses are expressed in (psi) unit. To create the Haigh's diagram according to Eq. (3.27) many data must be known from the experimental tests of the specimen.

3.3.2 Strain-Life (E-N) Prediction of Fatigue Failure

In Eq. (3.5) the Coffin-Manson strain life relation was indicated and improved by Morrow and others and deals with plastic strain. In the HCF regime-unlike in their method where local plastic deformation is also considered to assess fatigue failure-the total strain is dominated by the elastic strain compared to the contribution of the plastic strain as was illustrated in Fig. 3.4 (an E-N diagram). The strain life method has been presented in Sect. 3.2.2. However, it could be added that the influence of the mean stress σ_m on the elastic part of Morrow modified the Coffin-Manson equation into

$$\frac{\varepsilon(N)}{2} = \frac{(\sigma_f - \sigma_m)}{E}(2N)^b + \varepsilon_f(2N)^c \tag{3.30}$$

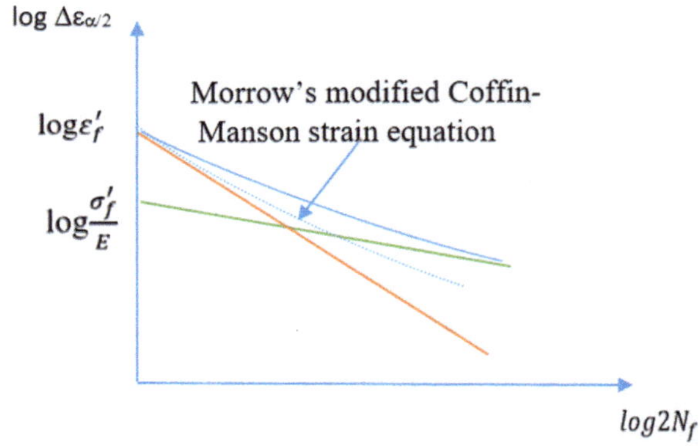

Fig. 3.6 Graphical representation of the Morrow's modification of the Coffin-Manson equation

(Note σ_f and ε_f, in Eq. (3.5) are σ_f' and ε_f'). In arbitrary loading modes and for different stress ratios, R, Smith, Watson, and Topper (SWT) instead of the mean stress, σ_m combined the cyclic strain range, $\Delta\varepsilon$ and σ_{max} into parameter Γ given as

$$\begin{aligned} \Gamma &= \sqrt{(\sigma_m + \sigma_a)\varepsilon_a E} = \sqrt{\sigma_{max}\varepsilon_a E} \text{for } a_m > 0. \\ \Gamma &= \sqrt{\sigma_a \varepsilon_a E} \text{for } a_m < 0. \end{aligned} \tag{3.31}$$

where $\varepsilon a = \frac{\Delta\varepsilon}{2}$ is the maximum principal strain and σ_{max} is the maximum tensile stress in the maximum principal strain plane. Multiplying Cofin-Manson relation of Eq. (3.5) by $\sigma_f E$, Eq. (3.32) is obtained for various R.

$$\sigma_f \frac{\varepsilon(N)}{2} E = \sigma_f^2 (2N)^{2b} + E\sigma_f \varepsilon_f (2N)^{b+c} \tag{3.32}$$

The SWT damage parameter in terms of the number of cycles to failure is

$$\Gamma(N) = \sqrt{\sigma_f^2 (2N)^{2b} + E\sigma_f \varepsilon_f (2N)^{b+c}} \tag{3.33}$$

Morrow modified the Coffin-Manson strain equation, (Eq. 3.3) and it is presented graphically in Fig. 3.6.

3.4 Very High Cycle Tests

In VHCF applications, 10^8 through 10^{11} (or more) cycles are experimentally observed for example in steel and Al alloys (Fig. 1.5). Figure 1.4 is redrawn schematically in an extended scale of N in Fig. 3.7 and an experimental illustration showing both cases the one with a fatigue threshold and one without it is shown in Fig. 1.5. In Fig. 1.5 represented here the S–N curves of Al and steel are shown. Note the N scale

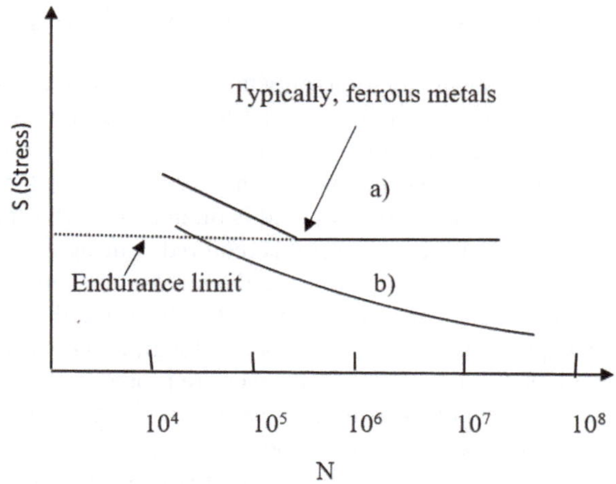

Fig. 3.7 S–N curves: **a** with a well-defined endurance limit, **b** without a definite fatigue limit

extends up to 10^{10} cycles. The failure probability increases with increasing numbers of cycles even in a region with assumed constant fatigue strength.

The requirement in actual applications of many components exceeds 10^8 cycles. Therefore, S–N data beyond 10^7 cycles are required, since components of aircraft, railways (high speed trains at 10^9 cycles), automobiles to mention some operate under such conditions. Despite the large amount of published data on fatigue S–N curves, these are limited to about 10^7 cycles because cost constrains rules out the use of conventional fatigue tests beyond 10^7 cycles. Nevertheless, since the fatigue life of such components is in the giga-cyclic range, it is essential to determine the stage of fatigue strength for ~10^9 cycles by experiments. Figure 3.8 is such an experimental S–N figure for Al and steel. It should be noted that accelerated testing is being

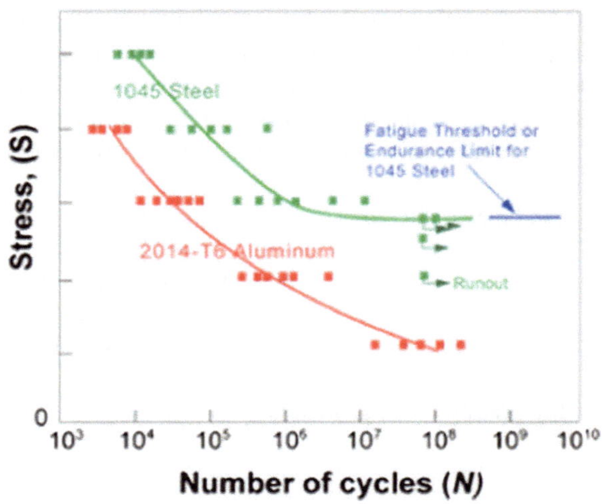

Fig. 3.8 S–N plot for 1045 steel and 2014-T6 Aluminum. Center for Nondestructive Evaluation, Iowa State University, Ames, Iowa 50,011, USA

considered by using high frequency cyclic loading by using piezoelectric fatigue systems.

The nature of vibrations effects the fatigue data. Ultrasonic fatigue test differs from the conventional fatigue tests in the nature of the vibrations. Such test is intended to reproduce free vibrations of the specimen vibrating with its own fundamental frequency, whereas in conventional testing the working frequency is set away from the fundamental frequency. and is often below. The specimen is subjected to forced vibration. In ultrasonic test it is required to design the specimen with a fundamental frequency that is tuned to matc the machine's. Ultrasonic fatigue machine is based on the concept of free vibration. Elastic wave theory can be used to explain this concept, in particular, the theory of longitudinal waves which propagate through a solid material. Longitudinal waves are propagated throughout the specimen and the displacement of its particles is parallel to the direction of the wave propagation. The speed of these waves is dependent on the material's properties and path geometry.

Materials tested are mainly high-strength steels and other materials containing heterogeneities in the form of pores, coarse second phase particles, which could act as internal fatigue crack initiation sites. Generally, UTS is related to fatigue strength, and the higher is the UTS, the steeper is the S–N curve in the VHCF (giga-cycle) range. Among the materials considered in VHCF are ferrous-, Ti -. Ni -, Al alloys and polycrystalline Cu which are extensively used in aeronautics, aerospace, automotive, railway and other industries. Tests are conducted to investigate the effect of specimen shape, size, surface roughness, inclusion size and load frequency on the VHCF testing results. The fatigue test results differ for different specimen shape and size, as the volume differs. Furthermore, VHCF is more sensitive to specimen size than LCF and HCF.

Regarding the crack, Paris et al. divides the total "fish-eye type fatigue life into three stages: (1) crack initiation; (2) crack growth from small to long and 3) crack growth from long to critical size. The crack starts from a defect (inclusion for example) size a_0 and becomes long at a_i. The growth rate is considered to be higher for small cracks than for long ones at the same driving force due to lack of crack closure. The transition point from small to long cracks is located at a factor X from the threshold corner of the stress intensity factor. X is a maximum for low load ratios ($R = 0, X = 3$) and minimum for high load ratios ($R > 0.8, X = 1$).

The total fatigue life is affected by the presence of crack, inclusions, pores etc. and it.

can be estimated as

$$N = N_{int} + N_{a_1 \to i} + N_{a_1 \to a} \tag{3.34}$$

The fatigue crack growth from small crack from $a \to a_i$ to long cracks from $a_i \to$

a,

Integrating crack growth law

$$\frac{da}{dN} = b\left(\frac{\Delta K_{\text{eff}}}{E\sqrt{b}}\right)^3 \tag{3.35}$$

and using stress intensity formula for internal circular cracks

$$\Delta K = \frac{2}{\pi}\Delta\sigma\sqrt{\pi a} \tag{3.36}$$

Paris et al. established the following expression for calculating fatigue life of small cracks from a_0 to a_i, as

$$N_{a_0 \to a_i} = \frac{\pi E^2}{2(\Delta\sigma)^2}\left[1 - \sqrt{\frac{a_0}{a_1}}\right] \tag{3.37}$$

And respectively fatigue life for long cracks from a_i to a, as

$$N_{a_i \to a} = \frac{\pi E^2}{2(\Delta\sigma)^2}\left[x^3\sqrt{\frac{a_0}{a_i}} - x^3\sqrt{\frac{a_0}{a}}\right] \tag{3.38}$$

Experimental results for different steels have shown that in the VHCF the fatigue life portion that corresponds to the crack growing from a_0 to a constitute only insignificant part (less than 1%) of the total fatigue life N.

The Paris' law is effective as it can characterize crack growth, which helps in assessing service life or inspection intervals for a structure. Even though the Paris' law is useful in many structural problems, it is involved with various sources of uncertainty such as initial crack length and material parameters.

Additional models for fatigue life prediction are presented in the publication of Deng et al.

References

1. J. Morrow, *"Fatigue Properties of Metals," Section 3.2 of Fatigue Design Handbook* (Society of Automotive Engineers, Warrendale, PA, 1968)
2. L.F. Coffin Jr., A study of the effects of cyclic thermal stresses on a ductile metal. Trans. ASME **76**, 931 (1954)
3. S.S. Manson, Fatigue-Complex subject—some simple approximations. Exp. Mech. **5**, 193 (1965)
4. K.N. Smith, P. Watson, T.H. Topper, A stress strain function for the fatigue of metals. J. Mater. **5**, 767 (1970)
5. I. Marines-Garcia, P.C. Paris, D. Lados, H. Tada, Fatigue crack growth from small to large cracks in gigacycle fatigue with fish-eye failures, in *9th International Fatigue Congress Atlanta, USA, May 2006*. Also, P.C. Paris, D. Lados, H. Tada, Eng. Fract. Mech. **75**, 299 (2008)
6. H. Deng, W. Li, T. Sakai, Z. Sun, Materials **8**, 8338 (2015)

Chapter 4
Fatigue Life

Abstract Strain life method is used to evaluate fatigue lifetime in service of structures exposed to certain loading conditions and to estimate the loading cycles a material can withstand before failure. Damage is induced in a component by crack initiation and propagation and their combination determines the total fatigue life. Cycles to failure in alumina SiC composite and yttria stabilized zirconia were applied for the determination of the fatigue limit. The SiC is reinforced by SiC fibers to improve high temperature strength. The moisture in air affect the cyclic behavior of SiC deteriorating its properties.

Keywords Strain life in alumina · In yttria stabilized zirconia · In SiC · Effect of moisture · Strength in air · Peak strain · Crack initiation · Crack propagation

4.1 Introduction

Two approaches are commonly used to express fatigue life. The traditional approach for safe life fatigue analysis (generally applied in aircraft structures to mention some), namely the stress life (S–N curves) and the other is the strain life method. The stress life method is the most common mode for failure analysis, in particular in metallic structures. Fatigue phenomena occur when a material is subjected to fluctuating stresses and strains, which lead to failure due to damage accumulation. It is also applied in HCF case -and as mentioned earlier- it is a graph of cyclic stress (S) range against the cycles to failure on a logarithmic scale (and is known as a Wöhler curve). S–N curves are derived from tests on samples of the material using a regular sinusoidal stress which is applied by a testing machine. The test is then repeated for stress cycles with different amplitudes and the outcomes are recorded and combined to construct a relation between stress and life. The progression of the S–N curve can be influenced by many factors such as corrosion, temperature, residual stresses, and the presence of notches. Statistical methods are used to construct the S–N curve for a certain probability of survival. The subject has been widely discussed in the previous chapter and no further consideration is given here for the topic.

J. Pelleg, *Cyclic Deformation in Oxides, Carbides and Nitrides*, Structural Integrity 22,
https://doi.org/10.1007/978-3-030-86118-6_4

4.2 Strain Life

The common notation of the curves is ε − N. The strain-life method is extensively used for crack initiation and propagation analysis. In the strain-life method of fatigue analysis, the elastic and plastic deformation of the material is used to determine how many cycles the material will last before failure. Failure is defined as the formation of a small crack (typically 1 μm) so the geometry of the material does not need to be considered provided the material properties did not change and the stress–strain tests are reliable. The equations for the strain-life diagram are used for calculating the number of cycles to failure. Summarizing the equations of the respective functions used for stress–strain and strain life are presented below:

Ramberg–Osgood equation:

$$\varepsilon_{tot} = elasticterm + plasticterm = \frac{\sigma}{E} + \left(\frac{\sigma}{K}\right)^{\frac{1}{n}} \tag{4.1}$$

Hysteresis curve equation:

$$\Delta\varepsilon = elasticterm + plasticterm = \frac{\Delta\sigma}{E} + 2\left(\frac{\Delta\sigma}{2K}\right)^{\frac{1}{n}} \tag{4.2}$$

Coffin-Manson equation:

$$\varepsilon_{tot} = elasticterm + plasticterm = \frac{\sigma_f}{E}\left(2N_f\right)^b + \varepsilon_f\left(2N_f\right)^c \tag{4.3}$$

Morrow Mean Stress Correction:

$$\varepsilon_{tot} = elasticterm + plasticterm = \frac{\sigma_f - \sigma_m}{E}\left(2N_f\right)^b + \varepsilon_f\left(2N_f\right)^c \tag{4.4}$$

Modified Morrow Mean Stress Correction:

$$\begin{aligned}\varepsilon_{tot} &= elasticterm + plasticterm \\ &= \frac{\sigma_f - \sigma_m}{E}\left(2N_f\right)^b + \varepsilon_f\left(\frac{\sigma_f - \sigma_m}{\sigma_f}\right)^{\frac{c}{b}}\left(2N_f\right)^c\end{aligned} \tag{4.5}$$

Smith–Watson–Topper Mean Stress Correction:

$$\begin{aligned}\varepsilon_{tot} &= elasticterm + plasticterm \\ &= \frac{\sigma_f^2}{\sigma_{max}E}\left(2N_f\right)^{2b} + \frac{\sigma_f\varepsilon_f}{\sigma_{max}}\left(2N_f\right)^{b+c}\end{aligned} \tag{4.6}$$

In these relations, variation of the elastic, plastic, and total strain amplitude as functions of the number of load reversals to failure ($2N_f$) are indicated.

Fig. 4.1 Strain Life Curve. Coffin-Manson relationship. Suria [1]

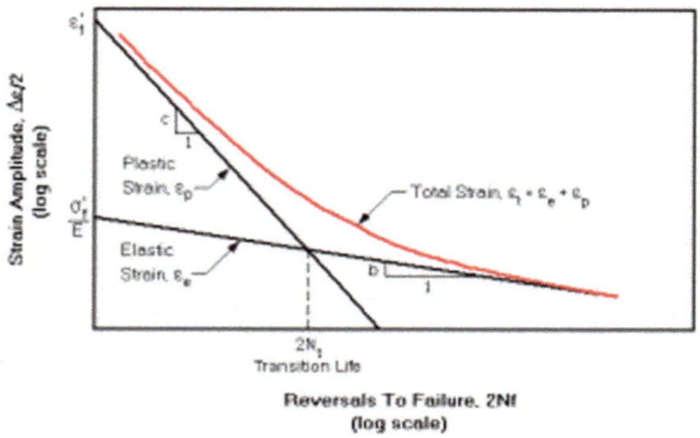

The strain-life relation equation is constructed from two separate curves (one for elastic life and one for plastic life) as shown in Fig. 4.1. An illustration of an $\varepsilon - N$ presentation based on Coffin-Manson relation of Eq. (4.3) is illustrated in Fig. 4.1. Similar experimental.

illustrations are also shown in Figs. 4.2 and 4.3 for high-strength-aluminum-alloy subjected to different manufacturing conditions and different test temperatures and steel material of medium static strength produced additively. Figure 4.3 is constructed by the known equation of the strain–life curve, Fig. 4.1, where the total strain amplitude $\varepsilon_{a,t}$ is divided into elastic $\varepsilon_{a,e}$ and plastic $\varepsilon_{a,p}$ components as such and seen in Eq. 4.3. Due to hardening or softening effects, materials subjected to cyclic loading are often unstable, producing stress amplitude changes during the tests. Thus, the stress amplitude corresponding to the stabilized state must be used, which occurs at the half of the number of cycles to crack initiation, which was determined at 10% stiffness loss of the stiffness over cycles signal. With this stress amplitude, the Young's modulus and the total strain amplitude the plastic part of the strain amplitude can be calculated as

Fig. 4.2 Strain-life curve according to Coffin and Manson, amended by the fatigue strength $\varepsilon_{a,E}$. Wächter et al. [2]. Open access

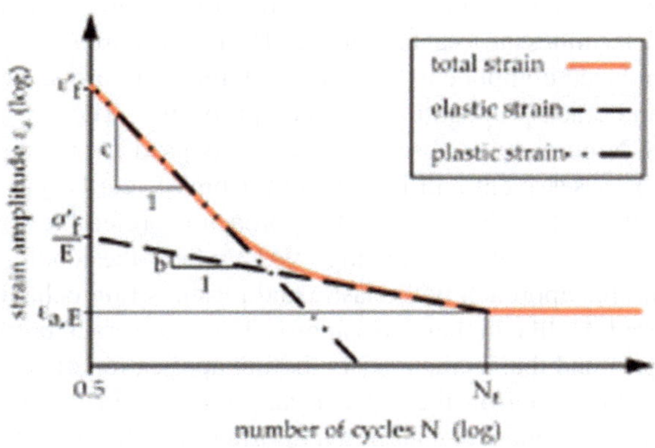

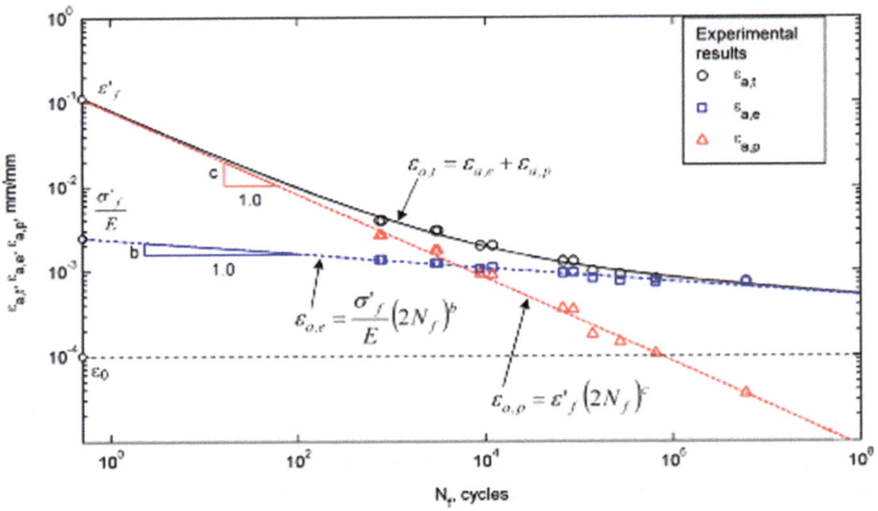

Fig. 4.3 Example of the Manson–Coffin–Basquin curve according to Eq. (4.3). Nieslorry et al. [3]. With kind permission of Elsevier

$$\varepsilon_{a,p} = \varepsilon_{a,t} - \varepsilon_{a,e} = \varepsilon_{a,t} - \sigma_a E \tag{4.7}$$

4.2.1 Strain Life in Alumina

Fatigue life usually refers to the number of stress or strain cycles that structures and components subjected to certain loading conditions can withstand before failure. The lifetime in service of a part is usually estimated from the number of loading cycles before cracks form. A combination of the initiation of cracks and their propagation determines the total fatigue life. The usual approaches commonly used for the damage induced components by crack initiation and propagation are: stress-life, strain-life, and crack propagation method approaches. The stress life approach in the form of S–N curves was discussed earlier, is used as the standard method for the fatigue life assessment and is suitable for high-cycle fatigue analysis that involves plastic deformation. The strain life approach is preferable when fatigue life is subjected to variable amplitude loading (VAL) where the loading sequence effect is essential. In this approach both elastic and plastic strain behavior of the material are considered. Unlike in these two approaches in the crack propagation method approach it is assumed that a crack exists (before application of stress or strain) in the material.

Few data on the subject can be found in the available literature. Despite the wide abundance of alumina, information relevant to the topic discussed here in pure alumina per se is missing. The industry is interested in alumina as a reinforcing

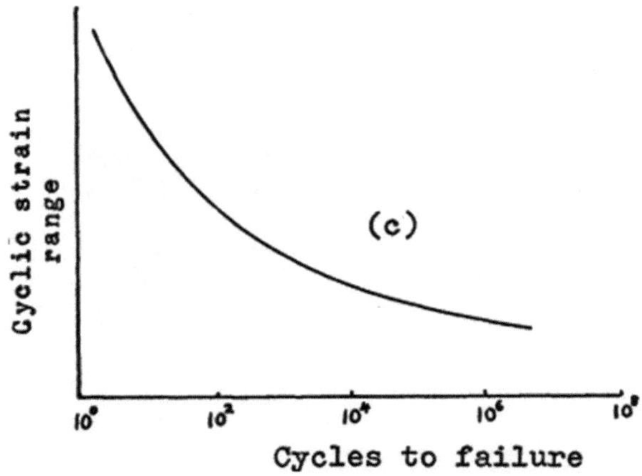

Fig. 4.4 $\varepsilon - \log N$ curve for strain cycling. After Kumar Sarkar [4]. Open access

and strengthening agent in various material and metals such as Al rather than as a monolithic pure unalloyed alumina, or even as a component in combination with other ceramics such as oxides carbides etc. Nevertheless, in an older publication an $\varepsilon - N$ curve, a cyclic strain vs. the cycles to failure has been found and reproduced here as Fig. 4.4.

4.2.2 Strain Life in Yttria Stabilized Zirconia (YSZ)

As known zirconia requires stabilization and the most popular stabilizer is yttria. Tetragonal zirconia polycrystals are stabilized at 3 mol % yttria addition although partially (PSZ). The strain life of PSZ uniaxial tension–compression fatigue behavior is presented below in Fig. 4.5. In the figure the accumulated plastic strain is plotted vs. fatigue cycle and the ratio, R refers to the ratio of the minimum to the maximum stresses during cycling. It is seen that the accumulation slows down with increasing the number of cycles. Also note that at the same ratio an increase in the stress causes more accumulated plastic strain, while an increase in ratio at the same stress results in higher damage.

4.2.3 Strain Life in SiC

In the framework of the effect of temperature and environment of SiC/SiC ceramic alloy,

strain-cycle curves were also included. In the SiC composite, SiC fiber-reinforcing was added to the SiC ceramic-matrix. As known ceramic materials possess a high strength and modulus at elevated temperatures. However, their use as structural

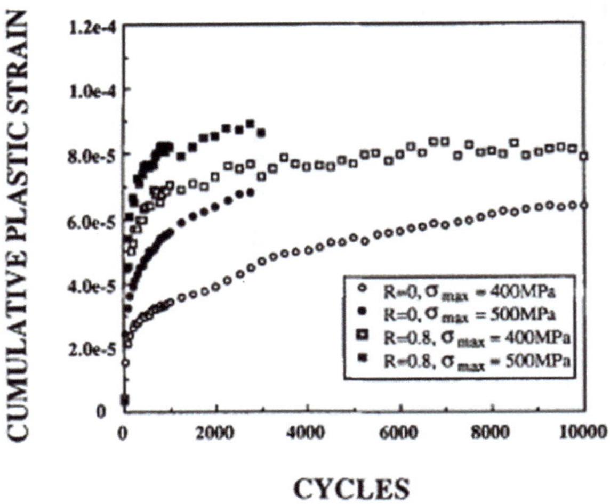

Fig. 4.5 Cumulative plastic strain versus cycle. Chen and Liu [5]. With kind permission of John Wiley and Sons

components are severely limited because of their brittleness. Continuous fiber-reinforced ceramic-matrix composites, by incorporating fibers in ceramic matrices, not only improves the high-temperature strength but also reduce their propensity for catastrophic failure. The fatigue tests shown in Fig. 4.6 represent test performed at different environment, i. e., air and steam and at various stresses. Moisture is known to affect the cyclic behavior of SiC deteriorating its properties, in particular, the fatigue life is decreased at the same stress level relative to that without moisture. In Fig. 4.6 a comparison at 750 °C in air and in a humid environment is made. In Fig. 4.7 the effect of stress on the peak strain is seen in air at 1000°. The stress levels are 80 and 100 MPa. The effect of moisture is illustrated in Fig. 4.8. Note the scale difference between Figs. 4.7 and 4.8 (in steam). In Figs. 4.9 and 4.10 the variation of peak strain at 1100 and 1200 °C in air is shown, while in Fig. 4.11 the

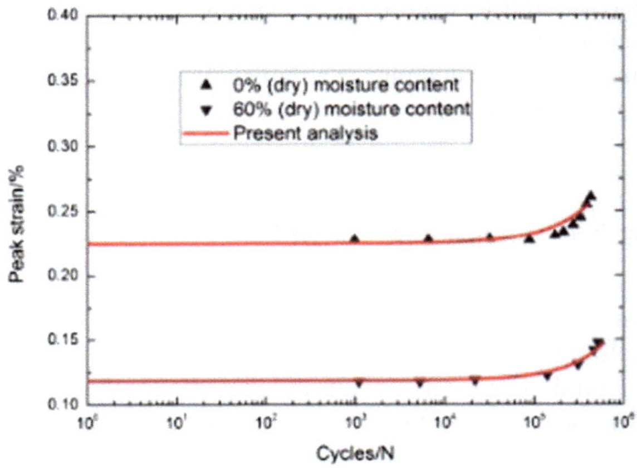

Fig. 4.6 The peak strain versus applied cycles with test environment of 0% moisture content condition and 60% moisture content. Li [6]. Open access

Fig. 4.7 The peak strain versus applied cycles at 1000 °C in air. Li [6]. Open access

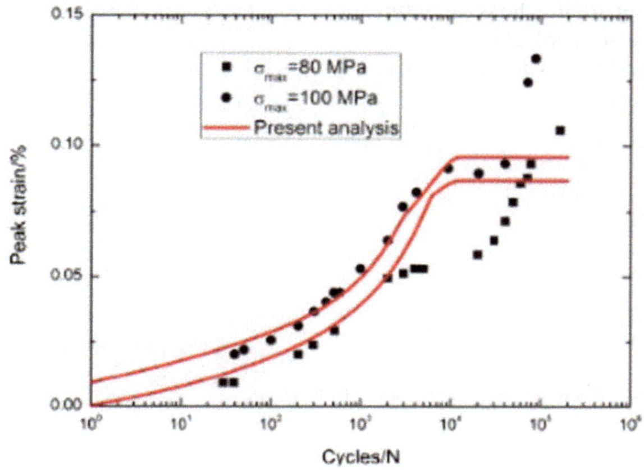

Fig. 4.8 The peak strain versus applied cycles at 1000 °C in steam. Li [6]. Open access

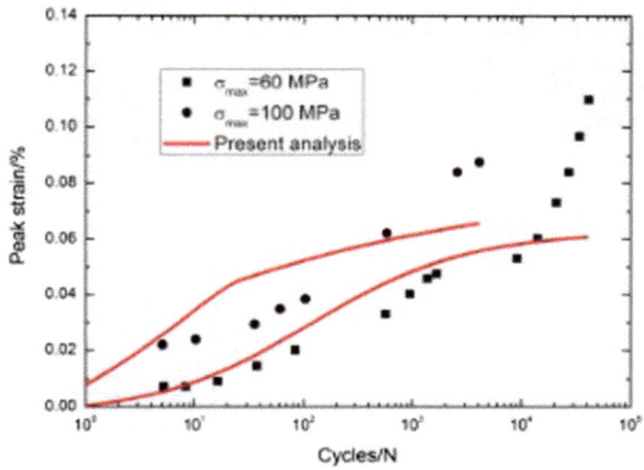

Fig. 4.9 The peak strain versus applied cycles at 1100 °C in air. Li [6]. Open access

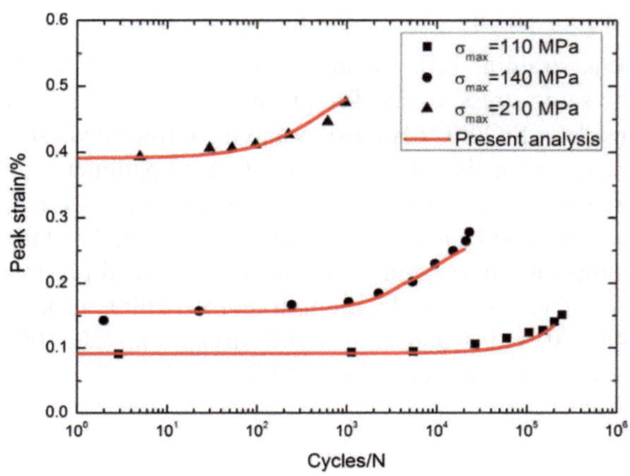

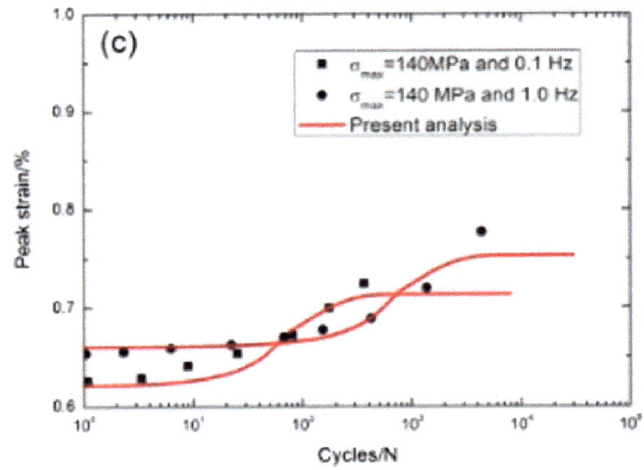

Fig. 4.10 The peak strain versus applied cycles at 1200 °C in air. Li [6]. Open access

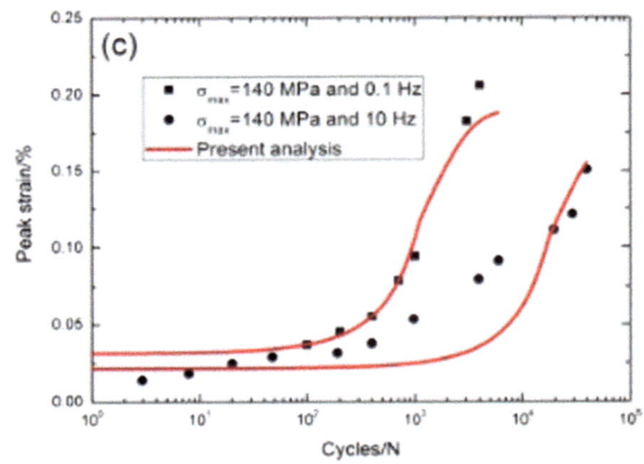

Fig. 4.11 The peak strain versus applied cycles at 1200 °C in steam. Li [6]. Open access

peak strain at 1200 °C in steam is shown and can be compared with Fig. 4.10 at the same stress levels. The presence of steam accelerated the damage development inside of SiC/SiC composites and the fatigue peak strain. Also, the damage development was analyzed through the damage parameters of fatigue hysteresis dissipated energy, fatigue hysteresis modulus, interface shear stress, S–N curves (which are not discussed and not shown in this section). The fatigue limit stress of 2D SiC/SiC composites have been predicted. The fatigue limit stresses approach 67%, 28%, 39% 17% and 28% tensile strength at 750, 1000, 1100, 1200 and 1300 °C in air, and 49%, 10%, 9% and 19% tensile strength at 750, 1000, 1200 and 1300 °C in steam conditions, are seen respectively.

References

1. O.V. Suria, A flexible lifting model for gas turbines: creep and low cycle fatigue approach M.Sc. Thesis, Cranfield University, U.K., 2006. Open Thesis, Also in My engineering world.net/2012/06/fatigue-analysis-in-turbo machinery
2. M. Wächter, M. Leicher, M. Hupka, C. Leistner, L. Masendorf, K. Treutler, S. Kamper, A. Esderts, V. Wesling, S. Hartmann, Appl. Sci. **10**, 5238 (2020)
3. A. Nieslorry, C. el Dsoki. H. Kaufmann, P. Krug, Int. J. Fatigue, **30**, 1967 (2008)
4. B. Kumar Sarkar, A Thesis, University of Surrey, (1967)
5. I-W. Chen, S -Y. Liu, J. Amer. Cer. Soc., **74**, 1197 (1991)
6. L. Li, Materials (Basel) **10**, 371 (2017)
7. V. Sharma, Air Force Institute of Technology Air University, Wright-Patterson Air Force Base, Ohio, Thesis, 2008
8. J. Delapasse, Air Force Institute of Technology Air University, Wright-Patterson Air Force Base, Ohio, Thesis, 2010
9. D. T. Christensen, Air Force Institute of Technology Air University, Wright-Patterson Air Force Base, Ohio, Thesis, 2009
10. T. Devon, Fatigue behavior of an advanced SiC/SiC composite at elevated temperture in air and steam. MS thesis, AFIT/GAE/ENY/09-D02. Schoool of Engineering and Management, Air Force Institue of Technology (AU), Wright-Patterson AFB, OH, December 2009

Chapter 5
Strengthening (Softening)

Abstract In static deformation, one of the work hardening occurs among other strengthening mechanisms when a material (mainly in metals and alloys) is strained beyond the yield point. An increasing stress is required to produce additional plastic deformation and the metal apparently becomes stronger and more difficult to deform. Dislocation activity is associated with hardening via pinning and hindering their motion. Grain boundaries are obstacles to dislocation motion by which they harden ceramics. Hardening in alumina, magnesia, pure zirconia, partially stabilized zirconia, SiC, B_4C and Si_3N_4 is discussed in this section. Also presented is in this section is the hardening of C/siC, Al/SiC and SiC/SiC composites. Dislocation motion by glide or slip, and the pinning by loops is considered. S–N fatigue data in some of the materials indicated above is discussed. Sand blasting improves surface properties of some of the listed substances. Fractured surface of B_4C showed twin formation during deformation by SEM and TEM.

Keywords Strengthening · Hardening in alumina · Hardening in magnesia · Zirconia · Partially stabilized zirconia · Hardening in SiC · B_4C · SiC/SiC · C/SiC · Composite · S–N · Experimental · Fiber · Degradation · Twins

5.1 Introduction

Work hardening in a metal occurs when a metal is strained beyond the yield point. An increasing stress is required to produce additional plastic deformation and the metal apparently becomes stronger and more difficult to deform. Work hardening is observed in conventional deformation as usually observed in static stress–strain plots, but is closely related to fatigue also. A few general words on strengthening in ceramics before considering the material of this book, namely, oxides, carbides and nitrides are listed below:

Several mechanisms by which materials may be strengthened are: (i) work (strain) hardening in ductile ceramics, (ii) solid solution strengthening due to pinning dislocations either by interstitial or substitutional atoms, (iii) strengthening by precipitates, (iv) hardening as a consequence of transformation (for example in YSZ) and (v) grain boundary strengthening. In general ceramics are brittle (there are exceptions) but at

J. Pelleg, *Cyclic Deformation in Oxides, Carbides and Nitrides*, Structural Integrity 22,
https://doi.org/10.1007/978-3-030-86118-6_5

high temperature plasticity is observed and the possibility of work hardening is likely, but not necessarily occur (since recovery might set in) and it depends on composition and test conditions. In superplastic oxides such as MgO strain hardening is absent (closely associated with superplasticity) and in β-Si_3N_4 under compression strain hardening was not observed. Often, little or no hardening at all occurs. Regarding (ii) and (iii) strengthening involves hindering dislocation motion. Similarly grain boundaries are obstacles to dislocation motion by which they harden the ceramics.

5.2 Hardening in Alumina

Polycrystalline alumina is brittle, does not undergo plastic deformation when deformed and there is an abrupt transition from elastic deformation to fracture and therefore polycrystalline aluminum oxide specimens do not undergo work hardening. However, sapphire crystals have been deformed in compression by basal plane (Pletka et al.) in the temperature range from 1400 to 1720 °C and show strain hardening. Beyond the yield-point stress–strain curves may be divided into three stages (see Fig. 5.1): (A) the work hardening rate, θ is at maximum, (B) where θ decreases continuously and (C) a plateau region in which θ is zero as illustrated schematically in Fig. 5.1. The experimental shear stress-shear strain curves at four temperatures are illustrated in Fig. 5.2. As can be seen and as would be expected the amount of hardening decreases with the increase in the temperature. As seen in Fig. 5.1, the work hardening rate, θ. is a maximum in stage B, decreasing continuously and at stage C the work hardening rate is zero.

The dislocation structure in addition to basal slip (0001) < $11\bar{2}0$ >, edge dipoles are formed by trapping dislocations in parallel slip planes. The dislocation loops are of two kinds formed by dipole breakup: The work hardening was attributed to the accumulation of these dipoles and loops. Following the steadily decreasing work

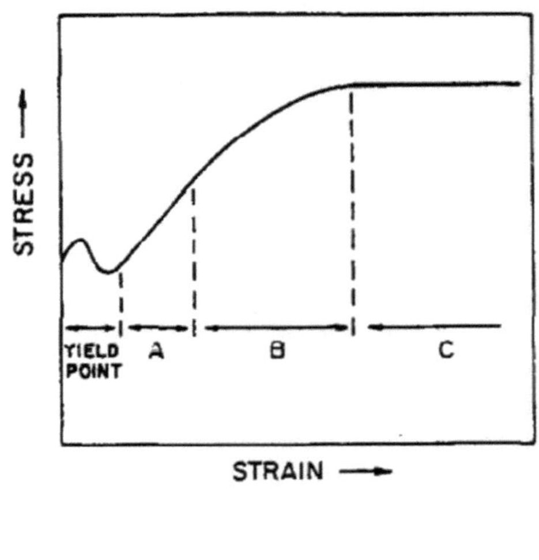

Fig. 5.1 Schematic stress–strain curve that has been divided into three regions beyond the lower yield point, Pletka et al. [1]. With kind permission of Elsevier

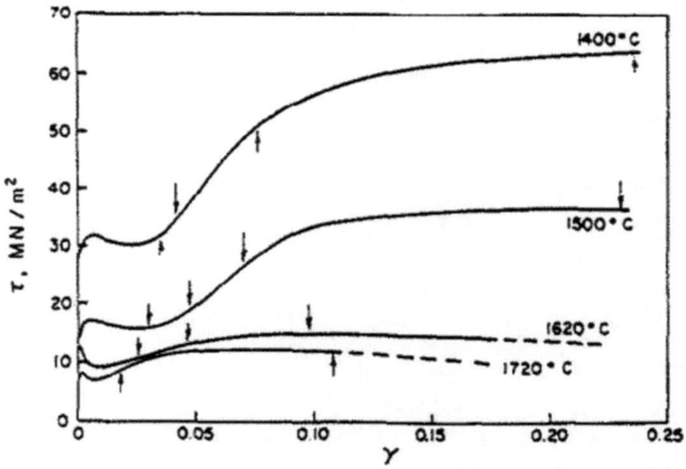

Fig. 5.2 Resolved shear stress-shear strain curves for sapphire deformed at temperatures from 1400–1720 °C at an initial strain-rate of 1.33 × 10^{-4} s^{-1}, Pletka et al. [1]. With kind permission of Elsevier

hardening rate observed in the decreasing stress–strain curves, a steady-state of zero work hardening is reached, when the rate of annihilation of the dipoles and loops by climb was equal to their rate of accumulation. The dislocation structure of the basal section of the sapphire deformed at 1500 °C is seen in Fig. 5.3. Dislocation densities were measured at several temperatures as a function of strain which were performed on specimens deformed beyond the lower yield point. The dislocation density in all theories show a parabolic dependence on stress as

$$\tau = x\mu b\rho^{1/2} \tag{5.1}$$

where τ is the flow stress, *x* a constant in the range of 0.2–0.5, μ the shear modulus, b the Burgers vector and ρ the total dislocation density. The square root of various dislocation densities is plotted as a function of the flow stress in Figs. 5.4, 5.5, 5.6, 5.7 and 5.8. The contribution of the dislocation-loop interaction, Δτ to the flow stress is

$$\Delta\tau = \frac{\beta\mu b}{\Lambda_l} \tag{5.2}$$

where Λ_l is the average spacing of the loops in the slip plane and β is a constant, of the order of unity and its magnitude is related to the strength of the dislocation-loop interaction. The values of β imply that loops are strong obstacles to the path of dislocations.

Dislocations emitted from a source glide and encounter a few loops acting as pinning points and bow between these pinning points until they break free and

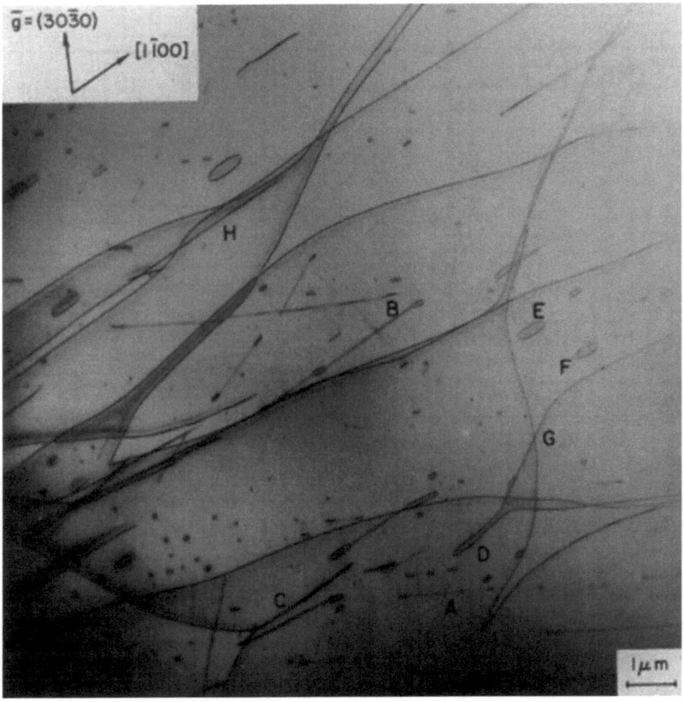

Fig. 5.3 Basal section of sapphire deformed at 1500 °C and an initial strain-rate of 1.33 × 10^{-4} s^{-1}. Examples of dipoles, glide dislocations and loops are illustrated in the micrograph, Pletka et al. [1]. With kind permission of Elsevier

Fig. 5.4 Square root of total dislocation density, ρ_t. for sapphire deformed at various temperatures, Pletka et al. [1]. With kind permission of Elsevier

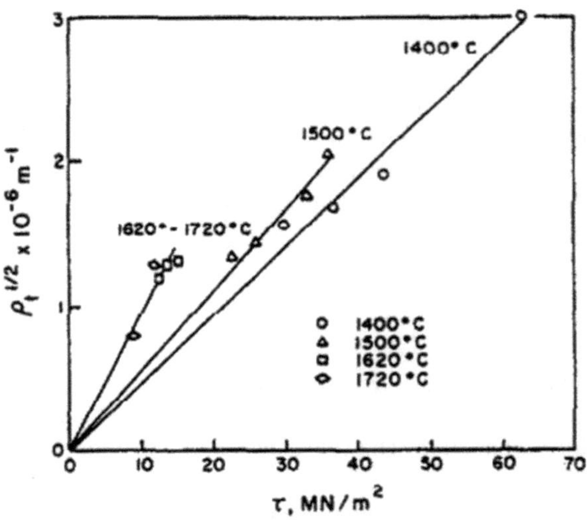

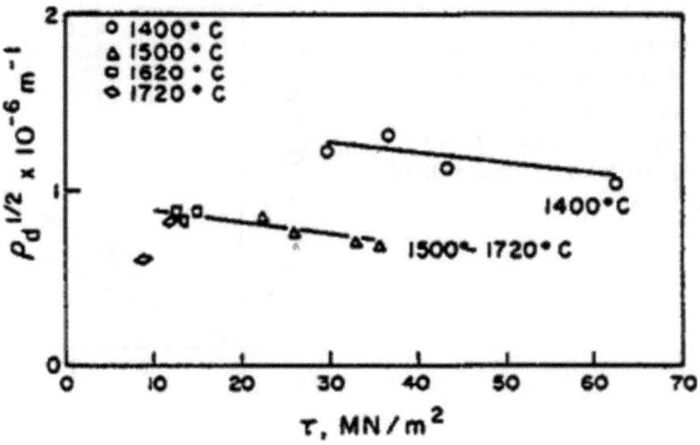

Fig. 5.5 Square root of dipole density, ρ_d versus τ for sapphire deformed at various temperatures, Pletka et al. [1]. With kind permission of Elsevier

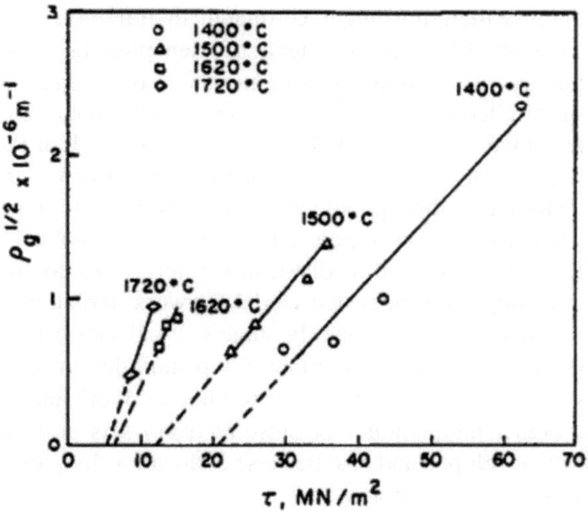

Fig. 5.6 Square root of glide dislocation density. ρ_g, versus τ for sapphire deformed at various temperatures, Pletka et al. [1]. With kind permission of Elsevier

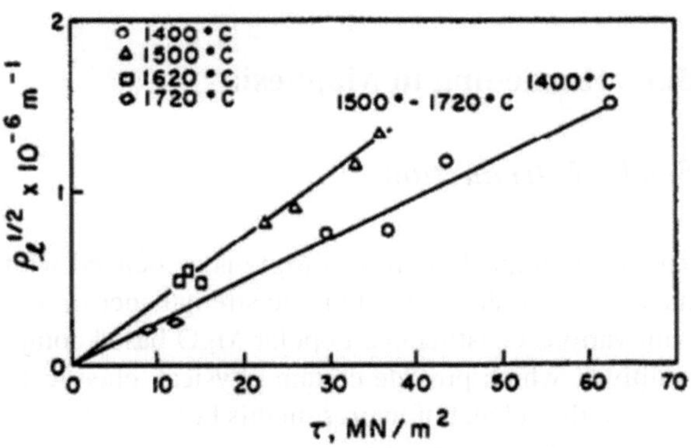

Fig. 5.7 Square root of loop density, ρ_l, versus τ for sapphire deformed at various temperatures, Pletka et al. [1]. With kind permission of Elsevier

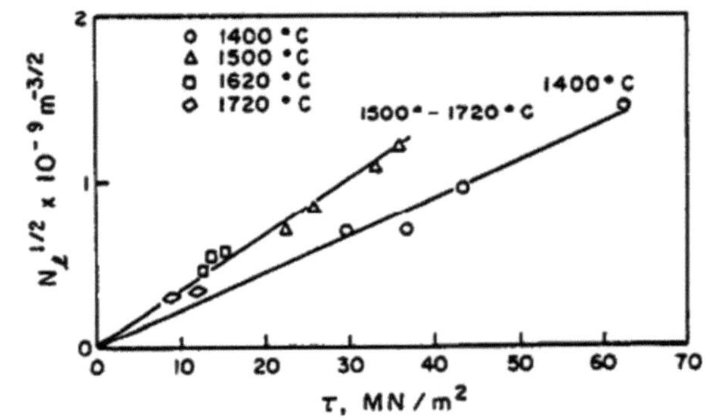

Fig. 5.8 Square root of the number of loops per unit volume, N_l, versus τ for sapphire deformed at various temperatures, Pletka et al. [1]. With kind permission of Elsevier

continue their movement. Ultimately their motion is stopped after encountering more and more obstacles until after a sufficient pinning points are intersected. Indicating the quantitative relations of work hardening is outside the scope of this book. However, for detailed description of the quantitative relations derived for sapphire crystals deformation, the reader should consult the work of Pletka et al.

In summary, the work hardening rate, θ is a maximum at stage A, decreasing continuously through stage B and reaching zero in stage C. The density of dislocations (glide dislocations, dipoles, and prismatic loops) and loops increase with increasing flow stress. Loop-glide dislocation interactions control the flow stress and the work hardening. The suggested model for work hardening is based on blocking the glide of dislocations by loops. In stage C a balance between recovery effects and work hardening becomes important. Loop annihilation is responsible for the recovery and work hardening is due loop accumulation. Work hardening rates (region A) are lower at higher temperatures because recovery sets in during yielding before stage A is fully developed and also because dislocation-loop interactions become less effective at higher temperatures.

5.3 Hardening in Magnesia

5.3.1 Introduction

The technological interest in MgO is associated with either as additives to various metallic, ceramic, and polymeric strengthener or as an alloy strengthened by itself with various constituents. Popular MgO based composites are those with C or SiC additives, which provide certain physical, chemical and/or mechanical properties. Clearly, the subject of interest in this book are the mechanical properties. One of the merits of MgO ceramics is its refractory character and its high temperature stability. Further attributes are its high thermal conductivity and low electrical conductivity. Therefore, a major use of MgO is as heating elements. MgO is also used as a basic refractory material for crucibles bricks and other refractories applicable at high temperatures. One has to emphasize the fire resistance, moisture resistance of MgO while maintaining high strength.

5.3.2 Strength of MgO

The MgO based composite chosen for illustrating the properties of MgO is MgO–C. The strength properties of 10, 15 and 20% composites are presented in this section. The relationship between the maximum compressive stress and the number of cycles to failure for the tree composites is shown in Fig. 5.9 for MgO. It is observed from the illustration that the curve becomes increasingly negative as the carbon content decreases, meaning that fatigue failure of the MgO composite occurs after less cycles or lower stresses. The load displacement curves evaluated experimentally with various C content are seen in Fig. 5.10.

The micrographs of MgO 10% C and 20% C are shown in Figs. 5.11 and 5.12, respectively. No large cracks are observed until the final stage, but some cracks were observed in the middle stage which were propagated and branched.

The crack progress behavior is evaluated according to

$$\frac{\mathrm{d}a}{\mathrm{d}t} = C\left(\frac{K_I}{K_{IC}}\right) \tag{5.3}$$

$$\frac{\mathrm{d}a}{\mathrm{d}N} = C\frac{\Delta K_I}{K_{IC}} \tag{5.4}$$

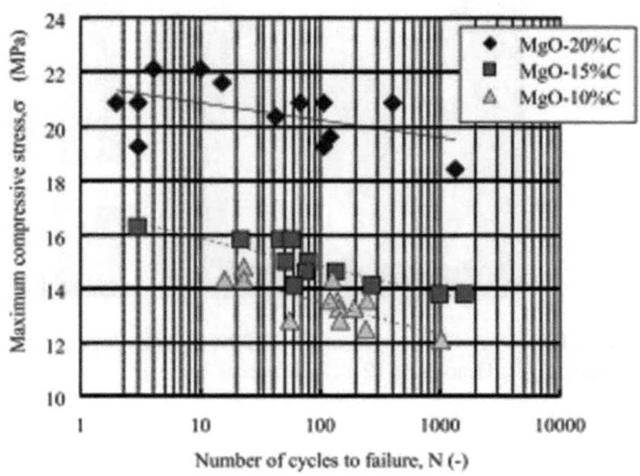

Fig. 5.9 Relationship between maximum compressive stress and number of cycles to failure in MgO–C bricks, Hino et al. [2]. Open access

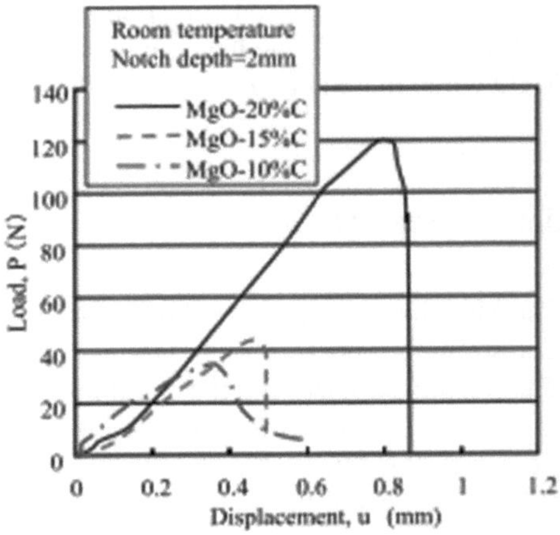

Fig. 5.10 Load–displacement curves of MgO–C bricks with different carbon contents, Hino et al. [2]. Open access

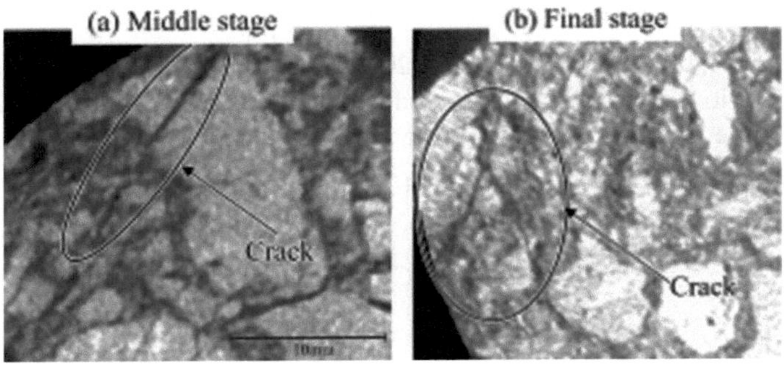

Fig. 5.11 X-ray CT images of MgO-10%C brick during fatigue failure test (**a** middle stage and **b** final stage), Hino et al. [2]. Open access. CT scan, is computed tomography scan

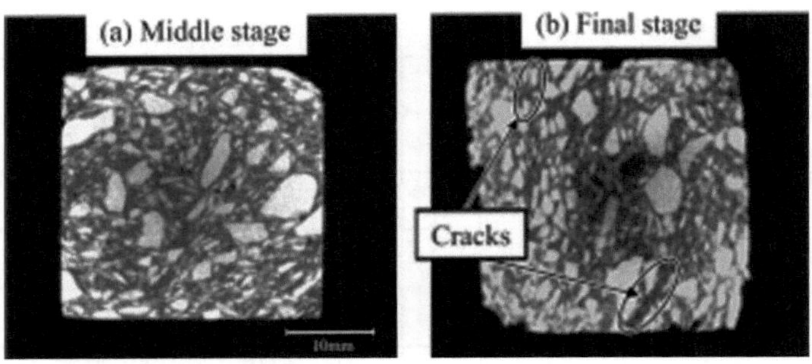

Fig. 5.12 X-ray CT images of MgO-20%C brick during fatigue failure test (**a** middle stage and **b** final stage), Hino et al. [2]. Open access

$$\log \Delta\sigma = -\frac{1}{n'}\log N + \frac{1}{n'}\log\left[\frac{2K_{IC}^{n'}}{CY\pi^{\frac{n'}{2}}(n'-2)}\left(a_0^{\frac{n'-2}{2}} - a_c^{\frac{n'-2}{2}}\right)\right] \tag{5.5}$$

n' is a material constant depending on experimental conditions, atmosphere and the temperature. $\Delta\sigma$ is the acting stress differential and N the number of loading (fatigue cycles). n′ can be considered as the degree of sensitivity to changes in the growth of the crack and in the stress intensity factor and is related also to the dynamic fatigue failure. A high value of n′ shows high resistance to dynamic fatigue failure and dynamic thermal fatigue. The value of n′ was determined from fatigue failure tests, assuming that the relation of Eq. (5.5) holds for each plot in Fig. 5.10. The value of n′ increases as the carbon content in brick increases. Clearly, K_{IC} is the fracture toughness of the material. K_I the stress intensity factor (in $MPa.m^{0.5}$) for model I where the fracture mode is according to Irwin's linear elastic theory.

In summary, the crack growth rate increases with decreasing the carbon content in MgO–C. X-ray CT scans indicate that no large cracks were observed until just before fracture in fatigue failure in specimens with large C content. However, at lower C content large cracks were observed even in in the middle of the tests. Thus, the C content in MgO–C affects the crack growth behavior from its appearance, propagation and growth until fatigue failure sets in.

5.4 Hardening in Pure Zirconia

Generally stabilized zirconia is the material of technological interest. However, few data on pure zirconia (ZrO_2) can be found sporadically when comparing alloyed zirconia with pure monolithic zirconia. Thus, for example the flexural strength of ZrO_2 versus indentation load is seen in Fig. 5.13 comparing it with 3Y-TZP/Ta, 3Y-TZP/Nb composites. Each of the data point represents the mean value of about twelve specimens tested at a given load. Linear fitting was applied. The fatigue limit for ZrO_2, ZrO_2/Ta and ZrO_2/Nb is 1200 ± 15 MPa, 860 ± 30 and 370 ± 30 MPa, respectively.

These data can be compared to the fatigue loading results. In fatigue loading a structure may fail at a small percentage of its fracture strength. The results for cyclic fatigue life for the specimens tested with a disc geometry are presented in a semi logarithmic form in Fig. 5.15 as peak stress versus cycles to failure (N). The tests were interrupted at $N = 10^7$ cycles, for the unfailed samples, which are marked with an arrow symbol. The number of stress cycles to failure is illustrated in Fig. 5.15. The indentation fatigue life for disc specimens is presented in semi logarithmic form as peak stress versus cycles to failure in Fig. 5.15. The indented specimens were tested under the same cyclic conditions as the un-indented ones. It is seen that fatigue life of the well-polished ceramic/metal composite without any artificial produced flaw is reduced when compared with monolithic zirconia (Fig. 5.14). Further, Fig. 5.15

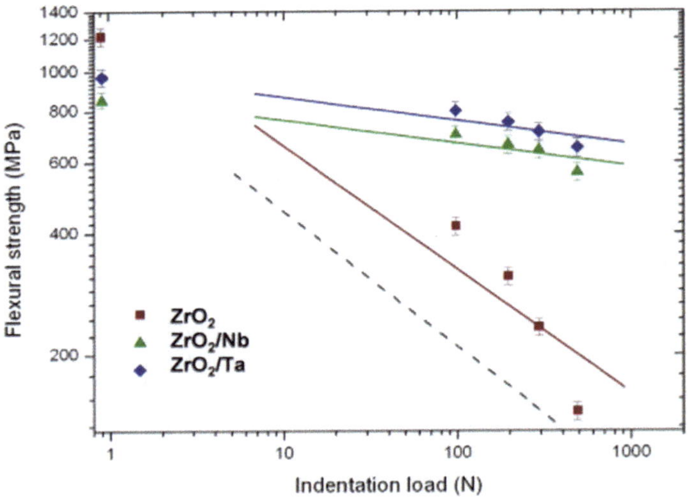

Fig. 5.13 Indentation load versus strength plots of 3Y-TZP/Ta, 3Y-TZP/Nb composites and zirconia ceramic, Smirnov et al. [3]. Open access

indicates that the fatigue limit changes much for the indentation fatigue. In this case the fatigue limit for ZrO_2/Ta, ZrO_2/Nb and ZrO_2 is 500 ± 25 MPa, 270 ± 10 MPa and 130 ± 40 MPa, respectively.

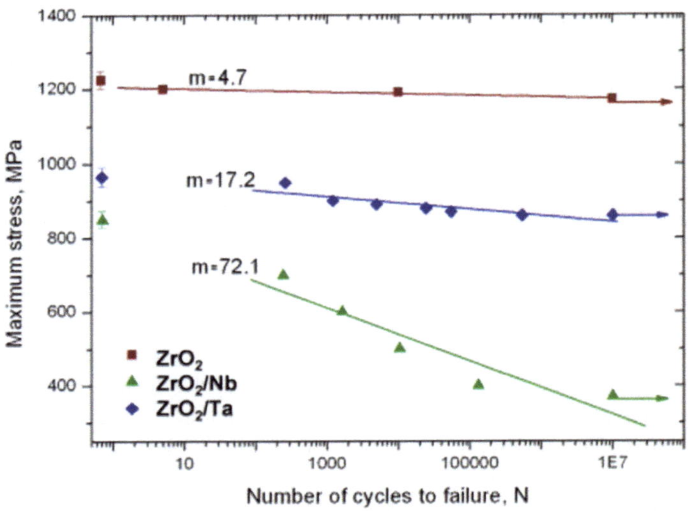

Fig. 5.14 Fatigue resistance S–N curves with values of slopes for un-indented polished specimens, Smirnov et al. [3]. Open access

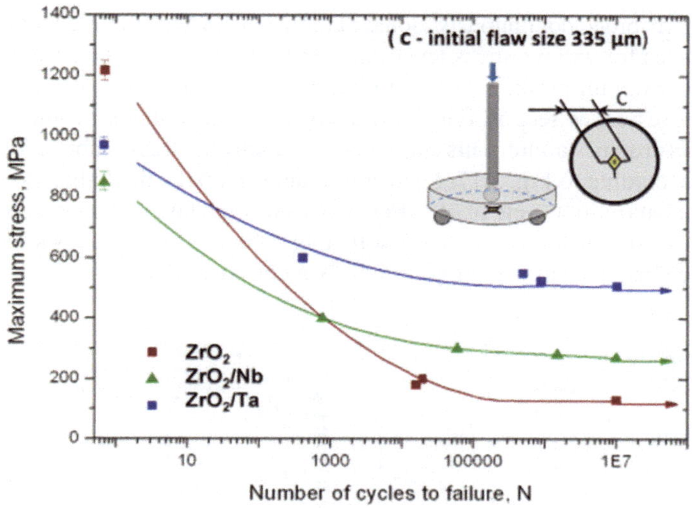

Fig. 5.15 S–N plot of the cycles to failure (N) of indented ZrO_2, ZrO_2/Ta and ZrO_2/Nb composites, Smirnov et al. [3]. Open access

The main issue of the work of Smirnov pivots around the composites ZrO_2/Ta, ZrO_2/Nb which are outside of the interest of this book. Regarding the monolithic ZrO_2, we can summarize the observations as: (1) The slope of ZrO_2 plot of σ_{max} with the number of fatigue cycles almost did not change up to 10^7 cycles· and is practically constant as observed in Fig. 5.14. (2) The fatigue performance of materials can be compared by the fatigue ratio, defined as the ratio of the fatigue limit to the static strength. For the monolithic ZrO_2 (shown in Table 2 of Smirnov et al. and not reproduced here), the ratio shows a significant decrease between un-indented and indented specimens. Therefore, the fatigue ratio of zirconia is extremely sensitive to surface condition and the presence of defects that act as stress raisers and, consequently, the chances of failure of monolithic ceramic is higher than in ceramic/metal composites. (3) Fatigue crack growth in artificially induced flaws in ZrO_2 at low applied loads—only of ~7% (80 MPa) of the flexural strength of un-indented specimens—This means that any flaw present in monolithic zirconia can lead to catastrophic failure.

5.5 Hardening in Partially Stabilized Zirconia (PSZ)

Addition of oxides such as CaO, MgO, CeO_2, Y_2O_3 to pure ZrO_2 can create partially stabilized zirconia (PSZ) which is stable at room temperature. Partial stabilized zirconia by MgO is shown below. To construct the plastic strain vs. number of cycles, the maximum plastic strain in tension and the plastic strain range of each cycle were recorded continuously. Some of these data in Mg-PSZ are plotted against fatigue cycles for various stress levels, as shown in Fig. 5.16. The accumulation of both, the maximum plastic strain and plastic strain range accelerates considerably at higher stresses, as seen in Fig. 5.17. Such strain accumulation could eventually lead to catastrophic failure. This should occur rapidly at stresses above 400 MPa in Mg-PSZ according to Fig. 5.17. Such stress, indeed, marks the fatigue tolerance of the material at run-out cycle of 10^3 (Fig. 5.16). Accompanying the increase in strain range is a tendency for cyclic strain softening (observed in hysteresis loop tilting toward the strain axis as cycling proceeds. Not considered here).

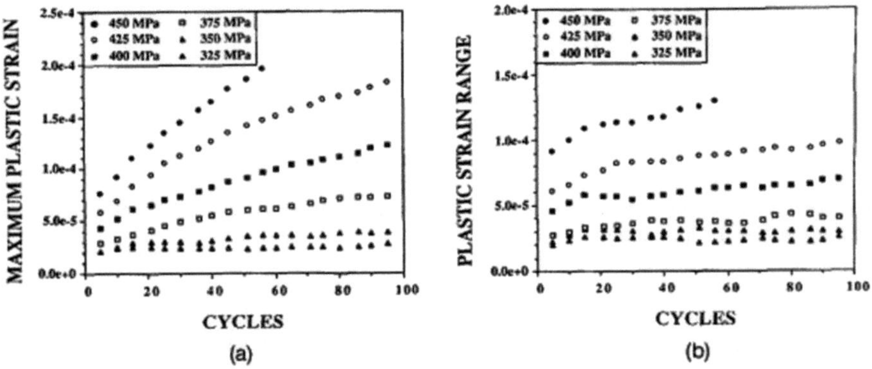

Fig. 5.16 **a** Maximum plastic strain versus fatigue cycles and **b** plastic strain range versus fatigue cycles, Liu and Chen [4]. With kind permission of John Wiley and Sons

Fig. 5.17 Plastic strain accumulated after 100 cycles versus maximum stress, Liu and Chen [4]. With kind permission of John Wiley and Sons

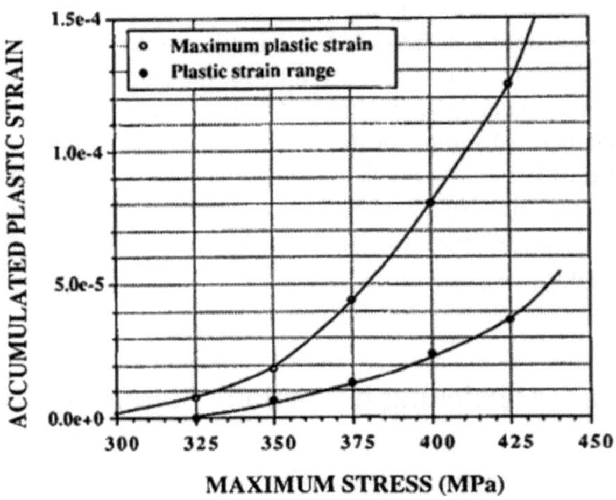

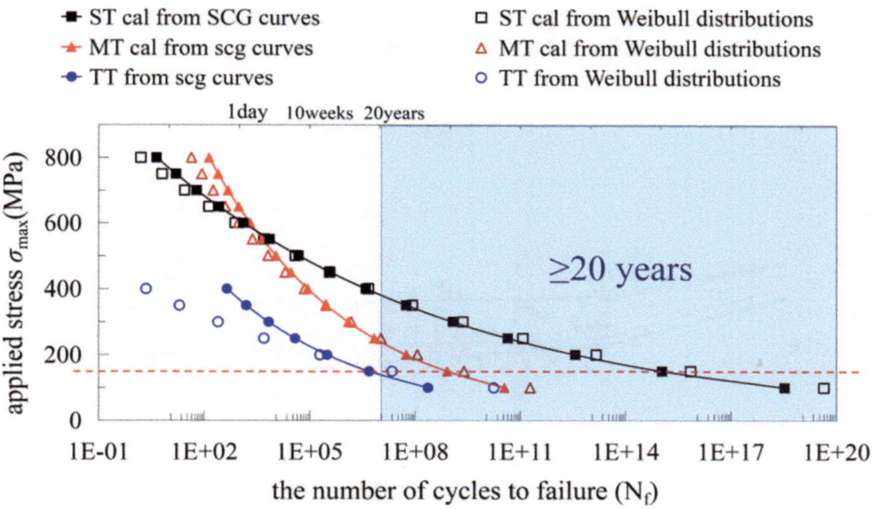

Fig. 5.18 Fatigue lifetime diagrams calculated from Weibull distributions and subcritical crack growth (SCG) curves without the effect of aging and without the effect of K_{I0}, Wei et al. [5]

5.5.1 *Hardening in Yttria Stabilized Zirconia (YSZ)*

Despite the extensive information on various topics related to yttria stabilized zirconia (YSZ), very limited data on cyclic strengthening of YSZ—the subject of this section—exists in the literature. However, in a very recent publication (Wei et al.) on lifetime assessment of yttria-stabilized zirconia dental ceramics, a relevant figure on the applied stress versus the number of cycles is included and shown in Fig. 5.18. In the figure three commercial zirconia dental materials are indicates and designated as super translucence (ST), medim translucence (MT) and tooth translucence (TT). TT is 5% yttria stabilized zirconia. For dental applications translucency is an important feature.

5.5.2 *Y-TZP Zirconia Fatigue Behavior in Water*

Recall that Y-TZP stands for yttria tetragonal zirconia polycrystal. It is extensively applied for fixed dental restorations due to its excellent mechanical properties, i.e., bending strength and toughness. It is thus exposed to saliva and its cyclic fatigue lifetime is of great interest. Therefore, to simulate actual performance in saliva, experiments were performed in water. Fatigue data for the "as received" and "CoJet sandblasted" zirconia are presented as S–N diagrams in Figs. 5.19a, b. The designation of the four specimens tested are: Zeno (ZW) (Wieland), Everest ZS (KV) (KaVo),

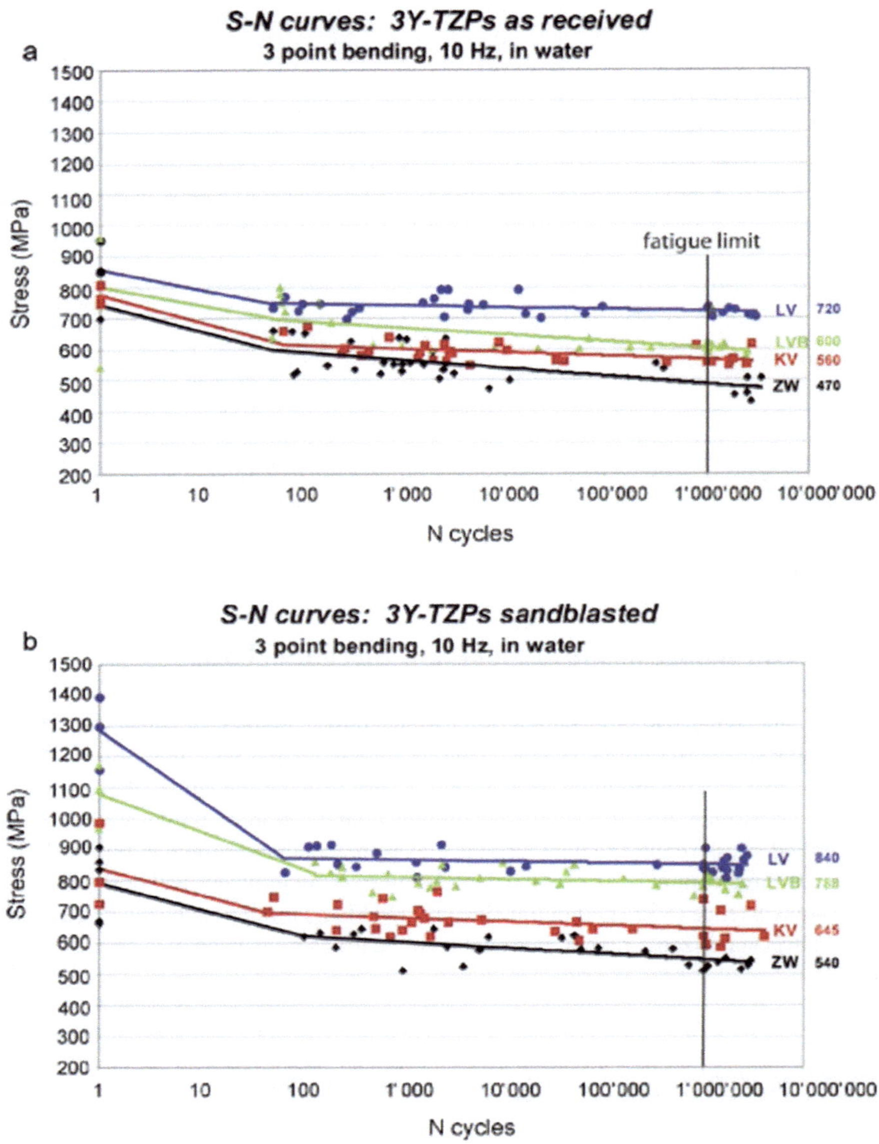

Fig. 5.19 **a** and **b** S–N fatigue data for Lava (LV) (3M Espe), Lava colored (LVB) (3M Espe), Everest ZS (KV) (KaVo), Zeno (ZW)(Wieland) "as received" and "after CoJet sandblasting". Fatigue limits at 106 cycles at 10 Hz in water were determined from trend lines using a least square linear regression fit. The fatigue limits "as received" were: LV = 720 MPa, LVB = 600 MPa, KV = 560 MPa, ZW = 470 MPa. The fatigue limits of "CoJet sandblasted" were: LV = 840 MPa, LVB = 788 MPa, KV = 645 MPa, ZW = 540 MPa. The initial strength for the "as received" was: LV = 854 ± 94 MPa, LVB = 749 ± 209 MPa (one outliner with major grinding flaw was responsible for dropping the mean initial strength), KV = 776 ± 31 MPa, ZW = 742 ± 39 MPa. The initialstrength for the "after CoJet sandblasting" was: LVs = 1282 ± 119 MPa, LVBs = 1077 ± 102 MPa, KVs = 836 ± 135 MPa, Scherrer et al. [6]. With kind permission of Elsevier

Lava white (LV) and Lava colored (LVB) (3M Espe). The fatigue limits (also indicated in Fig. 5.19) are: "as received" (ctr) LV = 720 MPa, LVB = 600 MPa, KV = 560 MPa, ZW = 470 MPa. The fatigue limits "after CoJet sand blasting" (CoJet stands for alumina–silica coated particle) were: LV = 840 MPa, LVB = 788 MPa, KV = 645 MPa, ZW = 540 MPa. The increase in fatigue limit after sandblasting was 15% for Zeno (ZW) and Everest (KV), 17% for Lava (LV) and 31% for Lava colored (LVB). As mentioned earlier yttria-stabilized zirconia ceramics is a high-performance material with excellent biocompatibility and mechanical properties, which suggest its suitability for posterior fixed partial dentures. The hypothesis is that the strength and reliability of Y-TZP zirconia ceramics are affected by the inner surface grinding of crowns and vary with the grinding parameter. Therefore, flexural strength, surface roughness, and fracture toughness were determined on samples machined by face and peripheral grinding with varied feed velocities and cutting depths to simulate the actual action of dental crowns while in general use. As seen in Fig. 5.20 an increase of 17% in fatigue limit is obtained in the sand blasted Y-TZP. Thus, sand blasting improves the fatigue resistance. Processing flows associated with fabrication and sintering are observed by SEM following fatigue tests. Grinding and pressing flaws are illustrated in the S–N diagram at Fig. 5.21.

The improved fatigue strength after sandblasting may be related to the presence of compressive stresses on the tensile surface. These could result from changes in the crystalline structure of the Y-TZP ceramics. X-ray diffraction data confirmed the

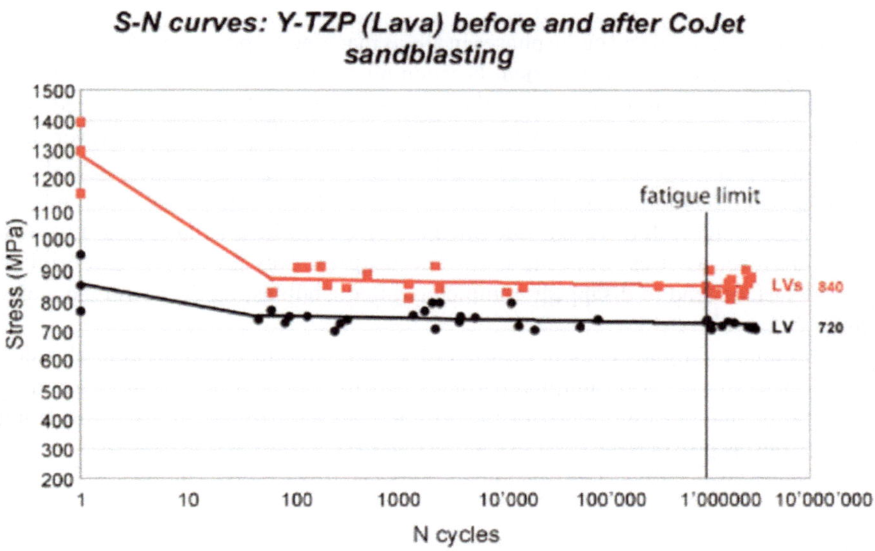

Fig. 5.20 S–N fatigue data for Lava (LV) "as received" (black, bottom) and "CoJet sandblasted" (LVs) (red, top). Fatigue limits at 10^6 cycles at 10 Hz in water were 720 MPa "as received" and 840 MPa for "CoJet sandblasted" which represented an increase of 17%, Scherrer et al. [6]. With kind permission of Elsevier

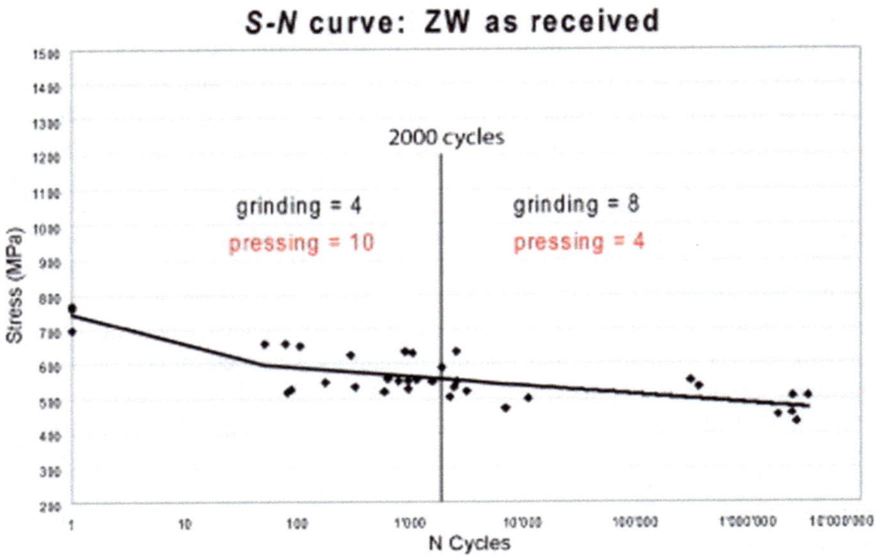

Fig. 5.21 Grinding and blank pressing failure flaws on the fatigue curve for ZW "as received" specimens. For specimens failing below 2000 cycles, the pressing flaws outnumbered the grinding flaws (4 vs. 10). After 2000 cycles, the grinding flaws prevailed (8 vs. 4), Scherrer et al. [6]. With kind permission of Elsevier

presence of tetragonal and cubic phases in all 'as machined' specimens. After blasting with CoJet™Sand, an asymmetrical broadening at the (1 0 1), (2 0 2) and (1 1 3) peaks and the reversal of intensities of peaks (0 0 2) and (2 0 0) were observed. The broadening could indicate the emergence of a new 'cubic 2' phase and the reversal of intensity of peaks (2 0 0) and (0 0 2) may be linked to the presence of ferroelasticity in tetragonal zirconia. An additional explanation for the improvement of the fatigue limits after sandblasting is the smoothening of the surface's machining grooves, which may act as stress concentrators. The deep grooves present on the original ground surface tend to disappear completely after sandblasting 20 s with the CoJet Sand.

In summary, fatigue behavior in water of Y-TZP zirconia ceramics after abrasion with 30 μm silica-coated alumina particles was investigated. Removal of grinding groves after sandblasting improves the fatigue resistance of Y-TZP used for dental applications as prostheses.

5.6 Hardening in SiC

Generally, SiC is used as a reinforcing agent to material, in particular to metallic materials. It should therefore not be surprising that only limited information exists on pure or monolithic SiC and literature quotes usually its composites. Very popular composites are Al/SiC and to a large extent those based on C/SiC. But before considering these composites let us look at the SiC/SiC composite.

5.6.1 Damage in SiC/SiC

In this section, the fatigue damage and lifetime of 2D SiC/SiC ceramic-matrix composites (CMCs) under cyclic fatigue loading at 750, 1000, 1100, 1200 and 1300 °C in air and in steam atmosphere is considered, presenting the fatigue peak strain and interface shear stress. The presence of steam accelerated the damage development inside of SiC/SiC composites. The fatigue limit stress values related to the tensile stress in air in the temperature range 750–1300 °C are compared with those in steam. Recall that ceramic materials have high strength and modulus at elevated temperatures, but their use as definite structural components are limited by their brittleness. Therefore, addition in form of fibers reinforces the ceramic-matrix, resulting in a composite with high strength and with reduced propensity for catastrophic failure. A life prediction model is discussed in detail by Li comparing with experimental and is not considered in this section. The respective figures in air and steam at various temperatures are illustrated in Figs. 5.22, 5.23, 5.24, 5.25, 5.26, 5.27, 5.28 and 5.29. The time dependent fiber strength will be controlled by surface defects resulting from oxidation and is given by Eq. (5.6) which was applied to Fig. 5.22, where the calculated and experimental data are compared.

$$\begin{aligned} \sigma_0(t) &= \sigma_0, t \leq \frac{1}{k}\left(\frac{K_{Ic}}{Y\sigma_0}\right)^4 \\ \sigma_0(t) &= \frac{K_{Ic}}{Y\sqrt[4]{kt}}, t > \frac{1}{k}\left(\frac{K_{Ic}}{Y\sigma_0}\right)^4 \end{aligned} \tag{5.6}$$

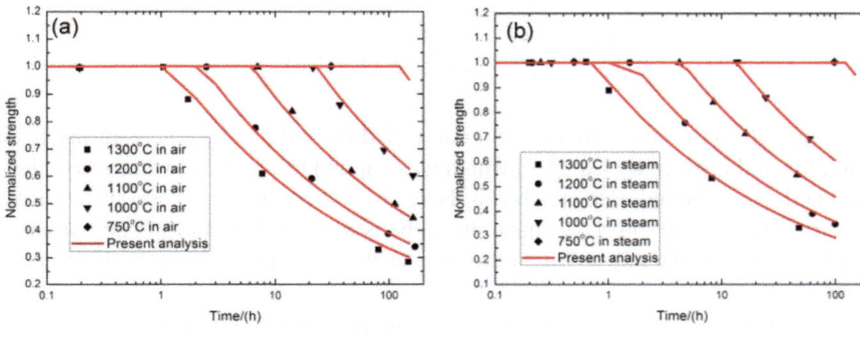

Fig. 5.22 The strength degradation of SiC fibers **a** in air and **b** in steam conditions at elevated temperatures, Li [7]. Open access

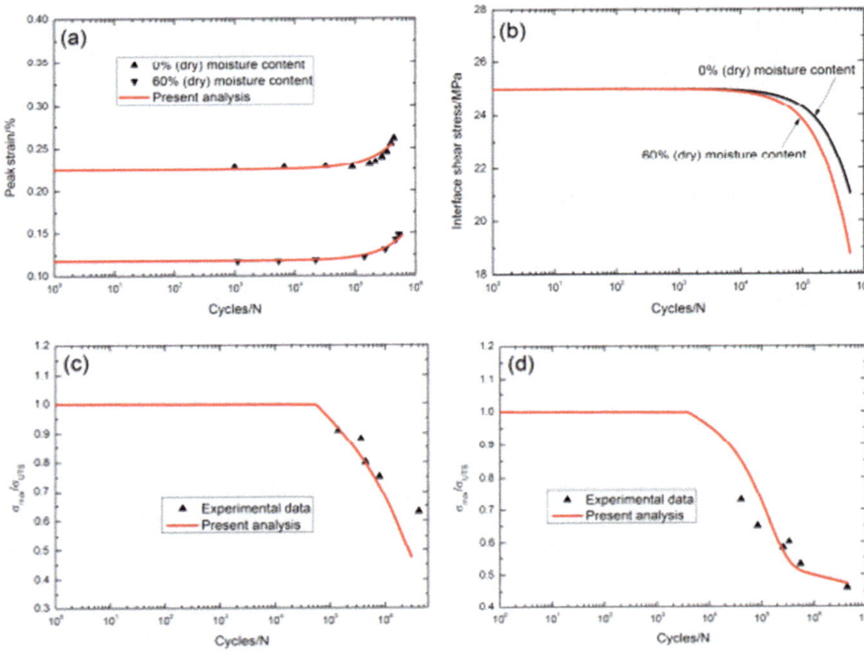

Fig. 5.23 **a** The peak strain versus applied cycles with test environment of 0% moisture content condition and 60% moisture content condition; **b** the interface shear stress versus applied cycle with test environment of 0% moisture content condition and 60% moisture content condition; **c** the fatigue life S–N curve with test environment of 0% moisture content condition; and **d** with test environment of 60% moisture content condition of 2D SiC/SiC composite at 750 °C in air, Li [7]. Open access

where K_{Ic} is the critical stress intensity factor, Y is a geometric parameter and k is the parabolic rate constant. The interface shear stress and the analysis according to Eq. (5.7)

$$\frac{(\tau_i(N) - \tau_s)}{(\tau_0 - \tau_s)} = (1 + b_0)\left(1 + b_0 N^j\right)^{-1} \tag{5.7}$$

were applied to the fiber strength and controlled by surface defects resulting from oxidation and is given by Eq. (5.7) which was applied to Fig. 5.22, where the calculated and experimental data are shown. Here, τ_0 is the initial interface shear stress, τ_s is the steady-state interface shear stress, b_0 is a coefficient and j is an exponent which determines the rate at which the interface shear stress drops with the number of cycles N.

In summary, with the increase of the fatigue peak stress, the degradation rate of the fatigue modulus and the interface shear stress increase and also the strain rate of the fatigue peak strain increases. With decreasing loading frequency, the degradation

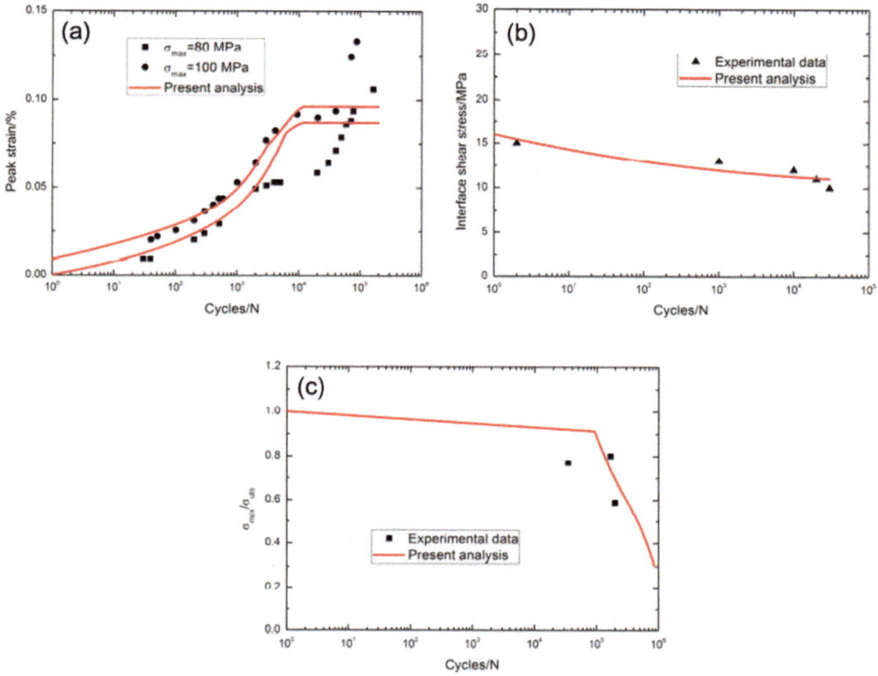

Fig. 5.24 **a** The peak strain versus applied cycles; **b** the interface shear stress versus applied cycles; and **c** the fatigue life S–N curves of 2D SiC/SiC composite at 1000 °C in air, Li [7]. Open access

of the modulus and the interface shear stress increase and the increasing rate of the fatigue peak strain and the fatigue hysteresis dissipated energy increases.

5.6.2 Hardening in Al/SiC

This section thus, concentrates on Al/SiC composites. The composite material is a 2124 powder aluminum alloy reinforced with SiC particulates. Three volume fractions (20, 25 and 30 vol.%) of SiC, were investigated. Following mixing the powders, bars (35 × 95 × 350 mm) were fabricated by hot compaction and extrusion. The machined surfaces of the test specimens were ground with 600 grit silicon carbide paper and then mechanically polished to remove all circumferential scratches and machine marks. Fully reversed, total strain controlled, tension–compression (R = 1) low cycle fatigue tests were performed at a constant strain rate of $1 \times 10^{-3}\ s^{-1}$.

The cyclic stress response curves of the composites having different volume fraction of SiC, are shown in Figs. 5.30, 5.31, 5.32, 5.33, 5.34, 5.35 and 5.36. At all plastic strain amplitudes, the composite material progressively hardened and softened only prior to failure. The hardening dominated the fatigue life up to 80–90%

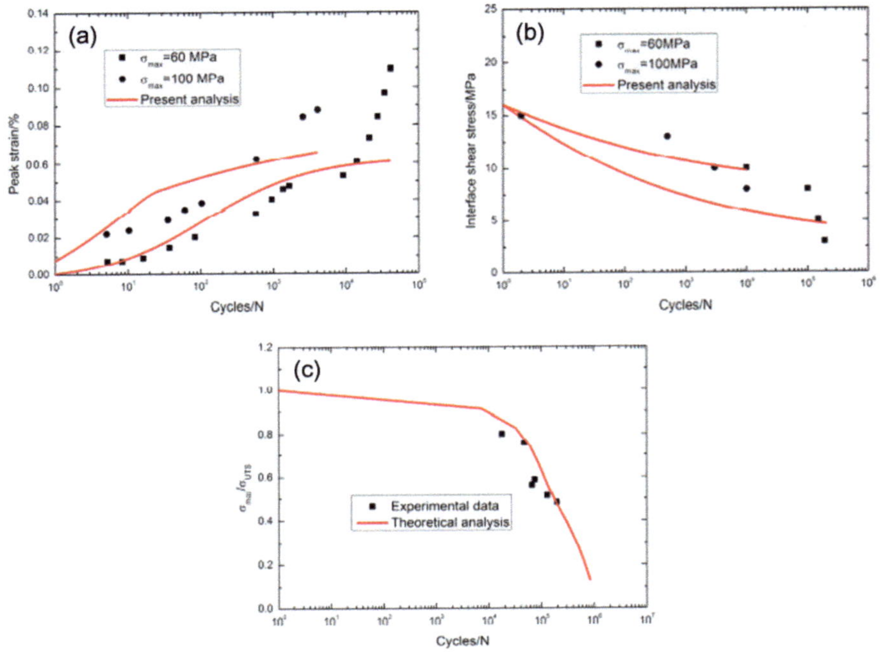

Fig. 5.25 **a** The peak strain versus applied cycles; **b** the interface shear stress versus applied cycles; and **c** the fatigue life S–N curves of 2D SiC/SiC composite at 1000 °C in steam, Li [7]. Open access

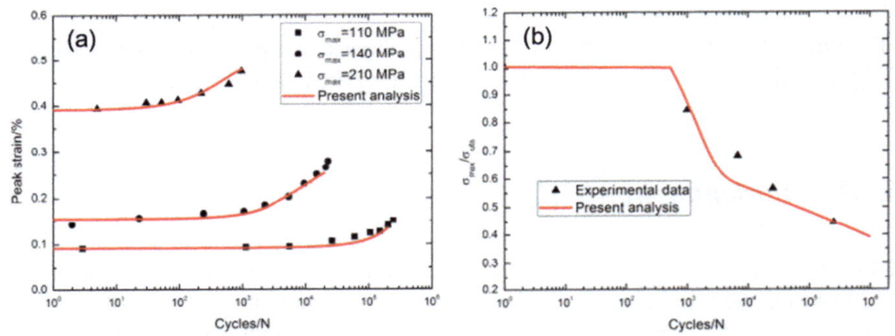

Fig. 5.26 **a** The peak strain versus applied cycles; and **b** the fatigue life S–N curves of 2D SiC/SiC composite at 1100 °C in air, Li [7]. Open access

but softening (drop load carrying capability) occurred at the onset of crack initiation and its growth. The effect of the cyclic strain amplitude is seen in Figs. 5.33, 5.34 and 5.35 at near equivalent strain. It is seen, that for a given volume fraction of SiC in the composite, the tendency of cyclic hardening decreased at lower cyclic strain amplitudes. Also it is observed that the overall degree of cyclic hardening during

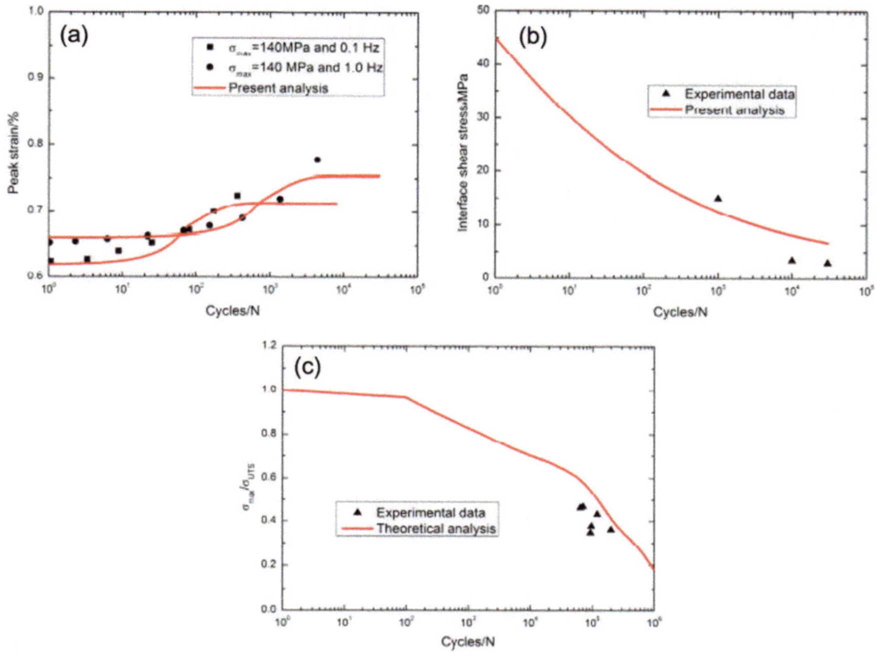

Fig. 5.27 **a** The peak strain versus applied cycles; **b** the interface shear stress versus applied cycles; and **c** the fatigue life S–N curves of 2D SiC/SiC composite at 1200 °C in air, Li [7]. Open access

straining decreased with the decrease in volume fraction of SiC reinforcing agent. Thus, the alloy containing 30 vol. % SiC has the greater cyclic strength and greater degree of hardening of the Al based composite. The hardening occurs by a progressive increase in stress in both aspects of the LCF test in tension and compression as seen in Fig. 5.36. Three stages can be observed in the cyclic hardening: (i) rapid hardening in the initial few cycles, followed by (ii) progressive hardening (strengthening) in most of the fatigue life and (iii) just prior to failure an instantaneous drop in the stress carrying capability occurs due to multiple crack initiation and rapid macroscopic crack growth. The cyclic stress–strain behavior is characterized by Fig. 5.37 where the stress amplitude is plotted versus the longitudinal strain amplitude which obeys Eq. (5.8).

$$\frac{\Delta\sigma}{2} = k'^{\left(\frac{\Delta\varepsilon_p}{2}\right)^{n'}} \tag{5.8}$$

where k′ is the cyclic strength coefficient and the exponential term (n′) is the cyclic strain hardening exponent, which provides a measure of the material's response to cyclic straining.

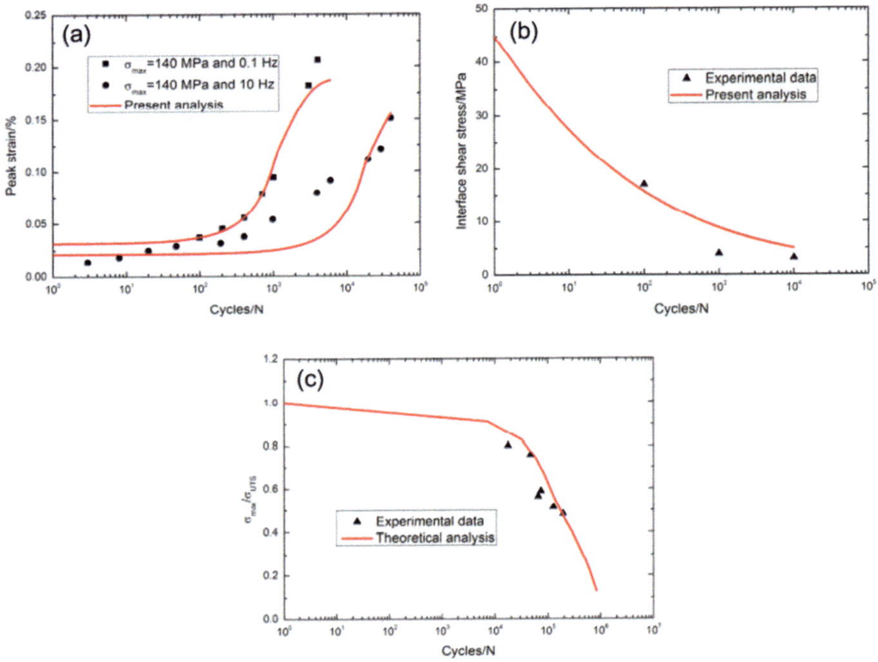

Fig. 5.28 **a** The peak strain versus applied cycles; **b** the interface shear stress versus applied cycles; and **c** the fatigue life S–N curves of 2D SiC/SiC composite at 1200 °C in steam, Li [7]. Open access

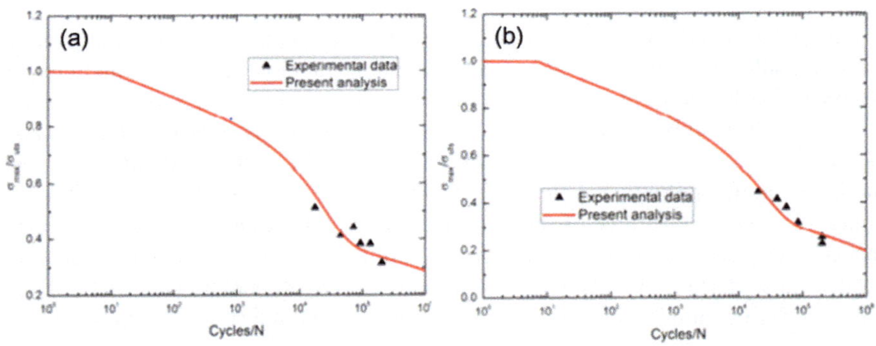

Fig. 5.29 **a** The fatigue life S–N curve with test environment in air; and **b** the fatigue life S–N curve with the environment in steam of 2D SiC/SiC composite at 1300 °C, Li [7]. Open access

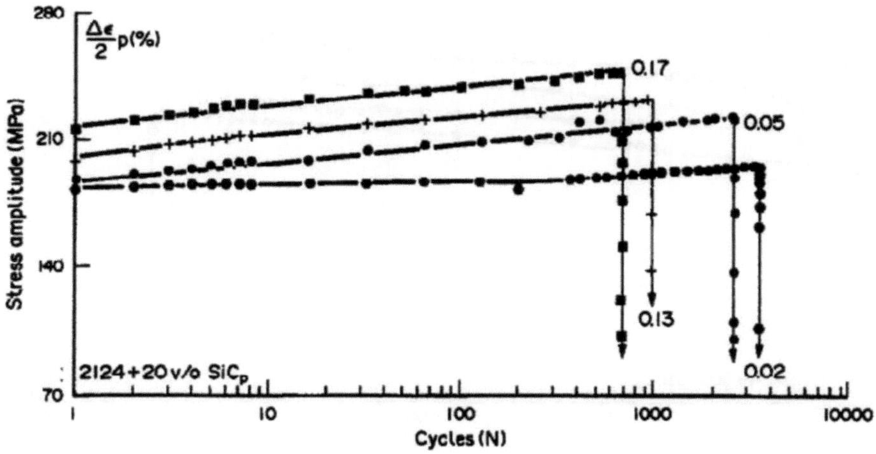

Fig. 5.30 Cyclic stress response curves for the powder metallurgy aluminum alloy 2124 + 2O v/o SiC_p (v/o stands for volume percent and SiC_p means SiC powder), Srivastan et al. [8]. With kind permission of Elsevier

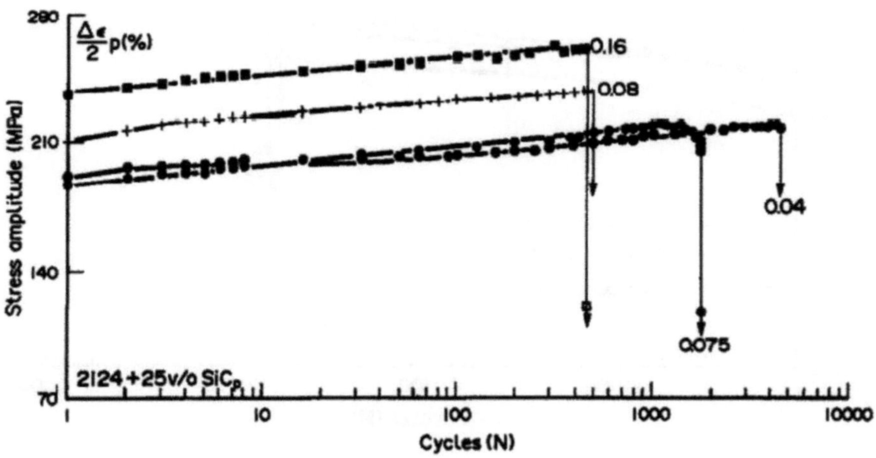

Fig. 5.31 Cyclic stress response curves for the powder metallurgy aluminum alloy 2124 + 25 v/o SiC_p, Srivastan et al. [8]. With kind permission of Elsevier

Crack initiation is seen in Fig. 5.38. The SEM illustration indicates that crack initiation starts with rupture of the hard SiC particle. With continued cyclic deformation many more SiC particles crack due to their inherent brittleness and decohesion at the matrix SiC particle interface. With further cyclic deformation the cracks at neighboring particles link to form a macroscopic crack, in the wake of which rapid softening occurs which is associated with a drop in the stress carrying capacity of

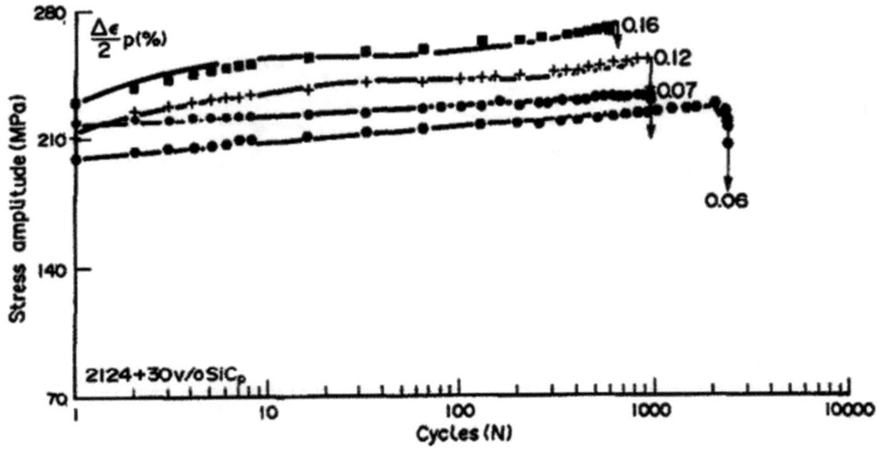

Fig. 5.32 Cyclic stress response curves for the powder metallurgy aluminum alloy 2124 + 3O v/o SiC_p, Srivastan et al. [8]. With kind permission of Elsevier

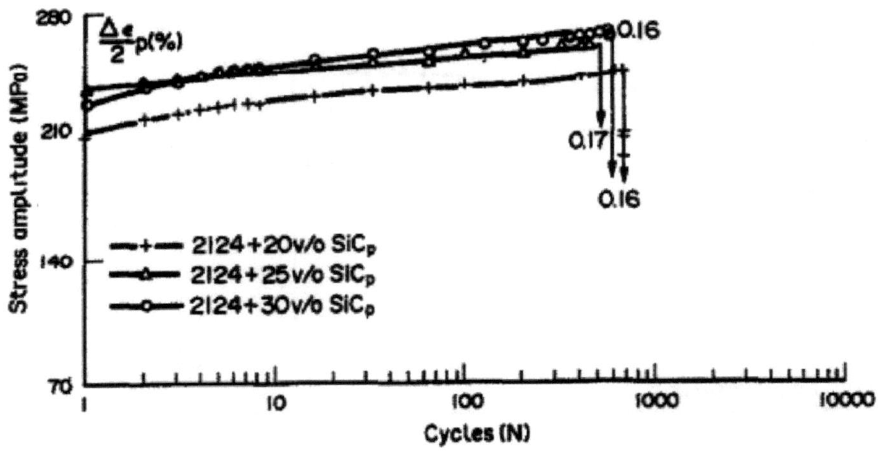

Fig. 5.33 Comparison of the effect of volume fraction of reinforcement on stress response behavior of the 2124 alloy at near equivalent strain amplitude ($\frac{\Delta\varepsilon_p}{2} = 0.16\%$), Srivastan et al. [8]. With kind permission of Elsevier

the composite prior to failure. At high plastic strain amplitude, fracture is observed to be brittle with low cyclic ductility and the fracture surface consists of microvoids homogeneously distributed and coalescence of microvoids as seen in Fig. 5.39. The macroscopic voids with those of the microscopic voids indicate ductile rupture. The sources of voids and dimples are attributed to fracture and decohesion of the SiC particles. It might be of interest to add Figs. 5.40 and 5.41 as illustrations of the frac-

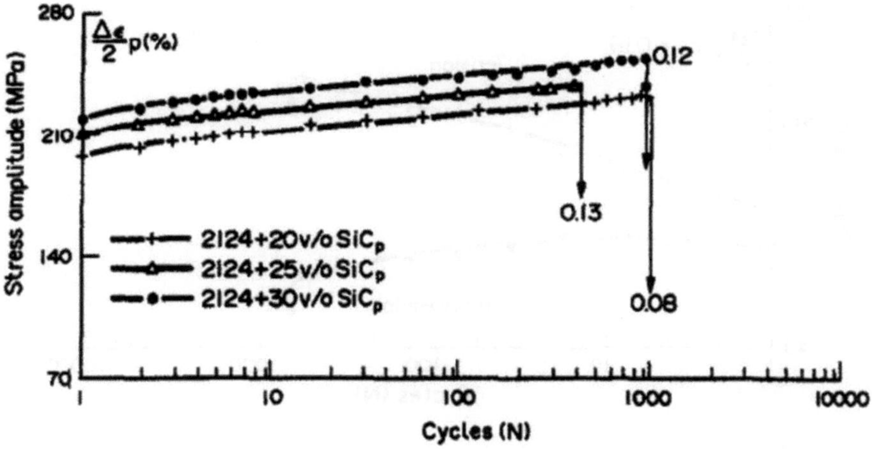

Fig. 5.34 Comparison of the effect of volume fraction of reinforcement on stress response behavior of the 2124 aluminum alloy at near equivalent intermediate strain amplitude ($\frac{\Delta \varepsilon_p}{2} = 0.12\%$), Srivastan et al. [8]. With kind permission of Elsevier

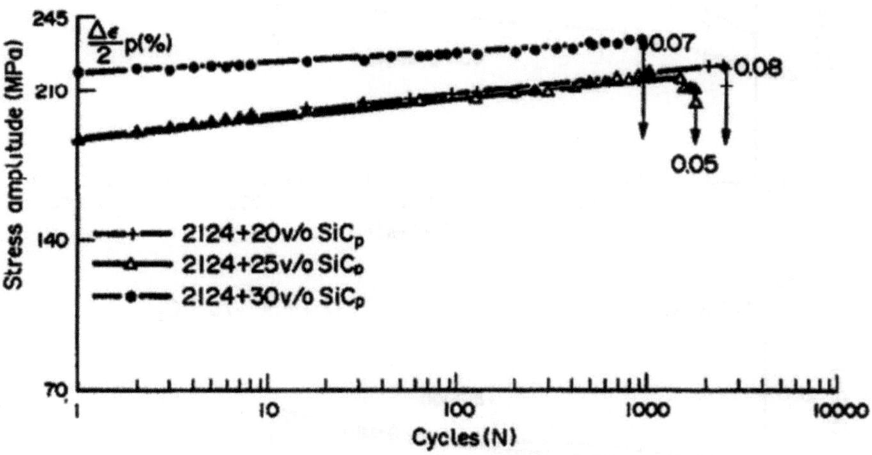

Fig. 5.35 Comparison of the effect of volume fraction of reinforcement on stress response behavior of the 2124 aluminum alloy at near equivalent strain amplitude ($\frac{\Delta \varepsilon_p}{2} = 0.07\%$), Srivastan et al. [8]. With kind permission of Elsevier

ture surface of the highest SiC content (30 vol. %) deformed to failure at different strain amplitudes.

In summary, the mechanisms associated with the cyclic stress response of the composite which controls the variation of stress amplitude with fatigue cycling deformations are dependent on the microstructure and the strain range. Since the

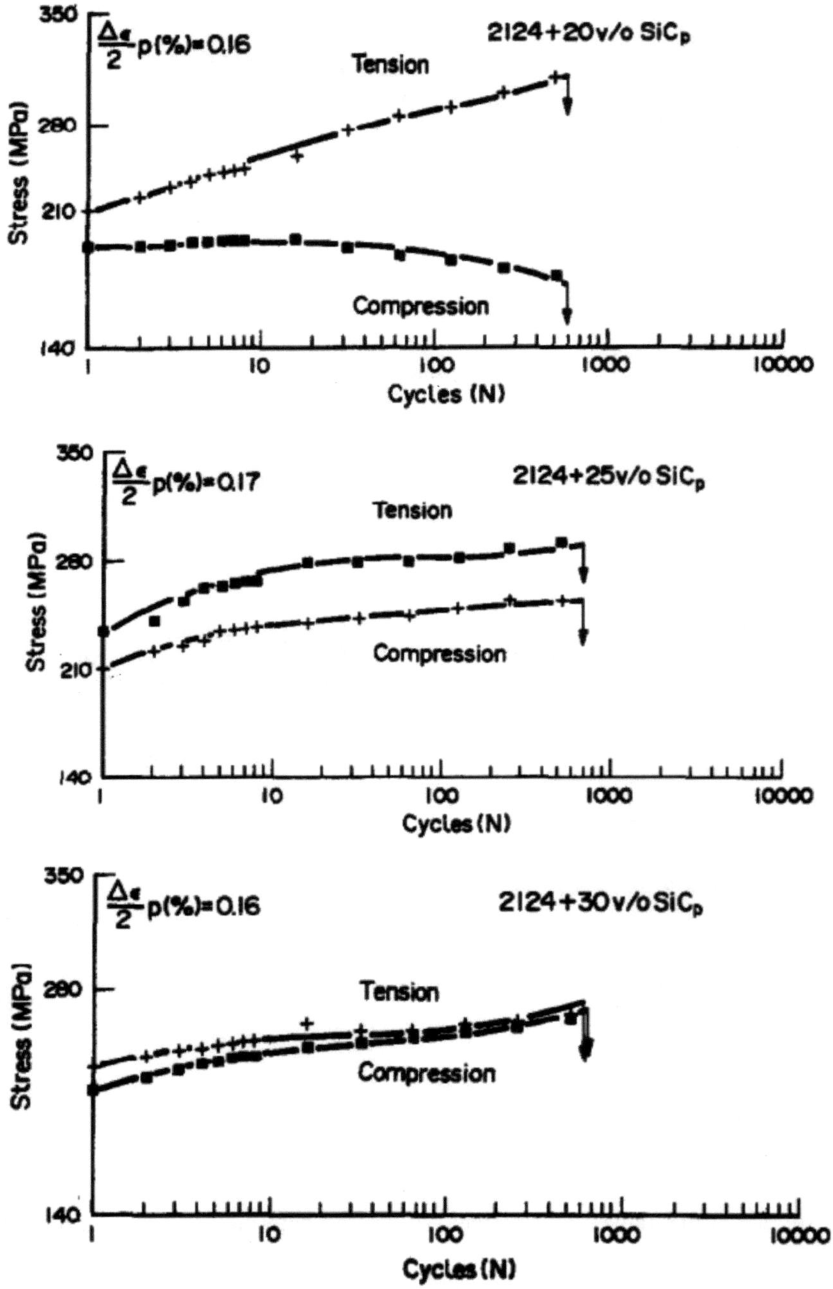

Fig. 5.36 Comparison of the effect of ceramic particle reinforcement on cyclic stress–strain curve for the powder metallurgy 2124 alloy, Srivastan et al. [8]. With kind permission of Elsevier

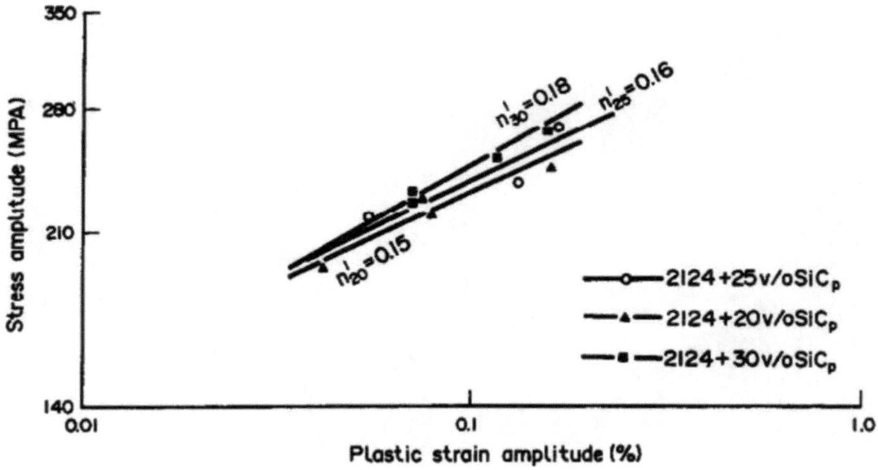

Fig. 5.37 Stress response curves showing the variation of tensile stress and compressive stress with cycles, at a fixed plastic strain amplitude, Srivastan et al. [8]. With kind permission of Elsevier

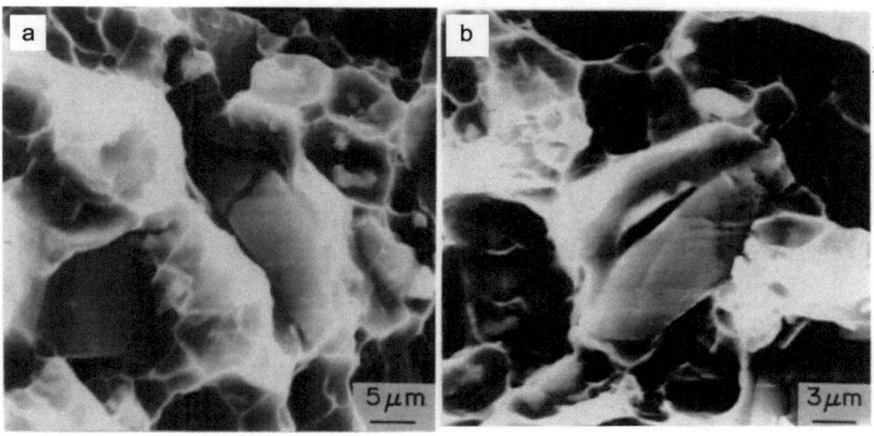

Fig. 5.38 Scanning electron micrograph showing crack initiation at ceramic (silicon carbide) particle. **a** 2124 + 25 v/o SiC, **b** 2124 + 3O v/o SiC, Srivastan et al. [8]. With kind permission of Elsevier

extruded microstructure was not heat treated by aging (no precipitation occurring for the strengthening), the hardening is possibly due to:

(i) the pre-existing high dislocation content in the matrix caused by the presence of SiC particles.
(ii) the dislocations due to the large thermal expansion coefficient mismatch between the aluminum matrix and SiC particles.

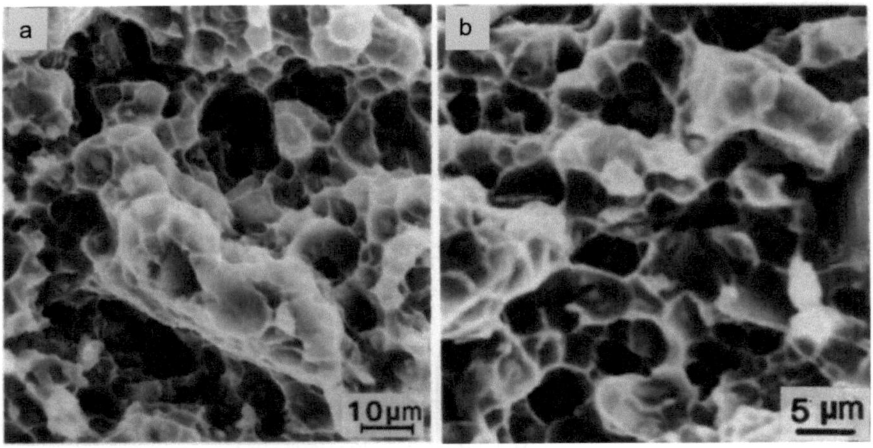

Fig. 5.39 Scanning electron micrograph showing voids and shallow dimples on the fatigue fracture surface, Srivastan et al. [8]. With kind permission of Elsevier

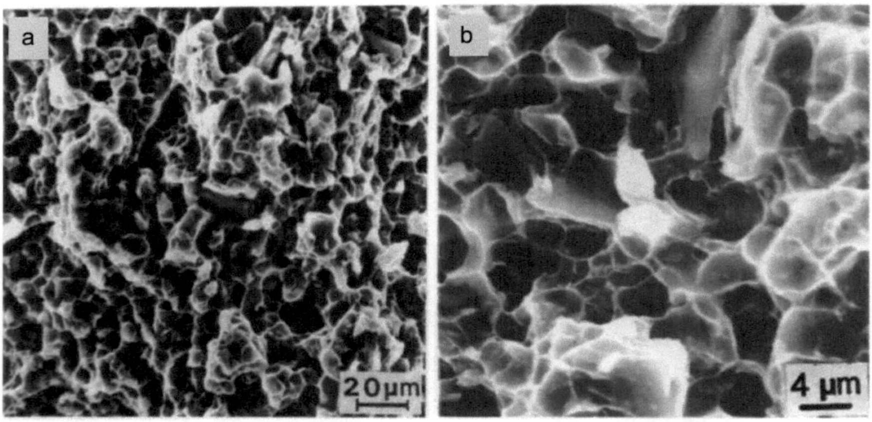

Fig. 5.40 Scanning electron micrograph of the fatigue fracture surface of the 2124 + 3O v/o SiC, sample deformed to failure at cyclic plastic strain amplitude of 0.15%, Srivastan et al. [8]. With kind permission of Elsevier

These dislocations are distributed throughout the matrix of the composite and the hardening during cyclic straining result from

(a) Multiplication of dislocation thus increasing the density of the already present dislocations,
(b) Interaction between dislocation—SiC particle and
(c) An increase in dislocation–dislocation interaction.

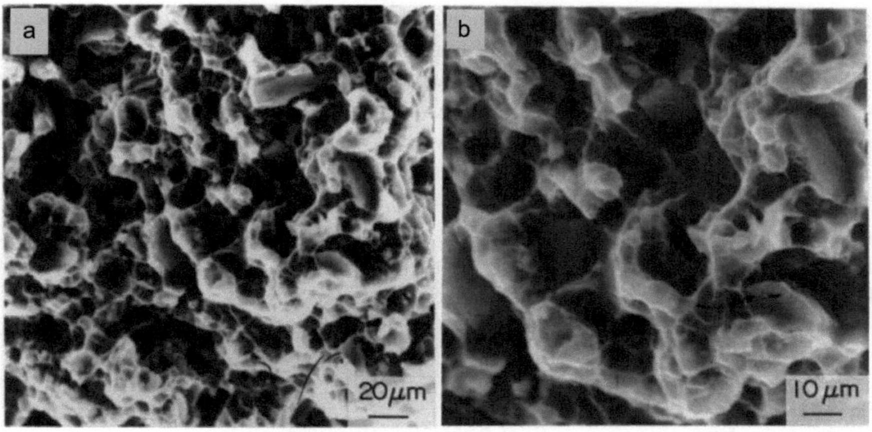

Fig. 5.41 Scanning electron micrograph of the fracture surface of the 2124 + 3O v/o Sic, sample deformed to failure at cyclic plastic strain amplitude of 0.07%, Srivastan et al. [8]. With kind permission of Elsevier

these dislocation interactions dominate the cyclic straining. When local stress contribution pile-up (build-up) at the SiC_p and exceeds a critical value, crack initiation through rupture of the brittle SiC sets in. With continued cyclic deformation more particles crack and deccohesion between particle and matrix occurs. With further deformation microcracks at neighboring particles link to form the macroscopic crack. Propagation of the macroscopic crack results in a rapid softening or drop in stress carrying capability of the material immediately prior to failure. On increase the volume of SiC_p the increased hardening is a consequence of (a) and (b) above. For the alloy containing a lower volume fraction of SiC reinforcement, fracture was dominated by ductile failure of the matrix material.

5.6.3 Hardening in C/SiC and C–C/SiC

In the composite Al/SiC discussed above the metal to be strengthened, Al is of three-valence, while hardening in the very common composite C/SiC or C–C/SiC the main constituent has four-valence. Therefore, cyclic hardening or the fatigue strength of these systems is presented below.

(i) C/SiC

The tension-tension fatigue behavior of unidirectional C/SiC composite at room temperature and 800 °C under air is considered here. The carbon fibers/SiC composite was manufactured by hot-pressing method providing a dense composite via a liquid phase sintering at low temperature. The volume fraction of the fibers was ~40%. The test specimens were further coated with SiC of ~20 μm thick by chemical vapor deposition to prevent oxidation at elevated temperature. For the tension-tension fatigue experiments dog bone-shaped specimens were used. The loading frequency of 10 Hz was applied to the fatigue tests using a sinusoidal wave front. The maximum number of the applied cycles is defined as 10^6 cycles. The stress levels in tension-tension fatigue tests were 90, 85, 80 and 70% of the tensile strength at room temperature, and 70, 65 and 60% of the tensile strength at 800 °C in air. The tension-tension fatigue life S–N curve at room temperatures is shown in Fig. 5.42. At 240 MPa no fatigue failure occurred at 10^6 cycles at room temperature. Compare this figure with that at 800 °C shown in Fig. 5.43. The test specimen failed already at a value of 77,215 cycles (below 10^5 cycles). The fatigue hysteresis modulus E is calculated by Eq. (5.9)

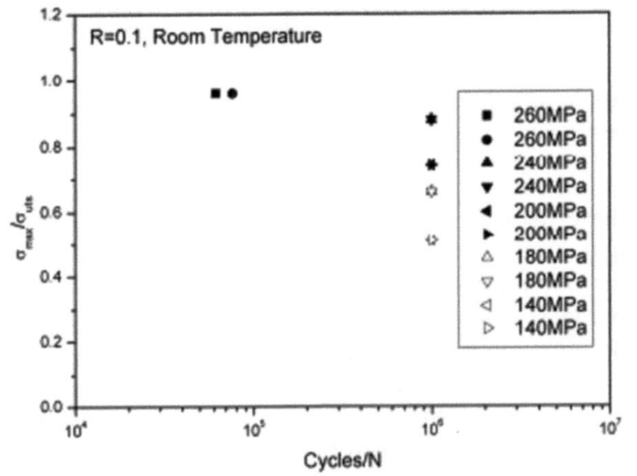

Fig. 5.42 The tension-tension fatigue life S–N curve of cross-ply C/SiC composite at room temperature, Li [7]. Open access

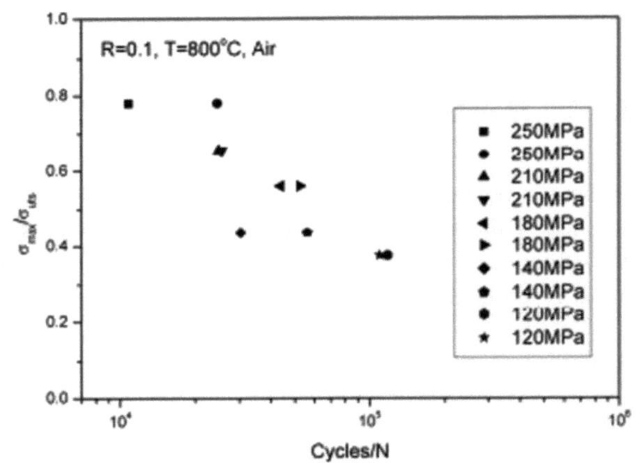

Fig. 5.43 The fatigue life S–N curve of unidirectional C/SiC composite at 800 °C under air, Li [7]. Open access

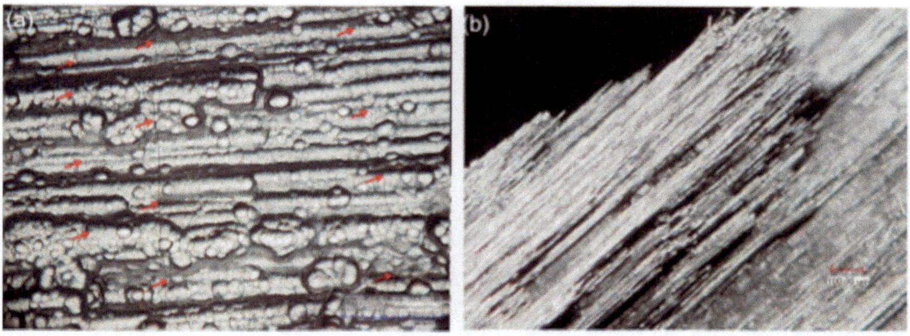

Fig. 5.44 **a** The multiple matrix crackings on the side surface; and **b** the fiber pullout at the fracture surface of the failure specimen observed under optical microscope, Li [7]. Open access

$$E = \frac{\sigma_{max} - \sigma_{min}}{\varepsilon_{max} - \varepsilon_{min}} \tag{5.9}$$

where σ_{max} and σ_{min} are the fatigue peak and valley stresses, and ε_{max} and ε_{min} denote the fatigue peak and valley strains, respectively. Specimens without fatigue failure were pulled by tension to failure in order to observe fracture surface. Multiple cracks were observed under optical microscope as seen in Fig. 5.44(a), and also fiber pullout is seen in (b) which is attributed mainly to low interface shear stress after fatigue loading. Fracture surface was inspected on specimens fatigued at 800 °C which is illustrated for comparison in Fig. 5.45. Multiple matrix cracking in front and side surfaces of failed specimens are seen in Fig. 5.45a, b and fiber pullout—longer than at room temperature—is seen in Fig. 5.45c, d. The pullout fibers are severely oxidized and interface shear stress degradation due to the oxidation of the C fibers is the main reason for the longer fiber pullout length under fatigue loading at 800 °C in air.

A model has been presented for the analysis—and not reproduced here (interested readers should consult Li)—of the various aspects of the fatigue test results. Figures 5.45 and 5.46 summarize the experimental and calculated data for room temperature and 800 °C fatigue tests, and Tables 5.1 and 5.2 summarize the relevant data for air and 800 °C fatigue tests respectively.

The experimental and predicted fatigue life S–N curves are illustrated in Fig. 5.46d, where the fatigue limit stress approaches 88% of the tensile strength. The predicted fatigue life S–N curve can be divided into two regions as shown in Fig. 5.46d, i.e., (1) the A-B part is affected by the degradation of interface shear stress and fibers strength; and (2) when the interface shear stress approaches steady-state, the B–C part is only affected by fibers strength degradation. On the other hand, the experimental and predicted S–N fatigue life curves of Fig. 5.47d where the fatigue life at 800 °C in air is greatly reduced compared with that at room temperature (Fig. 5.46d) is mainly attributed to oxidation of interphase of C fibers. The predicted fatigue life S–N curve can also be divided into two regions, as shown in Fig. 5.47d,

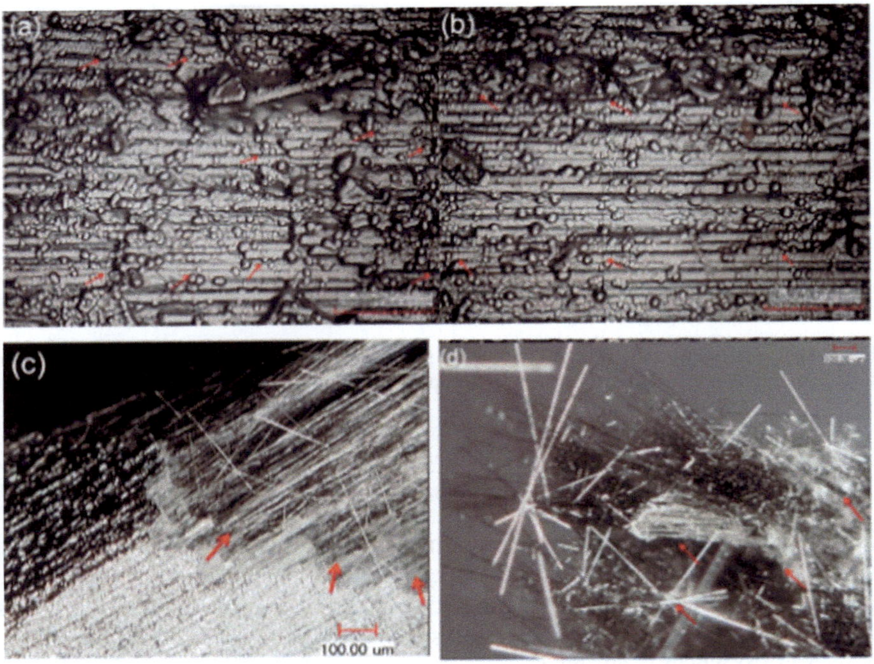

Fig. 5.45 **a** The multiple matrix cracking on the front surface; **b** the multiple matrix cracking on the side surface; **c** the pullout fibers at the fracture surface; and **d** the pullout fibers exhibit serious oxidation for fatigue failure specimen at 800 °C in air, Li [7]. Open access

namely: (1) the A–B part is affected by the degradation of interface shear stress and fibers strength; and (2) the B–C part is only affected by fibers strength degradation.

(ii) C/C-SiC

For high-performance C/C-SiC composites plain weave carbon fabric is used as the reinforcement for which is a 2D C/CSiC having high specific strength/modulus and good fracture toughness. This composite has been used in high performance braking systems and space structures. The cyclic stress could result in mechanical degradation and catastrophic failure, therefore investigating this system is important in order to get a better understanding of its performance. Widely used technique to fabricate C/C-SiC composites is the liquid silicon infiltration (LSI) method. The effect of the fractured surface of cyclically loaded specimens at 102, 104 and 106 cycles on the residual tensile stress (RTS) are presented here together with the microstructure. The tensile fatigue life of the C/C-SiC S–N curve is illustrated in Fig. 5.48. It is seen that the fatigue limit for 10^6 cycles is 57 MPa. Therefore, the maximum fatigue stress of 57 MPa was preselected for the cycles to eliminate fracture before the final quasi-static tensile test. The representative tensile stress strain curves of the virgin and post-fatigued C/C-SiC specimens is shown in Fig. 5.49.

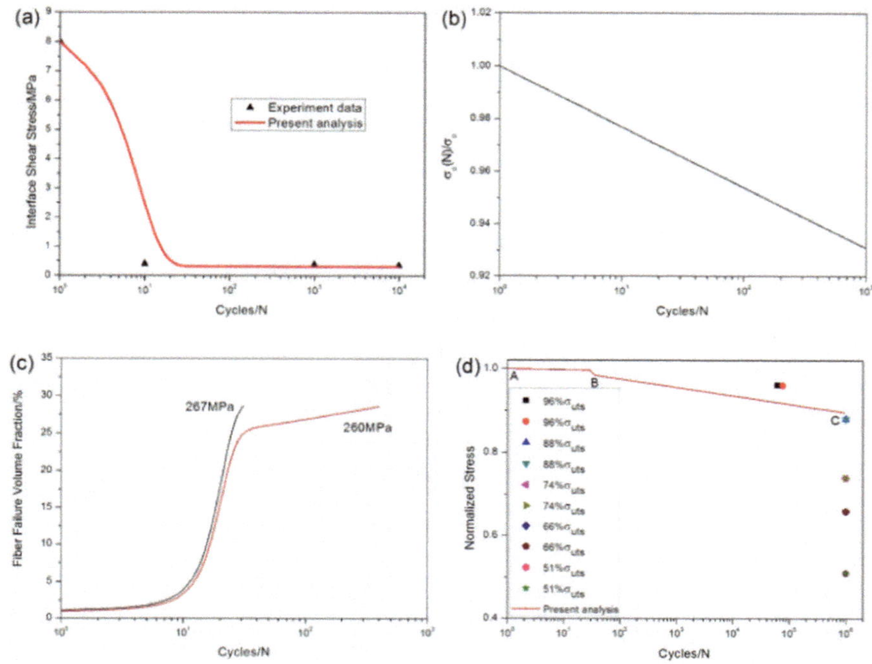

Fig. 5.46 **a** The interface shear stress versus applied cycles; **b** the fibers strength versus applied cycles; **c** the fibers failure probability versus applied cycles; and **d** the fatigue life S–N curve of experiment data and present analysis for unidirectional C/SiC composite at room temperature, Li [7]. Open access

Table 5.1 The fiber/matrix interface shear stress of unidirectional C/SiC composite corresponding to different number of applied cycles under fatigue loading at room temperature, Li [7]. Open access

Loading cycles	Experimental hysteresis loss energy/KPa	Interface shear stress/MPa
1	56 ± 3	8 ± 0.5
10	12.7 ± 0.5	0.40 ± 0.02
100	12.0 ± 0.5	0.38 ± 0.02
10,000	10.4 ± 0.5	0.35 ± 0.02
100,000	8.8 ± 0.4	0.32 ± 0.02
1,000,000	8.0 ± 0.4	0.30 ± 0.02

Table 5.2 The interface shear stress of unidirectional C/SiC composite corresponding to different applied cycles when $\sigma_{max} = 120$ MPa at 800 °C under air, Li [7]. Open access

Loading cycles	Experimental fatigue hysteresis loss energy/KPa	Interface shear stress/MPa
1	28 ± 2	6.1 ± 0.3
100	2.2 ± 0.5	0.30 ± 0.02
10,000	1.4 ± 0.5	0.22 ± 0.02
72,200	1.2 ±0.5	0.20 ± 0.02

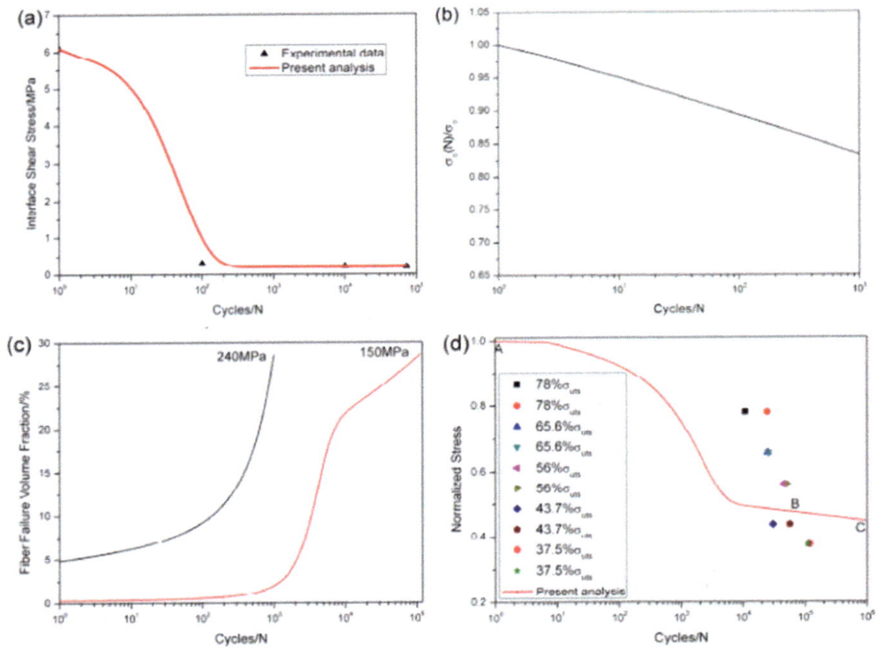

Fig. 5.47 **a** The interface shear stress versus applied cycles; **b** the fibers strength versus applied cycles; **c** the fibers failure probability versus applied cycles; and **d** the fatigue life S–N curves of experiment data and present analysis for unidirectional C/SiC composite at 800 °C under air, Li [7]. Open access

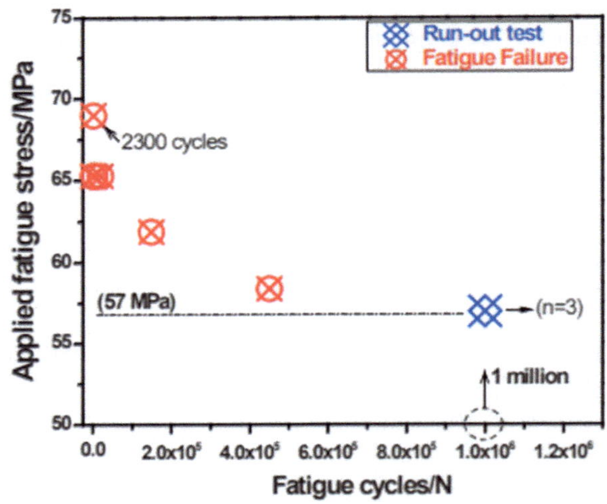

Fig. 5.48 Tensile fatigue life S–N curve of C/C-SiC composites, Zhang et al. [12]. Open access

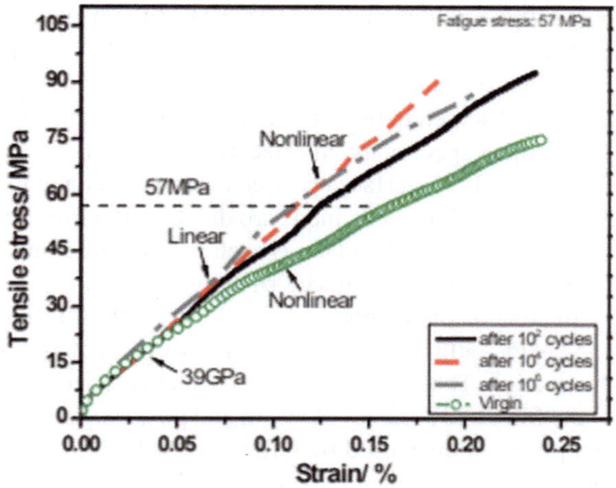

Fig. 5.49 Typical tensile stress–strain curves for the virgin and post-fatigue specimens, Zhang et al. [12]. Open access

The tensile stress–strain curves are non-linear associated with matrix cracking and fiber breakage, but the post-fatigue specimens are about linear when the tensile stress during the quasi-static test is lower than the maximum fatigue stress of 57 MPa. The almost linear behavior of the post-fatigue specimens is independent of the increasing fatigue cycles, which indicates that the cracks have been propagated already during the initial fatigue cycles. Only if the applied tensile stress is higher than 57 MPa. New damage sets in as indicated by cracks and interfacial degradation. The non-linear behavior of the post-fatigue seen in Fig. 5.49 occurs as mentioned with the beginning of the fiber rupture and cracking similar to the virgin sample. The RTS of the quasi-static tensile tests after 102, 104 and 106 cycles are summarized in Fig. 5.50. Figure 5.50 shows the distribution of the RTS for the virgin and the post-fatigue specimens. The fatigue induced enhancement for the RTS can be observed beyond 102 cycles and the RTS mean values for the post-fatigue specimens after 102, 104 and 106 loading cycles are about 90.9, 94,1 and 84.5, respectively (see Fig. 5.50). The microstructures of the virgin C/C-SiC and that of the post-fatigue tested after 10^6 cycles are displayed in Fig. 5.51. All microcracks in (a) and (c) are marked with red arrows. The cracks shown in Fig. 5.51a are the consequence of the mismatch of the coefficient of thermal expansion (CTE) between fibers and matrix during cooling down from LSI process temperature to room temperature. The 0° fibers suffer tensile thermal residual stress, while in the 90° fiber compressive stress is generated. The post-fatigue specimens after 10^6 cycles are seen in Fig. 5.51c, d. An increased number of long cracks are observed in (c) and they are bridging the adjacent 90° orientated fiber bundles, as observed in Fig. 5.51d. In Fig. 5.52, fully extended cracks after cyclic loading are stopped in front of 90° bundles, and these cracks propagate along the 90° fiber bundles under fatigue loading. Fracture surfaces of virgin and post-fatigue specimens after quasi- static tensile tests are displayed in Fig. 5.53. As seen in (a) only few single fibers are pulled out from the virgin specimens which are compact and limited detaching between fiber and matrix is observed. The post-fatigue specimens after 102, 104 and 106 cycles are seen in Fig. 5.53b–d. With increasing the number of cycles (beyond 102) crack propagation occurs not only in the matrix but also in the fiber as seen in (c) and (d). More pull-out of single fibers is observed in the interfacial debonded pulled out fiber bundle segments of the post-fatigue specimens. Interfacial degradation has occurred between the fiber and the matrix. The interfaces between carbon fiber and matrix in the virgin material and in the post-fatigue specimens are displayed in Fig. 5.54. Bonding at the interface in the virgin material is intact as seen in (a), while interfacial debonding between the fiber and matrix was induced by fatigue stress.

In conclusion the residual tensile strength (RTS) of C/C-SiC composite exposed to 57 MPa for three preselected loading cycles of 102, 104 and 109 was considered in this section. The fatigue damage was characterized by inspecting the microstructures of the fractured surfaces of virgin and post-fatigued specimens. Crack multiplication and propagation caused by fatigue stress both in fiber and matrix are involved with the decreased fatigue resistance of C/C-SiC composite leading ultimately to the final failure.

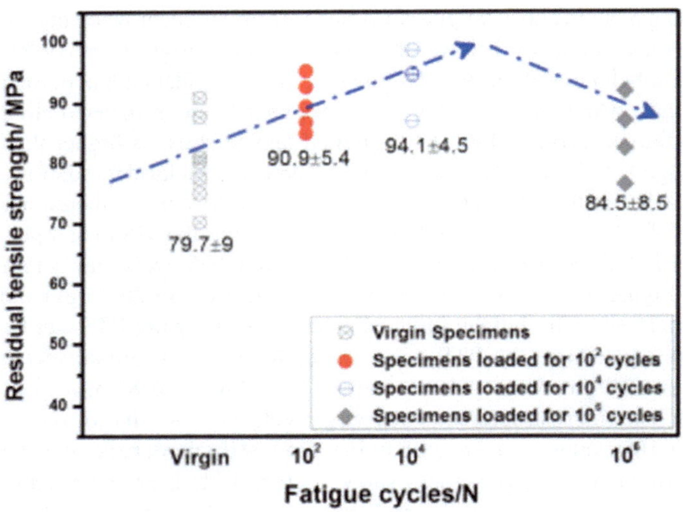

Fig. 5.50 Diagram of RTS for virgin and post-fatigue C/C-SiC, Zhang et al. [12]. Open access

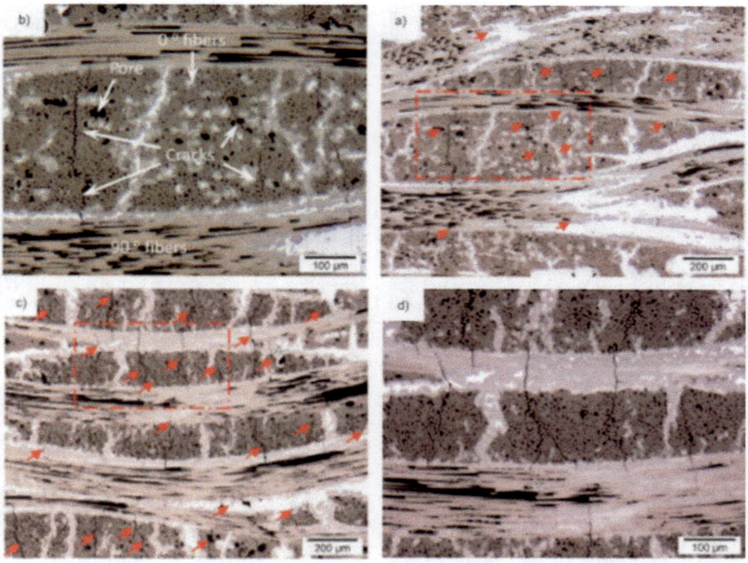

Fig. 5.51 Typical microstructure of C/C-SiC under OM; **a** virgin specimen; **b** magnification of the highlighted area in Fig. 5.51a; **c** post-fatigue specimen after 106 cycles; **d** magnification of highlighted area in Fig. 5.51c; cracks are marked by the red arrows; the loading direction is parallel to the 90° fibers, Zhang et al. [12]. Open access

Fig. 5.52 Crack propagation along the 90° fibers, Zhang et al. [12]. Open access

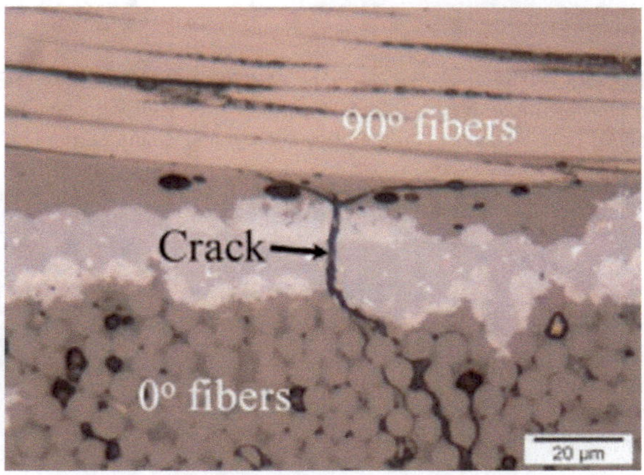

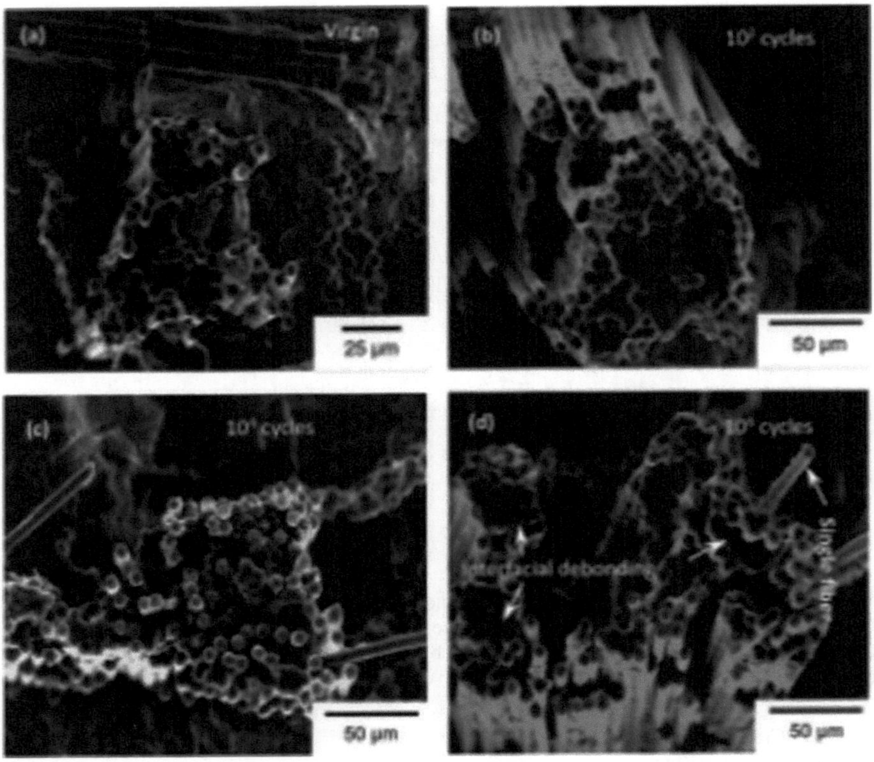

Fig. 5.53 The fractured surfaces of C/C-SiC composites (SEM images, topography contrast), **a** virgin sample; **b** post-fatigue specimen after 10^2 cycles; **c** post-fatigue specimen after 104 cycles; and **d** post-fatigue specimen after 106 cycles, Zhang et al. [12]. Open access

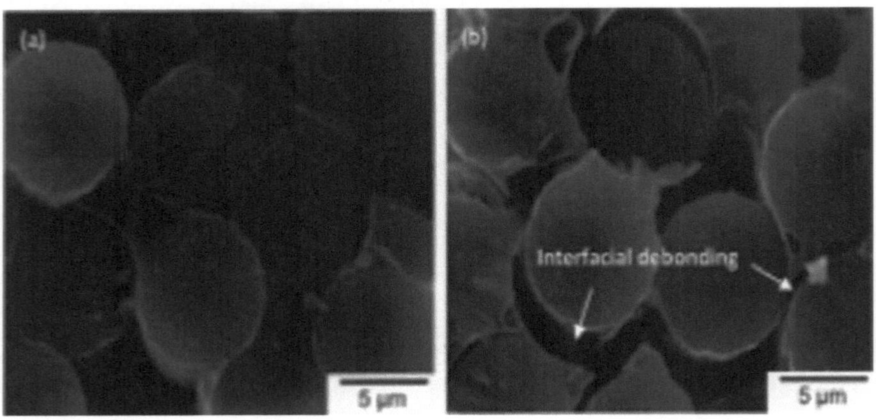

Fig. 5.54 SEM image of the interface between carbon fiber and matrix, **a** intact interface in the virgin sample; **b** interfacial debonding in the post-fatigue sample, Zhang et al. [12]. Open access

5.7 Hardening in B_4C

Boron Carbide (B_4C) is the hardest material after diamond and the cubic boron nitride. Room temperature fracture properties are of interest, but less information is available on the high temperature deformation. As expected, deformation-like in other materials-is produced by dislocation glide, however it is coupled with twins because formation of twins is energetically favorable. Thus, high temperature plasticity is coupled with twins. Bright field TEM images of as-processed fully dense polycrystalline B_4C are illustrated in Fig. 5.55. While the as-processed B_4C is dislocation free, twinned grains are seen in Fig. 5.55a, b, which is common in B_4C ceramic. Dark field TEM of dislocations in deformed B_4C obtained under weak-beam condition are shown in Fig. 5.56a and bright field images are seen in (b) and (c); the images result from dislocations interaction from multiple slip planes and with solute impurities (they are present in the original B_4C powder). The image shows substructure of complex nodes and curved lines, commonly associated with junction formation. Stacking faults (bound by partial dislocations) are also seen in Fig. 5.56a. The twins observed in Fig. 5.56b, c are also observed in the as-processed B_4C (Fig. 5.55) but no dislocations. The observations are a consequence of the high covalent character of B_4C, its low stacking fault energy (thus twin formation is promoted) and the dislocation-twin coupling occurring altogether at high temperatures. Now after characterizing B_4C and the observation of high temperature plasticity resulting from deformation, we can turn to its respond to cyclic deformation. The microstructure before and after high-current pulsed electron-beam treatment is illustrated in Fig. 5.57. B_4C exposed to high-current pulsed electron beam led to surface fragmentation due to a crack network shown in Fig. 5.57b. The formation of the crack network occurred due to high cooling rate of the surface layer observed during the electron beam treatment. The B_4C specimens exposed to electron-beam induced grain refinement (b). The fracture of the B_4C ceramic before and after the electron-beam treatment is fragile while the sintered ceramics undergoes an inter-crystalline fracture.

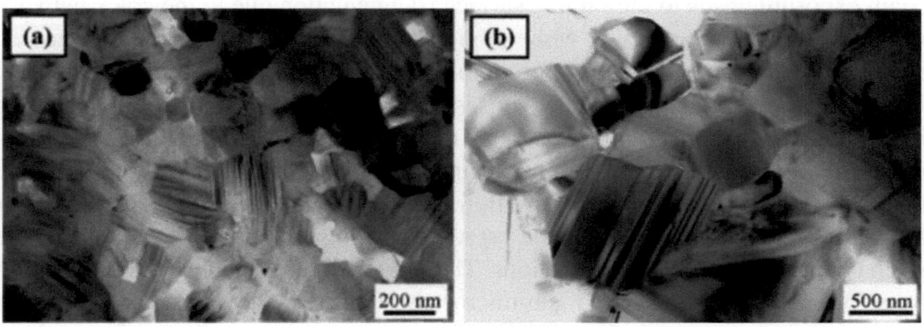

Fig. 5.55 Bright-field TEM images of as-processed B_4C ceramic: **a** n-B_4C and **b** sub-B_4C, Moshtaghioun et al. (2015). With kind permission of Elsevier (n-B_4C:130 nm, sub-B_4C:700 nm)

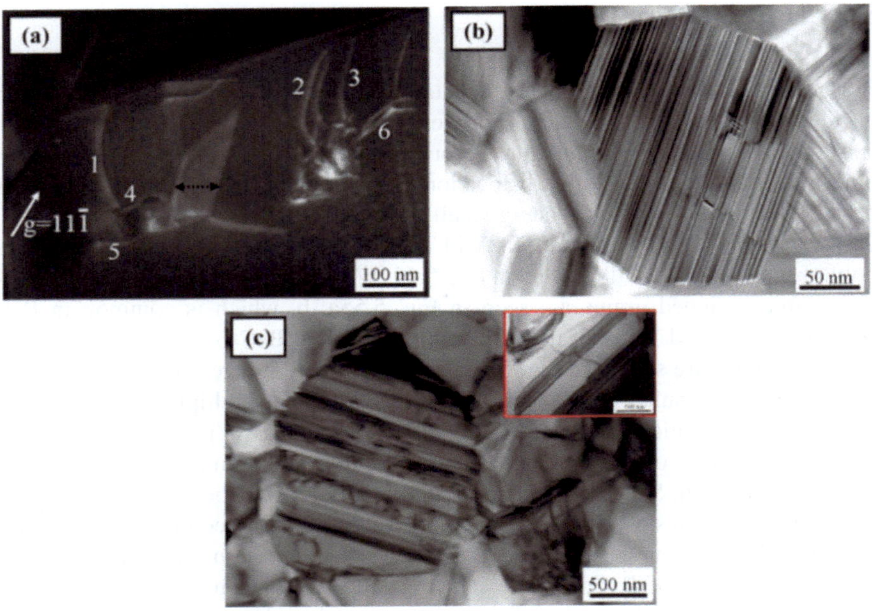

Fig. 5.56 **a** Dark-field TEM micrograph of deformed sub-B_4C ceramic (total strain ~35%) showing dislocations 1, 4 and 5 having b = [110], and dislocations 2, 3 and 6 having b = [101]. (The arrow indicates stacking fault.) Bright-field TEM micrographs of deformed B_4C ceramic: **b** n-B_4C and **c** sub-B_4C, showing extensive twin-dislocations interactions, Moshtaghioun et al. (2015). With kind permission of Elsevier

The fracture of the treated ceramics (modified by electron-beam treatment) occurs by inter-crystalline and trans-crystalline mixed mechanism. The observed size of trans-crystalline facets are 100–150 nm which indicates that the electron-beam treatment promotes nanostructure formation in the B_4C ceramics. The observed sub-grain structure formation seen in Fig. 5.57c, resulting from the high current pulsed electron-beam exposure, leads to the sample surface fragmentation due to the crack network which is shown in Fig. 5.57b, while at the same time it is possible that the mechanism of ceramics nanostructure formation is its mechanical twinning. Further, the formation of micro- and nanoscale twins is caused by elastic deformations occurring in the surface layer of ceramics due to the ultrahigh heating and cooling rates occurring actually under the high-current electron beam. TEM images of boron carbide thin films are presented in Fig. 5.59. Before and after the electron-beam treatment the layer is modified as seen by the presence of longitudinal interlayers along the grain boundaries and their intersections. These interlayers are indicated in Fig. 5.59b by arrows and have nano- and sub-microcrystalline structures. It is assumed that the surface layer shown in Fig. 5.59b is formed under the condition of two-phase state, that of liquid–solid. At a stage of the high rate heating of the sample the liquid phase is formed along the boron carbide crystalline interfaces, while at the cooling

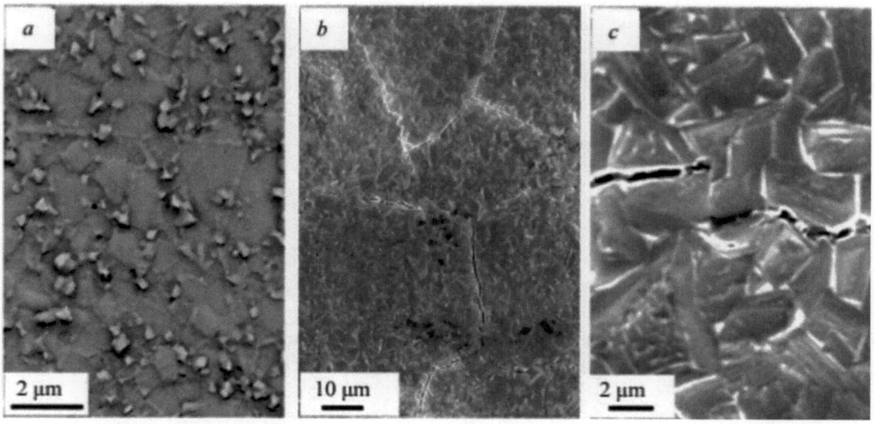

Fig. 5.57 Images of the surface structure of sintered boron carbide ceramics: **a** after etching with Murakami's reagent (optical image); **b, c** after the high-current pulsed electron-beam treatment (SEM images), Ivanov et al. [9]. Full access

stage the high crystallization rate of the liquid phase results in the formation of nanoscale interlayers along the boron carbide crystalline interface. SEM investigations of the structural phase states show that the increase of the fatigue life of boron carbide ceramics treated with the high-current electron beam is conditioned by (1) the formation of the micro-twins and (2) the formation of longitudinal interlayers along the grain-boundaries and their intersections having nano- and submicron crystalline structures. The hardness of the modified layer varies from 45 to 35 GPa and depend on the indentation load (from 5 to 50,300 mN). The hardness values before and after the electron-beam treatment is seen in Fig. 5.60. B_4C has low crack resistance, and its fatigue behavior under cyclic loading was tested by micro-hardness. 10 mN is the minimum indentation load, 5 s time exposure and the hardness change analysis is dependent on the number of cyclic loads. It was observed that the main changes in hardness are detected after 10–15 loading cycles. Therefore, the subsequent fatigue testing is carried out after 10 loading cycles. Averaged values for five areas of boron carbide ceramics are given in Fig. 5.60b.

In summary, high-current electron-beam treatment of sintered boron carbide ceramics leads to the formation of pore free polycrystalline structure. The fatigue life of the modified surface layer is improved without a decrease in the hardness value. SEM investigations show that the fatigue life of the high-current electron beam is associated with the formation of micro-twins and longitudinal interlayers along the grain boundaries, and their intersections having nano- and submicron crystalline structures.

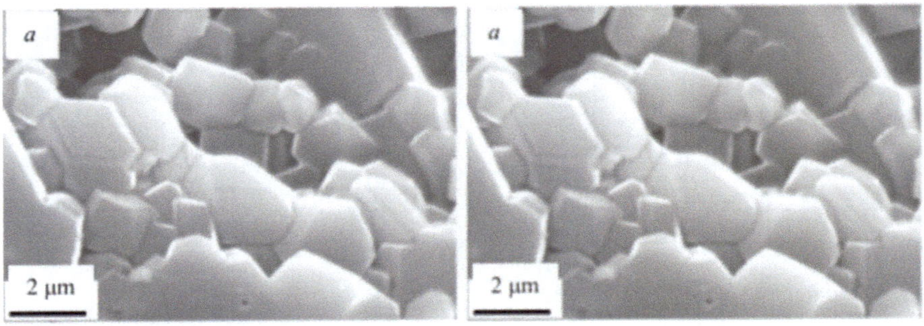

Fig. 5.58 SEM images of boron carbide ceramics: **a** sintered fracture surface; **b**, **c** after the high-current pulsed electron-beam treatment, Ivanov et al. [9]. Full access

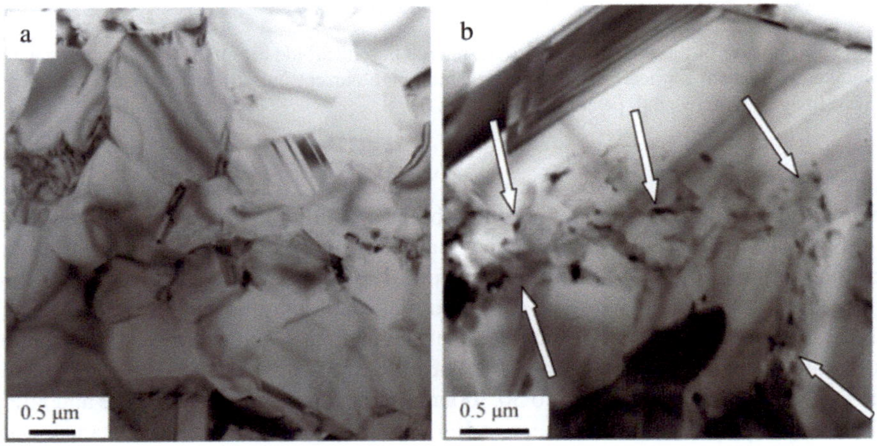

Fig. 5.59 TEM images of boron carbide ceramics: **a** sintered ceramics; **b** ceramics treated by high-current electron beam. Arrows indicate nano-crystalline interlayers, Ivanov et al. [9]. Full access

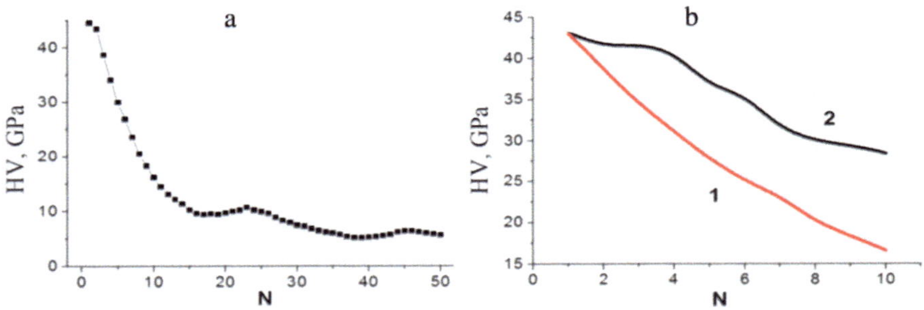

Fig. 5.60 **a** Hardness-cycle loading dependencies of B_4C ceramics: **b** before electron-beam treatment (curve 1); **b** after electron-beam treatment (curve 2). Fatigue testing, Ivanov et al. [9]. Full access

5.8 Hardening in Si_3N_4

The ceramic Si_3N_4 is a leading structural component for use in gas turbines, engine components and bearing parts because of its high mechanical strength and high heat resistance. The fatigue resistance is a key factor for its use, and understanding the resistance to failure of Si_3N_4 under cyclic conditions is important for its evaluation as an advanced structural material. Here fatigue tests are discussed, which were performed as accelerated tests by the resonant bending method, accumulating fatigue data up to 10^{10} cycles of sintered Si_3N_4. Also, the effect of frequency on cyclic fatigue at room temperature was included in the experiments. Further, tension–compression fatigue tests were performed at room temperature and high temperatures on sintered Si_3N_4 to compare their behavior to cyclic deformation. Two kinds of sintered Si_3N_4 were used in the experiments which are designated as SSN-A and SSN-B. Sintering additives were included in these specimens and their mechanical properties are listed in Table 5.3. Stress versus S–N curve under alternating load at frequencies 0.03–3 kHz for tapered SNNA2 specimens is seen in Fig. 5.61. Cyclic fatigue behavior of sintered Si_3N_4 appears to depend principally on the number of cycles rather than on time, because the fatigue strength at various frequencies did not show any discontinuities against the number of cycles. The solid curve suggests a decrease in the fatigue limit to about 40% of its initial strength. The stress amplitude σ_a versus the mean stress amplitude, σ_{mean} of specimen SSN-A1 at room temperature is shown in Fig. 5.62. Open circles represent specimens surviving 10^7 cycles and the closed circles represent the failed specimens. The dotted line represents tensile strength. The solid line from σ_e to s_t is the modified Godman line and is expressed as

Table 5.3 Mechanical properties of sintered Si_3N_4, Matsuda et al. [13]. With kind permission of Elsevier

Materials	*Density (g/cm^3)*	*Four-point bending strength*[a] *(MPa)*	*Fracture toughness*[b] *($MN/m^{3/2}$)*	*Young's modulus (GPa)*
SSN-A1	3·21	740	6·2	265
SSN-A2	3·24	820	6·3	270
SSN-A3	3·28	860	6·2	275
SSN-B	3·25	970	6·1	300

[a] According to JIS R1601.[11]
[b] Chevron notch method.

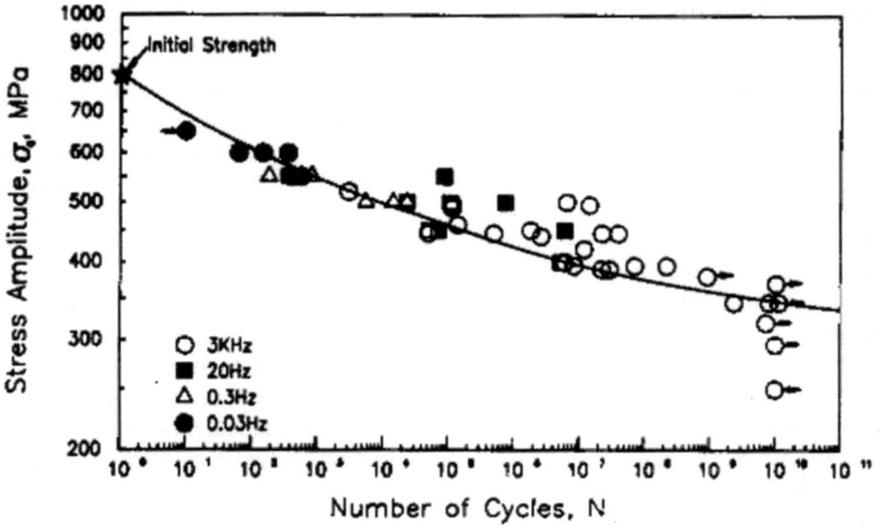

Fig. 5.61 S–N curve of cantilever beam specimens of SSN-A2 under alternating load, at frequencies from 0.03 Hz to 3 kHz, Matsuda et al. [13]. With kind permission of Elsevier

$$\sigma_a = \sigma_e\left(1 - \left(\frac{\sigma_m}{\sigma_t}\right)\right) \tag{5.10}$$

where σ_a is stress amplitude, σ_m mean stress, σ_t tensile strength and σ_e alternative fatigue strength at stress ratio $R = -1$. The effect of volume on cyclic fatigue at room temperature is shown in Fig. 5.63. Two SSN-B kinds of specimens were used with different volumes, namely, tensile specimens and cantilever beam specimens. The fatigue strength at the same number of cycles decreased with increasing effective volume. The slope of the S–N curve of the cantilever beam specimens—having a smaller effective volume—is larger than that of the tensile specimens having large volume (Fig. 5.63). Intersection of the lines in Fig. 5.63 (evaluated by extrapolation) provides the threshold stress of the cyclic fatigue. Log–log plot of stress versus the number of cycles to failure is shown in Fig. 5.64 at room temperature, 800 and 1000 °C with mean tensile stress of 585, 560 and 360 MPa, respectively. Despite the large scatter, it can be observed that the fatigue life of specimens greatly increased as stress amplitude decreased. The failed specimens at the onset of loading are marked with arrows pointing to the left while in specimens with arrows pointing to the right the test was terminated without failure.

The effects of mean tensile stress and static fatigue on cyclic fatigue of sintered Si_3N_4 (SSN-B) with a crystalline grain boundary phase at high temperature are shown in Fig. 5.65 obtained at a constant frequency of 20 Hz. The vertical axis represents a fully reversed fatigue condition, and the horizontal axis represents static fatigue strength and ultimate tensile strength. As temperature increased, static fatigue strength decreased. The mean stress effect can be approximate by Eq. (5.11) as

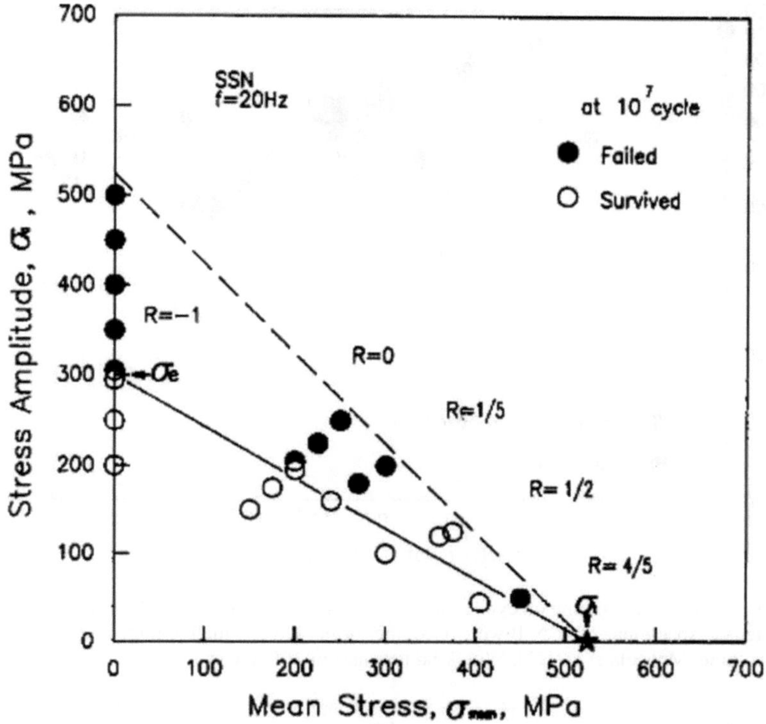

Fig. 5.62 Mean stress versus stress amplitude for fatigue failure in SSN-A1 at 10^7 cycles measured by high cycle fatigue test at room temperature. The solid line represents tensile strength, Matsuda et al. [13]. With kind permission of Elsevier

$$\left(\frac{\sigma_a}{\sigma_e}\right)^n + \left(\frac{\sigma_m}{\sigma_s}\right)^n = 1 \quad (5.11)$$

Here, σ_s is the static fatigue strength, and the other parameters were defined in Eq. (5.10). In order to determine the fracture origin and failure modes, fractured surfaces were examined by OM and illustrated in Fig. 5.66. Microcracks at grain boundaries are observed in specimens that failed under cyclic load, as shown in Fig. 5.67. Clusters of micro-cracks are seen sporadically at grain boundaries and some in other micro-cracks. Thus, it can be inferred form the experimental results that cyclic fatigue involves crack nucleation and crack propagation model. The micro-cracks are the result of stress concentration around the initial flaw. These (micro-cracks and initial flaw) form a fatigue damage zone. Growth of the damage zone to a critical value causes cyclic fatigue failure to occur. Crack propagation is faster under cyclic load than under static load. Pores and inclusions are preferential sites (see Fig. 5.66) for crack propagation as seen in Fig. 5.66c, d and likely at room temperature as seen in (a) and (b). No major defect as fracture origin was observed in the fracture surface

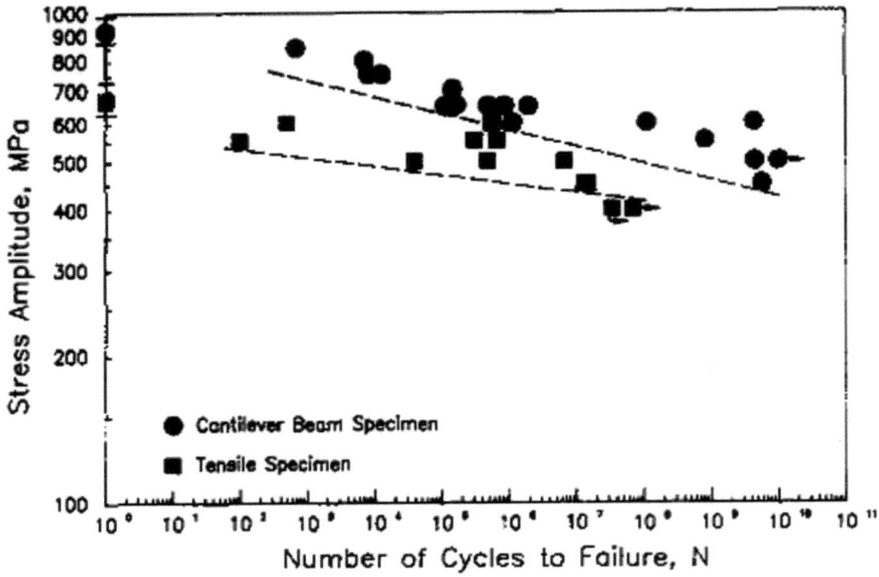

Fig 5.63 The effect of volume on cyclic fatigue at room temperature. Cantilever beam specimens and the tensile specimens of SSN-B were used; the former have a small volume and the latter have a large volume, Matsuda et al. [13]. With kind permission of Elsevier

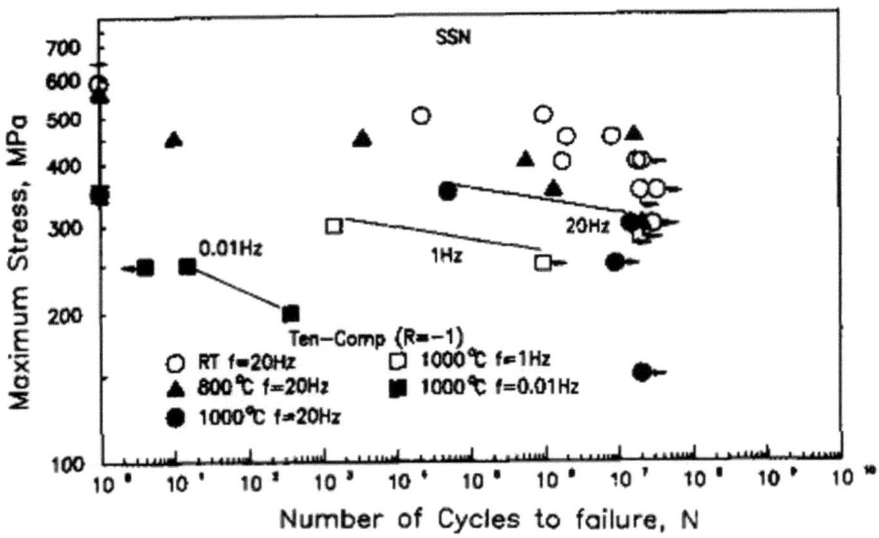

Fig. 5.64 S–N diagram of SSN-A3 using axial tension–compression load control at room temperature, 800 and 1000 °C. The symbols on the vertical axis represent the tensile strength of the materials used, Matsuda et al. [13]. With kind permission of Elsevier

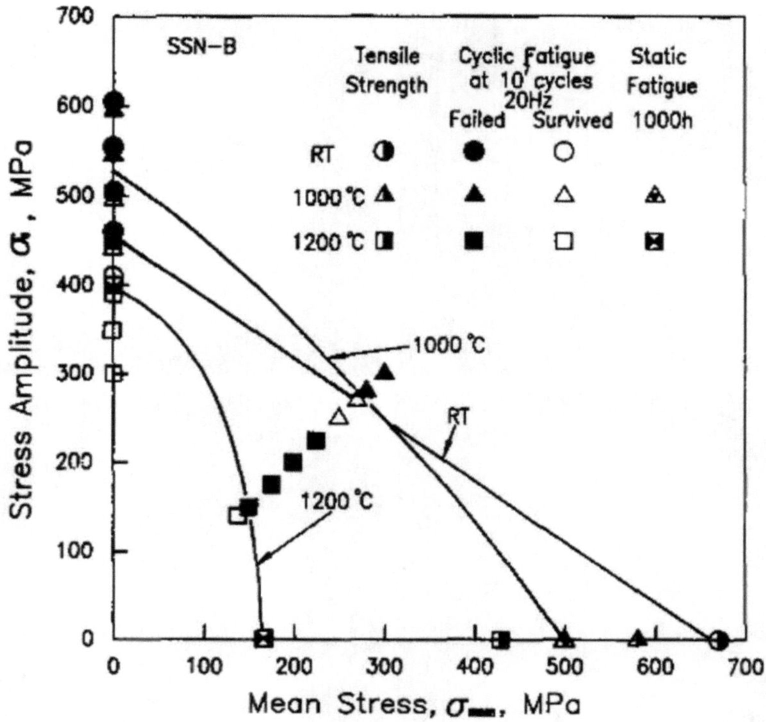

Fig. 5.65 Tensile mean stress effects on cyclic fatigue of SSN-B at room temperature, 1000 and 1200 °C, Matsuda et al. [13]. With kind permission of Elsevier

of low cyclic fatigue specimens as seen in Fig. 5.66e. The surface had the creep deformation region in such cases as static fatigue fracture, as shown in (f).

In summary, strength degradation of sintered Si_3N_4 in room temperature by cyclic fatigue depends on the number of cycles, while at high temperatures it depends on frequency, the lower the frequency the lower is the fatigue resistance. Microstructural observations suggest that cyclic fatigue is promoted by the formation of micro-cracks at grain boundaries. The progress of cyclic fatigue at high temperatures is caused by slow crack propagation and creep deformation.

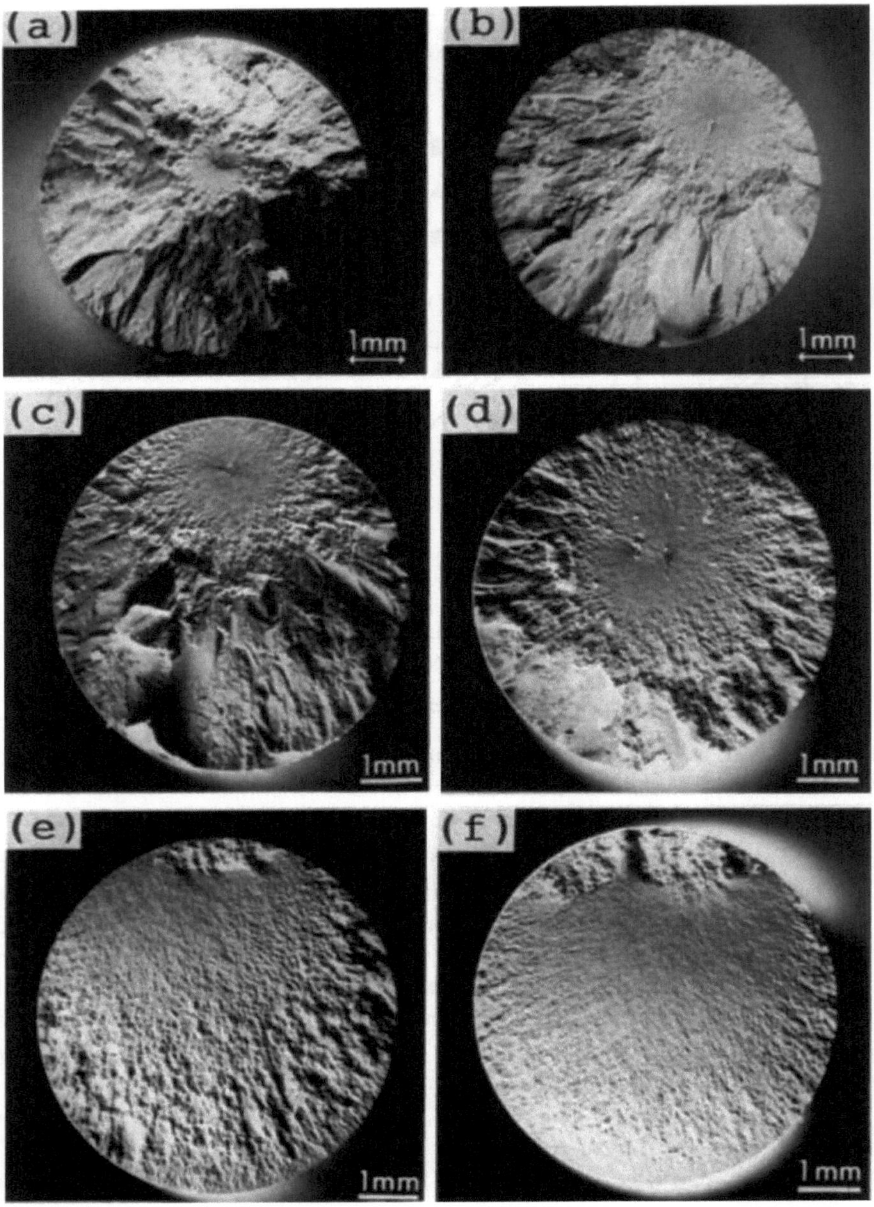

Fig. 5.66 Typical fracture surfaces of fatigue specimens of SSN-A3. **a** RT, 20 Hz, ±500 MPa, 2.2 × 10^4 cycles; **b** RT, 20 Hz, ±400 MPa, 1.8 × 10^7 cycles; **c** 800 °C, 20 Hz, ±350 MPa, 1.3 × 10^6 cycles; **d** 1000 °C, 20 Hz, ±350 MPa, 5.0 × 10^4 cycles; **e** 1000 °C, 1 × 10^2 Hz, ±200 MPa, 3.7 × 10^2 cycles; **f** 1000 °C, 150 MPa, 204 h, Matsuda et al. [13]. With kind permission of Elsevier

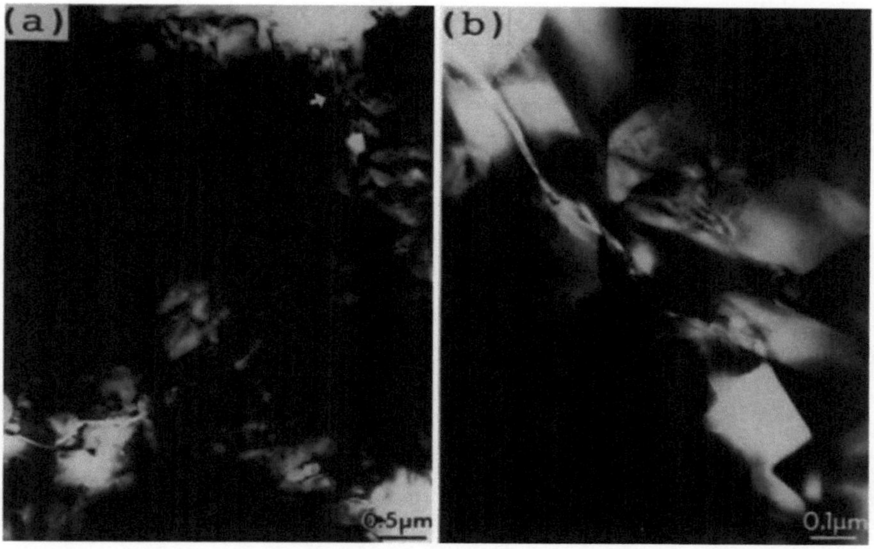

Fig. 5.67 Typical microstructures of SSN-A3 after fatigue test at room temperature. (a) ±450 MPa, 8.2 × 106 cycles; **b** enlarged view of area marked with arrow in (**a**), Matsuda et al. [13]. With kind permission of Elsevier

References

1. B.J. Pletka, A.H. Heuer, T.E. Mitchell, Acta Met. **25**, 25 (1977)
2. Y. Hino, K. Yoshida, Y. Kiyota, M. Kuwayama, ISIJ Int. **53**, 1392 (2013)
3. A. Smirnov, J.I. Beltrán, T. Rodriguez-Suarez, C. Pecharromán, M.C. Muñoz, J.S. Moya, J.F. Bartolomé, Sci. Rep. **7**, 44922 (2017)
4. S.-Y. Liu, I.-W. Chen, J. Amer. Cer. Soc. **75**, 1191 (1992)
5. C. Wei, X. Gong, C. Xie, Z.X. Chen, S.B. Li, L. Gremillard, J. Eur. Cer. Soc. **40**, 4647 (2020)
6. S.S. Scherrer, M.C.- Lorente, E. Vittecoq, F. de Mestral,' J.A. Griggs, H.W.A. Wiskott, Dental Mater. **27**, e28 (2011)
7. L. Li, Materials (Basel) **8**, 3316 (2015)
8. T.S. Srivastan, R. Auradkar, A. Prakash, Eng. Frac. Mech. **40**, 277 (1991)
9. Y. Ivanov, O. Tolkachev, M. Petyukevich, A. Teresov, O. Ivanova, I. Ikonnikiva, V. Polisadova, AIP Conf. Proc. **1698**, 030008 (2016)
10. Japanese Industrial Standard, Testing method for flexural strength (modulus of rupture) of high performance ceramics. JIS R1601 (1981)
11. G.R. Irwin, Handbuch der Physik. Springer Verlag, Berlin **6**, 551 (1958)
12. B.-G. Zhang, Y. Li, P. Xiao, W. Zhou, H. Luo, Z. Li, J. Ceram. Soc. Tech. **08**, 53 (2017)
13. M. Matsuda, T. Soma, M. Matusi, J. Eur. Ceram. Soc. **6**, 253258 (1990)

Chapter 6
Structural Features

Abstract Fatigue in materials is the result of cumulative damage processes that are usually induced by repeated loading cycles. Energy dissipation associated with irreversible damage, accompanied by the evolution of heat. The relation between stress and strain exhibits a lag between input and output termed hysteresis. Due to this lag, the forward path is different than the backward forming a loop. Evolution of stress-strain hysteresis in alumina is observed. Hysteresis as a structural feature has been recorded in magnesia, zirconia, SiC/SiC and Si_3N_4. The mean stress has been defined and its effect in alumina and yttria stabilized tetragonal zirconia inducing microcracks and damage is discussed in this chapter. Further, the mean stress effect is observed also in Si_3N_4 and the fatigue failure is expressed in terms of the relation between stress amplitude and the mean stress.

Keywords Structural features · Hysteresis · Alumina · MgO · Hysteresis in MgO/C Composite · Zirconia · Magnesia stabilized zirconia · Hysteresis in SiC/SiC · Monolithic Si_3N_4

6.1 Introduction

Structural ceramics can be manufactured by several techniques such as injection molding, uniaxial hot pressing, compression molding, slip casting and sintering. Structural ceramics exhibit desirable properties such as high temperature strength, high temperature stability wear resistance, corrosion resistance and chemical inertness. These desirable properties enable their use in a wide variety of applications ranging from turbine to aerospace systems including automotive engines. Heat exchangers, electronic devices and medical products are other fields of application.

The most common chemical bonds in ceramics are covalent and ionic which provide the high hardness, high compressive strength, and chemical inertness to them. These properties are also responsible for their low ductility sometimes up to complete brittleness. The absence of free electrons (characterizing metals) is responsible for the low electrical and thermal conductivity. The microstructure of ceramics can be entirely crystalline or glassy or their combination. In the lateral case the glassy phase usually surrounds small crystals, bonding them together. The crystal structures of

J. Pelleg, *Cyclic Deformation in Oxides, Carbides and Nitrides*, Structural Integrity 22,
https://doi.org/10.1007/978-3-030-86118-6_6

ceramics are many and varied and this results in a very wide range of properties which make ceramics such an important material for technological use. Control of the microstructure can overcome inherent stiffness resisting deformation to an applied force, to allow the production of ceramic springs, and ceramic composites.

The main compositional classes of engineering ceramics are the oxides, nitrides and carbides the subject of this chapter.

6.2 Hysteresis

Erber et. al. in "Hysteresis and Fatigue" already in 1993 has stated: "Fatigue in materials is the result of cumulative damage processes that are usually induced by repeated loading cycles. Since the energy dissipation associated with damage is irreversible, and the loading cycles are accompanied by the evolution of heat, the corresponding relation between stress and strain is not single valued, but rather exhibits a memory dependence, or hysteresis." As a **general** term, **hysteresis** means a lag between input and output in a system upon a change in direction. Thus **The meaning of hysteresis is "lagging"**. One can say that a system has some hysteresis when it follows a path to go from state A to state B, and a different path when going back from state B to state A. Principally on a graph two different curves connect at points A and B (one going out and one coming back) forming a loop as indicated in Fig. 6.1. The hysteresis is a borrowed concept from magnetism, where a lag of magnetization intensity, B occurs behind the magnetic field, H. as seen in Fig. 6.2. Looking at the graph if B is measured for various values of H and if the results are plotted in graphic forms then the graph will show a hysteresis loop.

6.2.1 *Hysteresis in Alumina*

Compromises has to be made in this section since limited information if at all exists in the open literature on the main concepts together, namely hysteresis in pure or

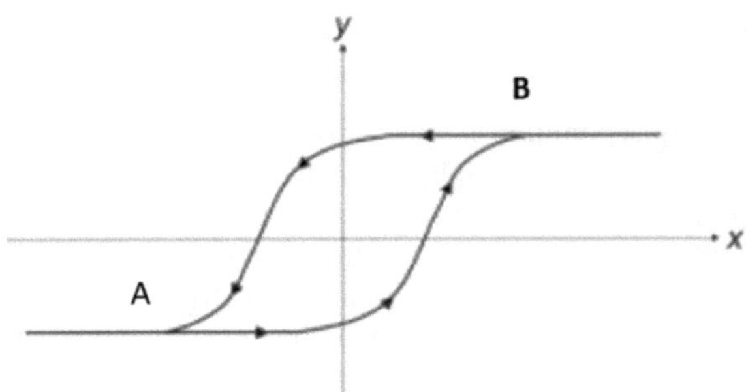

Fig. 6.1 A typical hysteresis curve

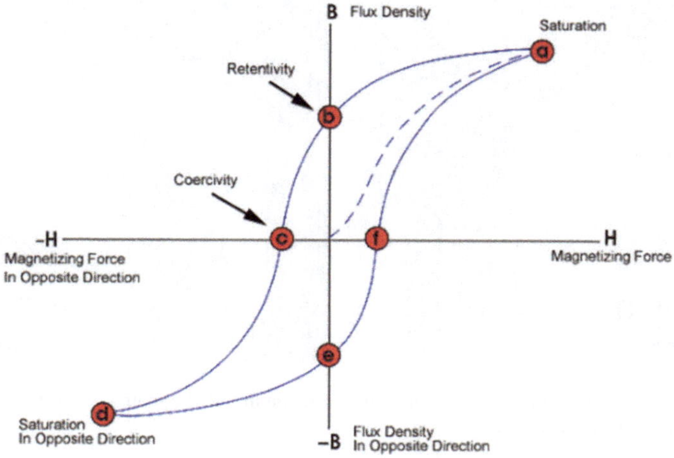

Fig. 6.2 A loop of hysteresis. After NDT resource center

monolithic alumina. Therefore, a composite Al_2O_3/NEXTEL™720 alumina will be considered first. And at a later stage other topics will be discussed. Alumina is non-magnetic and therefore, the very common hysteresis observed in magnetic material is rarely discussed in alumina-based materials.

Nextel ceramic fibers are an oxide and composite optic Inc. (COI) solution employed woven ™720 fibers within a matrix composed entirely of alumina (N720A). Nextel ceramic fiber 720 finds applications in ceramic matrix composites because of its high creep resistance. Oxide/oxide composites are being considered for use in high temperature aerospace applications where their inherent resistance to oxidation provides for better long-life properties at elevated temperature than most other ceramic matrix composites (CMCs). One promising oxide/oxide CMC is Nextel™720/Alumina (N720/A) as indicated above, which uses an 8-harness satin weave (8HSW) of Nextel™720 fibers embedded in a porous alumina matrix. The fiber fabric was infiltrated with the matrix in a sol–gel process. The laminate was dried with a "vacuum bag" technique under low pressure and low temperature, and then pressure-less sintered. The overall microstructure of the CMC is presented in Fig. 6.3. Previous research has shown N720/A to have excellent fatigue and creep-rupture properties in 1200 °C air. This oxide-oxide ceramic matrix composite was investigated in tension–compression fatigue behavior at 1200 °C in air and in steam. The composite has no interface between the fiber and matrix and relies on the porous matrix for flaw tolerance. Tension–compression fatigue behavior was studied for fatigue stresses ranging from 60 to 120 MPa at a frequency of 1.0 Hz. The R ratio (minimum stress to maximum stress) was −1.0. Fatigue run-out was defined as 10^5 cycles and was achieved at 80 MPa in air and at 70 MPa in steam. Steam reduced fatigue lives by an order of magnitude.

Fig. 6.3 As-received material. **a** Overview. **b** Porous nature of the matrix is evident, Lanser and Ruggles-Wrenn [1]. Free access

In all tests, a specimen was heated to test temperature at 1 °C/s and held at temperature 1200 °C for additional 45 min prior to testing. It should be noted that for a given stress level, the cyclic lives obtained in tension-tension fatigue were at least three orders of magnitude higher than those produced under tension–compression fatigue. The run-out stress in tension-tension fatigue was a high 170 MPa, more than twice the run-out stress of 80 MPa obtained in tension–compression fatigue. The S–N curves for tension–compression and tension-tension on the basis of maximum stress, σ_{max} reached in the cycle. are compared in Fig. 6.4. The evolution of hysteresis stress–strain relation with cycles at 1200 °C is seen in Fig. 6.5 for air and steam. The loops are nearly symmetric about the origin. The slopes of the stress–strain hysteresis curves of both tension-tension and tension–compression decrease with fatigue cycles which is accompanied with an increase in strain (and reduction of the tensile or compressive moduli). Clearly, lower strain accumulation with cycling means that less damage

Fig. 6.4 S–N curves for N720/A at 1200 °C in air and in steam. Arrow indicates that failure of specimen did not occur when the test was terminated, Lanser and Ruggles-Wrenn [1]. Free access, Tension-tension fatigue data from Ruggles-Wrenn et al. [2].

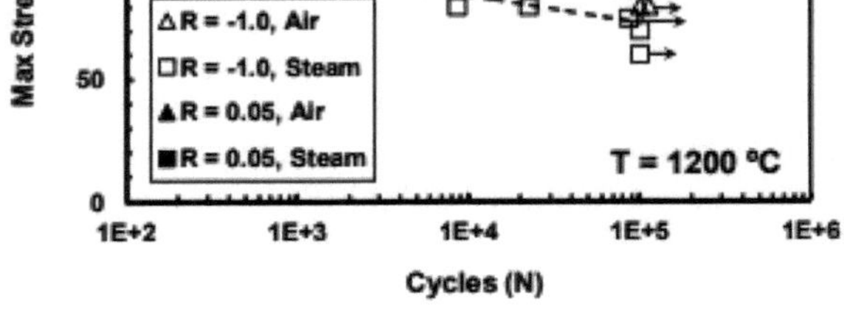

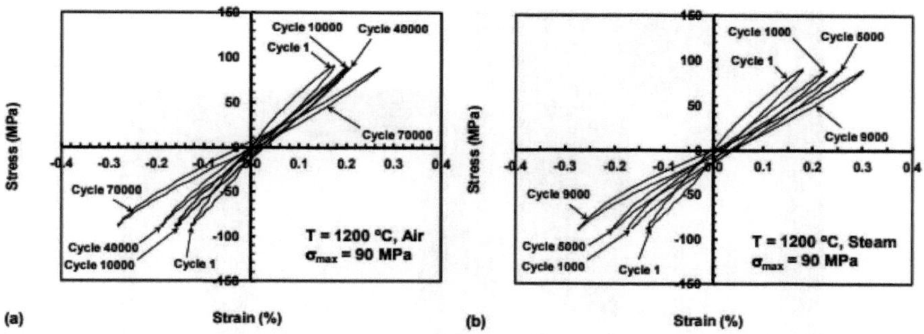

Fig. 6.5 Typical evolution of stress–strain hysteresis response of N720/A composite with fatigue cycles at 1200 °C. **a** In air. **b** In steam. $\sigma_{max} = 90$ MPa, Lanser and Ruggles-Wrenn [1]. Free access

has occurred, and that it is limited to some matrix cracking. Below, some optical and SEM micrographs of fracture surfaces are illustrated: Optical micrographs of tension–compression fatigue tests performed at 1200 °C in air and steam are shown in Fig. 6.6. In Figs. 6.7 and 6.8 N720/A specimens tested in air and steam at 1200 °C, respectively are illustrated under the stress and cycles indicated.

The fatigue test environment has a large influence on the fracture surface appearance. The fracture surface resulting for tests in air (Fig. 6.7) is fibrous with considerable regions of fiber pull-out and the individual fibers distinguishable. At higher magnification only small amounts of the matrix are bounded to the fiber surfaces. Contrary to this observation, the fracture surface of tests performed in steam is dominated by brittle failure. At high magnification as seen in Fig. 6.8c, increased fiber/matrix bonding occurred in steam. Also, loss of matrix porosity with subsequent matrix densification due to additional sintering and as a result, the N720/A composite exhibits decreased damage tolerance and reduced fatigue lifetimes at 1200 °C.

In summary, steam noticeably degrades tension–compression fatigue performance of N720/A and decreases tension–compression fatigue lives by ~ an order of magnitude. Prior fatigue in air does not reduce tensile stress (no damage occurred) while steam reduces the tensile stress suggesting that damage has occurred. The damage at 1200 °C in both air and steam are due to loss of matrix porosity and increased fiber/matrix bonding. Introducing compression into the load cycle promotes matrix densification and strengthening of fiber/matrix bond, which is accelerated in steam. Tension–compression cycling is considerably more damaging than tension-tension fatigue.

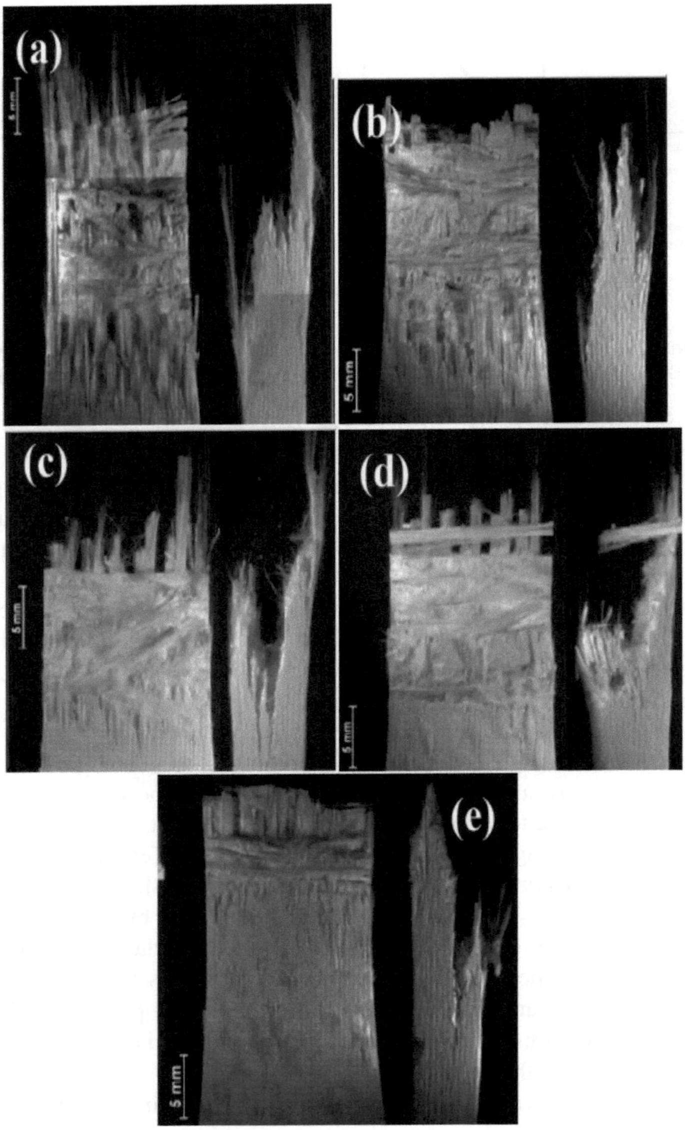

Fig. 6.6 Fracture surfaces of N720/A specimens tested at 1200 °C. **a** In compression to failure. **b** In tension–compression fatigue in air, σ_{max} = 120 MPa, N = 199 cycles. **c** In tension–compression fatigue in steam, σmax = 100 MPa, N_f = 450 cycles. **d** In tension–compression fatigue in air, σ_{max} = 80 MPa, N_f = 113,382 cycles. **e** In tension–compression fatigue in steam, σ_{max} = 75 MPa, N_f = 86,548 cycles, Lanser and Ruggles-Wrenn [1]. Free access

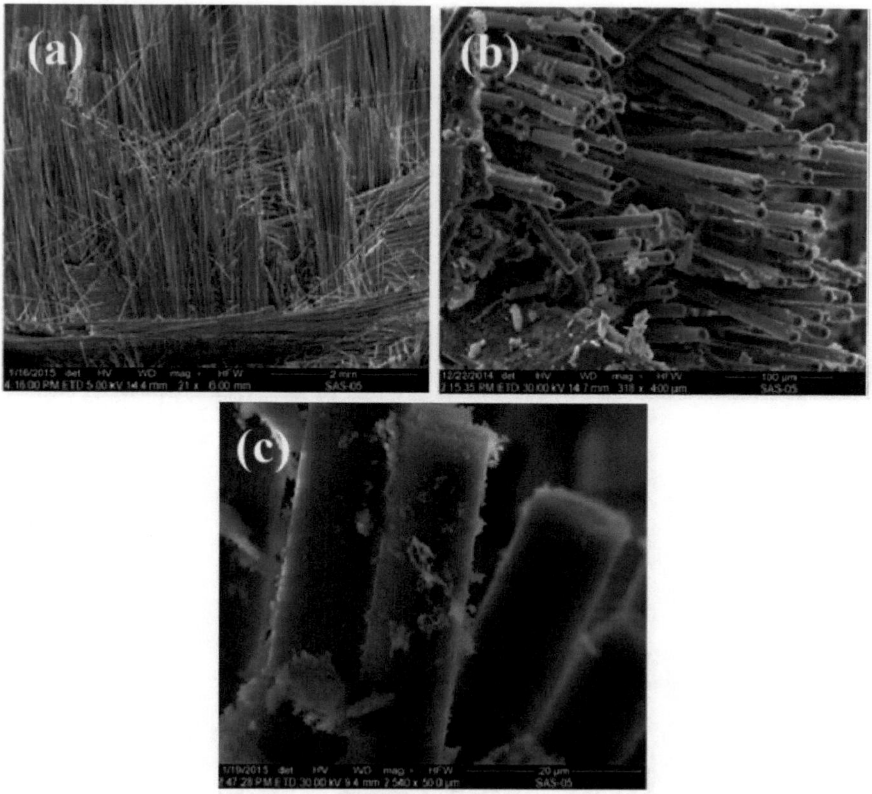

Fig. 6.7 Fracture surface of N720/A specimen tested in tension–compression fatigue at 1200 °C in air (σ_{max} = 80 MPa, N_f = 113,382 cycles), Lanser and Ruggles-Wrenn [1]. Free access

6.2.2 *Carbon/Al_2O_3*

Alumina possesses good mechanical properties such as high hardness, excellent wear resistance, and high chemical stability, but its drawbacks are its poor fracture toughness, low strength at elevated temperature as well as poor thermal shock resistance. Ceramics with carbon reinforced surfaces have been reported to exhibit improved wear resistance and a reduced friction coefficient. Thus, the addition of particulate carbon into oxide ceramics can significantly improve their durability and wear characteristics due to the lubricating properties of carbon.

Alumina nanoparticles and carbon nanoparticles (CNP) as additives were used to evaluated the properties relevant to this chapter. Indentation results are presented for the Al_2O_3-C specimens (CMS) in Fig. 6.9. For the experimental details see Tan et al. The indentation results show that pure alumina has the highest hardness while the hardness of CNP-Al_2O_3 composite samples decreases with larger Al_2O_3 particle (matrix) size. The distribution of CNP on the alumina depends on the surface area of

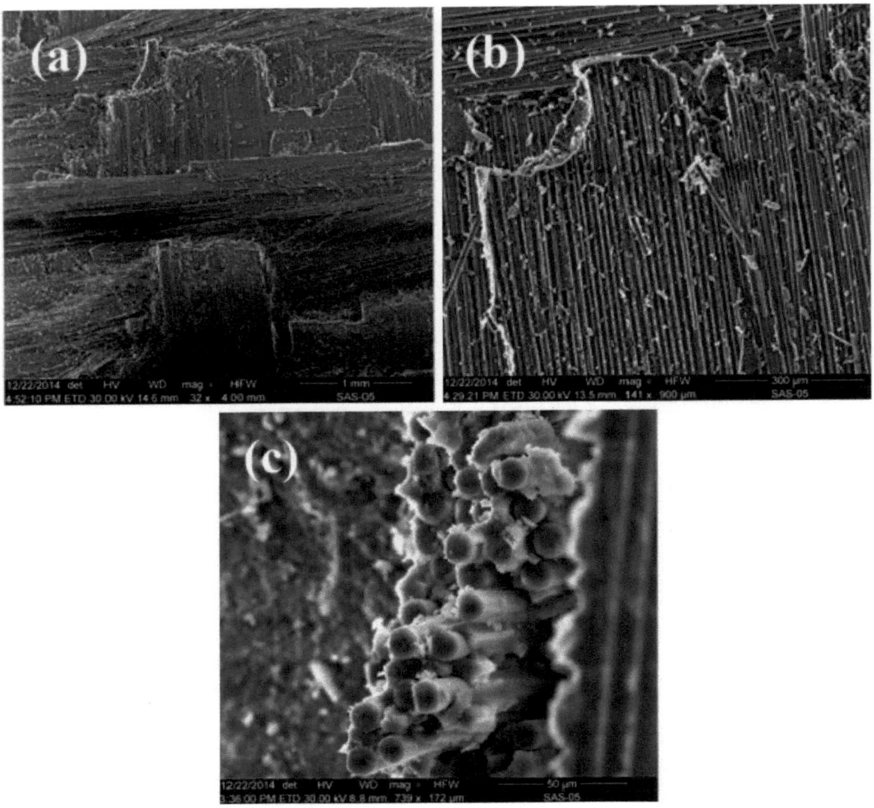

Fig. 6.8 Fracture surface of N720/A specimen tested in tension–compression fatigue at 1200 °C in steam (σ_{max} = 75 MPa, N_f = 86,548 cycles), Lanser and Ruggles-Wrenn [1]. Free access

the alumina and when it is larger a higher amount of CNP is adsorbed on the surface which leads to lower hardness, because it inhibits effective sintering between the Al_2O_3 interface. Also, the slip of the continuous carbon layer along the Al_2O_3 grain boundaries (see Fig. 6.11) might be the reason for the reduced hardness. Both reasons are affecting the microstructure. SEM images shown in Fig. 6.10 clearly indicate that the amount of CNP adsorbed onto the surface of alumina particle increase with a higher percent of CNP addition and the distribution is homogenous throughout the Al_2O_3 surface without agglomeration.

In summary, pure alumina exhibits the highest hardness value compared to the CNP-Al_2O_3 composite. The presence of higher amounts of CNP on the Al_2O_3 interface leads to reduced hardness due to inhibiting effective sintering.

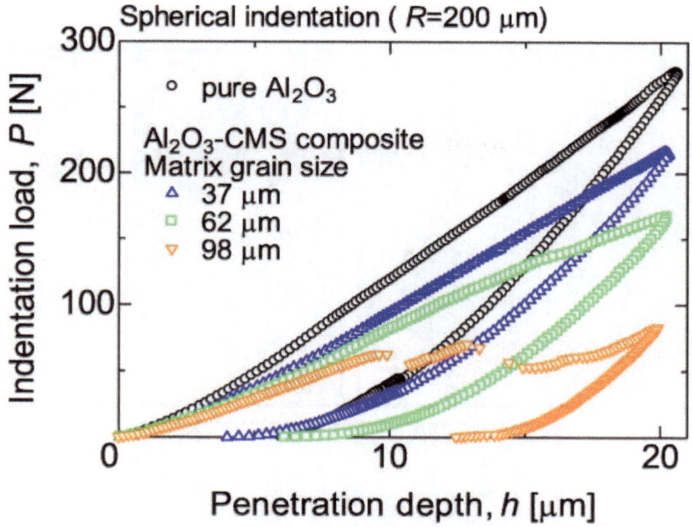

Fig. 6.9 Hysteresis curves of indentation load and penetration depth of 1.0 vol% CNP-Al_2O_3 composites, Tan et al. [3]. Open access. CMS stands for carbon molecular sieve

6.3 Hysteresis in Magnesia

MgO is among the few oxides which exhibits ferromagnetic behavior, consequently magnetic hysteresis might occur. However, at first hysteresis usually observed in testing some mechanical properties will be considered.

6.3.1 Mechanical Hysteresis (Cyclic Indentation)

Very limited information exists on the subject under consideration in this section for MgO. Nevertheless, MgO single crystal having the (001) orientation was indented by Berkovich indenter to a load of 50 mN to see if kinks formed in the plastic zone are responsible for the observed reversible hysteresis. In Fig. 6.12 reversible hysteresis is observed in single crystals during cyclic indentation of MgO (shown with other materials investigated), suggesting that this behavior can occur if there are insufficient slip systems to accommodate the strain around the indentation. These results show that reversible hysteresis is associated with conventional dislocation flow, without the need for unstable kinking.

It is suggested that the increase in work dissipated is due to the increase in the fraction of elastic material driving reverse flow. To investigate the idea that reversible hysteresis is caused by the build-up of elastic strains due to plastic anisotropy, rather than kinking, some MgO crystals were indented. MgO slips most easily on the {110}

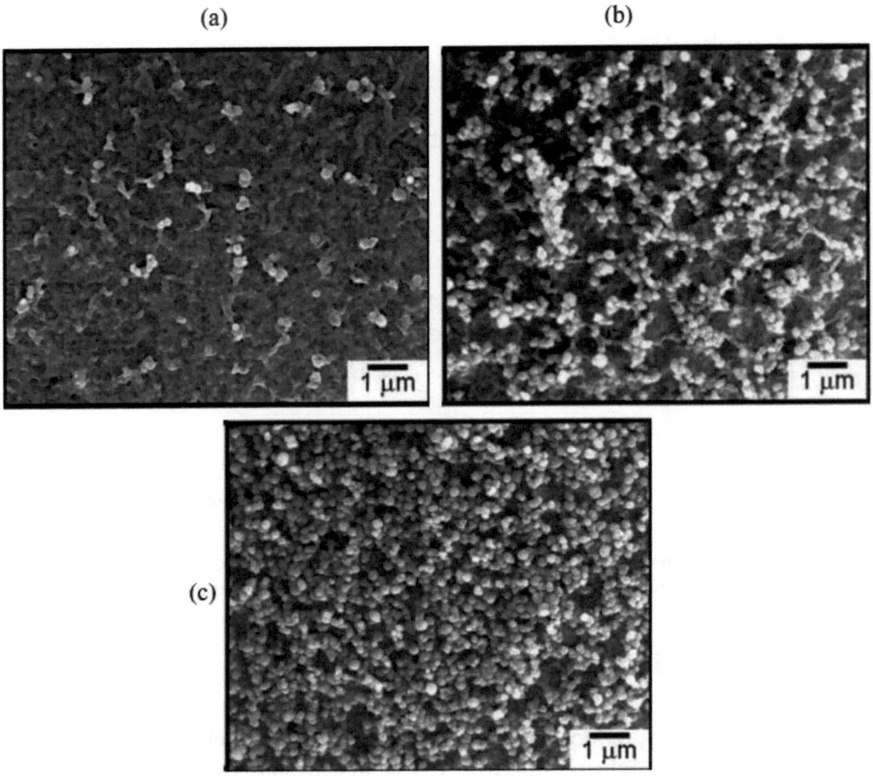

Fig. 6.10 SEM images of the **a** 0.3 vol%, **b** 0.6 vol%, and **c** 1.0 vol% CNP coated on Al_2O_3 granulation particle with average diameter of 62 μm, Tan et al. [3]. Open access

< 110 > slip system. However, this gives only two independent slip systems. Slip can also occur on the < 110 > system, which has a critical resolved shear stress much greater than the {110} < 110 > slip system at room temperature. Therefore, substantial elastic strains capable of driving reverse flow would be developed around the indenter, until slip is activated on the {0 0 1} < 110 > slip system.

In summary, it is suggested that the effect of reversible hysteresis in single-crystal indentation may arise in any material that has sufficient plastic anisotropy that there are insufficient soft slip systems to accommodate the deformation around the indentation. The role of this anisotropy is to enable the development of elastic strains that drive reverse plastic flow.

Further to the above, MgO is an interesting ceramic material owing to the fact that it is deformable at room temperature by conventional slip processes. In spite of having a large intrinsic strength characteristic of a ceramic material, with a NaCl-type structure, MgO presents a model system to understand plasticity in ceramics that have more than five independent {110} < 110 > -slip systems. Nano-indentation experiments were performed to investigate plasticity at low stresses. The depth and

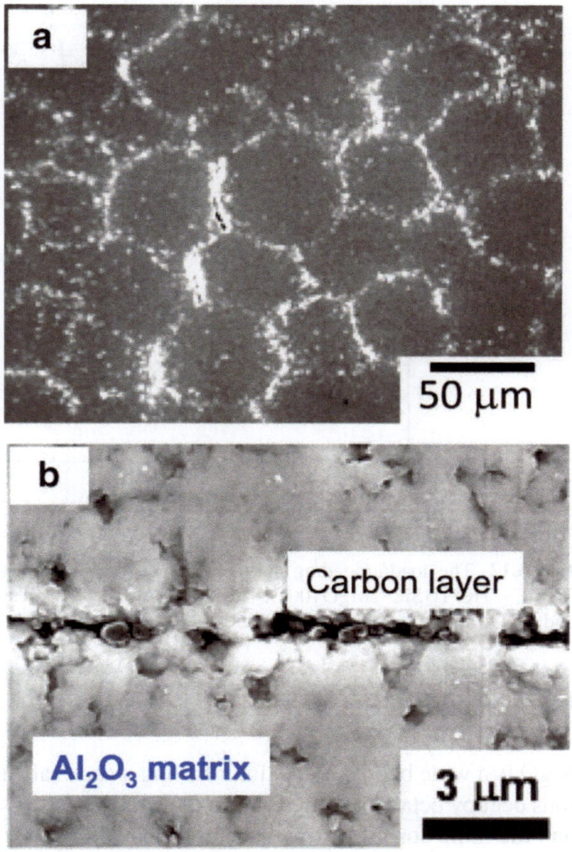

Fig. 6.11 **a** Microstructure of 0.6 vol% CNP-Al_2O_3 composite using Al_2O_3 with the average diameter of 62 μm. **b** Grain boundary of CNP-Al_2O_3 composite. Carbon layer could be observed at the interface between the Al_2O_3 matrix, Tan et al. [3]. Open access

rate of the indenter is based on the displacement readings form the experiments. A total of two loading cycles were applied to the MgO in the simulation and the displacements and load of the indenter in the simulations are shown in Fig. 6.13a. The simulated load and displacement vs. time is shown here. Finite element model was used for the calculation and the peak load from the simulation of 22 mN is close to the experimental 22 mN (Fig. 6.13). The corresponding load vs. displacement curves obtained are presented in Fig. 6.13b and c. The simulated curves obtained for two loading cycles show the presence of reversible hysteresis.

In conclusion, the depth and rate of the indenter is based on the displacement readings from the experiments. Simulation and the displacements and load of the indenter in the simulations agree. Hysteresis in MgO was observed through cyclic loading.

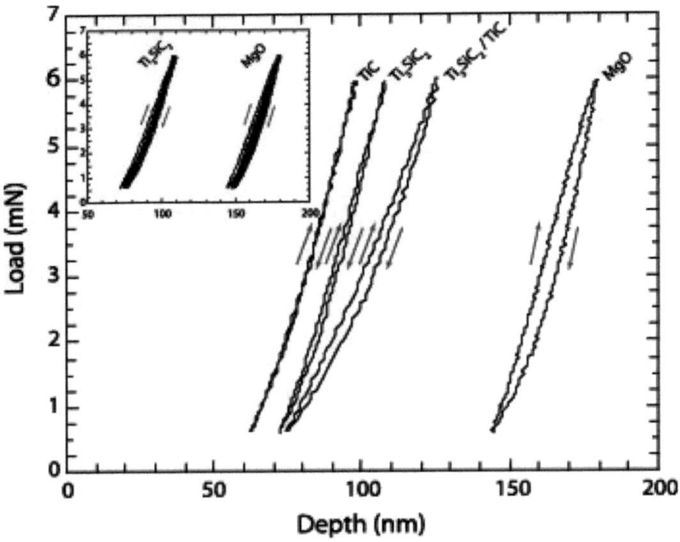

Fig. 6.12 The load/unloading curves obtained by nanoindentation experiments of TiC (111), Ti_3SiC_2, TiC/Ti_3SiC_2, MgO (001). The inset shows the hysteresis over four cycles for the multilayer and MgO, Jones et al. [4]. Open access

6.3.2 Magnetic Hysteresis

MgO is a wide band gap insulator having a rock salt structure (FCC), where the Mg ions occupy octahedral sites within the closed packed structure formed by the anions and the ionic constituents consist of a relatively small number of electrons.

Room-temperature ferromagnetic (RTFM) behavior has recently been reported in undoped metal oxides (MO) such as HfO_2, TiO_2, ZnO and MgO nanoparticles, powders and nanorods. The phenomenon is called d^0 magnetism to emphasize the fact that the observed magnetism is not a consequence of partially filled d-orbitals, but a result of moments induced by p-orbitals of the oxygen band. In other words, d^0-ferromagnetism arises due to completely-filled or unfilled d or f bands. Recall the well-known fact that magnetism arises due to exchange interaction between the electrons in partially-filled d-/f-bands in materials. Apart from this, magnetic behavior of certain materials with completely-filled or zero-filled d-/f-shells is also observed and has remained under debate among researchers.

MgO thin films ~0.1–0.5 μm were deposited at room-temperature on Si substrates by RF magnetron sputtering under an oxygen partial pressure of 1.3×10^{-4} mbar (in the mixture of Ar + O_2 with a total pressure of 40×10^{-4} mbar). Figure 6.14 shows the cross section of a 170 nm thick MgO film deposited at RT (Si substrate) and the films are uniform without observable pinholes. A protective layer of Pt was deposited on the MgO film to prohibit fabrication damage (sectioning). No magnetic impurities or elements are present in the deposited thin film as confirmed by XPS (x-ray photo

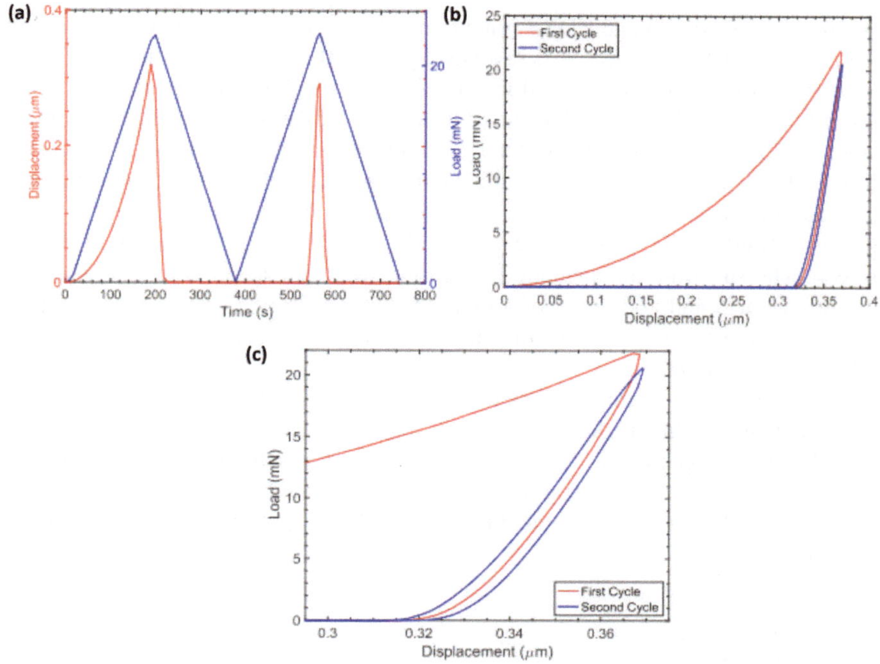

Fig. 6.13 **a** Simulated load and displacement of indenter with time. **b** Simulated load–displacement curve showing the hysteresis in the second and third cycle. **c** Magnified section of the loading–unloading part, Bhowmik et al. [5]. Open access

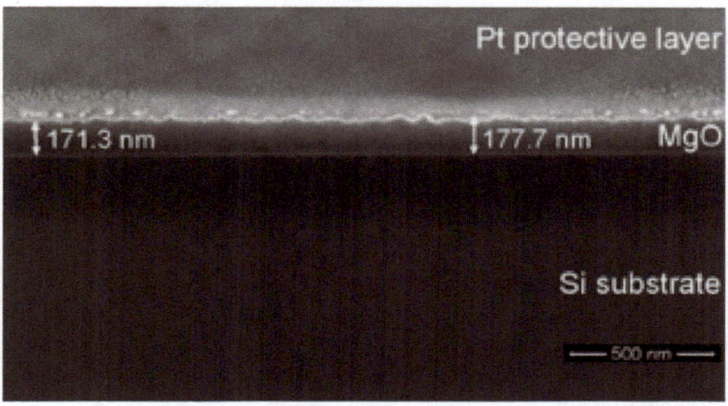

Fig. 6.14 Typical cross-section of the 170 nm thick MgO film (P_{O2}:1.3 × 10^{-4} mbar) using FIB showing a dense and pore free structure, Kapilashrami et al. [6]. Open access. FIB stands for focus ion beam

electron spectroscopy) and EDXS (energy dispersive x-ray spectroscopy) measurements. The RTFM in the film was revealed by SQUID magnetometer measurements and distinct variation in the saturation magnetization (M_S) with increasing film thickness was observed. The M_S value varied from 0.04–1.58 emu/g showing a maximum value at 170 nm thick film. Above the maximum M_S values decrease with increasing film thickness as see in Fig. 6.15, and eventually the MgO film becomes diamagnetic as in the case of the bulk (which is characteristic to oxide thin films showing magnetization). In Fig. 6.16 a typical RT M (H) loop for the 170 nm thick MgO film is seen. The substrate diamagnetic background was subtracted. The inset shows a low remanence, and coercivity of around 90 Oe, which is a typical soft magnetic behavior observed for oxide ferromagnetic semiconductors. Note in the inset the low field hysteresis loop.

In summary, the magnetic transition from ferromagnetism to para and then to diamagnetism is film thickness dependent and it is a common feature for oxide films. Further, magnetism is dependent on the oxygen stoichiometry of the film. The magnetic property of the 170 nm MgO film is enhanced initially with increasing oxygen content, however, continued increase in the oxygen pressure lowers the net magnetization.

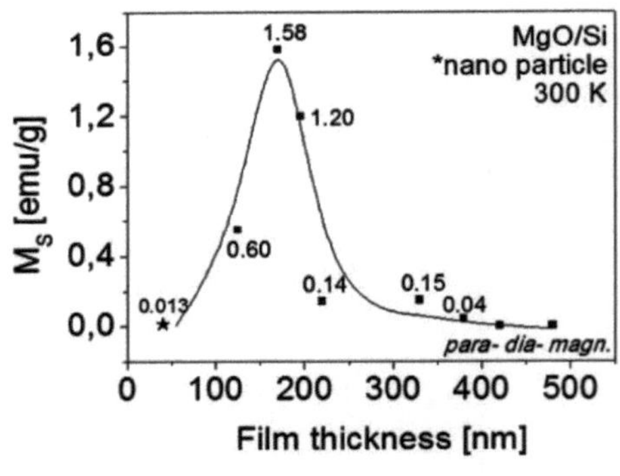

Fig. 6.15 Variation in M_S with increasing film thickness in MgO, Kapilashrami et al. [6]. Open access

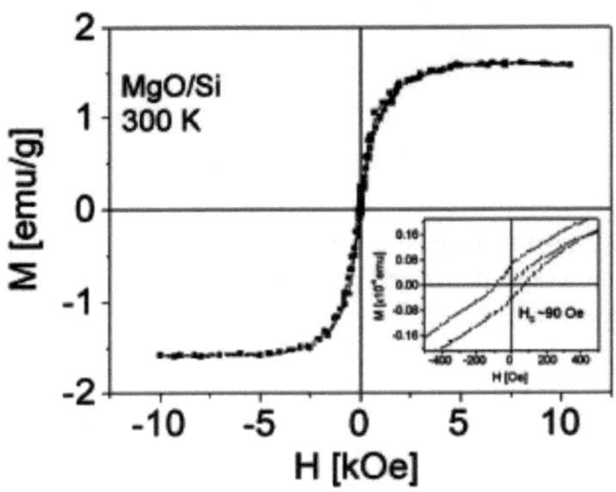

Fig. 6.16 M (H) loop at RT for the 170 nm thick MgO film, data has been corrected for the diamagnetic background. Inset shows low field hysteresis loop with low value for the coercivity, Kapilashrami et al. [6]. Open access

6.3.3 Hysteresis in MgO/C Composite

A very important magnesia-based composite is the magnesia-carbon refractory ceramic since it is exclusively used in the lining of steel converter. The lining is exposed to high temperature during steel making, exposed to thermal shock occurring during the heating and cooling temperature cyclic changes which can lead to failures reducing its lifetime. Refractories based on MgO/C exhibit a non-linear behavior even at extremely low strains which perturb the determination of their elastic properties such as Young's modulus.

The tensile and compression data were obtained by three-point bend tests and compression uniaxial crushing tests. The loading and unloading cycles are shown in Fig. 6.17 and the path revealed the existence of residual strain, hysteresis and reduction of stiffness due to degradation occurring in material B (as seen in Table 6.1 the material is bond by pitch binder). The reason is attributed to microcracks in the binder or created during the loading which propagated inside the binder and around the grains. Also, debonding at the binder-magnesia interface and in the

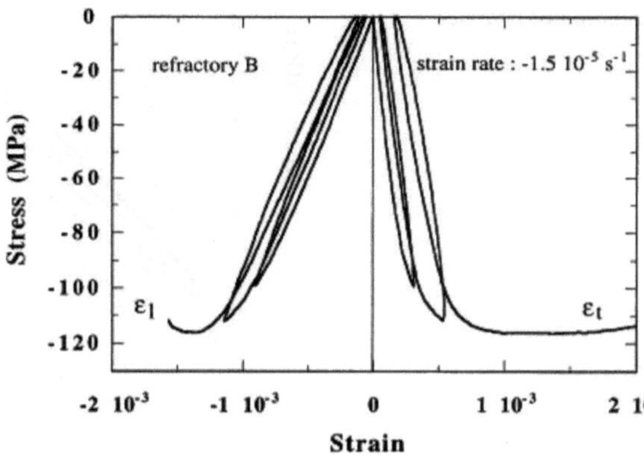

Fig. 6.17 Compression behavior for material B under load-deloading pathes, Schmitt et al. [7]. With kind permission of Elsevier

Table 6.1 Compression properties of magnesia-carbon refractories, Schmitt et al. [7]. With kind permission of Elsevier

Materials	MgO-C resin binder (A)	MgO-C pitch binder (B)
Young's modulus E_c (GPa)	10.6	102
Poisson's ratio ν	0.16	0.24
Compression strength σ_{cr} (MPa)	−27.2	−115.6
Maximal strain ϵ_{max}[a]	$-4.9\ 10^{-3}$	$-1.62\ 10^{-3}$

[a] Strain measured at the minimum stress of the curve.

magnesia, grains were observed. The stress–strain curves for refractories A and B are shown in Fig. 6.18. A linear behavior is observed at the beginning of the load, which is followed by hardening–softening regimes which arose from the initiation of microcracks around the aggregates, their propagation into the binder and then their coalescence to two or three macrocracks.

The load cycling, applied to material A during the test, is shown in Fig. 6.19. One can see that even for extremely small strains, typically less than 4×10^{-5}, the stress versus strain curve exhibits a hysteretic loop. In the loads shown in Fig. 6.19 higher strain levels were applied to specimen A. At first monotonic compression loading was performed (path AB), and non-linearity was observed with increasing strain. Unloading followed with tensile stress (path BC) and finally reloaded in tension (path DCE). The behavior remained linear up to C then became nonlinear (path CE). The true elastic modulus was given only by the very small elastic unloading path CD, which is higher than the one observed in compression load. The tensile properties

Fig. 6.18 Tensile behaviour for material A and B, Schmitt et al. [7]. With kind permission of Elsevier

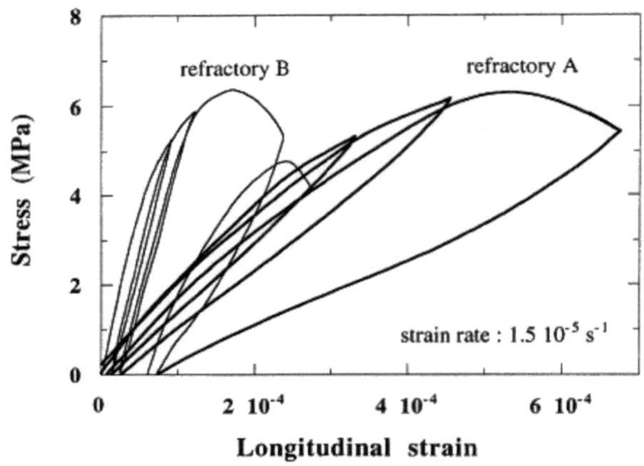

Fig. 6.19 Young's modulus definition in tension and compression for material. A, Schmitt et al. [7]. With kind permission of Elsevier

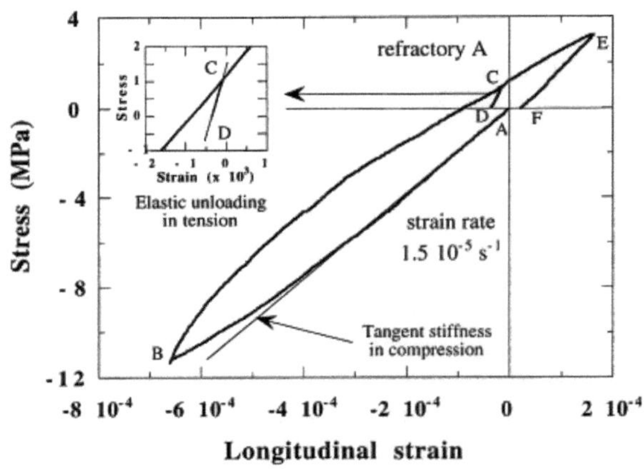

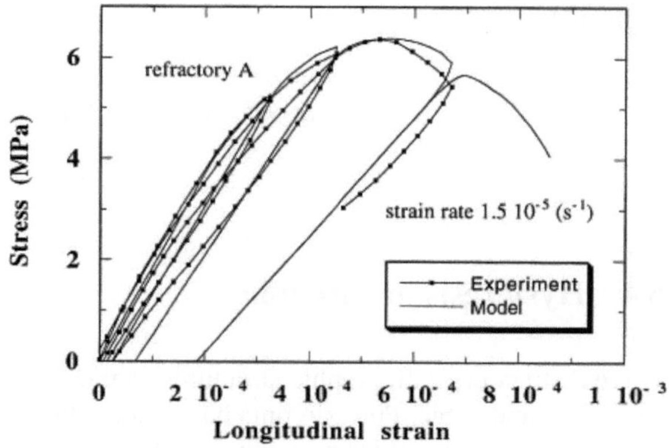

Fig. 6.20 Stress vs. strain curves under tensile load: model and experiment, Schmitt et al. [7]. With kind permission of Elsevier

were derived also by a three-point bend test to compare with those obtained by the indicated method above. Analysis was performed by finite element method. As an example, Fig. 6.20 can be illustrated where the model for uniaxial tension for sample A (with the resin binder) is considered. The strain rate used for the identification was equal to to 10^{-5} s^{-1}. The unloading the strain rate used for the identification was equal to to 10^{-5} s^{-1}. The unloading paths were well reproduced except for the hysteresis for which a more sophisticated model would be needed.

In the classical theory of elastic beam, under plane stress assumption and the hypothesis of St Venant, for the three-point bend test Young's modulus, E is related to force F and deflexion, f by the expression

$$E_f = \frac{Fl^3}{48If} \tag{6.1}$$

I is the moment of inertia of the cross section and l is the span $I = \frac{bh^3}{12}$. Expression (6.1) is no longer correct when the span to height ratio is short and the plane stress changes to curved surface condition due to distortion and a correction is required. The correction parameter, α is introduced to add the defluxions due to shearing and edge effects and is given as

$$E_f = \frac{Fl^3}{48If}(1+\alpha) \tag{6.2}$$

For prismatic beams, the correction α is given as

$$\alpha = 3(1+\upsilon)\left(\frac{h}{l}\right)^2 \tag{6.3}$$

A later correction is provided to take into account the warming established due to the loading force and since the shear deformation is not free. Thus, the corrected α is given as

$$\alpha = (2.85)^2 - 0.84\left(\frac{h}{l}\right)^3 \tag{6.4}$$

6.4 Hysteresis in Zirconia

Three distinct crystallographic structures characterize zirconia (ZrO_2): monoclinic, tetragonal and cubic. Pure zirconia has a monoclinic structure at RT which is stable up to 1170 °C. From this temperature up to 2370 °C the tetragonal phase is formed, while above this temperature the cubic zirconia forms and is stable. After processing and depending on cooling the tetragonal phase becomes monolithic. However, there is a large volume change (3–5%) during cooling to the monoclinic phase from the high temperature which limits its use because this change is sufficient to exceed elastic and fracture limits. In the wake of instability due to this volume change, cracks and flaws form in the zirconia ceramic. Therefore, various stabilizing agent are added such as yttria, magnesia etc., which will be discussed at a later stage of this chapter. These additives (CaO, MgO, Y_2O_3 or CeO_2) maintain the tetragonal phase at RT. The tetragonal or cubic phases can be partially or fully stabilized. The transformation from the tetragonal to the monoclinic phase is very rapid and martensitic in nature, thus it occurs by a shear process without diffusion. The transformation is thermodynamically reversible to ~1174 °C while the reversed transition, namely form tetragonal to the monoclinic occurs between 850 and 1000 °C depending on the strain energy; the change in atomic position curve shows a thermal hysteresis between heating and cooling. The martensitic transformation is influenced by shape, size and location of the particles of ZrO_2 (inter or intragranular), amount of the stabilizer and also by the difference in the thermal expansion.

6.4.1 Hysteresis—Zirconia Transformation

According to the International Union of Pure and Applied Chemists (IUPAC) classification there are six types of isotherms and four types of hysteresis loops. The hysteresis loops are shown in Fig. 6.21. The chemical nature of zirconia and its complex surface chemistry are considered under chromatographic conditions in order to better understand the chemical nature of zirconia. It has been observed that the physical and chemical properties of zirconia depend on the thermal treatment during the synthesis. H1 hysteresis loop (see Fig. 6.21) is considered to be the most favorable for chromatographic purposes.

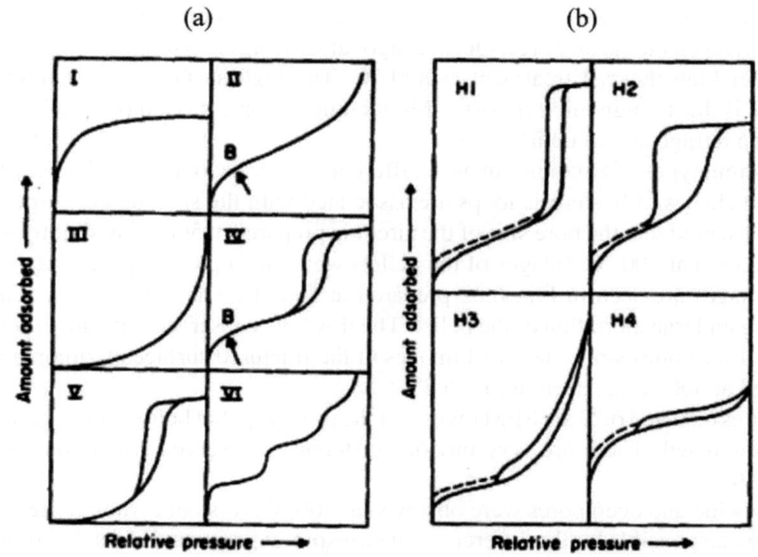

Fig. 6.21 Types of isotherms and types of hysteresis loops according to IUPAC classification, Anazia [8]. Free access

The method of preparation of zirconia as mentioned above strongly influences the properties of the precipitate and structure of the zirconia formed. A known method of preparation is by precipitation from zirconium or zirconyl salts or from zirconium alkoxides. In both methods of preparation, pH is an important factor that can influence the crystallinity of the precipitate formed. Tetragonal phases are most likely to form at pH < 6.5 or > 10.4 and monoclinic phase between the range 6.5 < pH > 10.4. Other techniques to prepare zirconia are by thermolysis of a zirconium salt and a sol–gel type process. (just to remind the reader, thermolysis is the dissociation of a substance as a result of heating, in other words thermal decomposition).

Two pellets of ZrO_2 were prepared at 500 °C and 700 °C. The steps of the synthesis are

$$ZrCl_{4(s)} + H_2O_{(l)} \rightarrow ZrOCl_{2(s)} + 2HCL_{(1)}$$

$$ZrOCl_{2(s)} + 2H_2O_{(l)} \rightarrow ZrO(OH)_{2(l)} + 2HCL_{(l)} \rightarrow \text{polymerization–}$$
$$\text{Condensation} \rightarrow \text{Zirconium(IV)oxide}$$

At temperatures below 400 °C during the synthesis, amorphous ZrO_2 is formed first; the monoclinic crystals are formed at 500 °C, and the tetragonal phase starts to precipitate at 1000 °C. The transformations (reversible) is expressed as

$$\text{Monoclinic} \overset{1170^\circ C}{\longleftrightarrow} \text{Tetragonal} \overset{2680^\circ C}{\longleftrightarrow} \text{cubic}$$

At temperatures below 400 °C zirconia contains micropores and the crystallinity has an effect on the surface area. Only tetragonal zirconia has a relatively high surface area after high thermal treatment at 800 °C. The high surface area of tetragonal zirconia is due to high microporosity. Monoclinic zirconia has a more stable surface area than tetragonal zirconia.

Different types of zirconia can have different hysteresis loops (see Fig. 6.21). The different shapes of hysteresis loops are associated with the specific pore structures. It was observed that the pore size of the zirconia prepared at 500 °C is slightly higher than zirconia at 700 °C. Images of the pellets were taken pre and post experiments. SEM images are seen in Fig. 6.22 prepared at 500 °C. The cracks are a result of handling and transportation of the pellet. The tiny holes observed indicate the porous nature of the compound. The SEM images of the fractured surface of zirconia pellet prepared at 500 °C are seen in Fig. 6.23.

Figures 6.22 and 6.23 are SEM images of the pristine pellet before any experiments were conducted. There are very obvious differences observed when compared to Fig. 6.23b.

Very wide and deep pores were observed at 700 °C. It is not certain if the experiments are responsible for the difference, since no pre-experiments of fractured surface at 700 °C were performed.

In summary, ZrO_2 exist in many crystallographic forms in addition to the amorphous phase. Because its unique surface chemistry which can be dynamically modified, zirconia can be used for chromatographic applications. Different shapes of hysteresis loops (see Fig. 6.21) were observed for various pore structures. Different types of zirconia can have different hysteresis loops and different adsorption. In particular, H1 hysteresis loop and type IV isotherm, are considered to be the most favorable for chromatographic purposes. See type IV isotherm among the other types in Fig. 6.21a.

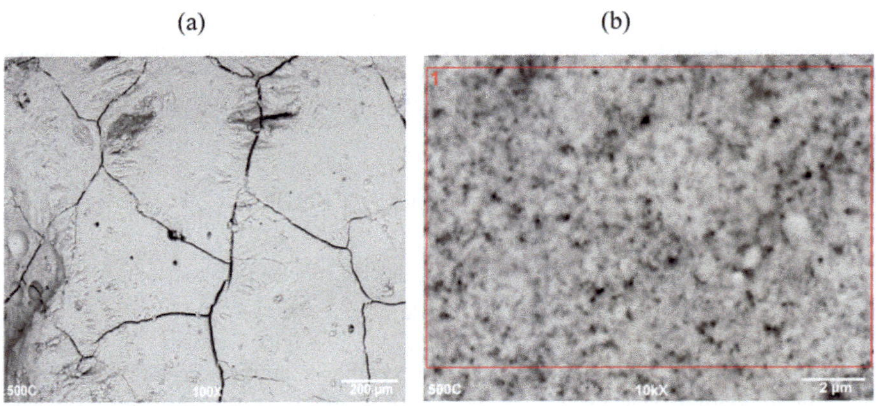

Fig. 6.22 **a** SEM image of the surface of the zirconia pellet prepared at 500 °C taken at 100 magnifications. **b** 10,000 magnifications, Anazia [8]. Free access

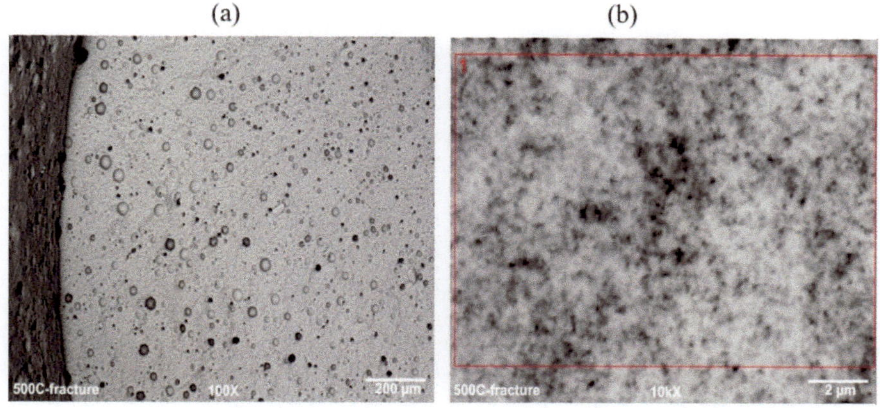

Fig. 6.23 **a** SEM image of the fractured surface of zirconia pellet prepared at 500 °C taken at a 100 magnification. **b** 10,000 magnifications, Anazia [8]. Free access

6.4.2 Stabilized Zirconia

As indicated earlier in this chapter, oxides such as CaO, MgO, Y_2O_3 or CeO_2 stabilize zirconia and maintain the tetragonal phase at RT. One of the most popular stabilizing agents is yttria, but also MgO is often used for stabilization of zirconia. Zirconia can be partially or fully stabilized depending on the amount of the additive, and in the case of yttria fully stabilized zirconia (YSZ) at least 8% Y_2O_3 is required, but generally 12 mol% yttria is used. About 3 mol% is sufficient for partially stabilized zirconia (PSZ), which is used for example to the topcoat of thermal barrier coatings in gas turbines and diesel engines.

6.4.2.1 Partially Yttria Stabilized Zirconia

Partially stabilized zirconia (PSZ) at a level of 3% yttria exhibiting hysteresis loops are considered below. The analysis of the hysteresis loops and curves under different stress conditions indicate anelastic behavior of the zirconia. Recall that anelastic behavior is time dependent as shown in Fig. 6.26. Thus, the anelasticity definition, is the property of a solid in which deformation depends on the time rate of change of stress as well as on the stress itself, or no definite relation between stress and strain. For 3% yttria tetragonal zirconia polycrystal (Y-TZP) the anelastic time dependent relationship is seen in Fig. 6.27. The stress vs. anelastic strain in cyclically loaded 3Y-TZP is shown in Fig. 6.28. In the figure ths first five cycles of hysteresis loops are indicated. The test speed is 5 MPa s^{-1} and the maximum stress is 100 MPa. The loops gradually shift to the right. If the sample is loaded and unloaded at a very high speed, the anelastic strain during the stress holding and after its removal, may be expressed by

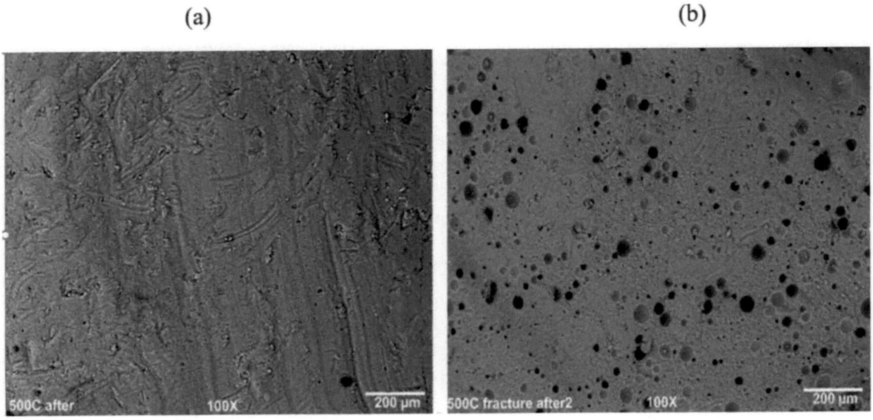

Fig. 6.24 **a** SEM image of the surface of zirconia pellet prepared at 500 °C taken at a 100 magnification (post experiments); **b** fractured surface at 100 magnification (Post experiments), Anazia [8]. Free access. International Union of pure and Applied Chemists (IUPAC)

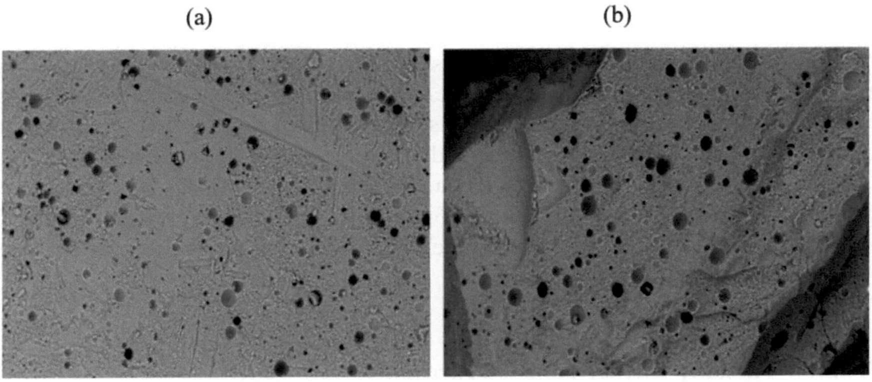

Fig. 6.25 **a** SEM image of the surface of zirconia pellet prepared at 700 °C taken at a 100 magnification. (Post experiments); **b** fractured surface at 100 magnification, Anazia [8]. Free access

$$\text{for stress holding } \varepsilon_a = C\left(\frac{\sigma}{\sigma_{CT}}\right)\left[1 - \exp(-k_1 t)\right] \tag{6.5}$$

$$\text{after stress is removed } \varepsilon_a = C\left(\frac{\sigma}{\sigma_{CT}}\right)\exp(-k_2 t) \tag{6.6}$$

Here, C is a constant characterizing the maximum and elastic deformation and σ is the applied stress which is smaller than the critical phase transformation stress σ_{CT}. The ease of deformation or recovery are characterized by k_1 and k_2. In other words, they present the extent of anelastic deformation and its recovery, respectively. The slope of the initial loading portion of the hysteresis loop (see Fig. 6.28) becomes

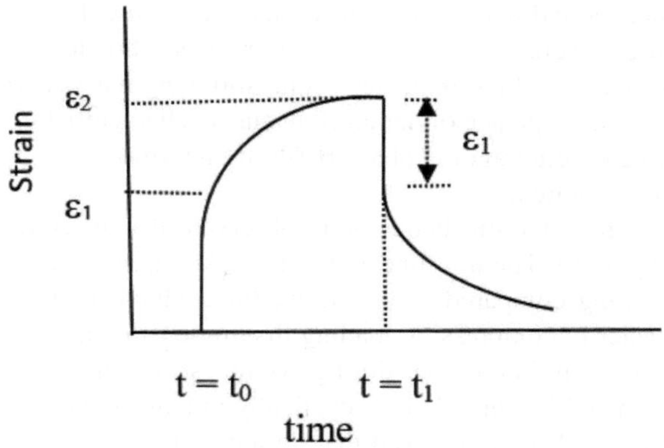

Fig. 6.26 Schematic illustration of anelastic behavior

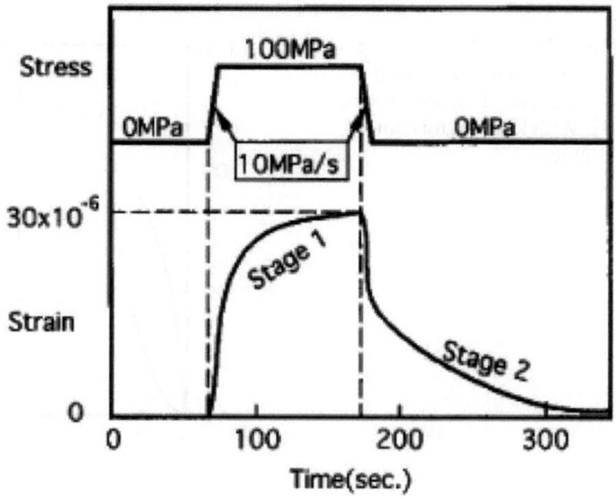

Fig. 6.27 The relationships of stress, non-linear strain vs. test time. This result reveals the anelastic behavior in 3Y-TZP ceramics, Pan and Horibe [9]. With kind permission of Elsevier

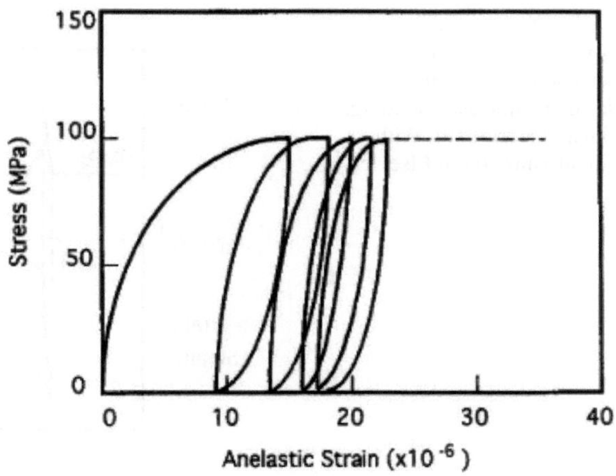

Fig. 6.28 A series of non-linear strain hysteresis loops within the first five cycles with a loading frequency of 0.025 Hz, Pan and Horibe [9]. With kind permission of Elsevier

steeper and steeper, an indication of hardening. However, this hardening is observed after several cycles only, and after several hundreds of cycles the non-linear strain increased which indicates cyclic softening has occurred as seen in Fig. 6.29. This is a consequence of damaged anelastic characteristic, due to microcracks formation when the tetragonal phase partly transformed (t → m) in areas where the stress is high enough.

The anelastic behavior is observed also in compressive loading as shown in Fig. 6.30. The non-linear strain quickly approaches saturation during compression loading compared to tensile loading, which implies that anelastic deformation is easier in compression loading than in tensile one. Figure 6.31 shows the stress and elastic strain relations during cyclic test (for the 5th and 5000th cycle). One loop of a 3Y-TZP sample loaded cyclically in tension and compression test is illustrated in Fig. 6.32. It is observed that, in a compressive half cycle, the sample seems much

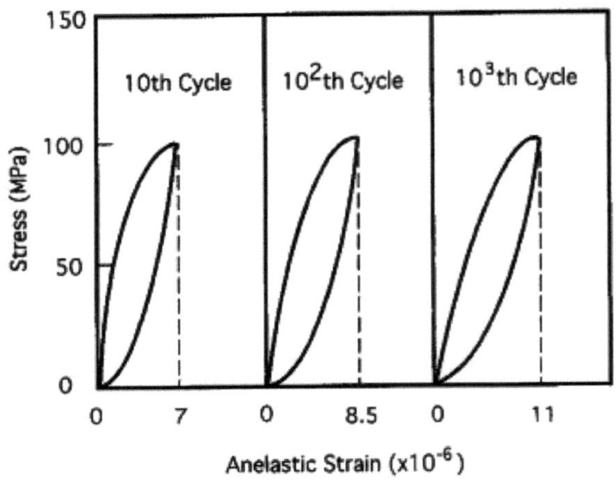

Fig. 6.29 Cyclic softening occurring with increasing stress cycles, Pan and Horibe [9]. With kind permission of Elsevier

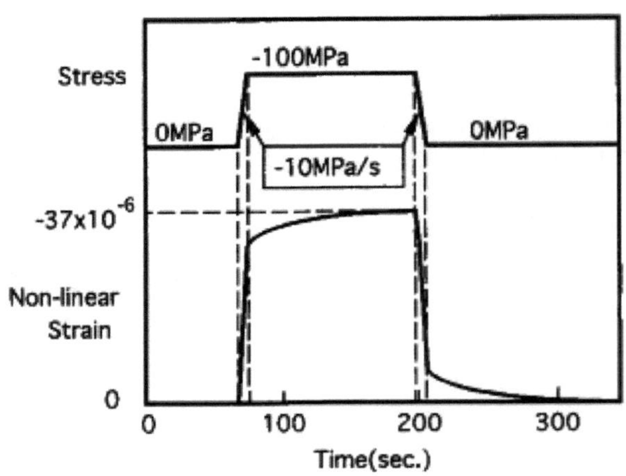

Fig. 6.30 The anelastic behaviour is also observed during compressive loading, Pan and Horibe [9]. With kind permission of Elsevier

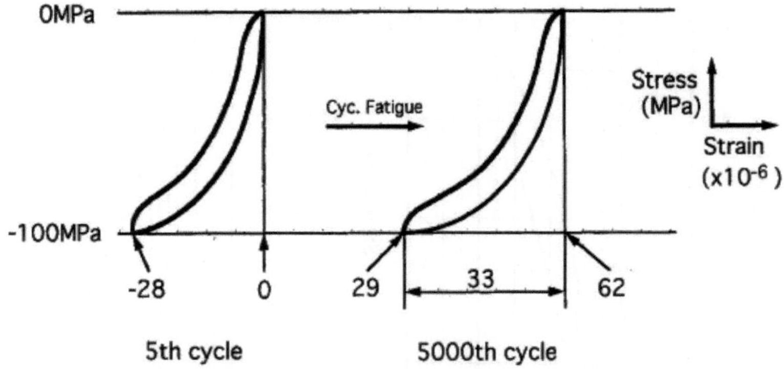

Fig. 6.31 The stress vs. non-linear strain curves in the axial compressive fatigue tests, Pan and Horibe [9]. With kind permission of Elsevier

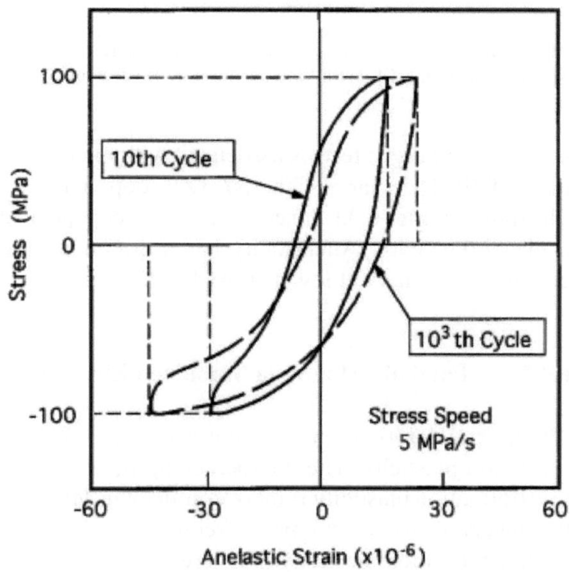

Fig. 6.32 The change of anelastic hysteresis loops under tension and compression fatigue testing, Pan and Horibe [9]. With kind permission of Elsevier

softer than in a tensile half cycle. In both cycles shown in the figure the strain in compression is higher.

In summary, 3Y-TZP exhibits anelastic behavior and the hysteresis loop change under both, tension and compression in fatigue test. Softening has been observed which is more pronounced in the compressive than in the tensile half cycle. The partially stabilized 3Y-TZP deforms more easily in compression stress than in the tensile stress at the same level. The non-linear deformation can almost be recovered after first loading unloading under a lower stress range. The anelastic characteristic of the material will be gradually damaged with the further cyclic loadings. Cyclic

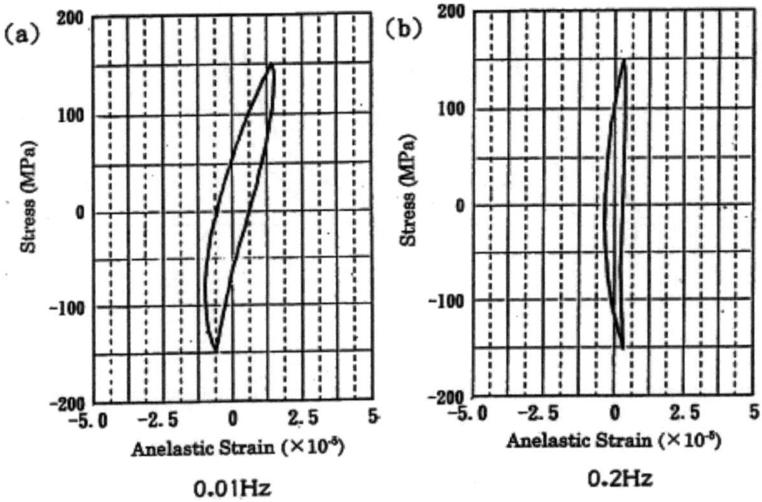

Fig. 6.33 Saturated stress–anelastic strain hysteresis loops in 3Y-TZP. **a** 0.01 Hz. **b** 0.2 Hz, Matsuzawa et al. [10]. With kind permission of Elsevier

softening in fatigue tests is associated with microcracking and phase transformation in 3Y-TZP. The shape of the 3Y-TZP loop changes as a function of frequency as illustrated in Fig. 6.33. It seems, that this dependence is caused by time-dependent anelastic behavior. Anelasticity has a dominant effect on deformation in Y-TZP ceramics especially at lower stress level.

6.4.2.2 Partially Magnesia Stabilized Zirconia (Mg-PSZ)

The fatigue deformation mechanism in Mg-PSZ occurs by microcracking or transformation plasticity. The microcracking mechanism is dominant in 3Y-TZP, while transformation plasticity is more dominant in Mg-PSZ. The relative portions of these two mechanisms depend on the frequency and stress amplitude, and the mechanism of transformation plasticity is favored at low frequency and high stress. The shape change of the hysteresis loop, which is an indication of cyclic hardening or softening, is the cyclic stress strain response under either stress controlled or strain controlled condition. After cycling, a steady state hysteresis loop develops under fully revised loading conditions at various stress or strain amplitudes which allows to assess the cyclically induced changes in the deformation resistance of the material. Four types of hysteresis loops are illustrated below schematically in wake of the mechanisms indicated above.

The steady-state hysteresis loop of dense, homogeneous material is shown schematically in Fig. 6.34. The figure consists of two linear portions AB and CD which occur after the elastic unloading and loading from the peak loads and two

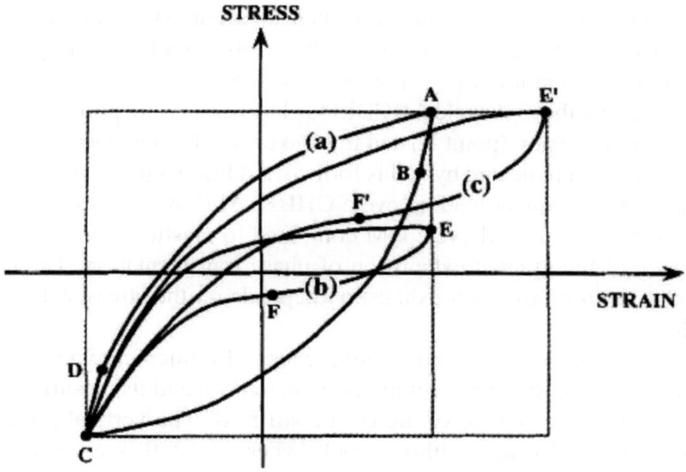

Fig. 6.34 Stress–strain response of **a** a dense, elastic-homogeneous plastic material, **b** a cracked material under constant strain-controlled condition, and **c** a cracked material under constant stress-controlled condition, Liu and Chen [11]. With kind permission of John Wiley and Sons

nonlinear portions BC and DA. The hysteresis is symmetric with respect to the origin with a negative (positive) curvature associated with the loading (unloading) portion DA (BC). The shape of the loop is retained until the appearance of fatigue cracks as seen in Fig. 6.34b from point A to E for the constant strain condition and in Fig. 6.34c from point A to E for the constant stress condition. The parts of unloading in (b) from point F to C and in (c) from point F' to C indicate negative curvature. Transformation induced hysteresis loop is shown in Fig. 6.35. The transformation

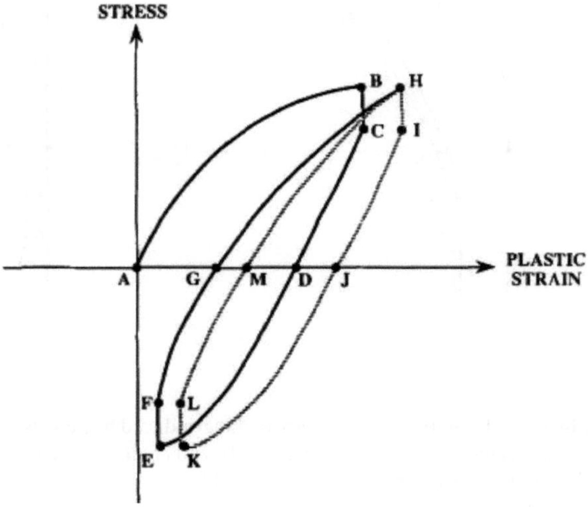

Fig. 6.35 Evolution of the transformation-induced hysteresis loops during stress cycling, Liu and Chen [11]. With kind permission of John Wiley and Sons

initiation (tetragonal → monoclinic) assumed to occur on loading and increases from point A to B, and after overcoming a back stress represented by the portion BC, reversed transformation (m → t) sets in. The process repeats itself after the load reversal to complete the first cycle ABCDEFG. The peak strain, point E, in compression is less than in tension (point B) and thus no closed loop is obtained and a finite residual plastic strain indicated by AG is formed. Additional phase transformation is triggered during subsequent loading cycle GHIJKLM (however less residual strain is accumulated in the second cycle GM compared to plastic strain in AG). This is a consequence of the gradual exhaustion of phase transformation. The number of fatigue cycles required to reach exhaustion depends on the stress state, stress rate and prior history.

The development of the hysteresis loop caused by microcracking is shown in Fig. 6.36. Let us consider crack accumulation in tension and its closure in compression which results in a change of the elastic stiffness. The vertical portion AB in (a) represents elastic loading without crack, where point B is assumed to be the threshold stress of microcracking and followed by BC plastic deformation from progressive microcrack formation, crack opening, and the release of residual strain. Elastic unloading indicated by CD is followed by a gradual recovery from point D until all crack surfaces are closed along DE. The process requires some compression point E. Further compression can be associated with fully recovered crack free specimen and is represented by the elastic compression indicated by the vertical line EF. Unloading the compressive stress from point F, the stress-plastic strain follows along the vertical line FG and at point G the crack reopens. The hysteresis loop ends at point H for the first cycle ABCDEFGH. The slope of the second cycle IJ is less than CD of the first cycle because of higher crack density. The strain accumulated in

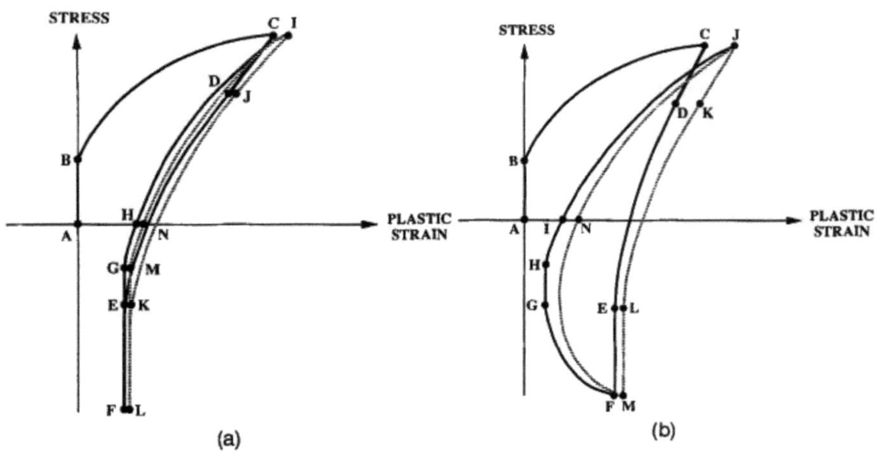

Fig. 6.36 Evolution of the microcracking-induced hysteresis loops during stress cycling **a** without crack surface friction and sliding, and **b** with crack surface friction and sliding, Liu and Chen [11]. With kind permission of John Wiley and Sons

the second cycle HN is smaller than AH in the first cycle. On reaching a microcrack saturation state within the first two cycles, a stable closed hysteresis loop forms in the third cycle as indicated by the dashed line NIJKLMN and retains its shape in subsequent cycling. This case of the hysteresis loop is entirely a consequence of microcracks opening and closing.

A variation of the case of microcracking discussed above concerning the schematic hysteresis loop formation, in Fig. 6.36b the effect of sliding operating on the microcrack surfaces during deformation is included. The sliding related deformation may be regarded as anelastic in nature. Following the peak compressive load an additional compressive strain can accumulate during unloading resulting in curve FG with a negative slope. When the anelastic strain is at last exhausted, a vertical unloading portion GH appears and the hysteresis loop continues. The steady-state hysteresis loop NJKLMN is represented by dashed lines.

The hysteresis loop representing the case when both mechanisms, the phase transformation and microcracking are operational is illustrate in Fig. 6.37. The curvature of tensile unloading may vary from negative to positive when the dominant mechanism is changing from microcracking to phase transformation. When the contribution of both mechanisms is about the same a straight unloading line may result.

Now, after considering the possible various schematic shapes of the hysteresis loops representing their association with the causes of their formation, we can turn to Mg-PSZ, the partial stabilized zirconia of 8–9 mol% Mg oxide (magnesia). Coarse grained Mg-PSZ was used to evaluate the hysteresis loop development in prefatigued specimens at a 5 Hz frequency which were then stressed in the range ± 300 MPa for 10^5 cycles. As the stress range becomes larger the tensile unloading curvature becomes less negative as seen in Fig. 6.38 and contrary to the observed case in

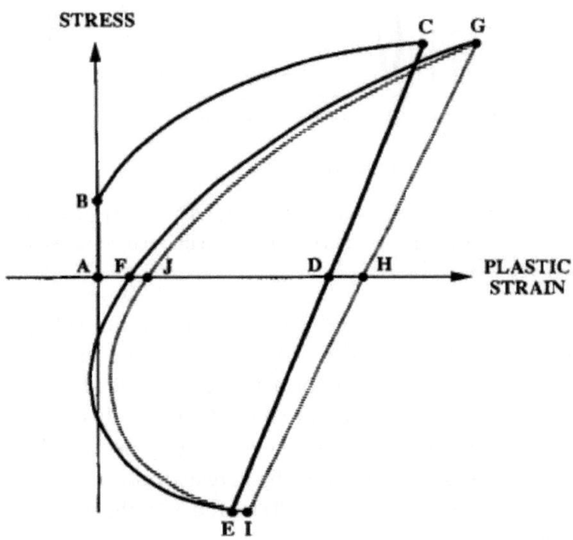

Fig. 6.37 Evolution of the hysteresis loops caused by both phase transformation and microcracking, Liu and Chen [11]. With kind permission of John Wiley and Sons

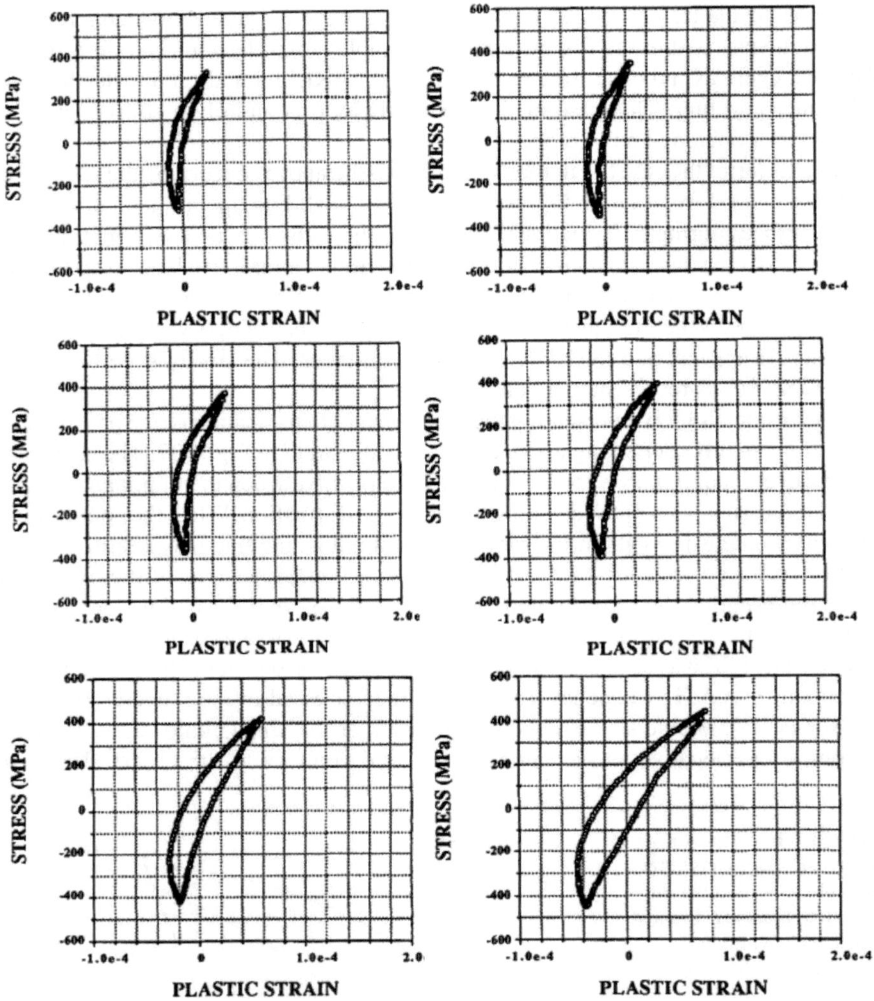

Fig. 6.38 Hysteresis loops of Mg-PSZ at 0.25 Hz. The specimen has been prefatigued at ± 300 MPa for 100,000 cycles, Liu and Chen [11]. With kind permission of John Wiley and Sons

3Y-TZP. At the highest stress range the unloading portion is nearly straight (the last illustration in Fig. 6.38), which indicates that both mechanisms, namely nicrocracking and phase transformation contribute about equally to the cyclic deformation (the illustration is similar to the schematic illustration of Fig. 6.37). From these plots a cyclic stress-plastic strain curve was constructed which is shown in Fig. 6.38.

For the development of hysteresis loops in the stress rang of 100–375 MPa and at strain rates of 20 and 200 MPa s^{-1} the interested readers should consult the publication of Liu and Chen.

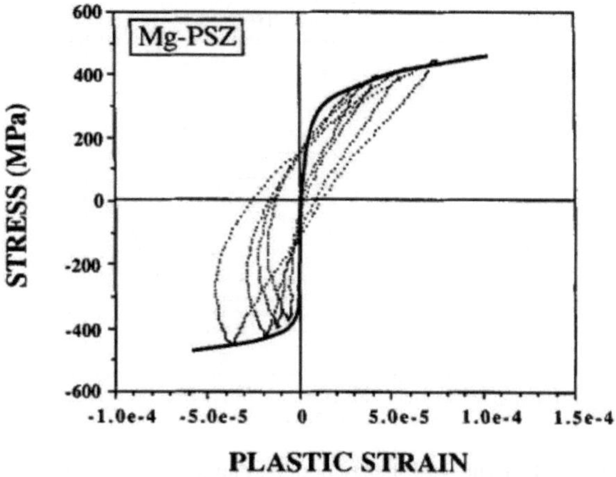

Fig. 6.39 Cyclic stress-plastic strain curve of Mg-PSZ at 0.25 Hz. The specimen has been prefatigued at ± 300 MPa for 100,000 cycles. (Dasd lines are hysteresis loops at different stress ranges), Liu and Chen [11]. With kind permission of John Wiley and Sons

In summary: (1) two distinct fatigue deformations were considered, namely, that of microcracking and transformation plasticity. They are operational at all stress ranges. (2) Microcracks form in the tensile portion of the cycle, and they partly close in the compressive half cycle. (3) Microcrack deformation of anelastic nature has been observed in the compressive half cycle after unloading from the peak compressive load. (4) Cyclic deformation is dependent on the strain rate or frequency which affects both mechanisms. At lower strain rates transformation plasticity is more prominent, which can be suppressed at very high strain rates and microcracking mechanism remains the dominant one. Evidence for fatigue damage is the cyclic softening which has been observed regardless of which deformation mechanism is dominant.

6.4.2.3 Fully Stabilized Zirconia (8Y-FSZ)

Pure ZrO_2 transforms from monoclinic to tetragonal at ~1170 °C as stated in Sect. 6.4 (more accurately 1173–1174 °C) which is accompanied with a volume change of about 9%. Due to this volume change the transition may be accompanied with crack formation. It is required to stabilize the tetragonal zirconia structure of over wider temperature range which is accomplished by adding yttria. The stabilization is a consequence of substituting some of the Zr^{+4} ions with the somewhat larger Y^{+3} ions to form the yttria stabilized zirconia (YSZ). The very good mechanical properties of YSZ are associated with the tetragonal phase. The zirconia stabilized with 8 mol % yttria is strong, wear resistant and chemically stable at elevated temperatures and is known as fully yttria stabilized zirconia (8YSZ).

(a) Most of the work in yttria stabilized zirconia is related to partially stabilized one, and almost no experimental research was performed on the fully stabilized zirconia, namely to 8Y-FSZ or of higher yttria content zirconia. However, a communication where hysteresis is indicated in an 8% yttria stabilized zirconia, but not related directly to fatigue in its common sense, but rather to thermal cycling has been

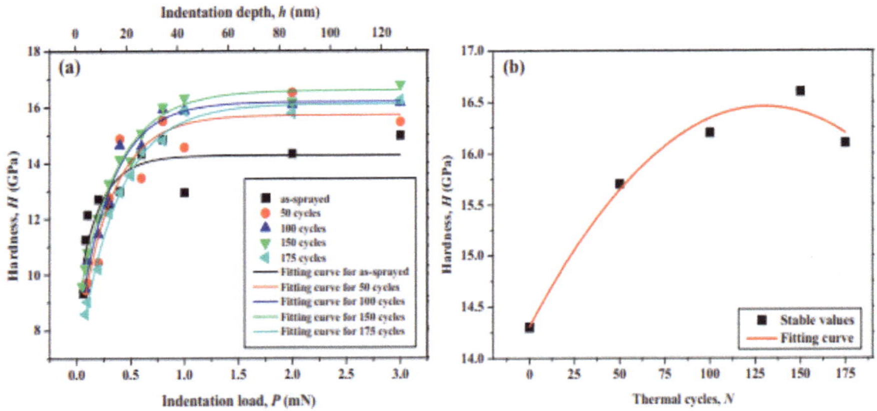

Fig. 6.40 **a** The RISE behavior of hardness of 8YSZ at small scales under different thermal cycles. **b** The variation of hardness during the steady state with thermal cycles, Maoa et al. [12]. With kind permission of Elsevier

published and is considered below. Since fatigue is a classical deformation where the loading represents changes observed in a material under the influence of stress generated during cyclic loading, it was felt that the resulting hysteresis from thermal cycling is sufficiently representative of hysteresis observed in 8YSZ.

The mechanical properties of air plasma sprayed 8 mol % yttria stabilized zirconia (8YSZ) coating were explored by nanoindentation with a Berkovich indenter. Such coatings are widely used in aircraft, gas turbines, etc. as thermal barriers as protection at high temperatures and against oxidation, thereby increasing the lifetime of the respective components exposed to such conditions. The hardness increases for all 8YSZ samples tested with the increase of the indentation load, P within the load range P < 1 and when P > H gradually falls to saturation at a hardness level of 14.8 as seen in Fig. 6.40a. The true hardness, H_{true} is calculated on the basis of the energy of indentation contact given as

$$U_r = \frac{1}{3}\sqrt{\frac{1}{\alpha_0 \tan^2(\Psi) H_{true}}} P_{\max}^{3/2} \tag{6.7}$$

where $\alpha_0 = 3\sqrt{3}$ and $\Psi = 67.3°$ are geometrical constants and the semi apex angle of a Berkivitch indenter, respectively. The loop energy U_r of the elastic–plastic deformation can be calculated by the area difference between the loading and unloading paths, U_r and the corresponding $P_{\max}^{3/2}$ providing good fits to the straight lines as shown in Fig. 6.41. From the slope of the linear relationship, H_{true} can be determines and the results are shown in Fig. 6.41b. As seen H_{true} increases from a value of 73 to 80 GPa for less than 100 cycles, which is followed by a decrease from 75 to 175 GPa

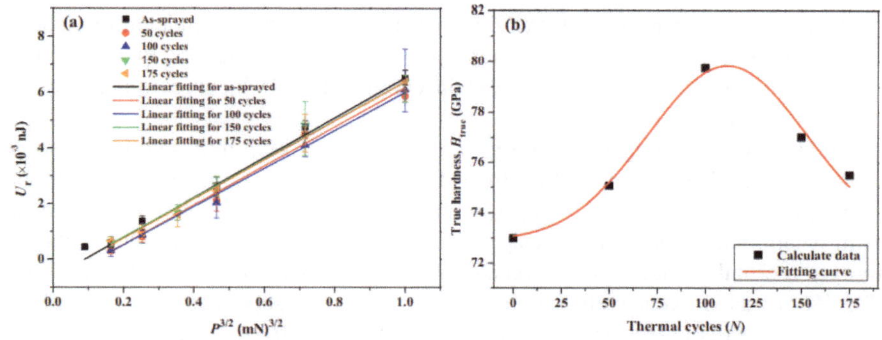

Fig. 6.41 **a** Plot of U_r–$P^{3/2}$ max for 8YSZ before and after thermal cycling. **b** Evolution of true hardness with thermal cycles, Maoa et al. [12]. With kind permission of Elsevier

after 175 thermal cycles. Further, additional indentation tests were performed under $P_{max} = 0.08$ mN and the result are shown in Fig. 6.42a and as seen, the P–h curves match the elastic Hertzian theory and no hysteresis is observed which indicates a perfectly elastic behavior during nanoindentation. Based on the Hertzian theory the P–h relation is generally expressed as

$$P = (4/3)E_r R^{1/2} h^{3/2} \tag{6.8}$$

Assuming R ≈ 402 nm, Er for the test can be calculated ~207 GPa with P = 1.75×10^8 $h^{1.5}$. The details from the elastic to elasto-plastic transition of 8YSZ after thermal cycling carried out at P_{max} in the range 0.1–6 mM is shown in Fig. 6.42b.

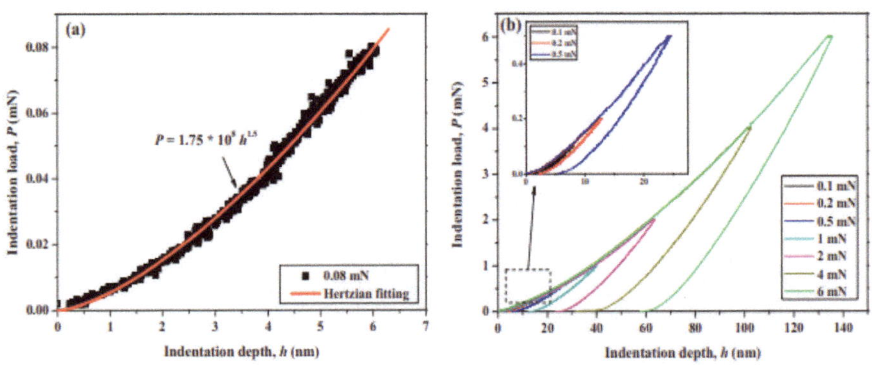

Fig. 6.42 **a** Fully elastic loading and unloading-depth curves of 8YSZ after thermal cycling. **b** Onset of elasto-plastic transition of 8YSZ after thermal cycling under different indentation peak loads. Inset is magnified from the small indentation load regime, Maoa et al. [12]. With kind permission of Elsevier

In summary, he hardness of 8YSZ increases with increase in indentation load P. The hardness at first increases with increasing the thermal cycles and then decreases as seen in Fig. 6.41b. No hysteresis was observed and the P–h curves was expressed by the Hertzian elastic theory.

(b) However, contrary to the absence of a hysteresis loop indicated in (a), some hysteresis was observed in YSZ coatings. For example, time dependent deformation of air plasma sprayed 7 wt.% Y_2O_3 stabilized heat treated ZrO_2 coating loaded using a nanoindenter show hysteresis loss during cyclic loading. Atmospheric plasma sprayed thermal barrier coatings (APS-TBC) based on 7 wt.% Y_2O_3 stabilized ZrO_2 (7YSZ) are used extensively for thermal protection of superalloys in gas turbine engines. The thicknesses of the coatings are in the range from 300 μm to a few mm. The microstructure contained features such as porosity, lamellar interfaces, micro and macro cracks which were deliberately introduced in the stabilized zirconia to reduce the elastic modulus of the coating to provide compliance to add in their integration with the underlying bond coat and super alloy.

Hysteresis loss in cantilever bending experiments were performed to investigate hysteresis loss per loading–unloading cycle. The load was applied by an indenter—as mentioned—of 300 μm tip radius. Residual stresses are built up in the 7YSZ coating either during deposition or during thermal cycling (due to mismatch in the thermal expansion coefficients of coating and the adjoining materials) leading to the build-up of strain energy,

The microstructural features of the 7YSZ indicated above result in non-linear elastic properties of the coating including observation of hysteretic behavior during loading and unloading cycles.

The load–deflection curve is shown in Fig. 6.43a and the hysteresis loss per cycle (area under the load–deflection curve) as a function of the load is shown in Fig. 6.43b.

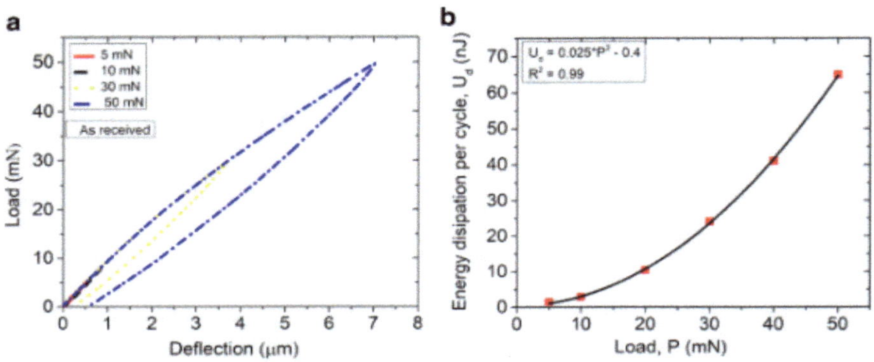

Fig. 6.43 **a** Variation in load–deflection curve with increase in load (cantilever dimension 0.3 × 0.2 × 2 mm^3). **b** Variation of energy dissipation or the hysteresis loss per cycle with maximum applied load in cantilever (The datum points represent the calculated values), whereas the solid line shows the best fit curve, where the equation along with regression parameter (R) given in the legends, Lal et al. [13]. With kind permission of Elsevier

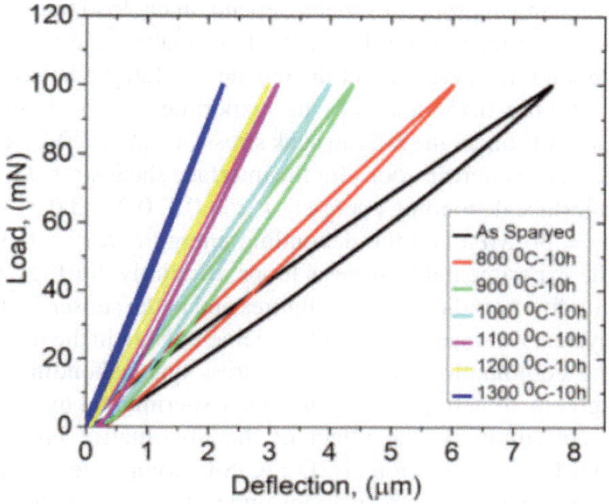

Fig. 6.44 Load–deflection curves of sintered coatings, loaded up to 100 mN. The sintering conditions of each curve is given in the legend, Lal et al. [13]. With kind permission of Elsevier

It was observed that the hysteresis loss increases proportionately with the square of the maximum load in the cantilever. The variation of the load deflection curves with sintering are shown in Fig. 6.44. It is seen that with an increase in the temperature of exposure at fixed time, the total deflection is reduced and also the area under the load–displacement curve (i.e., total hysteresis loss per cycle).

In summary, it has been indicated that hysteresis was observed in fully stabilized YSZ also, not only in partially stabilized 3YSZ. It is shown that air plasma sprayed 7YSZ coating shows a hysteresis loss during the loading–unloading cycle. The variation of energy dissipation or the hysteresis loss per cycle with maximum applied load in cantilever loading is illustrated graphically. The miniaturized cantilevers prepared from free-standing heat-treated coatings were loaded using a nanoindenter.

6.5 Hysteresis in SiC/SiC

Since SiC is either being reinforced with other components or it acts as a reinforcing agent for various materials, metals and ceramics alike, SiC/SiC composite is presented below. Thus, seldom one can find publications on hysteresis or other relevant information on pure SiC and the choice of SiC/SiC is clearly a consequence of the similarity of matrix and reinforcing agent. The material is a 2D woven SiC/SiC composite and the effect of fiber interface on the tensile and fatigue properties is of interest. Ceramic matrix composites (CMCs) are widely used at high temperature as a light and high-performance structural composite material, which makes it a candidate to replace superalloys for aero-engines. CMC materials are used for additional applications, such as combustion chambers, turbines, exhaust nozzles and other components of aero-engines. As mentioned regarding fiber-reinforced CMCs, the mechanical properties much depend on the load transfer at the fiber/matrix interface. The interface properties of the fiber/matrix interface shear stress and the interface debonding energy affect the tensile and fatigue behavior of the CMC.

Figure 6.45 illustrates the experimental and predicted hysteresis loops of 2D SiC/SiC under the fatigue peak stress of $\sigma_{max} = 200, 240$ and 275 MPa. The predicted fatigue hysteresis loops for the interface shear stress ($\tau_j = 30, 40$ and 50 MPa) and the interface debonding energy ($\zeta_\delta/\zeta_\mu = 0.1, 0.2$ and 0.3) using the interface shear stress $\tau_j = 50$ MPa and the debonding energy $\zeta_d/\zeta_m = 0.3$ are in good agreement with the experimental hysteresis loops. Similarly, for Figs. 6.46, 6.47 and 6.48 under the conditions indicated at the figures and in the subscript the predicted and experimental hysteresis loops show good agreement. Note in the illustration under what conditions the combination of the shear stress and debonding energy ratio are for the best agreement between predicted and experimental hysteresis loops.

In summary, the effect of the fiber/matrix interface properties of the tensile and fatigue behavior of 2D SiC/SiC composite was presented. The predicted and experimental hysteresis loops under different conditions are illustrated.

6.6 Hysteresis in Si_3N_4

A variety of applications are in use of monolithic Si_3N_4 and its composites, staring with pH sensors devices based on insulator/electrolyte interfaces, followed by gate insulator with high quality and low leakage current in MOSFET (and other devices) and clearly in mechanical applications. In the following we shall indicate some of its uses including a much-used composite and clearly the hysteresis in stress–strain relation which is the main interest in this chapter.

6.6.1 PH-Sensitive Coating

The use of devices based on insulator/electrolyte interfaces is limited by a slow response which manifests itself as a hysteresis loop. Use of Si_3N_4 to test the time taken to perform a complete hysteresis loop is of interest because of the good insulator qualities of Si_3N_4. The devices used in the measurement are light-addressable potentiometric sensors (LAP) with Si_3N_4 insulator which was deposited by low pressure chemical vapor deposition (LPCVD). For the measurement setup see details in the work of Luc Bousse et al. It has been shown that pH-sensitive insulators have a memory to pH changes, which manifests itself as slow response or as hysteresis. The results depend on the time taken to perform a complete pH loop. Figure 6.49 shows the hysteresis loops obtained with a Si_4N_4 sample at times of 1, 2, 4 and 6 min per pH step (loop times of 960–5760 s; in the figure it is given in minutes). Figure 6.50 shows data for the same sample for times of 10, 20, 40, and 80 min per pH step (loop times of 9600–7.68×10^4 s; in the figure it is given in minutes). It can be noted

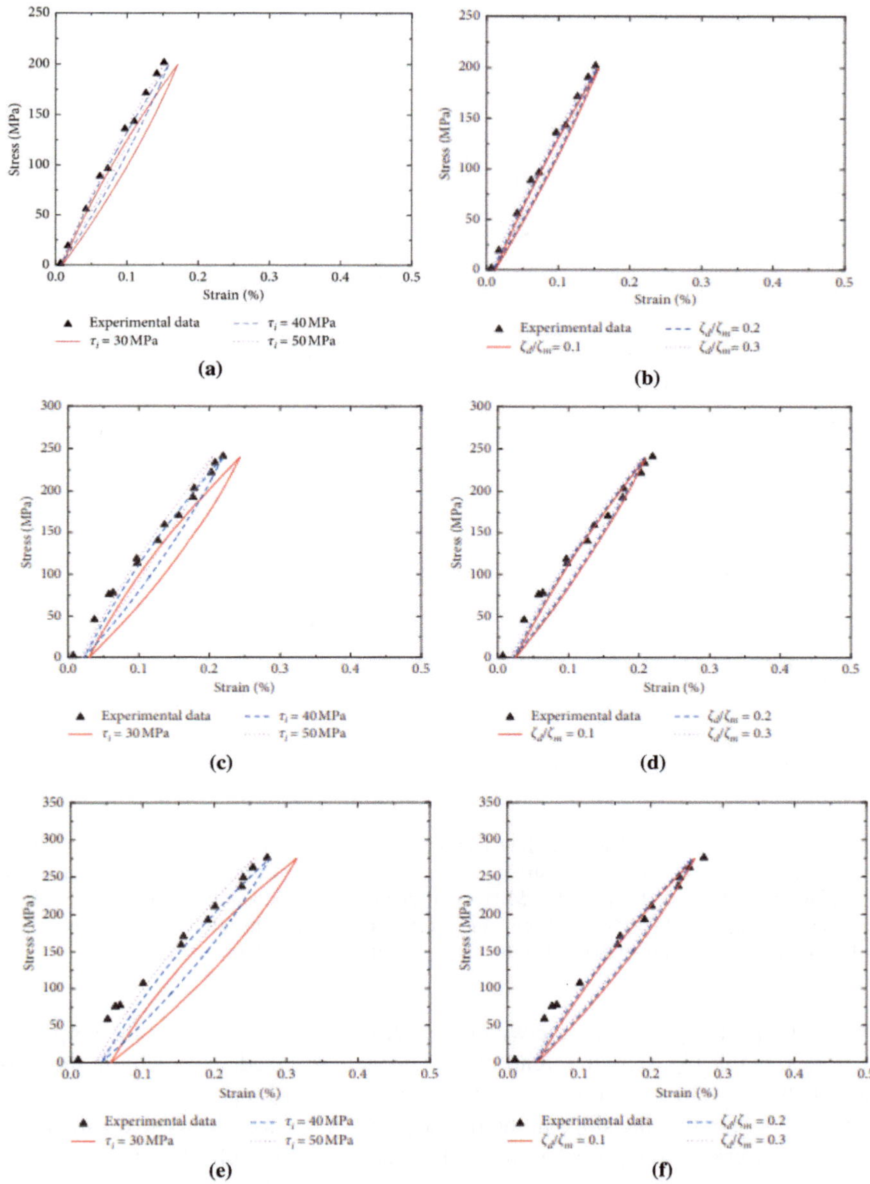

Fig. 6.45 The experimental and predicted hysteresis loops of 2D Sylramic™ SiC/SiC composite under. **a** The fatigue peak stress of $\sigma_{max} = 200$ MPa for different interface shear stress. **b** The fatigue peak stress of $\sigma_{max} = 200$ MPa for different interface debonding energy. **c** The fatigue peak stress of $\sigma_{max} = 240$ MPa for different interface shear stress **d** The fatigue peak stress of $\sigma_{max} = 240$ MPa for different interface debonding energy. **e** The fatigue peak stress of $\sigma_{max} = 275$ MPa for different interface shear stress. **f** The fatigue peak stress of $\sigma_{max} = 275$ MPa for different interface debonding energy, Longbiao [14]. Open access

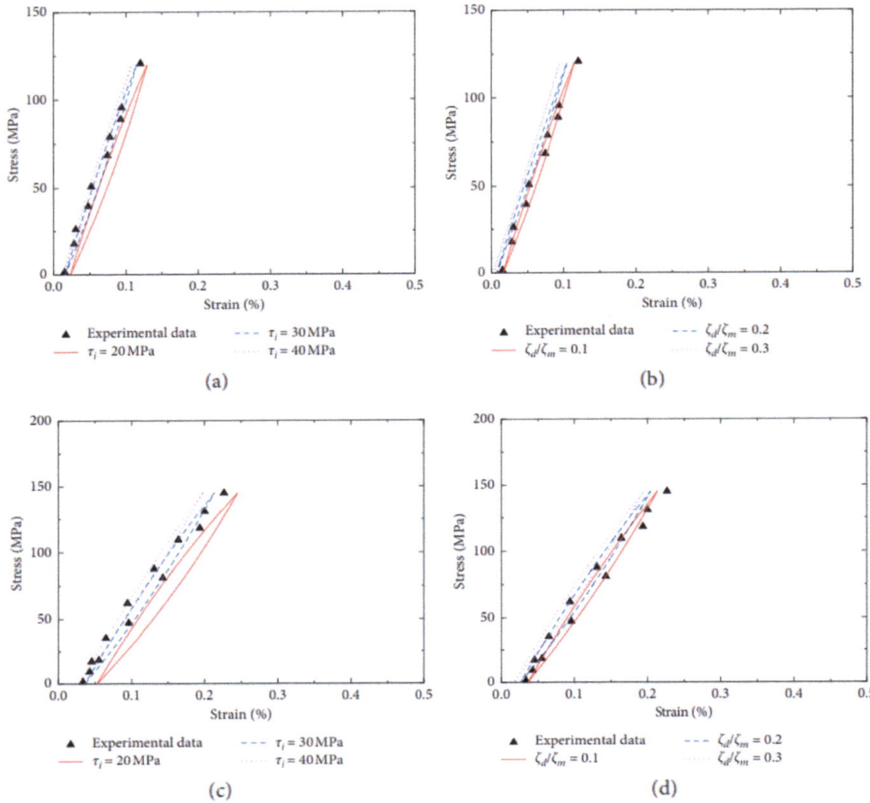

Fig. 6.46 The experimental and predicted fatigue hysteresis loops of 2D Tyranno™ SiC/SiC composite under. **a** The fatigue peak stress of σ_{max} = 120 MPa for different interface shear stress. **b** The fatigue peak stress of σ_{max} = 120 MPa for different interface debonding energy. **c** The fatigue peak stress of σ_{max} = 145 MPa for different interface shear stress. **d** The fatigue peak stress of σ_{max} = 145 MPa for different interface debonding energy, Longbiao [14]. Open access

that the loops in Fig. 6.3 become wider, i.e., show more hysteresis, as the loop time increases, but in Fig. 6.50 the trend of the curves is reversed, especially in the acid half of the pH range. Here the hysteresis diminishes at loop times above 9600 s. Further it is seen that the slope in the pH range 8–11 tends to decrease compared to the overall average slope, as the loop time is increased. The results depend on the time taken to perform a complete pH loop. The accuracy of pH sensor devices based on insulator/electrolyte interfaces is limited by a slow response, which manifests itself as hysteresis and drift. With Si_3N_4, the degree of loop closure sets an upper limit to the average background drift: it was less than 1.7 mV in 1.536 × 10^5 s, or l.1xlO^{-8} V/s. A comparison of the amplitude of the hysteresis of three materials is listed in Table 6.2. Note that this result is not only a material property, but also depends on the deposition method of the insulator films, in our case Si_3N_4.

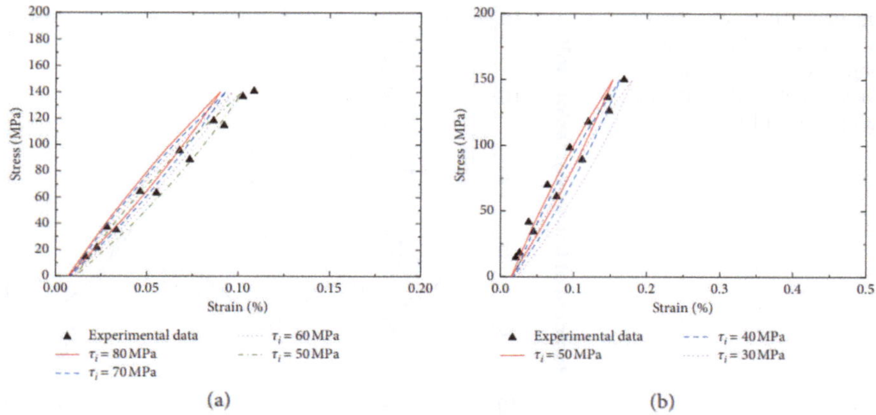

Fig. 6.47 The experimental and predicted fatigue hysteresis loops of 2D Hi-Nicalon™ SiC/SiC composite under. **a** The fatigue peak stress of $\sigma_{max} = 140$ MPa for N = 13,000 cycles. **b** The fatigue peak stress of $\sigma_{max} = 150$ MPa for N = 27,000 cycles, Longbiao [14]. Open access

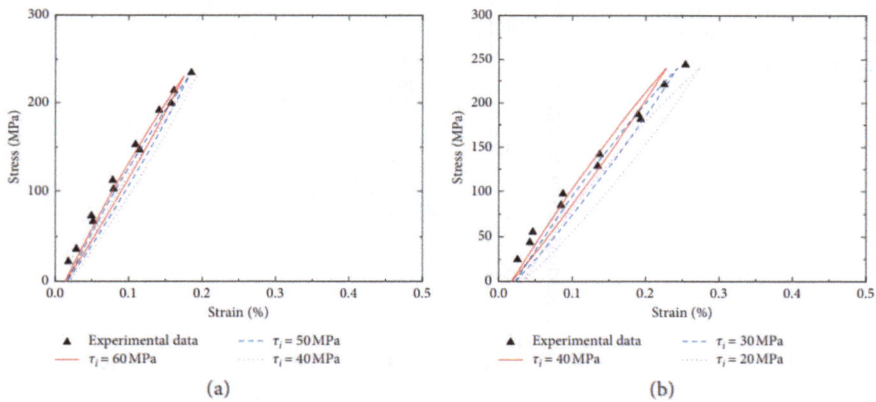

Fig. 6.48 The experimental and predicted fatigue hysteresis loops of 2D Tyranno™ SiC/SiC composite under. **a** The fatigue peak stress of $\sigma_{max} = 230$ MPa for N = 16,000 cycles. **b** The fatigue peak stress of $\sigma_{max} = 240$ MPa for N = 85 cycles, Longbiao [14]. Open access

In summary, the data in the LAP sensor using Si_3N_4 as insulator show that there is little influence of any background drift on the data, namely on the hysteresis loop enclosure. The results depend on the time taken to perform a complete pH loop. The data also show that hysteresis is larger than the effect of background drifts, which are at most 3×0^{-8} V/s (0.9 mV/8 h).

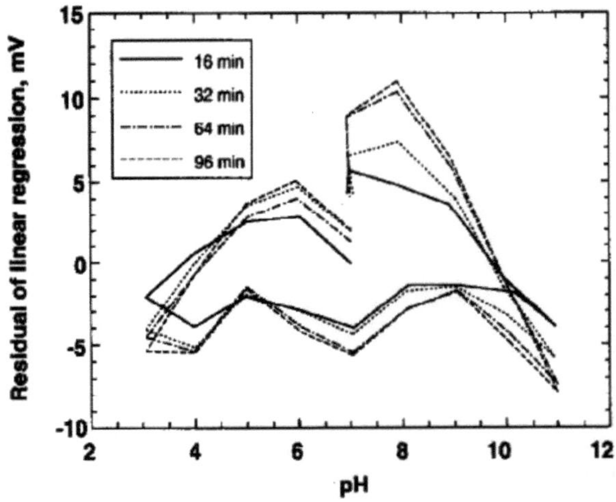

Fig. 6.49 Hysteresis curves for a Si_3N_4 surface for short loop times, 16–96 min, Bousse et al. [15]. With kind permission of Elsevier

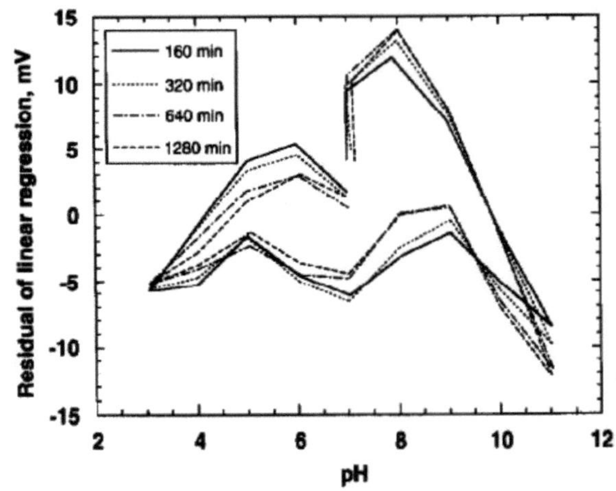

Fig. 6.50 Hysteresis curves for a S_3N_4 surface for long loop times, 160–1280 min, Bousse et al. [15]. With kind permission of Elsevier

Table 6.2 Comparison of hysteresis widths of Ta_2O_3, AI_2O_3 and Si_3N_4 insulators for a loop time of 32 min. The data for Al_2O_3 are from Ref. 13. The acid-side hysteresis is the difference at pH 6 (pH 7 for Al_2O_3) after an excursion to pH 3; the basic-side hysteresis is the difference at pH 7 (pH 8 for $A1_20_3$) after an excursion to pH 11, Bousse et al. [15]. With kind permission of Elsevier. Note Ref. [16] refers to L. Bousse, H. H. van den Vlekkert and N. F. de Rooij

Material	Ta_2O_5	Al_2O_3	Si_3N_4
Hysteresis acid side (mV)	1.68	3.73	6.29
Hysteresis basic side (mV)	1.72	4.40	10.92

6.6.2 Electrical Properties (Hysteresis in C-V Curves)

Si_3N_4 is a promising candidate for use as a gate insulator for MOSFET. Recall that MOSFET stands for Metal Oxide Semiconductor Field Effect Transistor. The MOSFET is a capacitor operated transistor device. The capacitor plays an essential role for operating a MOSFET. Low leakage current gate is required, and one way to reduce leakage current is by using high-k gate dielectric. To remind the reader, the term high-κ dielectric refers to a material with a high dielectric constant (κ, kappa), as compared to silicon dioxide. High-κ dielectrics are used in semiconductor manufacturing processes where they are usually used to replace a silicon dioxide gate dielectric. Silicon nitride is an attractive candidate for this purpose due to its high relative dielectric constant (7.5) and low interface states densities. Another advantage is to use Si_3N_4 between high-κ dielectric and silicon substrate. The interface of Si_3N_4/Si is correlated with electric properties is discussed below. The 1.5-nm-thick Si_3N_4 films were grown in microwave excited high-density Xe/NH_3 mixture gas plasma with a pressure of 6.6 Pa. The microwave frequency and power were 2.45 GHz and 5 W/cm^2, respectively. Si_3N_4 were fabricated by direct nitridation on N-type silicon substrate. The Al gate was deposited by thermal evaporation. C-V curves were measured for MIS capacitors having Si_3N_4 films. Recall a MIS capacitor is a capacitor formed from a layer of metal, a layer of insulating material and a layer of semiconductor material. It gets its name from the initials of the metal–insulator-semiconductor structure.

The films were grown at 400 and 600 °C without and with annealing at 600 °C for 30 min. Hysteresis could be seen in C-V curve for Si_3N_4 grown at 400 °C followed by annealing in N_2 atmosphere, but could not be detected for Si_3N grown at 600 °C followed by annealing. The C-V curves of MIS capacitor in the Si_3N_4 films grown by 400 and 600 °C are show in Fig. 6.51.

In summary, sub-nitride could be detected at Si_3N_4/Si interface structure and correlate with the width of hysteresis in C-V curve. The sub-nitrides consist of Si^+, Si^{2+} and Si^{3+}. Here, Si^{n+} denotes Si atom bonded with n N atoms and $(4 - n)$ Si atoms.

6.6.3 Inelastic Deformation in Monolithic Si_3N_4

Ceramics exhibit inelastic deformation at elevated temperatures which effects fatigue and other mechanical properties such as creep at high temperatures, however only limited information is available because the difficulty in measurement of axial strain at these temperatures. Understanding of the mechanical properties of monolithic ceramics is important for application as machine components and for design.

Extensometer equipped with the closed loop type push–pull cyclic loading test system was developed for high temperature use which aided the measurements at high temperature. The stress–strain response in monolithic Si_3N_4 was examined after

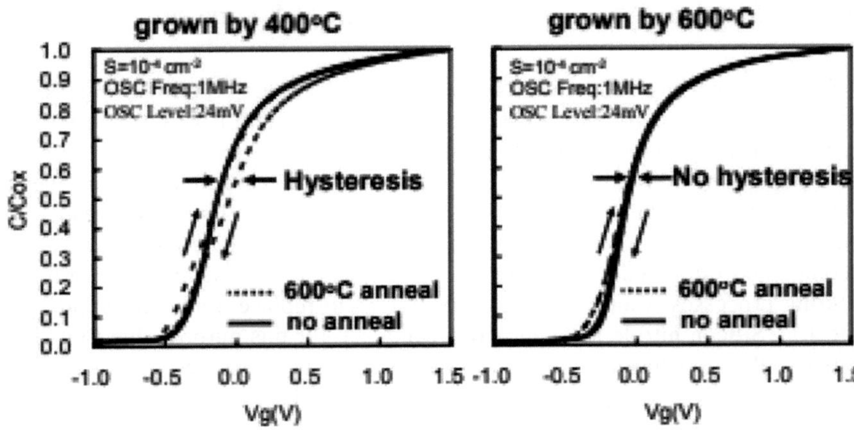

Fig. 6.51 The C-V curves of MIS capacitor in the Si_3N_4 films grown by 400 and 600 °C, Higuchi et al. [17]

axial strain was measured. Si_3N_4 was prepared by mixing Y-α SiAlON and Si_3N_4 particles at the ratios of 40 and 60 wt.%, and then the mixed particles were sintered in N_2-gas atmosphere at 2023 K. (Note that SiAlONs are ceramic alloys based on the elements silicon (Si), aluminum (Al), oxygen (O) and nitrogen (N) and were developed to solve the problem of Si_3N_4 being difficult to fabricate.) This ceramic is denoted below as Si_3N_4 (A). The strain or stress rate-controlled push–pull cyclic loading tests were performed by employing the closed loop type-electrohydraulic test systems for the monolithic-Si_3N_4(A) at 1573 K in the air. The variation of the cyclic stress–strain at a stress rate of $\dot{\sigma}$ =1.0 Mpa s^{-1} and stress amplitude $\Delta\sigma/2 =$ 250 MPa at 1573 K is shown in Fig. 6.52. Note that the inelastic strain detected at N = 1 cycle is gradually reduced with the progress of loading cycles, and then nearly disappears at N = 807 cycles, which is an indication of the increasing resistance against cyclic loading. The effect of stress rate on the cyclic inelastic deformation is seen in Fig. 6.53. The inelastic strain (arising from cyclic loading) increases with decreasing the controlled stress rate.

In summary, the inelastic strain during cyclic loading at 1573 K is greater at the smaller stress/strain rate (smaller on the tensile stress side than on the compressive stress side). The inelastic strain decreases with increasing the number of stress/strain cycles at 1573 K showing increasing resistance to cyclic deformation in the cyclic loading process.

6.7 The Mean Stress

The definition of the mean stress has been given in Chap. 1, page 10 and is reindicated here as: **The mean stress** is the arithmetic **mean** of the maximum stress and the

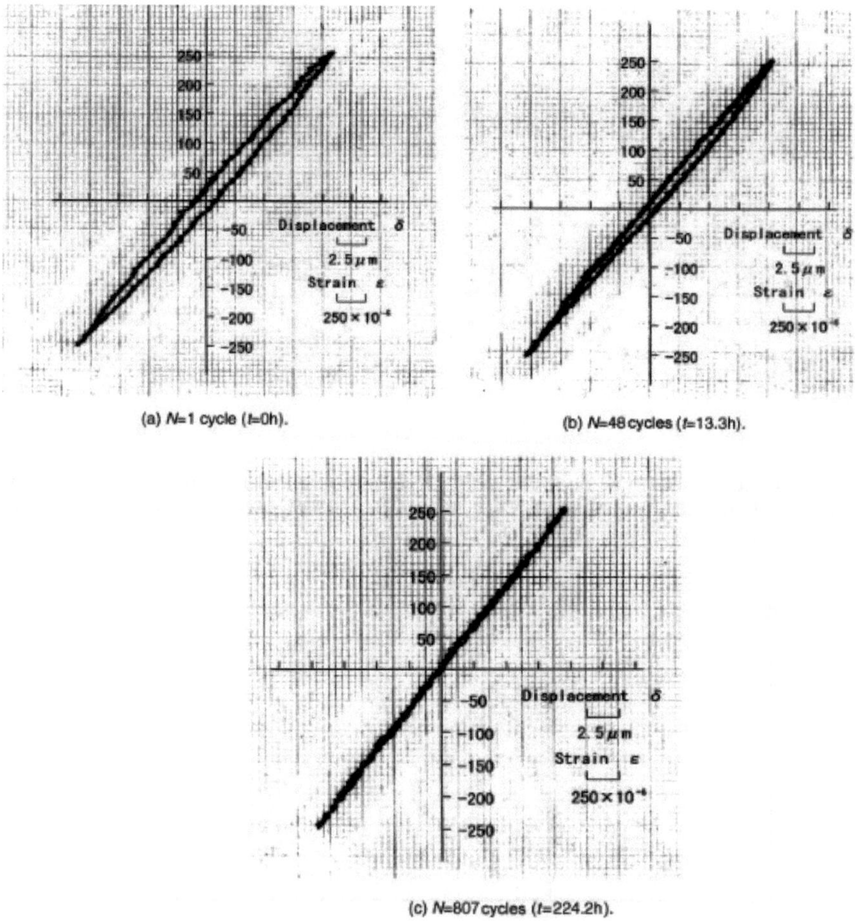

Fig. 6.52 Change in cyclic stress–strain response with increasing number of triangular stress wave cycles at $\Delta\sigma/2 = 250$ MPa and $\dot{\sigma}$=1.0 MPa/s at 1573 K in monolithic-Si_3N_4(A): **a** at $\dot{\sigma} = 7.2$ MPa/s, **b** at $\dot{\sigma}$ =3.6 MPa/s, and **c** at $\dot{\sigma}$ =0.72 MPa/s, Hatanaka et al. [18]. With kind permission of Elsevier

minimum stress as seen in the redrawn illustration of Fig. 1.3 in Fig. 6.54. Another illustration of the cyclic stress variation with time is seen in Fig. 6.55. Further to the mean stress, the alternating stress is defined (see Eq. 1.2 in Chap. 1) as the force changing direction or magnitude with respect of time. The alternating stresses are an important aspect of machine design since they are responsible for fatigue in a component which may cause it to fail at a stress value lower than its yield stress after a large number of cycles. The mean stress effect plays an important role in the overall fatigue strength of engineering materials, particularly under uniaxial fatigue loading. The fatigue damage increases as the applied tensile superimposed static stress, $\sigma_{x,\,m}$, increases. Many attempts have been made to propose sound engineering

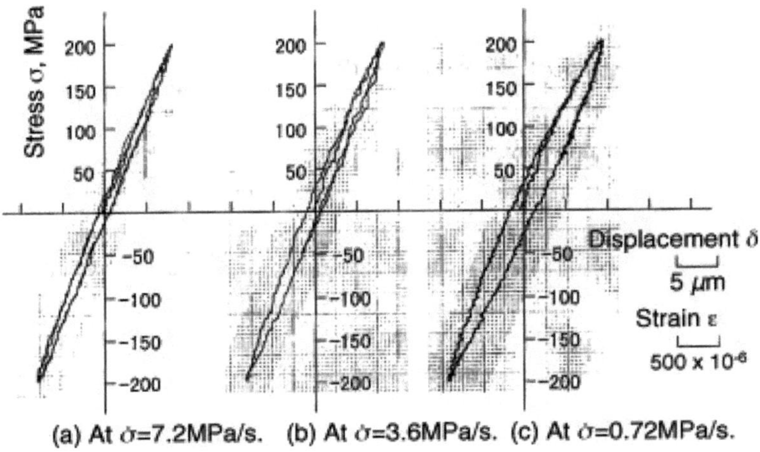

Fig. 6.53 Effect of stress rate on cyclic stress–strain response under the triangular stress wave loading at 1573 K in the monolithic-Si_3N_4(A) at $\Delta\sigma/2 = 200$ MPa: **a** $t_h = 30$ s and **b** $t_h = 10$ min, Hatanaka et al. [18]. With kind permission of Elsevier

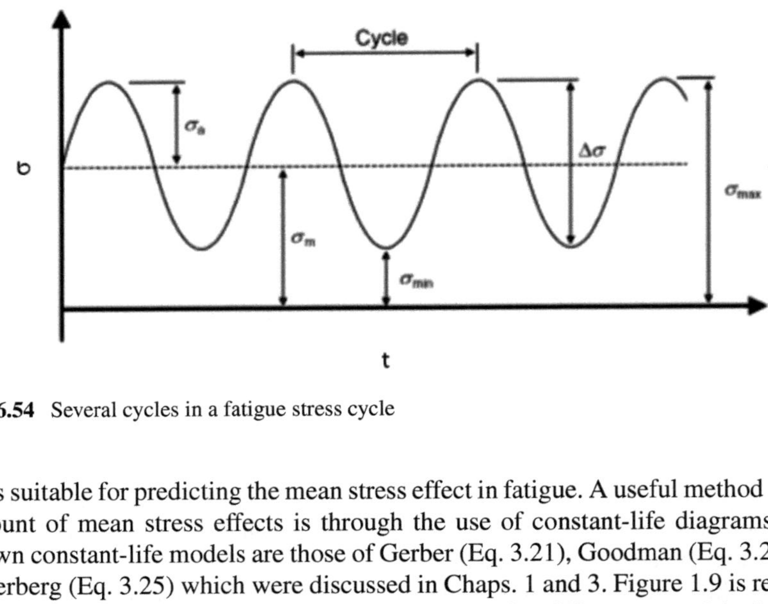

Fig. 6.54 Several cycles in a fatigue stress cycle

rules suitable for predicting the mean stress effect in fatigue. A useful method to take account of mean stress effects is through the use of constant-life diagrams. Well known constant-life models are those of Gerber (Eq. 3.21), Goodman (Eq. 3.22) and Soderberg (Eq. 3.25) which were discussed in Chaps. 1 and 3. Figure 1.9 is redrawn here with experimental data at different conditions for different materials. In these models, different combinations of the stress amplitude and mean stress are plotted to provide a constant fatigue life. Almost all the proposed criteria can be summarized by using Marin's general equation

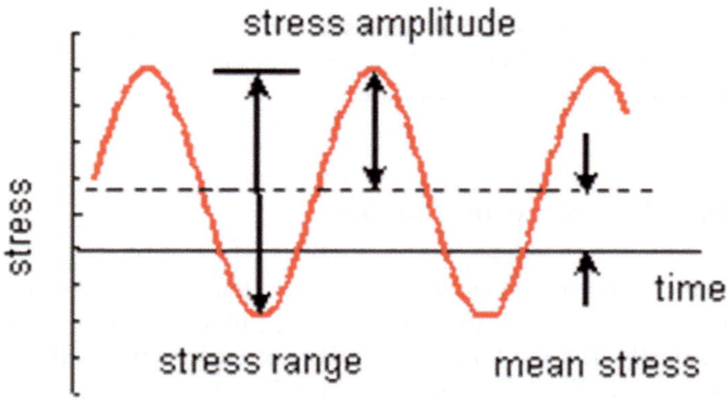

Fig. 6.55 The Stress variation verses time

$$\left(\frac{\sigma_0}{\sigma_{0,R=-1}}\right)^n + \left(f\frac{\sigma_{x,m}}{\sigma_{UTS}}\right)^m = 1 \tag{6.9}$$

where $\sigma_{0,\ R=-1}$ denotes the fatigue limit generated under fully reversed loading, σ_{UTS} is clearly the ultimate tensile strength, f, m and n are constants which take on different values according to the considered fatigue damage model. Figure 6.56 also indicates that the experimental points are mainly located within the region delimited by the

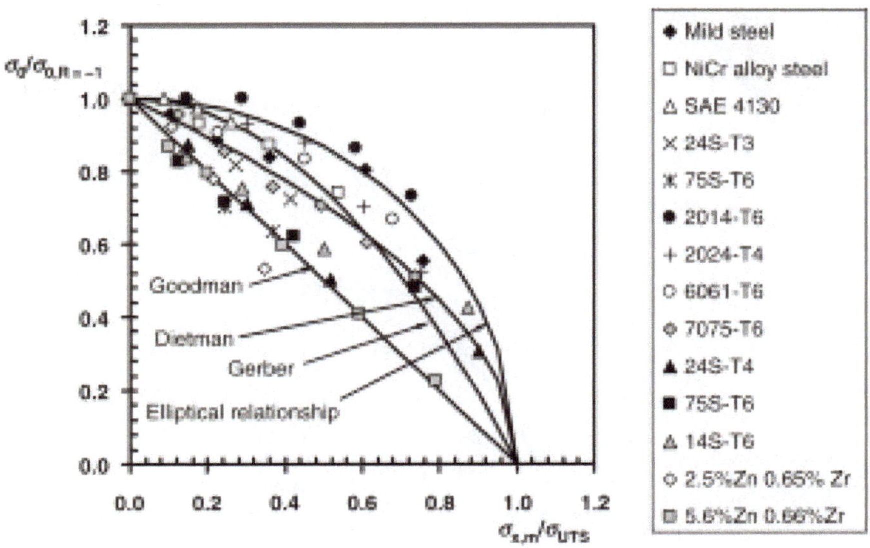

Fig. 6.56 Accuracy of the considered criteria in predicting the mean stress effect under uniaxial fatigue loading. Comparison of mean stress equations

elliptical relationship and by Goodman's straight line. Typically, more damage is accumulated when the product is under a tensile mean stress, while less damage is accumulated under compressive mean stress.

6.7.1 The Mean Stress in Alumina

Plasma sprayed alumina coatings were subjected to four-point bending (4 PB) tests. Cyclic tensile and compressive loading of the samples were performed mechanically and thermally. The non-linear observation and the hysteresis induced, indicate inelastic behavior during deformation. Figure 6.57 is a plot of the mean stress vs. mean strain of the plasma deposited alumina on Ti6Al4V substrate at three different settings. Ti6Al4V was chosen for its relatively low modulus and high elastic limit that permits testing of the coatings up to relatively large strains. The three loading modes used consequently for each specimen is listed below:

1. loading up to stepwise increasing loads, with full unloading (mode I),
2. loading up to the maximum load, with full unloading (mode II), and
3. loading up to stepwise increasing loads, with partial unloading (mode III).

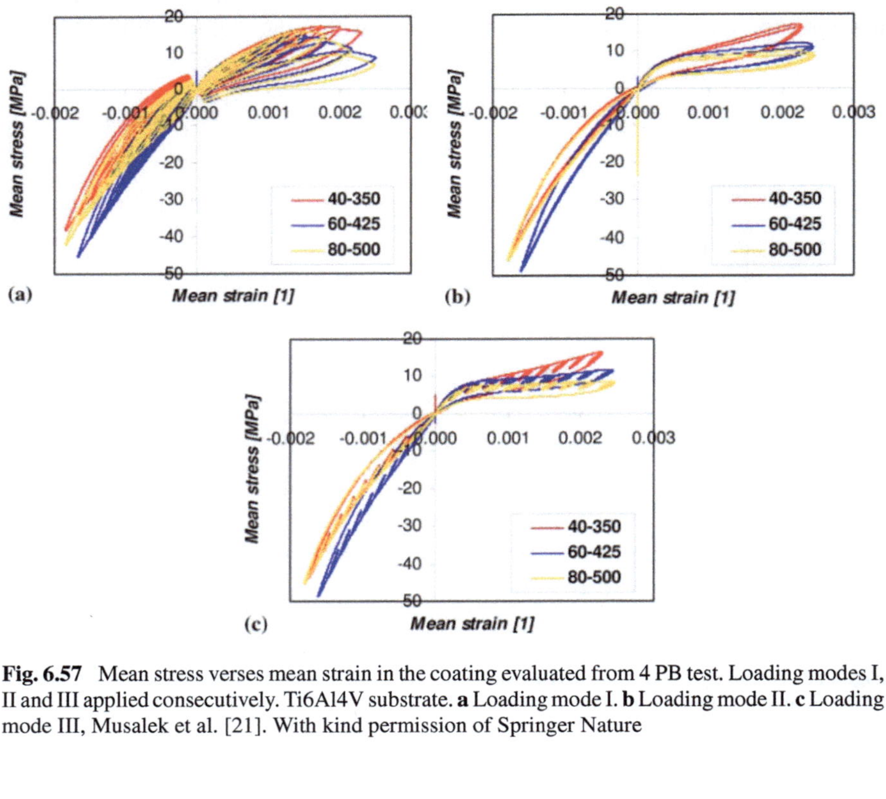

Fig. 6.57 Mean stress verses mean strain in the coating evaluated from 4 PB test. Loading modes I, II and III applied consecutively. Ti6Al4V substrate. **a** Loading mode I. **b** Loading mode II. **c** Loading mode III, Musalek et al. [21]. With kind permission of Springer Nature

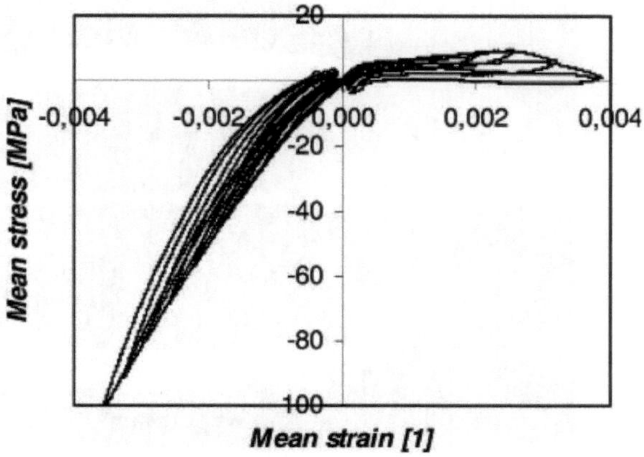

Fig. 6.58 Mean stress verses mean strain in the coating evaluated from 4 PB test of half specimen. Coating 40–350 on Ti6Al4V substrate, Musalek et al. [21]. With kind permission of Springer Nature

The illustration in Fig. 6.57 shows the load–deflection curves obtained from 4 PB test which were transformed to plots showing evolution of mean stress mean strain curves. From the stress–strain curves maximum (Fig. 6.57a) one can estimate the tensile strain level causing large-scale cracking of the coating and loss of coating stiffness (ca. 0.15% corresponding to effective mean stress approx. 15 MPa). The maximum loading (1000 N) admissible by the 4 PB fixture was insufficient to achieve macroscopic failure of the coating on Ti6Al4V substrates. Therefore, previously tested specimens were cut longitudinally in two halves to reduce their stiffness (thereby doubling the maximum strain achieved) and tested again in mode I (Fig. 6.58). Specimens coated on 99.5Al substrate were loaded using 4 PB test both in tension and compression until failure of the coating as illustrated in Fig. 6.59. Residual stresses in the coating resulted in a noticeable curvature of the specimens after spraying as seen in Fig. 6.59a. Due to low yield stress of the substrate, plastic deformation occurred and coating failure in tension and compression resulted as seen in Fig. 6.59b and c.

In summary, sprayed alumina coatings exhibited non-linear, inelastic, anisotropic and inhomogeneous behavior during deformation by 4 PB indentation which is irreversible. Permanent coating deformation and cracking were observed. Gradual decrease in coating stiffness occured during tensile loading, which finally led to the failure of the coating. In compression, significantly higher capability to withstand extensive deformation was observed.

6.7.2 The Mean Stress in Y-STZ

In uniaxial tension and compression fatigue, the behavior of 3 mol % yttria stabilized tetragonal zirconia (Y-STZ) polycrystals are considered and the effect of the mean stress on inducing microcracks and damage at some mean stress level. Preexisting processing flaws are the fracture origins and the dominant mechanism of fatigue

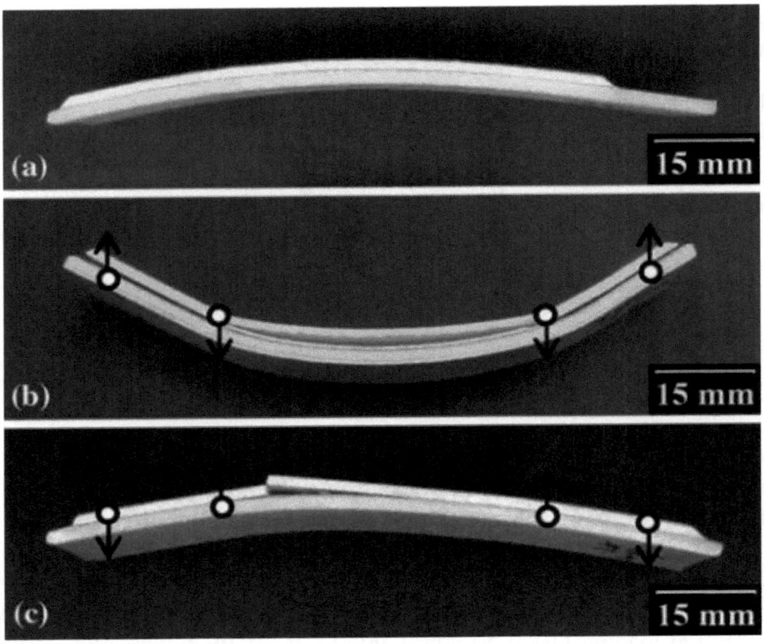

Fig. 6.59. 4 PB test of specimen coating 60–425 on Al99.5 substrates. **a** Specimen prior to the loading. **b** Compression loading-failure strain $\varepsilon_c = 1.36\%$. **c** Tensile loading-failure strain $\varepsilon_c = 0.26\%$

damage. The fatigue lifetime is controlled mainly by the propagation of cracks which is sensitive to the maximum stress.

Following unloading to zero stress a residual strain remains in the hysteresis loop. Although the accumulated plastic strain slows down (levels off) with the increase of the number of cycles, actually it increases with the number of cycles. At the same ratio (namely that of the minimum stress to the maximum stress) during cycling, an increase in maximum tensile stress, i.e., that of the mean stress, an increase in the number of cycles induces more accumulated plastic strain. An increase of R ratio at the same maximum stress causes more damage. Thus, a higher mean stress is more damaging than higher stress amplitude. A plot of the cumulative plastic strain against the mean stress is seen in Fig. 6.60. The scatter is mostly due to temperature fluctuation during the long-time test. Figure 6.61 shows failure loci for runout cycle of 10^4 as the stress amplitude against the mean stress. The R ratios are indicated on the graph. In the figure, numbers next to the symbols indicate the number of specimens tested. The three groups are indicated in the figure. The line with the negative slope indicates a failure locus which can be expressed by Goodman's relation Eq. (3.22), rewritten her in a different way as

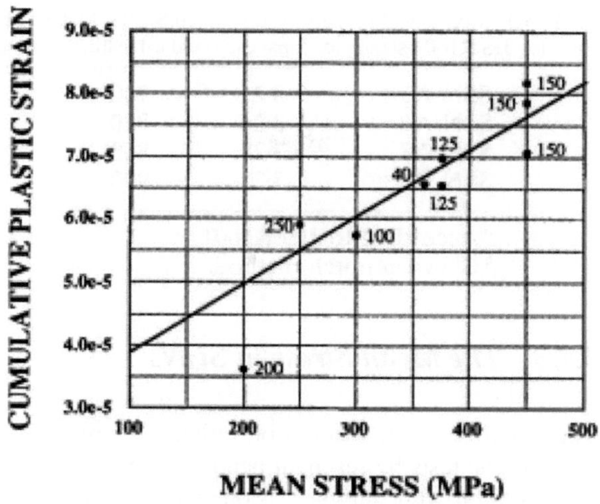

Fig. 6.60 Stress amplitude-failure cycle (S–N) curve of 3Y-TZP. Tests terminated at 10^5 cycles if failure did not occur, Liu and Chen [19]. With kind permission of John Wiley and Sons

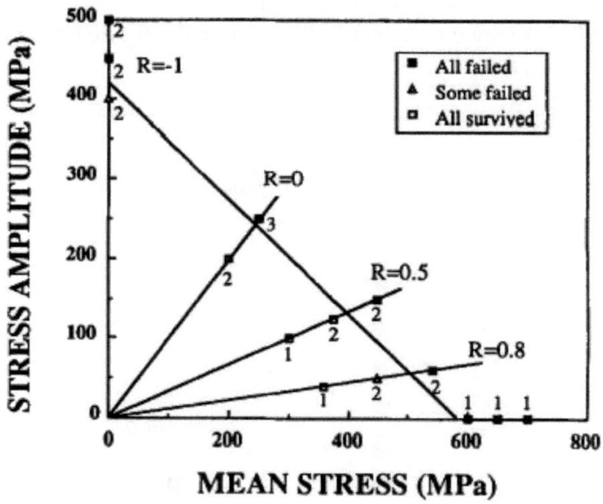

Fig. 6.61 Goodman diagram of 3Y-TZP for specimens tested for 10^4 cycles. Numbers next to the symbols indicate the number of specimens tested, Liu and Chen [19]. With kind permission of John Wiley and Sons

$$\sigma_a = \sigma_e\left[1 - \left(\frac{\sigma_m}{\sigma_u}\right)\right] \tag{6.10}$$

where σ_a is the stress amplitude, σ_e is the fatigue strength at $R = -1$, σ_m is the mean stress and σ_u is tensile strength.

In summary, mean stress and stress amplitude dependence of failure follows the Goodman relationship. Fatigue failure originates from internal, preexisting flaws and fatigue lifetime is likely to be controlled by crack propagation.

Table 6.3 Mechanical properties of sintered Si_3N_4, Matsuda et al. [20]. With kind permission of Elsevier. JIS R1601 stands for Japanese standard testing method

SSN-A1	3·21	740	6·2	265
SSN-A2	3·24	820	6·3	270
SSN-A3	3·28	860	6·2	275
SSN-B	3·25	970	6·1	300

[a] According to JIS R1601.[11]
[b] Chevron notch method.

6.7.3 The Mean Stress in Si_3N_4

The effect of cyclic loading on fatigue failure in Si_3N_4 was evaluated by accelerated fatigue tests by resonant bending method and also to investigate the effects of frequency on cyclic fatigue at room temperatures. Cyclic fatigue at room temperature and high temperatures of sintered Si_3N_4 having the mechanical properties shown in Table 6.3 are compared. The stress amplitude vs. mean stress is shown in Fig. 6.62 of SSN-A1 specimen (see Table 6.3). In the figure open circles represent specimens that

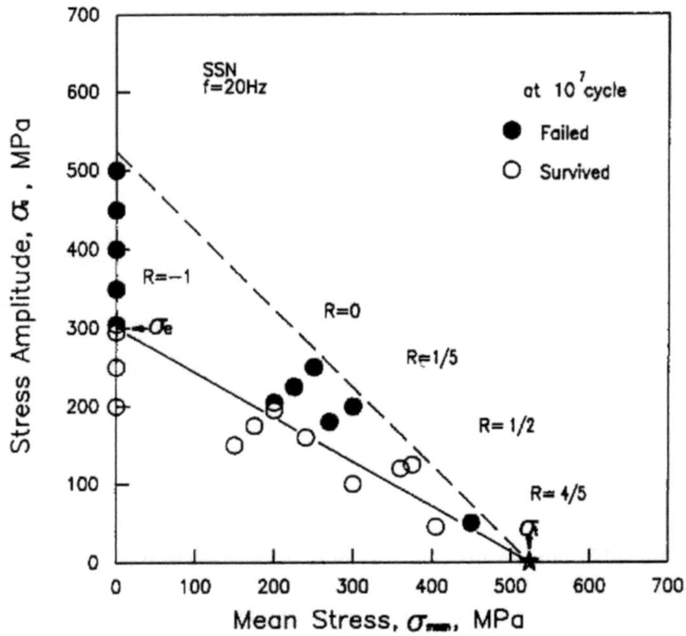

Fig. 6.62 Mean stress versus stress amplitude for fatigue failure in SSN-A1 at 10^7 cycles measured by high cycle fatigue test at room temperature. The solid line represents tensile strength, Matsuda et al. [20]. With kind permission of Elsevier

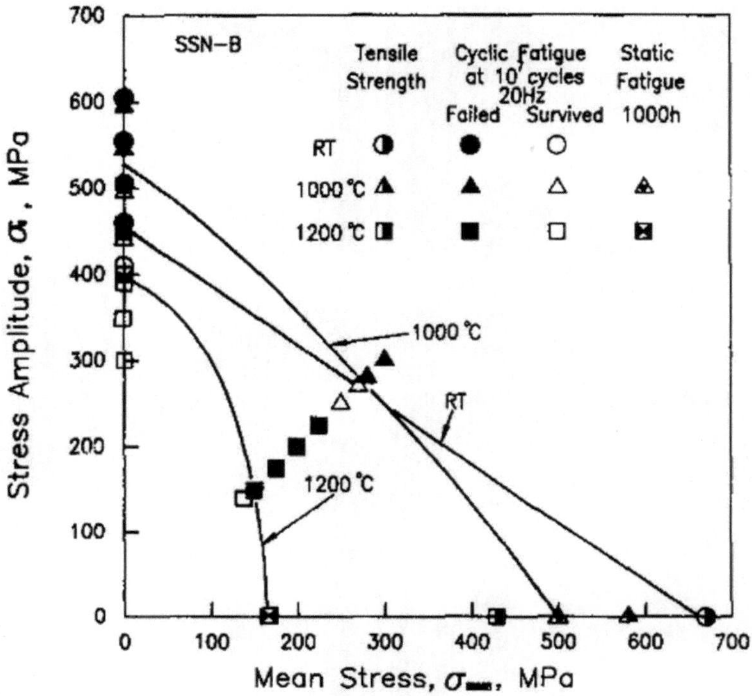

Fig. 6.63 Tensile mean stress effects on cyclic fatigue of SSN-B at room temperature, 1000 and 1200 °C, Matsuda et al. [20]. With kind permission of Elsevier

survived 10^7 cycles and the closed ones represent failed specimens. The dashed line in the figure represents tensile strength. The effect of mean tensile stress and static fatigue on cyclic fatigue of sintered Si_3N_4 (specimen SSN-B) is shown in Fig. 6.63. The results were obtained under load control at a constant frequency of 20 Hz. It is seen that with increasing temperature the fatigue strength decreases. An approximation for the mean stress effect is show in Eq. (6.11) which is similar to Eqs. (3.22) and (3.24)

$$\left(\frac{\sigma_a}{\sigma_e}\right)^n + \left(\frac{\sigma_m}{\sigma_s}\right) = 1 \tag{6.11}$$

where σ_a is stress amplitude, σ_m mean stress, σ_e alternating fatigue strength and σ_s static fatigue strength. The n value increased with increasing temperature.

Fractured surfaces of several SSN-A3 specimens (see Table 6.3) are shown in Fig. 6.64. The purpose is to see the origin of the fracture, the failure mode and defects (such as pores and inclusions) which are preferential sites for crack nucleation and their propagation. In the figure various stages of the cyclic fatigue failure is indicated. Thus, microcracks are formed as a result of stress concentration around initial flaws

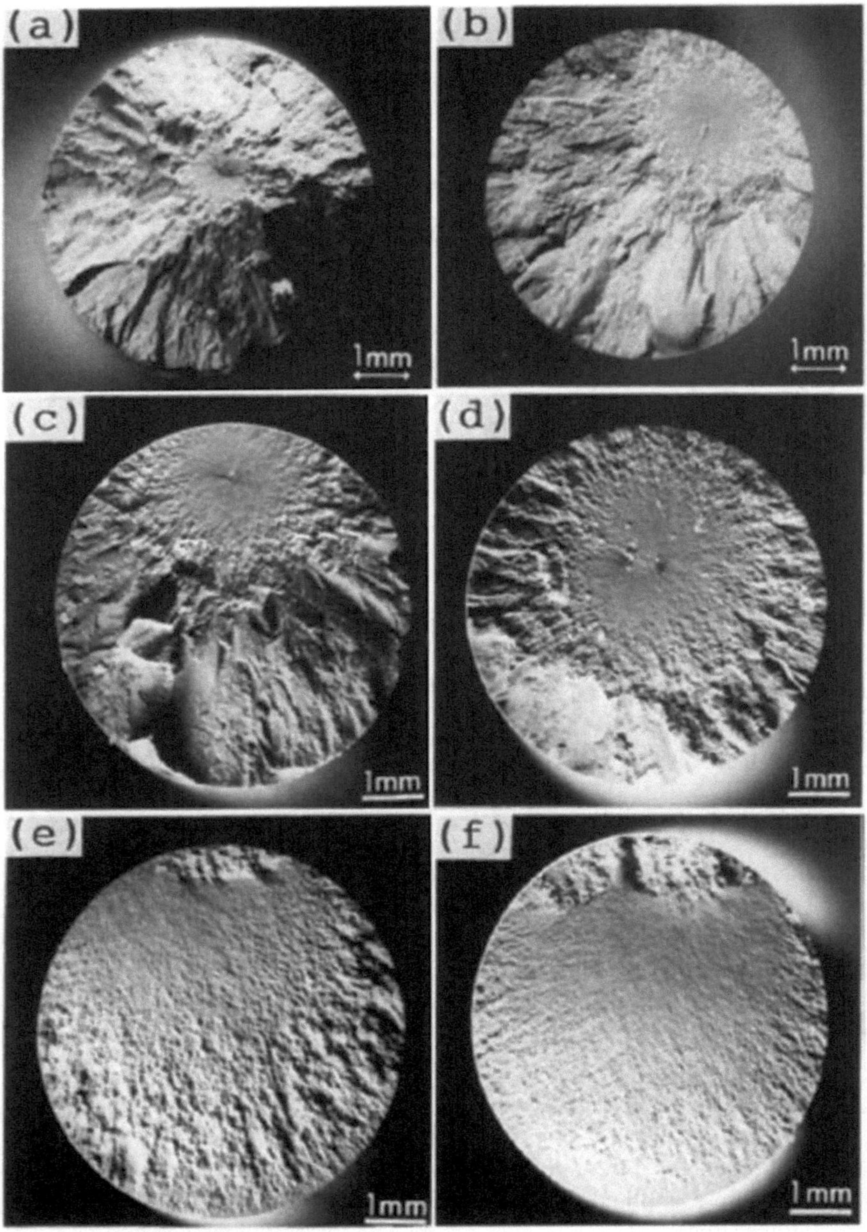

Fig. 6.64 Typical fracture surfaces of fatigue specimens of SSN-A3. **a** RT, 20 Hz, ± 500 MPa, 2.2 × 10^4 cycles, **b** RT, 20 Hz, ± 400 MPa, 1.8 × 10^7 cycles, **c** 800 °C, 20 Hz, ± 350 MPa, 1.3 × 10^6 cycles, **d** 1000 °C, 20 Hz, ± 350 MPa, 5.0 × 10^4 cycles, **e** 1000 °C, 1 × 10^2 Hz, ± 200 MPa, 3.7 × 10^2 cycles, **f** 1000 °C, 150 MPa, 204 h, Matsuda et al. [20]. With kind permission of Elsevier

in the microstructure, growth, the damaged zone gets larger and at a critical value, acyclic fatigue failure occurs. Generally, crack propagation is faster under cyclic load than under static load. Major defects are pores and inclusions which are preferential nucleation and crack propagation sites.

In summary, the effect of the mean stress on fatigue strength, and an equation expressing the allowable stress under cyclic loading was suggested. Microstructural observations by optical microscope indicates that cyclic fatigue of sintered Si_3N_4 occurs by the formation of microcracks at grain boundaries, they grow and progresses, leading to failure at a critical value. The cyclic fatigue life depends on the number of cycles at room temperature and on the frequency at high temperatures. The lower the frequency the less resistance is Si_3N_4 to fatigue.

.

References

1. R.L. Lanser, M.B. Ruggles-Wrenn, in *10th International Conference on Composite Science and Technology*
2. M.B. Ruggles-Wrenn, G. Hetrick, S.S. Baek, Int. J. Fatig. **30**, 502 (2008)
3. W.K. Tan, N. Hakiri, A. Yokoi, G. Kawamura, A. Matsuda, H. Muto, Nanoscale Res. Lett. **14**, 245 (2019)
4. N.G. Jones, C. Humphrey, L.D. Connor, O. Wilhelmsson, L. Hultman, H.J. Stone, F. Giuliani, W.J. Clegg, Acta Mater. **69**, 149 (2014)
5. A. Bhowmik, J. Lee, T.B. Britton, W. Liu, T.-S. Jun, G. Sernicola, M. Karimpour, D.S. Balint, F. Giulian, Acta Mater. **145**, 516 (2018)
6. M. Kapilashrami, J. Xu, K.V. Rao, L. Belova, Proc. Appl. Cer. **4**, 225 (2010)
7. N. Schmitt, Y. Berthaud, J. Poirier, J. Eur. Cer. Soc. **20**, 2239 (2000)
8. O. Anazia, Thesis, The Faculty of the Department of Chemistry, Western Kentucky University, Bowling Green, Kentucky
9. L.S. Pan, S. Horibe, Acta mater. **45**, 463 (1997)
10. M. Matsuzawa, E. Fujimagari, S. Horibe, Mater. Sci. Eng. A **314**, 105 (2001)
11. S.-Y. Liu, I.-W. Chen, J. Am. Cer. Soc. **75**, 1191 (1992)
12. W.G. Maoa, J.M. Luoa, C.Y. Dai, Y.G. Shen, App. Surface Sci. **338**, 92 (2015)
13. D. Lal, P. Kumar, S. Sampath, V. Jayaram, Acta Mater. **194**, 394 (2020)
14. L. Longbiao, Adv. Mater. Sci. Eng. **2020**, 1 (2020)
15. L. Bousse, S. Mostarshed, B. van der Schoot, N.F. de Rooij, Sens. Actuators, B **17**, 157 (1994)
16. L. Bousse, H.H. van den Vlekkert, N.F. de Rooij, Sens. Actuators B **2**, 103 (1990)
17. M. Higuchi, A. Teramoto, M. Komura, S. Shinagawa, E. Ikenaga, H. Nohira, K. Kobayashi, T. Hattori, S. Sugawa, T . Ohm, in *208th ECS Meeting, Abstract* p. 748, (2006).
18. K. Hatanaka, Y. Ishiga, R. Kawazoe, M. Hasui, J. Ohgi, H. Ogawa, Mater. Sci. Eng. A **468**, 273 (2007)
19. S.Y. Liu, I.W. Chen, J. Am. Ceram. Soc **74**, 1197 (1991)
20. M. Matsuda, T. Soma, M. Matsui, J. Eur. Cer. Soc. **6**, 253 (1990)
21. R. Musalek, J. Matejicek, M. Vilemova, O. Kovarik, J. Thermal Spray Technology, **19**, 422 (2010)
22. T. Erber, S.A. Guralnick, S.C. Michels, Ann. Phys. **224**, 157 (1993)
23. J. Marin, Interpretation of fatigue strengths for combined stresses, in *Proceedings of the International Conference on Fatigue of Metals*, Institution of Mechanical Engineers, London, UK (1956), pp. 184–195.
24. N. Pathak, S.K. Gupta, C.L. Prajapat, S.K. Sharma, P.S. Ghosh, B. Kanrar, P.K. Pujari, R.M. Kadam, Phys. Chem. Chem. Phys. **19**, 11975 (2017)

Chapter 7
Underloading, Overloading and Cumulative Damage

Abstract Components in service under cyclic conditions might undergo underloading and overloading under common loading conditions usually below the ultimate tensile strength. Failure can occur suddenly without warning and therefore of prime concern even at room temperature applications. Fatigue is a major failure mechanism in structural components, about 90% of all mechanical failure in moving parts of automotive, marine and aircraft components. Overloads are known to retard crack growth, while underloads accelerate crack growth. Underloads are known to have a detrimental effect on both fatigue-crack initiation and growth from notches. Tensile overload retards crack growth. Crack growth behavior under single tensile overload conditions in alumina is presented. Fatigue failure occurs in localized weak spots of a material following its deterioration after some number of cycles which limits and even prohibits safe use. Cumulative damage in alumina, magnesia, zirconia and SiC/SiC is considered in details in this chapter.

Keywords Underloading · Overloading · Cumulative damage · Alumina · Magnesia · Zirconia · SiC/SiC · Variable amplitude · Experimental observations

7.1 Introduction

It has been observed that machine components working under cyclic conditions during service may experience overstressing or understressing under common loading conditions. Fatigue failure occurs under cyclic loading when a component is subjected to a certain number of cycles, normally below the ultimate strength. Fatigue also usually occurs without warning after a progressive degradation of the material subjected to cyclic loadings. Fatigue is a major failure mechanism in structural components, about 90% of all mechanical failure in moving parts of automotive, marine and aircraft components. Performance under cyclic loads is of prime concern when applying materials not only at high temperatures, but also at RT. Therefore, understanding fatigue and life assessment of engineering applications is of utmost importance. Many factors affect the fatigue life of structural components among them microstructure, grain size, surface finish of parts exposed to motion, loading condition, environment, etc. This subject has been discussed wide in the past mainly in

J. Pelleg, *Cyclic Deformation in Oxides, Carbides and Nitrides*, Structural Integrity 22,
https://doi.org/10.1007/978-3-030-86118-6_7

regard to metallic materials, in which plastic deformation before failure is a key factor leading to fracture. Oxides, carbides and nitrides are mostly brittle at room temperature, fail almost completely without deformation, but are, nevertheless, exposed to fatigue fracture. Some of the mentioned factors will be considered in this chapter.

7.2 Underloading

Cyclic loading might be performed under different conditions. Figure 7.1 illustrates some cycles, reproduced from Fig. 1.1, defining certain important parameters, which might be relevant to loading conditions and in actual service. Of the many reports on loading conditions, some principal experimental results will be discussed.

Designers wish to increase the fatigue resistance of machine components exposed to cyclic or dynamic stresses. The end results of such exposure may be failure, which begins with a crack. Hence, the effect of various loading sequences and modes are usually related to crack initiation and growth, which are inherent steps in the progress to failure. Overloads (OLs) are known to retard crack growth, while underloads (ULs) accelerate crack growth relative to the background rate. It has been stated that

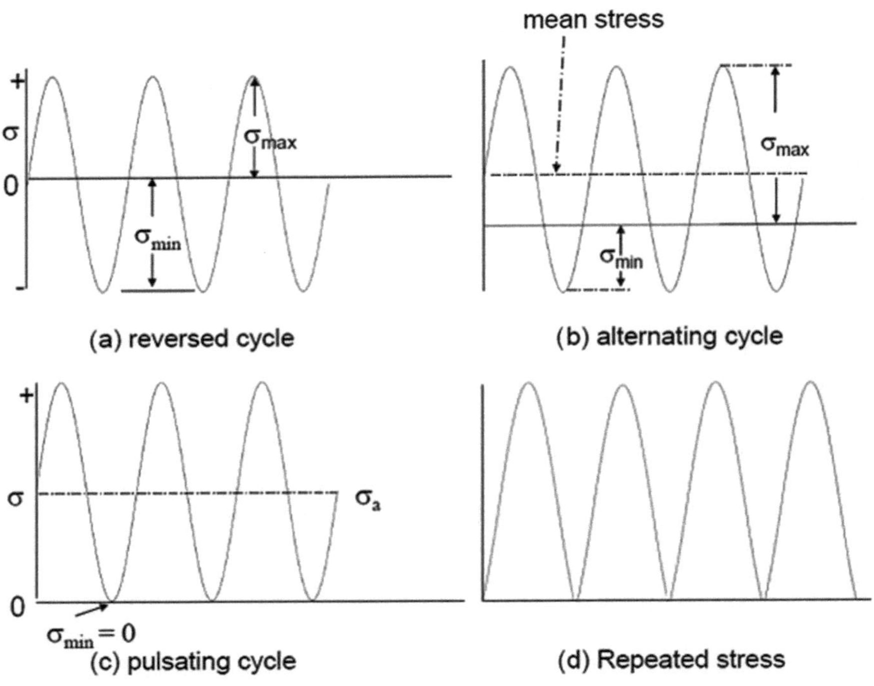

Fig. 7.1 Various forms of cyclic stresses. **a–c** are sinusoidally varying cycles; **d** represents a repeating cycle

fatigue ULs are known to have a detrimental effect on both fatigue-crack initiation and growth from notches. These effects are thought to be due to the production of a tensile, plastic zone in the crack tip and the flattening of the crack tip surface roughness leading to a reduction in crack-closure effects (Geary). The loading of test specimens may be done by single or periodic ULs. In both cases when applied to aluminum alloys, crack growth acceleration is observed. In aluminum testing, similar behavior is also observed (Stephen et al.). Apparently, the reduced contribution to the crack-closure mechanism is involved in the accelerated crack growth. A single UL gives only a temporary increase in the crack growth rate: however, it has been found that periodically repeated ULs result in noticeable crack growth acceleration. Repeated ULs keep the crack growth mechanism active. Thus, if in fatigue there may be frequent ULs, such as occur in airplane wings, then crack growth acceleration will continue promoting premature failure. Furthermore, when UL occurs during an otherwise constant-amplitude loading, the crack growth following the UL will be greater than that occurring during constant amplitude loading. The period of the crack-growth acceleration is short and the normal, constant amplitude growth rate is quickly restored, unless, as indicated, the machine element for the test specimen is exposed to repeated UL. Note that if UL immediately follows overloading (OL), the degree of crack retardation due to OL is reduced, but not eliminated. An UL applied prior to an OL, on the other hand, has little effect on the degree of crack retardation.

An example of schematic OL/Ul spectra is shown in Fig. 7.2. An experimental program was conducted to evaluate the effects of periodic overloads and underloads on the fatigue crack growth behavior of 2024-T3 aluminum alloy. The specimens were cycled to failure using a load sequence that consisted of single repeated block. The block began with 2500 constant amplitude cycles with a maximum stress of S_{max} – 10 ksi and a stress ratio of $R = 0.02$. The constant amplitude loading was followed by a single spike overload at a stress of S_{OL} that ranged from 1.125 to 3.4 S_{max}. The

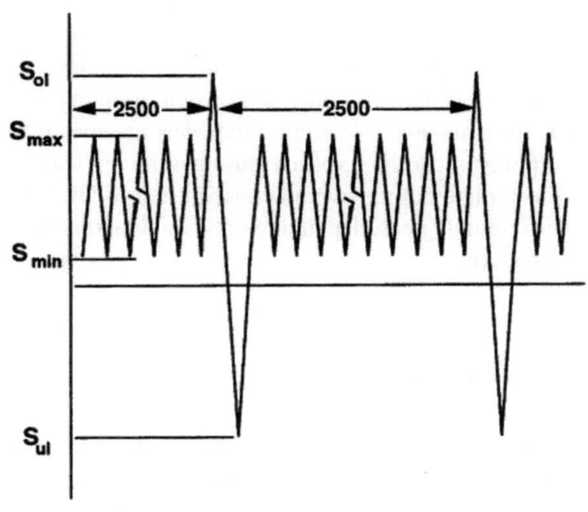

Fig. 7.2 Schematic of overload/underload spectra, Stephens et al. [1]. NASA Contractor Report 201668. Distribution: unclassified—unlimited

overload was followed by a single spike underload at a stress of S_{UL} that ranged from 0 to $-3\ S_{max}$. In each test, the number of cycles to failure was measured.

Tests done on aluminum 7010 and 8090 alloys (Zitounis and Irving) have indicated that fatigue crack-growth rates under loading sequences containing tensile UL show crack acceleration effects, compared with growth rates calculated from the linear summation of constant amplitude growth rates. In aluminum 7010 the acceleration of cracking was in the range of 30%, whereas in Al 8090 alloys it was up to a factor of 12 (i.e. 1200%). The number of reversed cycles determine crack acceleration. The experiments on these alloys were performed at large cycles of $R = 0$ and small cycles of $R = 0.9$ to create load spectra similar to those observed in helicopter lift frame and rotor heads. These parts contain large numbers of small amplitude cycles of R approaching zero.

Reports over the last decades state that the compression part of the loading under tensile-compression tests (negative R ratios), may have a significant effect on the results of crack initiation and propagation. Examples of components with this type of loading include those of helicopter and gas turbines. Topper studied crack-opening stresses under spectra consisting of original constant amplitude small cycles of a range of R ratios, followed by tension–compression ULs. He found that the crack-opening stress level was reduced immediately after the application of the UL and then gradually increased with subsequent constant amplitude cycling, reaching a steady state level after a large number of cycles. Thus, observed crack-opening had a significant acceleration effect when the R ratio was low. At high R ratio (0.8), changes in opening stress had no effect on the crack growth rates, because the crack opening stress was always less than the minimum stress of the loading cycle.

Several mechanisms have been suggested for crack-growth rate, among them fatigue-crack closure, strain hardening ahead of the crack tip (caused by the ULs) and the influence of mean stress, etc. Topper and Yu suggested that the compressive ULs caused by squeezed crack tip to have decreased crack-closure stress, thus increasing the effective stress-intensity factor range of the baseline constant amplitude cycles at a low R ratio.

To conclude this section, note that the loading conditions of a particular material greatly influences the outcomes when applying ULs to specimens of that material. Experimental—loading conditions are meant to simulate the loads to which components can be exposed during service. Various loading-UL sequences have been studied, but the overall consequence of UL is that crack growth is accelerated to different degrees, depending on several factors commonly encountered during the test pattern.

7.3 Overloading

Fatigue behavior in service condition in general differ significantly from those prevailing in laboratory tests because many factors influence fatigue behavior. To insure component reliability in service conditions, one has to consider these factors when designing a machine part. The single most important factor often considered is the stress concentration since most of fatigue failures are initiated at a stress concentration. Stress concentrations are present in machine parts and their geometrical shapes frequently induce stress concentration. For example, compressor valves are valves used within a compressor to allow gas flow to and from the cylinder area. The valves function as spring-loaded non-return valves, with the gas only being able to flow in one direction. Fatigue is a primary cause of compressor valve failure due to the repeated bending. The application of an overload can significantly improve fatigue behavior by increasing the fatigue limit or reducing the growth rate of an existing fatigue crack. Several theories have been proposed to explain the transient fatigue crack growth decelerations and accelerations which follow overloads (transient fatigue crack growth is studied for a material system in which the crack tip is shielded due to crack bridging. The process of material separation during fatigue crack growth is described by the use of an irreversible cohesive zone model). The mechanisms that have been proposed to explain retardation after a tensile overload, for example, include residual stress, crack deflection, crack closure, strain hardening, and plastic blunting/resharpening. It is suggested that no single mechanism can be expected to represent observed effects over the entire range of crack growth rate, da/dN. A relation of $\frac{\mathrm{da}}{\mathrm{dN}}$ versus ΔK, e.g., the behavior ranging from the near threshold region to the Paris region is presented below:

$$\frac{\mathrm{d}a}{\mathrm{d}N} = A(\Delta K)^n \tag{7.1}$$

Crack growth rate essentially indicates a linear relationship between log da/dN and log ΔK, which correspond to the formula

$$\ln\left(\frac{\mathrm{d}a}{\mathrm{d}n}\right) = \ln A + n(\Delta K) \tag{7.2}$$

Here K is the strain hardening coefficient, and n is the strain hardening exponent. Here, ΔK is the stress intensity factor range and α is a function of the applied load to be identified based on the experimental data. In Eq. (7.2) n is the slope of the line, and A is the coefficient found by extending the straight line to ΔK = 1 MP $\sqrt{m}$. The load dependency of α enables the description of the load sequential effects. The equation of Paris is

$$D = \left(\frac{n}{N}\right)^{\alpha} \tag{7.3}$$

In Eq. (7.3) D is the fatigue damage of the material. Fatigue damage increases with the applied loading cycles. Under different loadings the damage accumulation is different, and the linear damage accumulation rule is known as the Palmgren–Miner's rule expressed as

$$D = \frac{n_1}{N_1} + \frac{n_2}{N_2} + \frac{n_3}{N_3} + \cdots + \frac{n_i}{N_i} = \sum_{1=1}^{k} \frac{n_i}{N_i} \tag{7.4}$$

In Eq. (7.4) as stated D is the fatigue damage of the material, n_i is the number of applied loading cycles corresponding to the *i*th load level, and N_i is the number of cycles to failure for the *i*th load level from constant amplitude experiments.

A plastic zone is assumed to be present in front of the crack tip during the application of the overload, which depends on the magnitude of the overload and the yield strength. In the interior the plastic zone is developed under planes strain conditions and on the surface plane stress conditions are approached. The plastic zone calculation is based on the assumption that the material (in the present case Ti and Al alloys) responses were perfectly plastic. Table 7.1 shows the results indicating that the use of the plastic zone approach to calculate the plastic zone sizes is reasonably acceptable for Al and Ti alloys.

Tensile overload retards crack growth. The intensive exploration on retardation was pivoted around the effect of residual stress and crack closure. It should be noted that the plane stress conditions at the surface cannot be fully realized because of restraint from the interior material. It has observed that for thin specimens the computed plane stress and measured values were in good agreement. For thick specimens, however, he observed that the measured values were between the computed plane stress and the plane strain values. A measure of the duration of retardation is the distance that

Table 7.1 Mechanical properties, Carlson et al. [2]. With kind permission of Elsevier

Material	Plane stress plastic zone* (mm)	Plane strain plastic zone** (mm)
Titanium alloy[10]	0.37	0.12
Aluminium alloy[11]	1.13	0.38
Low carbon steel[12]	4.03 (1.95)	1.34 (0.65)

*$\frac{1}{\pi}\left(\frac{K_{OL}}{\sigma_y}\right)^2$ (plane stress)

**$\frac{1}{3\pi}\left(\frac{K_{OL}}{\sigma_y}\right)^2$ (plane strain)

K_{OL} is the overload stress intensity factor and σ_y is the yield strength.

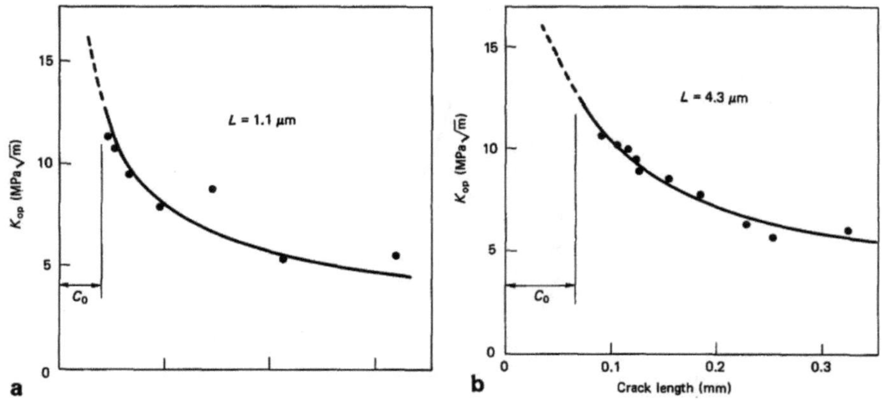

Fig. 7.3 Opening stress intensity factor versus crack extension **a** after 100% overload and **b** after 150% overload, Carlson et al. [2]. With kind permission of Elsevier

a crack advances during the transient behavior following an overload. This has been called the 'delay distance' and has been measured for the materials listed in Table 7.1. The compressive residual stress would be expected to be eliminated when the crack has passed through the overload plastic zone.

Measurement of the delay distance reveals that compressive residual stress mechanism cannot be the primary reason for the retardation for the alloys indicated above, although it may be an operative mechanism for Al alloys and steel alloys. The behavior following immediately an overload (i.e. post-overload) is attributed to a crack tip stretching with the consequent reduced closure contact pressures and by effectively increasing the stress intensity factor range. This resulted in a brief crack growth acceleration period. Ultimately, the crack extended through the stretched ligament and formed an asperity which resulted in the K_{OP} data points shown in Fig. 7.3 (note Kop is opening stress intensity factor). In the tests obstruction to closure was observed, and the opening values of the stress intensity factor, Kop, were determined for crack extension after overloads were applied. The data in Fig. 7.3a are for an overload of 100%. Those of Fig. 7.3b are for an overload of 150%. As plotted, the overloads occurred at the origins. During the interval between the origin and the first data point, a transient crack growth acceleration or delayed retardation was observed. The character of the obstruction form resulting from the tensile overload, intended to examine closure obstruction is illustrated in Fig. 7.4. It was observed that even though crack closure obstruction prior to an overload was of the plain strain for, obstruction after an overload occurred under plane stress conditions. The Kop data obtained refer to an Al alloy.

The variation of the stress intensity factor for opening, Kop as the crack extends is presented in Fig. 7.5. Constant ΔK is maintained from A to B. At B, a tensile overload is applied and the crack tip stretching decreases the misfit obstruction of asperities formed previously, and the Kop has a reduced value as shown between B

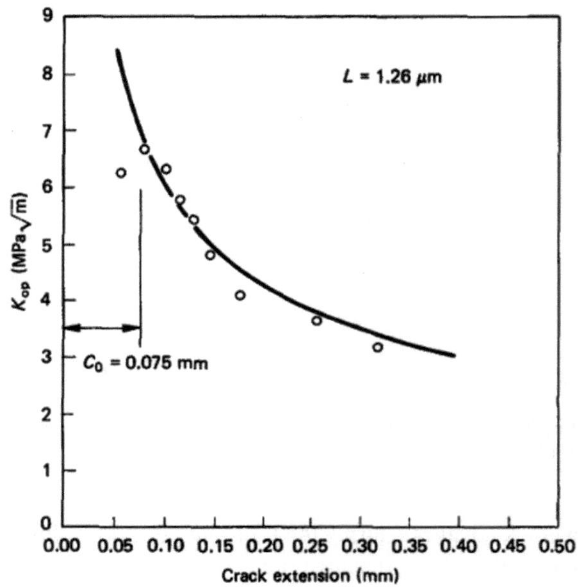

Fig. 7.4 Opening stress intensity factor versus crack extension, Carlson et al. [2]. With kind permission of Elsevier

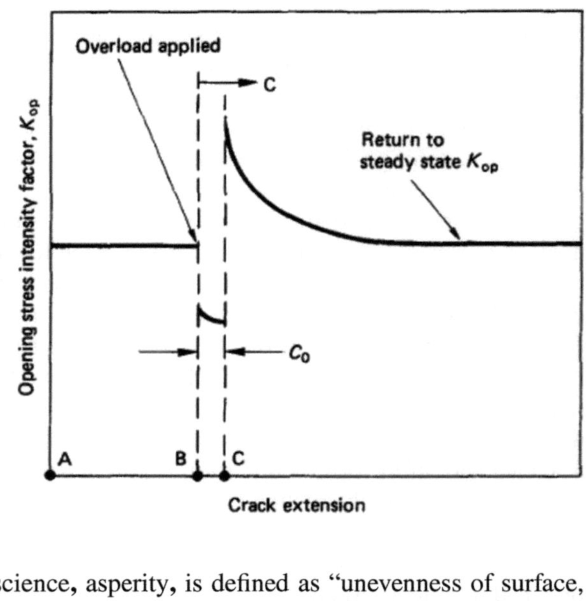

Fig. 7.5 Comparison of crack growth rates after tensile overload, Carlson et al. [2]. With kind permission of Elsevier

and C. (Note: in materials science, asperity, is defined as "unevenness of surface, roughness, ruggedness"). In the transition between Band C the stretched crack tip ligament is fractured to form a stretched asperity which, by increasing the obstruction misfit, results in an abrupt increase in Kop at C. As this asperity recedes from the crack tip, Kop decreases and returns to a steady state. Crack tip stretching is involved in the formation of asperities.

In summary, a crack growth method is presented to assess its limits. Fatigue life estimation attributes crack growth retardation to a single mechanism. The variation in the opening stress intensity factor after a tensile overload results in obstruction of the closure. A discrete asperities model is used as a diagnostic tool. This model ascribes the acceleration transient after a compressive overload and the difference in crack growth rates in inert and active atmosphere. When obstruction to closure develops after a tensile overload, retardation can continue beyond the crack tip plastic zone. This indicates that in addition to the residual stress fracture surface microstructural features have to be considered.

Following Borrego et al. on the evaluation of overload effects of fatigue crack growth and closure, based on the Paris' et al. partial closure model, suggesting the effective range of K between its real minimum and maximum. Paris et al. evaluated ΔK for 6061-T6 and 2024-T3 aluminum alloys considering the fatigue crack rate as a function of the applies intensity range $\Delta K = K_{max} - K_{min}$, where K_{max} and K_{min} are respectively the maximum and minimum stress intensities in the loading cycle. Accordingly

$$\Delta K = K_{\max} - \frac{2}{\pi} K_{op} - \sigma_{nom}\sqrt{\frac{\pi d}{2}} \tag{7.5}$$

where K_{op} is the stress intensity factor at opening load, σ_{nom} is the nominal uniform stress that would be present at minimum load if the crack were absent, and d is the distance between the crack tip and the contact zone behind the crack. The effective range of K could be bounded by:

$$K_{max} - \frac{2}{\pi}K_{op} - \left(1 - \frac{\pi}{2}\right)K_{min} \leq \Delta K_{eff} \leq K_{max} - \frac{\pi}{2}K_{op} \tag{7.6}$$

An approximate result, thus can be given by

$$\Delta K_{eff} = K_{\mathrm{MAX}} - \frac{2}{\pi} K_{op} \tag{7.7}$$

Experimental results showed that, for several aluminum alloys, this equation produces very good correlation of fatigue crack growth data under load-reduction (threshold) simulations. Note, that the partial closure model was physically established assuming that, under ΔK reduction simulations, the crack is open at tip and closed near the load reduction location, at loads below the crack opening load. For the experimental details the reader should consult the work of Borrego et al.

Fatigue crack growth tests under constant ΔK conditions were performed [on AlMgSi (6082) aluminium alloy with T6 heat treatment], resulting in controlled constant-crack-wake history, which is more sensitive to changes in fatigue crack growth rates associated with changing driving force mechanisms than constant-ΔP tests. All experiments were performed in a servo-hydraulic, closed-loop mechanical test machine with 100 kN load capacity, interfaced to a computer for machine control and data acquisition. All tests were conducted in air, at room temperature and with a load frequency of 20 Hz. The tests were performed under constant ΔK and stress ratio R conditions. Single tensile overload tests were performed at several ΔK baseline levels. The overload ratio, OLR (1.5 and 2) is defined as

$$OLR = \frac{\Delta K_{OL}}{\Delta K_{BL}} = \frac{K_{OL} - K_{min}}{K_{max} - K_{min}} \tag{7.8}$$

where K_{max}, K_{min}, and K_{OL} are the maximum, minimum and peak overload intensity factors, respectively. From the load–displacement records, variations of the opening load, P_{op}, were derived. The fraction of the load cycle for which the crack remains fully open, parameter U, was calculated by the following equation

$$U = \frac{P_{max} - P_{OP}}{P_{max} - P_{min}} \tag{7.9}$$

The values of the effective stress intensity factor range, ΔK_{eff}, are given by the expression:

$$\Delta K_{eff} = K_{max} - K_{op} = U \Delta K \tag{7.10}$$

A typical transient crack growth behavior obtained when a specimen is subjected to a single tensile overload in a constant ΔK test is shown in Fig. 7.6. In Fig. 7.6a the crack length from the overload event, $a–a_{OL}$, is plotted against the number of cycles from the point of overload application, $N–N_{OL}$, where a_{OL} and N_{OL} are the

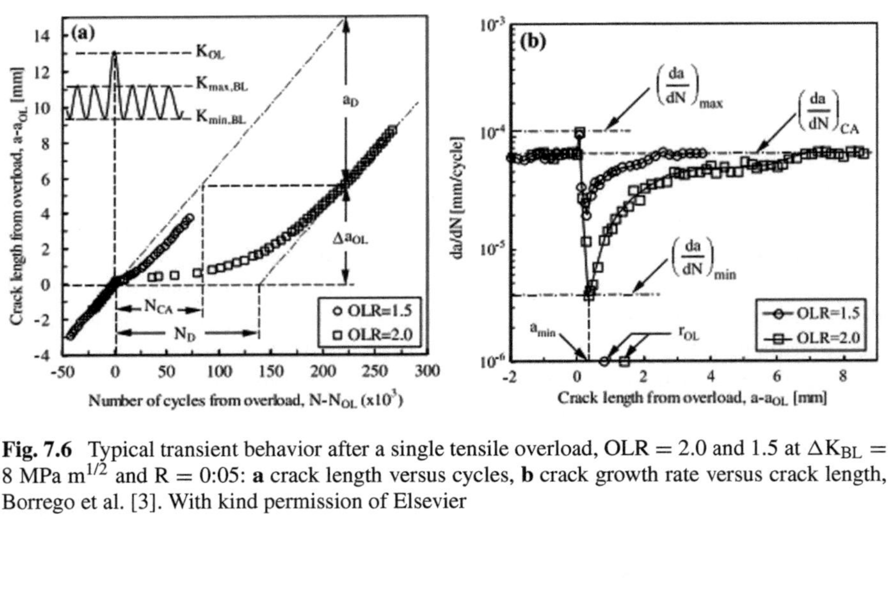

Fig. 7.6 Typical transient behavior after a single tensile overload, OLR = 2.0 and 1.5 at ΔK_{BL} = 8 MPa $m^{1/2}$ and R = 0:05: **a** crack length versus cycles, **b** crack growth rate versus crack length, Borrego et al. [3]. With kind permission of Elsevier

crack length and the number of cycles at which the overload is applied, respectively. Differentiation of this data gives the respective crack growth which may be plotted in function of crack length from the overload event, a–a_{OL}, as illustrated in Fig. 7.6b where da/dN is plotted versus a–a_{OL}. Following a brief initial acceleration of crack growth rate after the overload, subsequent crack growth rate decreases to a minimum value and then approaching to the base line steady state. This is seen in Fig. 7.6b where the experimental points are for OLR = 1.5 and 2.0. The observed behavior is usually referred to as delayed retardation of crack growth.

The size of overload monotonic plastic zone, r_{OL}, was also calculated, being presented in Figs. 7.6b and 7.7 for comparison. r_{OL} was evaluated, for a plane stress condition, from the following equation:

$$r_{OL} = \frac{1}{\pi}\left(\frac{K_{OL}}{\sigma_Y}\right)^2 \tag{7.11}$$

where K_{OL} is the stress intensity factor at peak load during the overload cycle and σ_Y is the yield stress. This plastic zone superimposes the plastic zone created during regular crack growth at constant amplitude fatigue. Figure 7.8 presents the influence of the ΔK baseline level ΔK_{BL}, at which the overload is applied, in terms of the normalized crack growth ratio, da/dN/da/dN_{CA}, against a–a_{OL}. Figure 7.8a, b present the results obtained at R = 0.05 for OLR = 2 and 1.5, respectively. The Figure 7.8 show that crack growth retardation increases significantly with ΔK_{BL}. The values of parameters Δa_{OL}, a_{min} and N_D increase as ΔK_{BL} is increased from the lower to the higher ΔK_{BL} analyzed, corresponding to an increase in fatigue life of approximately four and three times for OLR = 2 and 1.5, respectively. Further, the influence of the

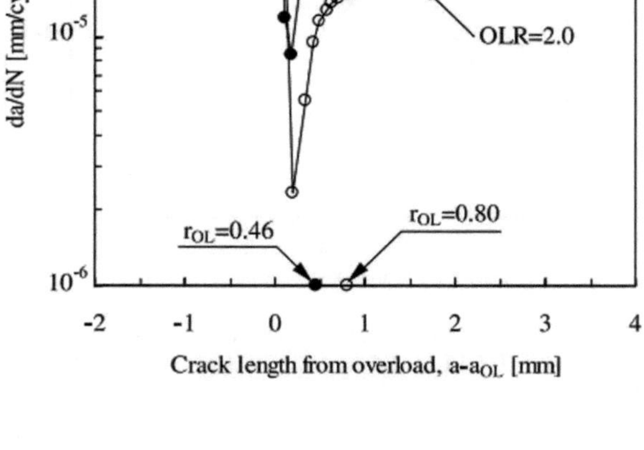

Fig. 7.7 Crack growth rate as a function of crack growth distance from the overload event for OLR values of 1.5 and 2 at DK_{BL} = 6 MPa $m^{1/2}$ and R = 0:05, Borrego et al. [3]. With kind permission of Elsevier

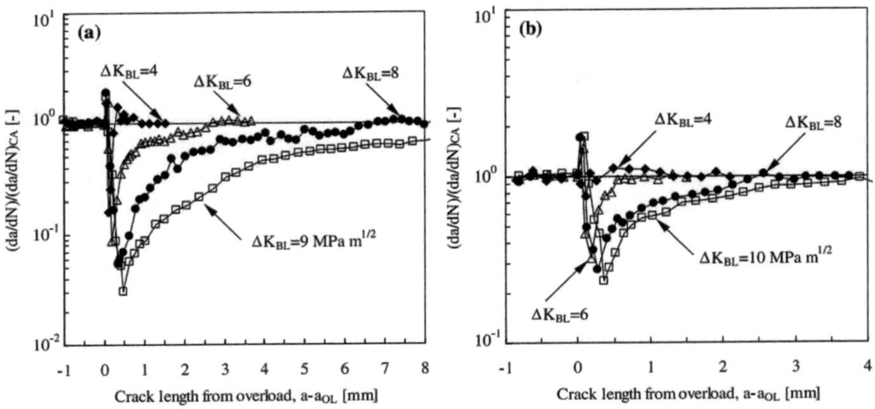

Fig. 7.8 Effect of ΔK baseline level on the normalized crack growth rate, R = 0.05: **a** OLR = 2.0, **b** OLR = 1.5, Borrego et al. [3]. With kind permission of Elsevier

stress ratio on the transient crack growth behavior following a single tensile overload can be seen in Fig. 7.9 for different OLR and ΔK_{BL} levels. Fig. 7.9a, b present the results obtained for OLR = 2 at ΔK_{BL} = 6 MPa $m^{1/2}$ and for OLR = 1.5 at ΔK_{BL} = 10 MPa $m^{1/2}$, respectively.

More information on the effect of overload under various conditions, their analysis and the resulting illustration are extensively analyzed in the work of Borrego et al., and the interest readers should consult this publication.

In summary, the crack growth behavior under single tensile overload conditions in 6082-T6 aluminum, alloy was presented. (a) It was observed that the influence of OLR and baseline ΔK level on the crack growth transients is significant and with

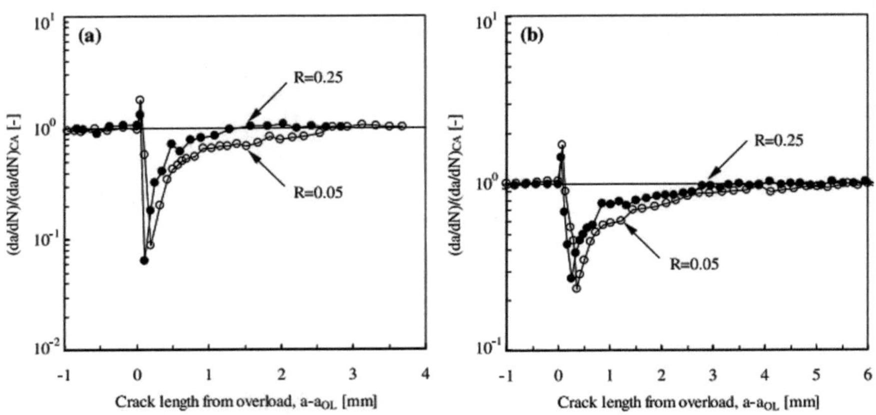

Fig. 7.9 Effect of stress ratio on the normalized crack growth rate: **a** OLR = 2.0 at ΔK_{BL} = 6 MPa $m^{1/2}$, **b** OLR = 1.5 at ΔK_{BL} = 10 MPa $m^{1/2}$, Borrego et al. [3]. With kind permission of Elsevier

increase OLR and ΔK_{BL} crack growth retardation increases while decreasing R ratio. (b) The length of the transient region over which crack growth rates are affected by the overload and its variables. (c) The OLR is a major factor in determining the influence on the retardation effect of ΔK_{BL}. (d) The partial closure model is not able to completely account the observed transition period from full closure to partial closure. The suggested modification of this model improved significantly the inferred growth rates for all the post-overload transients.

7.4 Cumulative Damage

As indicated several times in this book, fatigue is the major cause of the catastrophic failures of machine elements or parts. Such structural components in service endure cyclic loads with varying intensity until fracture may set in. Note that fatigue failure occurs in localized weak spots of a material following its deterioration after some number of cycles which limits and even prohibits safe use. Understanding the damage process in fatigue is required to predict and evaluate fatigue life of components exposed to cyclic loads. Essentially, the fatigue damage is associated with crack nucleation (initiation) and propagation and with local flaws on micro-scale, such as extrusions, intrusions and with imperfections on an atomic level such as dislocations, vacancies, slip bands, etc. One of the methods to address fatigue problems is based on the linear damage rule which is used for practical engineering design because calculation efforts are reduced.

The usual way of analyzing and predicting fatigue life of materials (originally formulated for metallic components) is to plot stress amplitude against the number of loading cycles to failure, namely S–N diagrams. The basic stress-life relation is given by a power law as (following Peng et al.)

$$\sigma^m N_f = C \quad \text{or} \quad C = \sigma_f^{'}\left(2N_f\right)^h \tag{7.12}$$

here N_f is the number of loading cycles to failure at a given stress level σ_i, m and C are material constants, $\sigma_f^{'}$ and h denote the fatigue strength coefficient and fatigue strength, respectively. This relation is known as Basquin's power law. The equation can be rewritten as a linear function on log–log scale and presented as shown in Fig. 7.10.

$$\log(\sigma) = a + b\log(N_f) \tag{7.13}$$

a being the intercept of the curve, and b is the slope given as $b = -1/m$.

Initial damage is induced by the stress amplitude σ_1 for n_1 cycles and residual number of cycles to fracture, namely the residual life N_r, at the same stress amplitude is given as

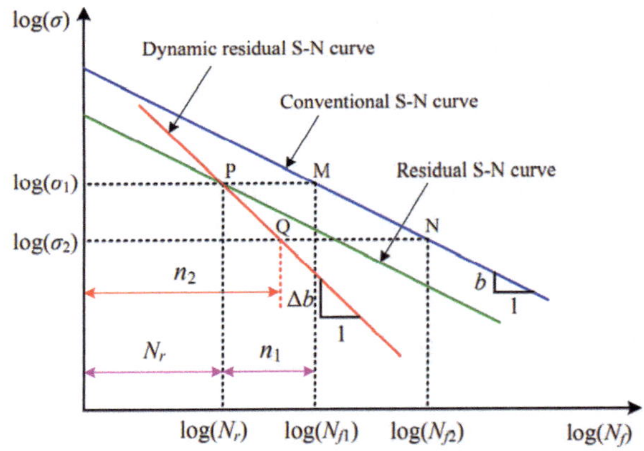

Fig. 7.10 Schematic representation of conventional S–N curve, residual S–N curve, and dynamic residual S–N curve, Peng et al. [4]. Open access

$$N_r = N_{f1} - n_1 \tag{7.14}$$

seen in Fig. 7.10. For the undamaged specimen, the residual life corresponds to the fatigue failure cycles determined from the conventional S–N curve. Assuming that the damaged specimen can be considered as an undamaged specimen, the stress-life relation of the damaged specimen can be described by a residual S–N having similar expression as the conventional one. Thus, the residual S–N curve can have the same form and the same slope as Eq. 7.13, i.e.

$$\log(\sigma) = a' + b\log(N_r) \tag{7.15}$$

where a′ is the intercept of the residual S–N curve.

For variable amplitude loading tests, the simplest cases are the two-stage cycle loading. In such tests first a pre-cycle at a stress σ_1 for n_1 cycles is applied, then the specimen is cycled at another stress, σ_2 for n_2 cycles to failure. The relation between Eqs. 7.14 and 7.15 can be expressed in a linear cumulative damage rule as

$$\frac{n_1}{N_{f1}} + \frac{n_2}{N_{f2}} = 1 \tag{7.16}$$

Equation (7.14) is called Milner's residual S–N curve and the slope is the same as that of Eq. (7.13). In Eq. 7.16 the effect of loading history is not indicated, and the slope of Eq. 7.15 may be a dominant factor of describing loading effects on fatigue. The approach taken is that the residual S–N curve slope is considered as a variable with respect to previous fatigue loadings instead of a basic material constant. Then, a dynamic residual S–N curve, as shown in Fig. 7.10, can be expressed as:

$$\log(\sigma) = a'' + \Delta b \log(N_r) \tag{7.17}$$

Equation (7.17) can be described as responsible for the virgin material without initial damage, Δb is identical to the original slope b; With continued fatigue loading, the absolute value of Δb increases with the progressive damage; at fracture it tends to be infinite. As a consequence, the slope ratio $b/\Delta b$, defined as b_r, decreases with the loading cycles or the expended life fraction and should range from 1 to 0. In the dynamic residual S–N curve method, the change in slope is appropriate to present the fatigue failure process and the evolution law of damage accumulation. It is essential to quantify b_r when accumulating fatigue damage.

For further analysis of the cumulative damage at various stages and conditions and the comparison with experimental results for Al and steel (outside the scope of this book), the reader should consult the work of Peng et al., who propose a new cumulative fatigue damage rule based on dynamic residual S–N curve and material memory concept.

In summary, a phenomenological cumulative damage rule is proposed by incorporating a dynamic residual S–N curve and material memory concept to describe damage accumulation behavior. The model follows a linear trend in accumulating damage and also takes the load-level dependence and loading sequences into account. It predicts the damage and fatigue life with a small amount of data necessary from the conventional S–N curve.

7.4.1 Cumulative Damage in Alumina

Very little information can be found in the literature specifically on the subject of interest in this section. However, an older publication discussing surface damage in alumina was explored by Warren et al. The essence of this publication is the mechanical properties of brittle structural ceramics which decreases drastically in the presence of cracks which are inevitably introduced in the preparation of ceramic material via machining and polishing. In addition to the mechanical properties optical and electronic properties may also be affected by the presence of such cracks. Two contrasting methods are illustrated below, both utilizing the mechanical probing of a surface to evaluate the damage.

When the surface of a brittle material (a glass, a semiconductor or a ceramic) is machined (via a hard diamond grit, for example) then a variety of structural changes in the near-surface region occur. Usually, a groove is left behind the particle, producing a rough surface; beneath the groove is a region of plastic zone of intense plastic deformation, which leaves the surface in a residual stress and finally may result in microcracks at the region induced by the machining. The crack patterns observed are median/radial cracks which are predominantly perpendicular to the surface and lateral cracks which are parallel to the surface. In some cases, (semiconductor material) the final stage of surface preparation is usually a chemo-mechanical polish. As indicated two methods were used to evaluate the damage, ultrasonic force method and acoustic emission indentation.

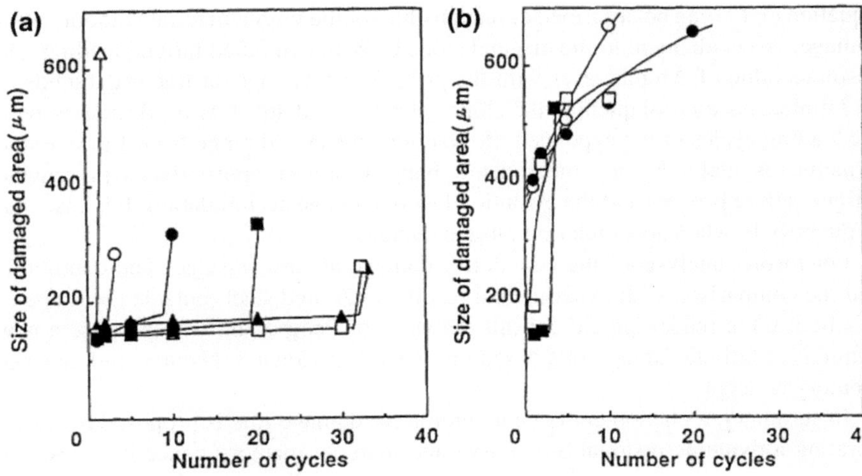

Fig. 7.11 Development of the damaged area under cyclic indentation; peak applied load = 196 N. **a** Covalently bonded materials: SiC (O) transgranular, (●) intergranular; Si_3N_4 (Δ) transgranular, (▲) intergranular; sialon (□) 0.3 μm, (■) 1.0 μm. **b** Ionically bonded materials. MgO (O) 10 μm, (●) 25 μm; Al_2O_3 (□) 10 μm, (■) 2 μm, Takakura and Horibe [5]. With kind permission of Springer Nature

Indentation fatigue damage accumulation in the surface layer of alumina caused by local cyclic loading is presented here. Two types of indentation fatigue behavior were observed. Type I is characterized by indentation induced damaged zone at the surface associated with chipping with increased indentation cycles. Type II behavior is characterized by significant increase in damaged zone already in the early cycles associated with cracking. The relations between lateral cracking, microcracks and the resultant damaged zone in alumina is of major interest and is discussed below:

In Fig. 7.11 the size of the damaged area as a function of the number of cycles is illustrated by the open squares. Alumina is indicated by the squares (open and closed as indicated in the subscript of this figure. Included are SiC, MgO, Si_3N_4 and SiAlON. Recall that SiAlON ceramics are a special class of high-temperature refractory materials, with high strength at ambient and high temperatures, good thermal shock resistance and exceptional resistance to wetting or corrosion. The SiAlONs are based on the elements Si, Al, O and N. Table 7.2 list the relevant data on this material. Figure 7.12 illustrates the dependence of indentation fatigue type upon the grain size and ductility/brittleness. From this figure, it is clear that Type I materials are characterized by brittleness and/or small grain size, whereas Type II materials are plastic and/or have a large grain size. The effect of grain size upon cyclic indentation in alumina seem to be considerable as seen in Fig. 7.13. Coarse-grained alumina behaves as type II whereas fine-grained behaves as type I. In type I indentation fatigue, lateral cracking propagates during cyclic indentation which might not be initiated in coarse-grained alumina. However, more tests with scanning acoustic microscopy (SAM) clearly indicates the presence of lateral crack in coarse grained alumina as

Table 7.2 Fabrication process, additives, Vickers hardness, grain size and fracture appearance in the conventional flexural strength testing for the materials tested, Takakura and Horibe [5]. With kind permission of Springer Nature

Material	Final fabrication	Additives	Grain size (μm)	Hardness (GPa)	Fracture appearance
MgO	Pressureless	None	25	5	Transgranular
MgO	Pressureless	None	10	5	Transgranular
Al_2O_3	HIP	None	10	13	Intergranular
Al_2O_3	HIP	None	2	18	Intergranular
SiC	Pressureless	B, C	5	24	Transgranular
SiC	Pressureless	Al_2O_3	1	19	Intergranular
Si_3N_4	HIP	None	2	18	Transgranular
Si_3N_4	Pressureless	Y_2O_3, $MgAl_2O_4$	0.5	13	Intergranular
Sialon	HIP	Y	1	19	Intergranular
Sialon	HIP	Y	0.3	19	Intergranular

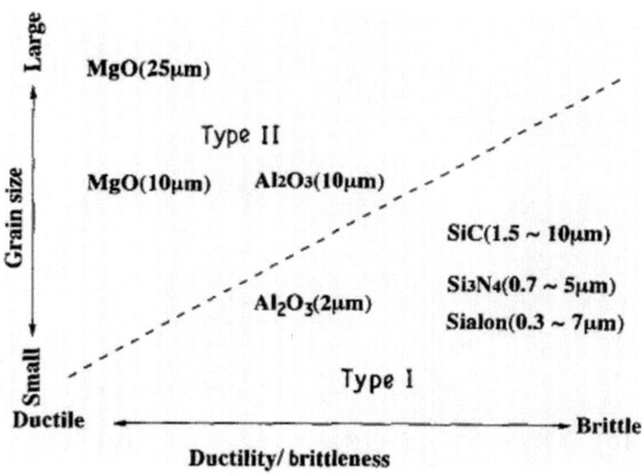

Fig. 7.12 The dependence of the type of indentation fatigue behavior on grain size and plasticity, Takakura and Horibe [5]. With kind permission of Springer Nature

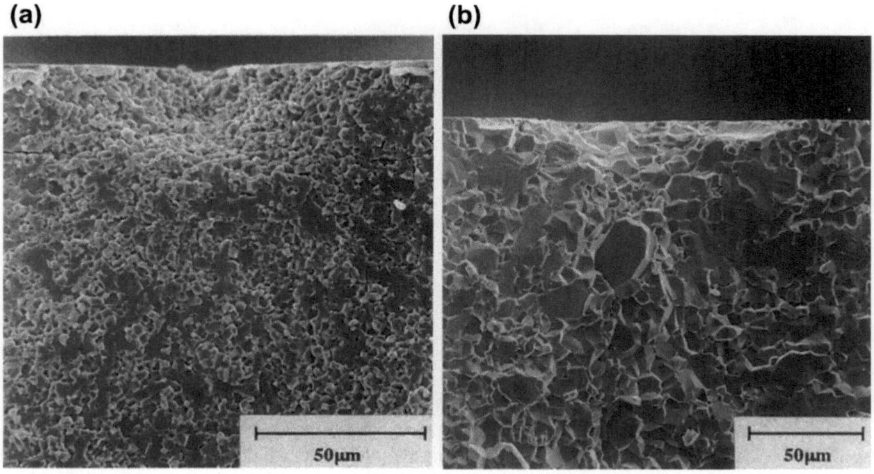

Fig. 7.13 Fractured surfaces produced by flexural strength testing in alumina: **a** fine-grained, **b** coarse-grained, Takakura and Horibe [5]. With kind permission of Springer Nature

well, as seen in Fig. 7.14c even after a single indentation in coarse-grained alumina. Thus, irrespective of the grain size, lateral cracks develop and propagate. The damage accumulation in coarse grained alumina is shown schematically in Fig. 7.15 and occurs as follows: The first indentation induces a lateral crack around an indent as seen schematically in Fig. 7.15a. SEM observation of the indentation damage in the coarse-grained alumina seen in Fig. 7.16, indicates that the fracture is characterized mainly by intergranular cracking. Grain bridging specific to coarse grained alumina allows bridging grains to rotate or slide as seen in Fig. 7.15b with the consequent secondary microcrack formation at grain boundaries and inside grains. The process

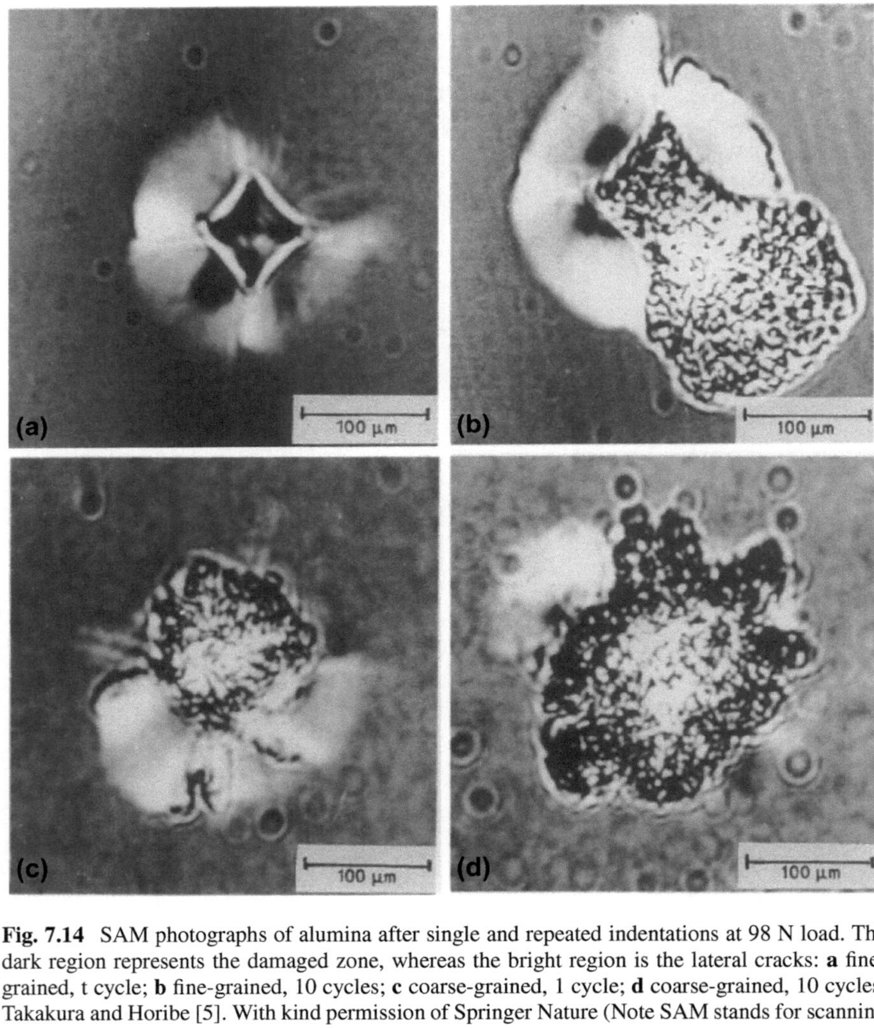

Fig. 7.14 SAM photographs of alumina after single and repeated indentations at 98 N load. The dark region represents the damaged zone, whereas the bright region is the lateral cracks: **a** fine-grained, t cycle; **b** fine-grained, 10 cycles; **c** coarse-grained, 1 cycle; **d** coarse-grained, 10 cycles, Takakura and Horibe [5]. With kind permission of Springer Nature (Note SAM stands for scanning acoustic microscopy)

Fig. 7.15 Schematic illustration of the indentation fatigue mechanism in coarse-grained alumina, Takakura and Horibe [5]. With kind permission of Springer Nature

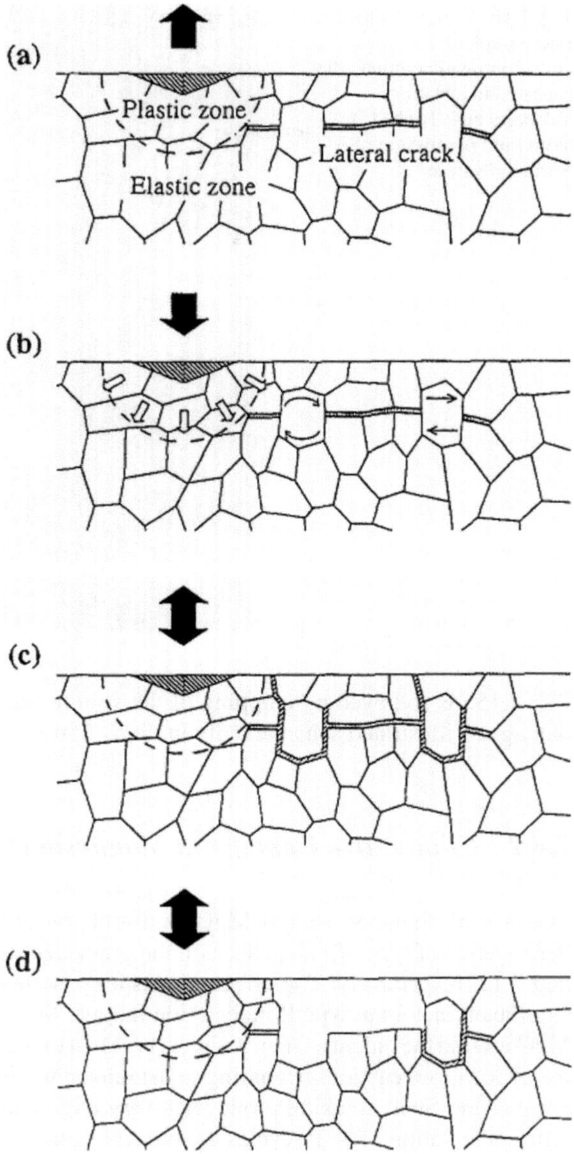

is associated with lateral crack growth propagation prevention during subsequent cycles by the residual stress developed and which is relieved by microcracking. The residual stress before being relieved has the role of the driving force for lateral crack propagation. The cracks on reaching the surface of alumina, as illustrated in

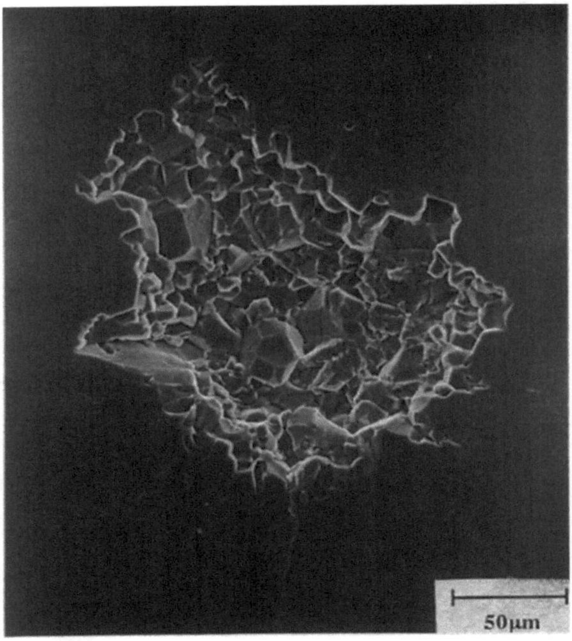

Fig. 7.16 Scanning electron micrograph of coarse-grained alumina after repeated indentation, Takakura and Horibe [5]. With kind permission of Springer Nature

Fig. 7.15c is followed by chipping of the volume surrounded by cracks as its final damage as illustrated schematically in Fig. 7.15d.

7.4.2 Cumulative Damage in Magnesia (MgO)

The size of damaged area of MgO as a function of the number of cycles and the dependence of type of indentation on the grain size were also included in Figs. 7.11 and 7.12, respectively. The effect of plasticity is considered to be one of the most significant factors in type II indentation fatigue. Many microcracks were observe in MgO around the indent as seen in Fig. 7.17. Most of these microcracks are the result of the dislocation reactions according to the pile up model. Also, some cleavage cracks are present. Such cracking produces a very large damaged zone around an indent. After indentation of 1–10 cycles the damage zone around an indent is illustrated as SAM photographs in Fig. 7.18. The dark region around the indent in Fig. 7.18a, b is the damaged zone. The bright region surrounding it is the microcracking area. As seen a plastic characteristic is seen in magnesia at cyclic indentation. It would be difficult for a lateral crack to propagate continuously during cyclic indentation because the residual stress which is the driving force crack propagation is relieved by plastic deformation or microcracking. It is suggested that microcracking and the resulting local chipping occur gradually with repeated indentation, followed by saturation.

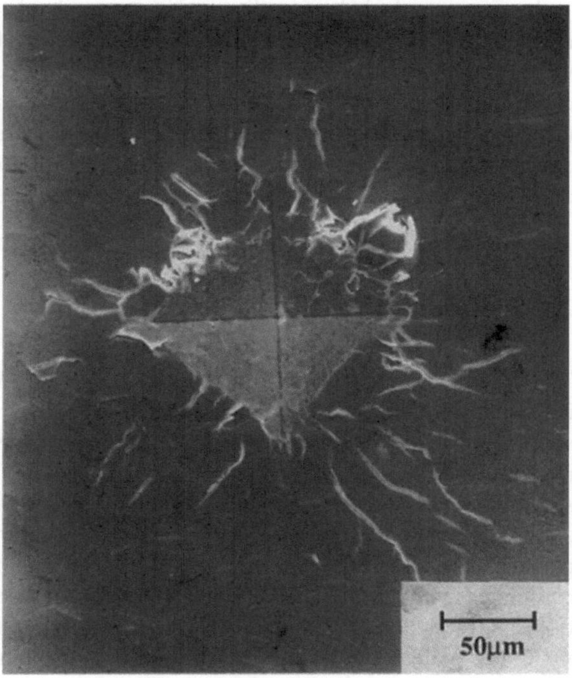

Fig. 7.17 Scanning electron micrograph of magnesium oxide after a single indentation, Takakura and Horibe [5]. With kind permission of Springer Nature

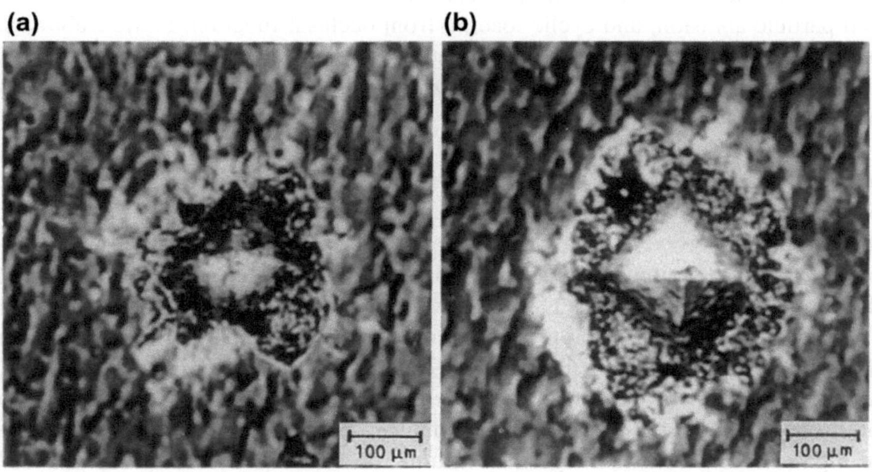

Fig. 7.18 SAM photographs of magnesium oxide after repeated indentation: **a** 1 cycle; **b** 10 cycles, Takakura and Horibe [5]. With kind permission of Springer Nature

In summary, the behavior observed in repeated indentation, namely local chipping is caused by successive initiation and propagation of microcracks. This behavior where indentation induced damaged zone occur in early stages of the indentation cycles has type II fatigue indentation characteristic and was observed in ceramics having grain-bridging or plastic-deformation characteristics. The formation of microcracks during cyclic indentation relieves the residual stress which is involved in lateral crack propagation. The relevant properties of MgO are listed in Table 7.2.

7.4.3 Cumulative Damage in Zirconia

The significant interest in yttria stabilized zirconia (3-YSZ) ceramics is related to its use in dental crowns and hip implants. They may fail in cyclic use due to the sub-critical fatigue damage at the load levels that are significantly lower than those required for damage 3-YSZ under monotonic loading. Damage from particle abrasion, not necessarily immediately apparent, weaken or harm the fatigue strength of zirconia (and alumina ceramics) in crown-like structures. Small flaws introduced by particle abrasion can outweigh any resistant strengthening effect from compression associated with surface damage or, in the case of zirconia, with phase transformation. The clinical survival of all-ceramic crowns is a complex combination of factors, including material properties such as strength, modulus, and toughness, damage caused by shaping and handling procedures such as casting, machining, grinding, and particle abrasion, and cyclic loading from occlusal function. Particle abrasion is of special interest because of its countervailing influences on crown performance. Damage accumulation and fatigue life of zirconia which was abraded by particles is presented in this section.

In Figs. 7.19 and 7.20 the flexural stress in Y-TZP (recall the abbreviation refers to Yttria Tetragonal Zirconia Polycrystal) as a function of number of cycles is presented. Since both zirconia and alumina are clinically relevant ceramics the figures compare their stress properties. To see the effect of the surface quality, polished and sand-blasted particle (to simulate the effect of particles abrasion) specimens were used in the comparison. It is seen in Fig. 7.19 that the as polished specimens endure higher stress level, and also the stress level of zirconia versus the number of cycles is higher than that of alumina (included in Figs. 7.19 and 7.20). The Y-TZP (also the alumina) are fine grained. Particle abrasion by cyclic fatigue loading degrades the strength of zirconia (and alumina). The damage induced by particle-abrasion is illustrated in Fig. 7.21. Basically, damage in a ceramic seen in Fig. 7.19 that has polished specimens endure higher stress level, and also the stress level of zirconia versus the number of cycles is higher than that of alumina (included in Figs. 7.19 and 7.20). The Y-TZP (also the alumina) are fine grained. Particle abrasion by cyclic fatigue loading degrades the strength of zirconia (and alumina). The damage induced by particle-abrasion is illustrated in Fig. 7.21. Basically, damage in a ceramic undersurface may result from several reasons as illustrated schematically in Fig. 7.22. Note that blunt trauma is physical trauma to a body part, either by impact, injury, or physical attack,

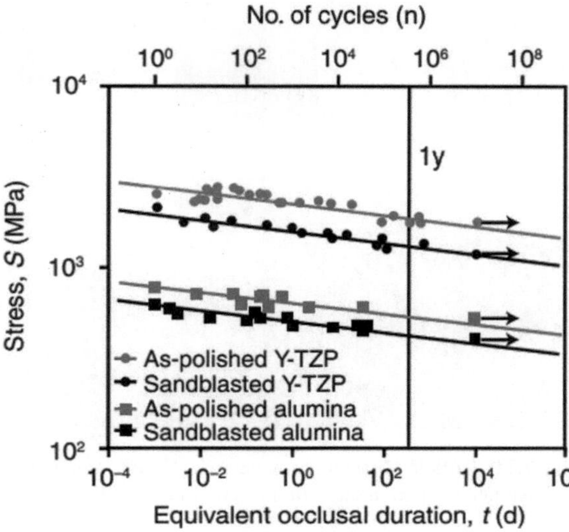

Fig. 7.19 Flexural stress needed to produce radial cracks as a function of number of cycles n or equivalent occlusal contact cycles in fatigue tests, comparing Y-TZP and alumina bilayer specimens (see Fig. 1) with as-polished and particle-abraded surfaces. Data are individual test results. Solid lines are linear best fits.14 arrows indicate test "runouts." Zhang et al. [17]

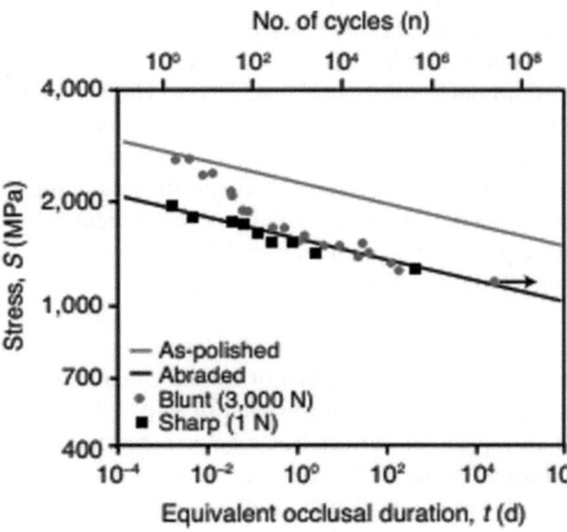

Fig. 7.20 Flexural stress as a function of number of cycles n or equivalent occlusal contact duration needed to produce radial cracks in Y-TZP for bilayer specimens with different ceramic cementation surface treatments. Data points are individual test results. Solid lines are linear best fits from Fig. 7.28 of their paper. Arrow indicates test "runouts." Zhang et al. [17]

which can result also from high velocity impact. In Fig. 7.22c it was caused by a tungsten carbide (WC) indentation.

A critical load is required for the onset of radial fracture as a number of loading cycles, which can be expressed as

$$S = 0.75\left(\frac{P}{d^2}\right)\log\left(\frac{E_c}{E_s}\right) \tag{7.18}$$

Fig. 7.21 Particle-abrasion damage in a Y-TZP specimen. Substantial damage to a depth of approximately 4 μm in the ceramic is revealed. Section and part-surface views, Zhang et al. [17]

Here, P is the load to initiate fracture, d is the ceramic thickness, E_c is the elastic modulus of the ceramic and E_s is the elastic modulus of the substrate (it is polycarbonate). For Y-TZP $E_c = 205$ GPa (for alumina $E_c = 372$ GPa) and for the polycarbonate substrate $E_s = 2.3$ GPa. The material strength in single cycle loading is presented in Table 7.3. The experiments are intended to simulate ceramic crowns on dentin. The contact loading experimental setup is illustrated schematically in Fig. 7.23. It induces flexural tensile stresses in the lower half of the ceramic layer, which results in radial cracking at the cementation surface. The initiation and the subsequent propagation of radial fractures were recorded by a video system located beneath the specimen (see Fig. 7.23).

In summary, damage accumulation and fatigue life of Y-TZP by particle abrasion was discussed in this section. The damage created by particle abrasion can reduce the strength of Y-TZP (and that of alumina also), and in fatigue cycling test this reduction can be 20–30% greater than in as-polished specimens. (Recall that for fatigue testing specimen surface has to be prepared immaculately, generally by polishing to a flawless surface). Y-TZP is considered to be a structural ceramic and is a major ceramic for use as dental crowns and hip implants. The experimental setup had a configuration for simulating occlusal loading of a monolithic crown on dentin.

7.4.4 Cumulative Damage in SiC/SiC

Some information on cyclic deformation in SiC has been included in Figs. 7.11, 7.12 and Table 7.2. The data were obtained by cyclic indentation and are related to the size of the damaged zone (Fig. 7.11) as a function of number of cycles and to the grain size as a function of plasticity (Fig. 7.12). The failure is transgranular, the size of damage

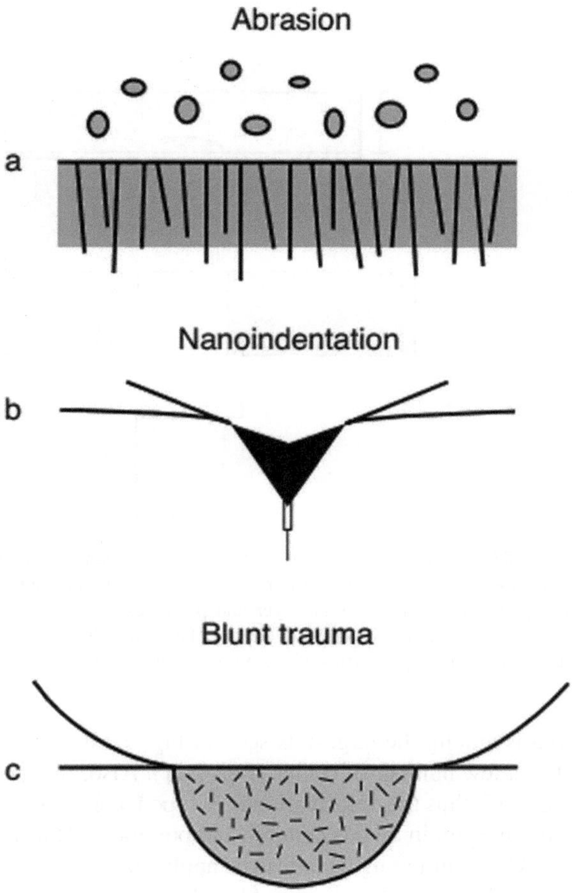

Fig. 7.22 Damage treatments in the ceramic undersurface: **a** particle abrasion, producing a quasiplastic damage zone of microcracks; **b** sharp diamond indentation at low load (1 N), producing an intense deformation zone with individual, well-developed cracks; and **c** WC sphere indentation, producing a blunt trauma quasiplastic zone with microcracks, Zhang et al. [17]

Table 7.3 Material strength in single-cycle loading*, Zhang et al. [17]

Material class	Strength (MPa)		Brand name	Manufacturer
	As polished	Particle abraded		
Alumina	754 ± 60	691 ± 69	AD995	CoorsTek
	706 ± 46		Procera Alumina	Nobel Biocare
Zirconia	2486 ± 88	2078 ± 60†	Pro-TZP	Norton
	1654 ± 152		Procera Zirconia	Nobel Biocare
	1475 ± 92		Lava Zirconia	3M ESPE
	1267 ± 116		VITA Zirconia	VITA Zahnfabrik

*As-polished strength of commercially available dental ceramics is provided for comparison.
†Significantly different from as-polished strength.

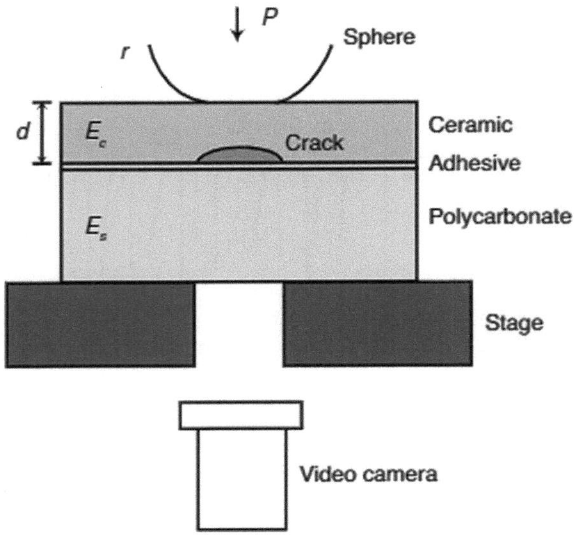

Fig. 7.23 Schematic of the flat bilayer test specimen, consisting of a ceramic plate of thickness d and modulus E_C bonded with epoxy to a polycarbonate substrate of modulus E_S. Loading with the sphere flexes the ceramic layer and produces a radial crack at the cementation surface. This configuration simulates occlusal loading of a monolithic crown on dentin. Cyclic loading is used to determine fatigue properties of the bilayer system, Zhang et al. [17]

area is among the largest as seen in Fig. 7.11b and sets in early in the fatigue test, after a few number of cycles (Fig. 7.11). Also, the tendency to brittleness is seen in Fig. 7.12, thus the deformation is of type I. Generally, however, SiC is used as a reinforcing agent in metals ceramic and polymers. Thus it is not surprising that this role of SiC is dominating its industrial application, also in SiC matrix, where it is an additive in the form of a Nicalon fiber. Fatigue damage accumulation in 3-dimensional SiC/SiC composites is presented in this section. The damage propagation, the interaction between the different damage modes and the resulting damage accumulation promotes critical damage modes within the composite structure, alters the material stiffness and damping characteristics and reduces the strength together with the fatigue life of the composite laminates. Particularly, subcritical fatigue loading of ceramic matrix composites (CMCs) appears whenever the maximum applied cyclic stress exceeds the matrix strength for cracking. Then, the cyclic opening and closing of matrix cracks is the basic responsible mechanism of fatigue for CMCs. In Fig. 7.24 the variation in the loading direction of the normalized effective dynamic modulus of elasticity is presented versus the fatigue cycles in both applied maximum stress levels. The closed symbols stand for the maximum applied fatigue stress of 0.7 of the UTS, while the open symbols relate to 0.75 of the UTS. The Eigen frequencies (natural frequencies) are indicated in the figure. As seen a very steep reduction in the effective dynamic modulus of elasticity occurs for both cases of loading at the first 25 kcycles. This sharp drop is associated with the formation of matrix crack

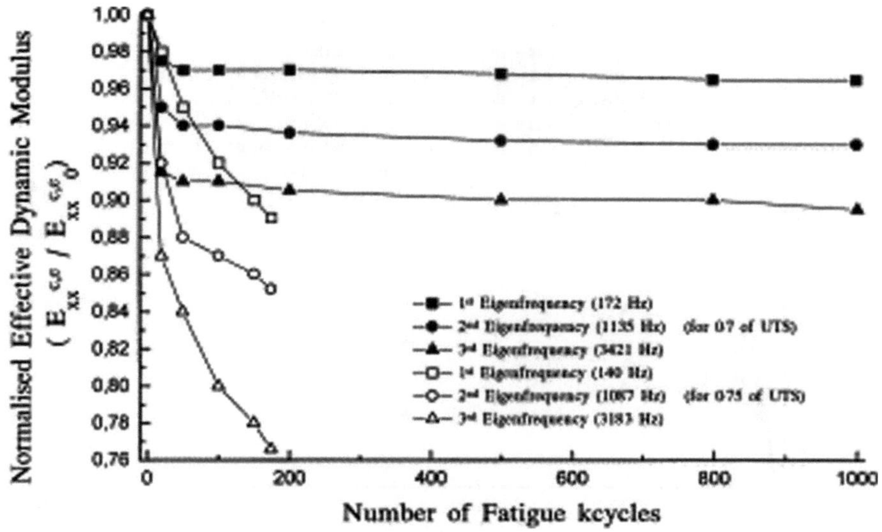

Fig. 7.24 Normalised effective dynamic modulus in the loading direction versus fatigue cycles of 3-D SiC/SiC for maximum applied stress of 0·7 (solid symbols) and 0·75 (open symbols) of UTS, respectively, Kostopoulos et al. [18]. With kind permission of Elsevier

network, it extends up to saturation where fiber-matrix debonding sets in with failure of some of reinforcing fibers in the loading direction. The stiffness reduction is associated with this failure. Further, after ~30 kcycles the effective dynamic modulus of elasticity remains almost constant in the case of applied load 0.7 of UTS, while it keeps decreasing in the case when the maximum applied load was 0.75 of UTS. The variation of the normalized damping coefficient (indicates loss factor) versus the number of the fatigue cycles is illustrated in Fig. 7.25 for both cases and the symbols represent the same stress levels as in Fig. 7.24.

As seen in this plot the damping coefficient increases very fast within the first 25 kcycles but in the case of applied load of 0.75 of UTS was higher in all cases. The increase in the case of applied load of 0.7 of UTS, the increase is at a lower rate, almost constant until the end of the fatigue experiment.

Acoustic emission (AE) is one of the nondestructive techniques to monitor micromechanical-macromechanical fatigue behavior of materials. In Figs. 7.26 and 7.27, AE events corresponding to fiber breakage as a function of fatigue cycles for maximum applied stress of 0.7 and 0.75, respectively are shown. Three stages can be seen. Up to 25 kcycles, fiber fracture events occur during the application of fatigue mean load and the first cycles of the fatigue test. The second stage is the steady state phase, represents an increase in damage, which weakens the material structure and causing degradation of its performance and an increase of internal damping. At the final stage (third stage) crack saturation phase approaches and an unstable damage propagation initiates. It is associated with an increase rate of fiber fracture, which lead to the material final failure. This stage is not indicated in Fig. 7.26. The tensile and

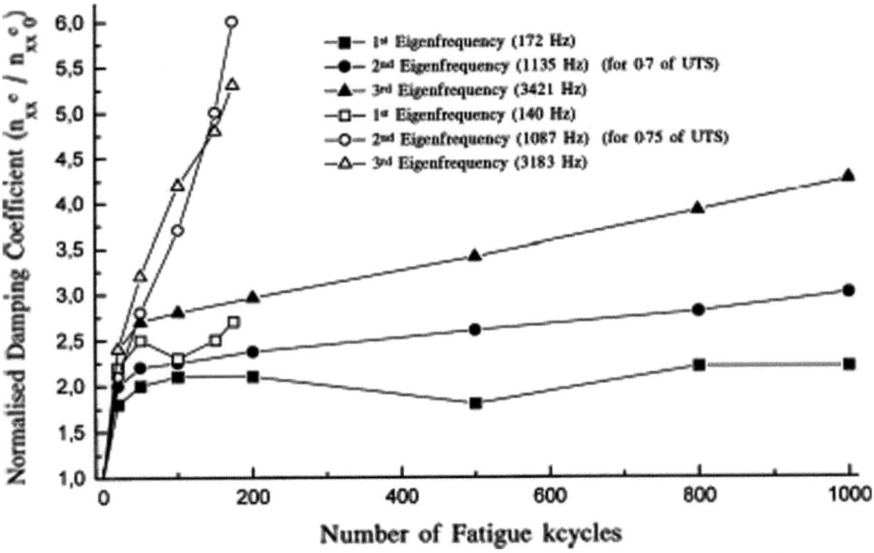

Fig. 7.25 Normalized damping coefficient (loss factor) versus fatigue cycles of 3-D SiC/SiC for maximum applied stress 0.7 (solid symbols) and 0.75 (open symbols) of UTS, respectively, Kostopoulos et al. [18]. With kind permission of Elsevier

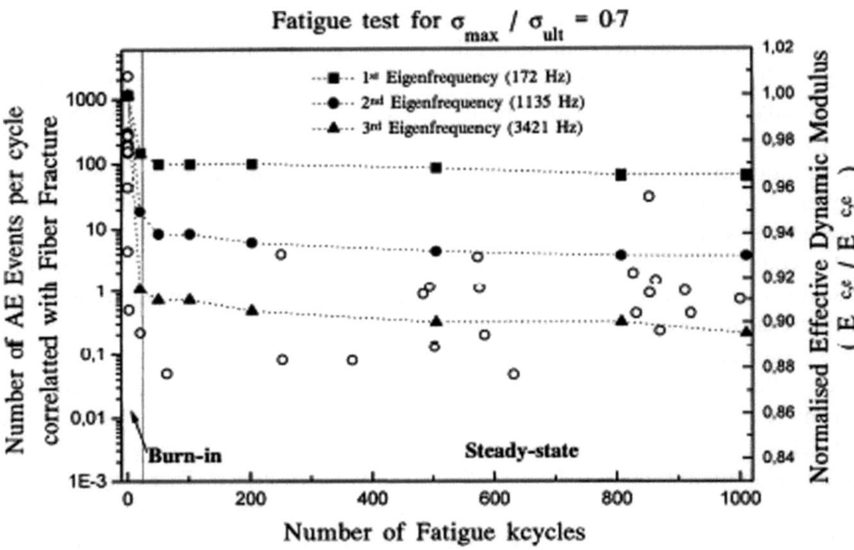

Fig. 7.26 Correlation between AE activity related to fiber fracture and normalized effective dynamic modulus for fatigue loading of 3-D SiC/SiC up to 0.7 of UTS, Kostopoulos et al. [18]. With kind permission of Elsevier

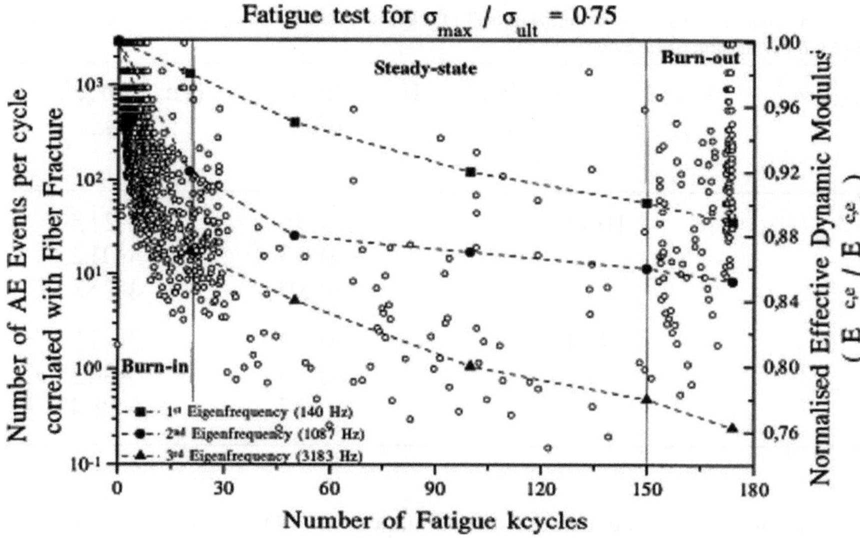

Fig. 7.27 Correlation between AE activity related to fiber fracture and normalized effective dynamic modulus for fatigue loading of 3-D SiC/SiC up to 0.75 of UTS, Kostopoulos et al. [18]. With kind permission of Elsevier

fatigue properties of 3-D SiC/SiC composites are listed in Table 7.4. Whenever the maximum applied stress exceeds the strength of the material during the fatigue test, an extensive matrix crack network is established within the 3-D SiC/SiC. This, together with the processing induced crack accompanied with fiber debonding consists of the first group of the damage mechanism developed during the first fatigue cycle. Table 7.5 summarizes the maximum normalized variations measured by using the dynamic response monitoring during the fatigue of 3-D SiC/SiC composites under two different maximum applied stress levels.

Table 7.4 Tensile and fatigue properties of 3-D SIC/SiC composites, Kostopoulos et al. [18]. With kind permission of Elsevier

	Mean value	*SDD*
Tensile strength (UTS) (MPa)	161	0·7
Tensile tangent modulus of elasticity (GPa)	27	0·7
Endurance fatigue limit ($R = 0{\cdot}1, f = 10$Hz) (% UTS)	70	0·02

Table 7.5 Dynamic response monitoring of 3-D SiC/SiC of composites, Kostopoulos et al. [18]. With kind permission of Elsevier

$\sigma_{max}/\sigma_{ult}$	*Maximum applied fatigue cycles (kcycles)*	$(E_{xx}^{c,e}/E_{xx}^{c,e}{}_o)$ *min*	$(n_{xx}^{e}/n_{xx}^{e}{}_o)$ *max*
0·7	1000	f_1: 0·96	2·213
		f_2: 0·93	3·012
		f_3: 0·89	4·258
0·75	174	f_1: 0·89	2·716
		f_2: 0·85	5·342
		f_3: 0·77	6·084

In conclusion, fatigue damage accumulation in 3-D Si/SiC composite was presented in this section. A non-destructive macromechanical approach was applied to evaluate the damage state by recording its dynamic response as function of fatigue cycles. The micromechanical fatigue behavior was monitored by acoustic emission approach. The fatigue damage produces a progressive reduction in the dynamic modulus in the first 30 kcycles. The damping coefficient (loss factor) is a sensitive and reliable measure of the fatigue damage state in the case of CMC such as SiC/SiC composite. In Table 7.4, n_{xx}^{e} is the effective loss factor and $E_{xx}^{c,e}$ is the effective complex bending modulus of elasticity. These values are defined in the derivation of the vibration of a linear viscoelastic beam (Kostopoulos et al. [18]).

7.5 Variable Amplitude Loading

In many practical applications, complex loading conditions occur and, in service, no fixed loading is encountered. Therefore, fatigue tests are performed under various conditions and cyclic patterns in order to understand the effects of non-uniform amplitude on fatigue life. To improve fatigue design of ceramic components it is necessary to conduct tests under variable stress amplitude loadings also not only under constant stress amplitude as has been done in most of the earlier investigations. Evaluation of fatigue life under variable stress amplitude is presented below and the analysis is based on the stable crack growth from an inherent flaw which is the fracture origin.

The analysis (following Hoshide et al.) is based on the postulate that stable crack growth per cycle, da/dN is correlated to the stress intensity factor, K as

$$\frac{da}{dN} = CK^n \tag{7.19}$$

$$K = \sigma(\pi a)^{1/2} M_0 \geq \mathrm{Kth} \tag{7.20}$$

Where a is a half crack length, σ is an applied stress, M_0 is a magnification factor determined by the crack geometry and loading system and K_{th} is a threshold stress intensity factor. Introduce a crack length, l defined as a M_0^2 and rewrite Eq. (7.19) as

$$\frac{\mathrm{d}l}{\mathrm{d}n} = C' K^n \tag{7.21}$$

where $C' = CM_0^2$. For the analysis to apply for small cracks also (not only large cracks) the material constant l_0 should be added for the evaluation of the stress intensity factor. Integrate Eq. 7.21 for the range from l_i to l_f leads to the relation of fatigue life N_f given as

$$\sigma^n N_f = A\left[(l_i + l_0)^{-q} - (l_i + l_0)^{-q}\right] \tag{7.22}$$

where A is given as

$$A = 1/(\pi^{n/2} C' q) \tag{7.23}$$

and

$$\mathrm{q} = (\mathrm{n} - 2)/2 \tag{7.24}$$

In Eq. (7.25) l_f presents the equivalent crack length at failure and is related to the critical stress intensity K_c as

$$l_f + l_0 = K_c^2/\pi\sigma^2 \tag{7.25}$$

It is presumed in the derivation of Eq. 7.22 that M_0 does not change during the process of crack growth to its final length of l_f. Now Eq. (7.22) is applicable to estimate crack propagation period to an arbitrary range of l such that $l_i \leq l \leq l_i$.

Variable tests are performed by applying a combination of various patterns of loading cycles. The combination of overloading (OL) and underloading (UL) cycles is such a pattern. It has been observed in metals for example that an UL applied immediately after OL reduces the post OL retardation more significantly than an UL which immediately precedes an OL. Usually, faster, transient fatigue-crack growth rate and, therefore, shorter crack propagation lives are observed. Figure 7.28 is an illustration of a plot of the crack length versus the applied cycles. Note that for a given crack length, the alloy can endure much larger cycles when an OL precedes loading marked by (a), with the next best case being (b), where an OL precedes an UL, followed by a baseline loading sequence. Coming back to the above relations, which are discussed for a repeated two-step loading condition. In one loading block

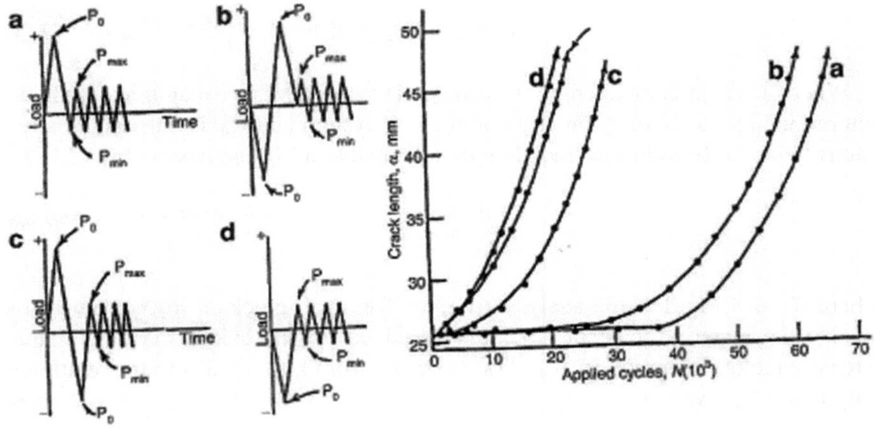

Fig. 7.28 Effect of UL position on post-OL retardation in 2024-T3 Al alloy (Scorupa 1998). With kind permission of John Wiley and Sons. From Joshua Pelleg, Mechanical Properties of Materials. Springer 2013, p. 396

described by a combination of low stress σ_L and subsequent high stress σ_H ($>\sigma_L$), the numbers of cycles for the low and high stresses are denoted by n_L and n_H, respectively. Let us consider several cases:

Case A: At the start of the crack growth, Eq. (7.20) is satisfied by both low and high stresses. When the initial and final values of equivalent crack length in the *k*th block are denoted by l_k and l_{k+1} as illustrated in Fig. 7.29, the fatigue life given by Eq. (7.22)

Fig. 7.29 Crack growth model under variable stress amplitude, Hoshide et al. [6]. With kind permission of Elsevier

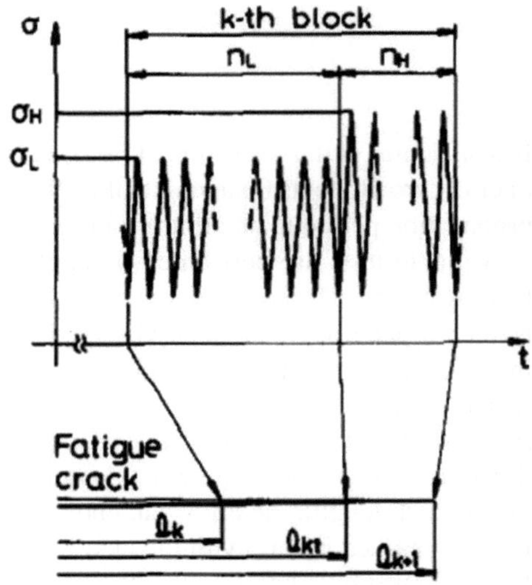

reduces to the number of cycles n_L and n_H for the crack growth corresponding to the two ranges of crack length, i.e., $l_k \sim l_{kt}$ and $l_{kt} \sim l_{k+1}$, as Eqs. (7.26) and (7.27) written in terms of the new ls, as

$$\sigma_L^n n_L = A\left[(l_k + l_0)^{-q} - (l_{kt} + l_0)^{-q}\right] \tag{7.26}$$

and

$$\sigma_H^n n_H = A\left[(l_{kt} + l_0)^{-q} - (l_{k+1} + l_0)^{-q}\right] \tag{7.27}$$

which is similar to Eq. (7.22) in terms of the current ls.

Summing Eqs. (7.26) and (7.27) results in

$$\sigma_L^n n_L + \sigma_H^n n_H = \mathrm{A}\left[(\mathrm{l_k} + \mathrm{l_0})^{-\mathrm{q}} - (\mathrm{l_{k+1}} + \mathrm{l_0})^{-\mathrm{q}}\right] \tag{7.28}$$

The number of blocks is denoted by Z. The number of blocks to failure, Z_f can be evaluated from Eq. (7.27) if each side is summed from 1 to Z with respect to k as

$$(\sigma_L^n n_L + \sigma_H^n n_H)\mathrm{Z_f} = \mathrm{A}\left[(\mathrm{l_i} + \mathrm{l_0})^{-\mathrm{q}} - (\mathrm{l_f} + \mathrm{l_0})^{-\mathrm{q}}\right] \tag{7.29}$$

where $l_i \equiv l_1$, and $l_f \equiv l_{Z_f+1}$ or l_f may be defined as the length of the final failure in the Z_fth block.

Another case where Eq. (7.20) is satisfied by the high stress but not by the low stress condition is at the start of crack growth. Two possibilities should be considered:

Case B: When the low stress is able to satisfy Eq. (7.20) during the crack growth process only due to high stress. When Z_L is defined as the number of blocks immediately before Eq. (7.20) is satisfied by the low stress, Eq. (7.22) gives Z_L in the relation (7.30)

$$\left(\sigma_H^n n_H\right) Z_L = A\left[(l_i + l_0)^{-q} - (l_L + l_0)^{-q}\right] \tag{7.30}$$

where

$$l_L + l_0 = K_{th}^2/(\pi\rho_L^2) \tag{7.31}$$

The period of the process after the low stress becomes active for the crack growth, $Z_f - Z_L$, is given as

$$\left(\sigma_L^n n_L + \sigma_H^n n_H\right)\left(Z_f - Z_L\right) = A\left[(l_L + l_0)^{-q} - \left(l_f + l_0\right)^{-q}\right] \tag{7.32}$$

Note that in Eq. (7.32) Z_f and l_i of Eq. (7.29) are replaced by $Z_f - Z_L$ and l_L, respectively. Eliminating in Eqs. (7.30) and (7.32) the term $(l_L + l_0)$ leads to Eq. 7.33

given as

$$\left(\sigma_H^n n_H\right) Z_f = A\left[(l_i + l_0)^{-q} - \left(l_f + l_0\right)^{-q}\right] - \left(\sigma_L^n n_L\right)\left(Z_f - Z_L\right) \quad (7.33)$$

Equation (7.33) can be expressed as (7.34)

$$\left(\sigma_L^n n_L + \sigma_H^n n_H\right) Z_f = A\left[(1 + \xi\eta)(l_i + l_0)^{-q} - \xi\eta(l_L + l_0)^{-q} - \left(l_f + l_0\right)^{-q}\right] \quad (7.34)$$

where $\xi = (\sigma_L/\sigma_H)^n$ and $\eta = n_L/n_H$. It could be noted that when the term $\xi\eta$ is eliminated one obtains Eq. (7.29).

Another case when Eq. (7.20) is not satisfied by the low stress during the whole crack growth process, the failure life is obtained by eliminating the second term on the right side of Eq. (7.33).

A further situation might occur if the initial size of the flaw is very small. When Eq. (7.20) is not satisfied even by high stress, the specimen is anticipated not to be broken until a sufficiently large number of cycles for example 10^7–10^8 cycles.

7.6 Experimental Observations

7.6.1 Partially Stabilized Zirconia (PSZ)

The fatigue life of partially stabilized zirconia ceramics is considered under variable stress amplitude loadings. The fatigue life formula was derived based on an assumption that the fatigue life is identical to the period required for the crack growth from a predominant natural flaw included in a specimen. The formulas were applied as described above (Hoshide et al.) to investigate the characteristics in fatigue life distribution under a variable stress amplitude condition. All fatigue tests were performed under 3-point bending with the span length of 20 mm in an ambient atmosphere. The loading pattern was chosen as a repeated two-step type. The cumulative probabilities, F the number of blocks to failure under several conditions is plotted in Fig. 7.30 for PSZ. The symbols with arrows represent unbroken specimens before 10^7 cycles of the total low and high cycles $(n_L + n_H)$Z, and the number with the arrow indicates the number of unbroken specimens at a specified test condition. The stress, $\sigma_L =$ 600 MPa is identical to a constant stress amplitude test. In Fig. 7.31 the Weibul distribution of the summation of the high cycles to failure, $n_H Z_f$ are indicated to investigate the effect of low stress set to be the fatigue limit on the fatigue life. The dot-dash lines present the fatigue life distributions under constant stress amplitudes equal to those of the high stress levels. The fatigue limit stress appears to affect the life distribution (this situation corresponds to case B in the above analysis), while σ_L

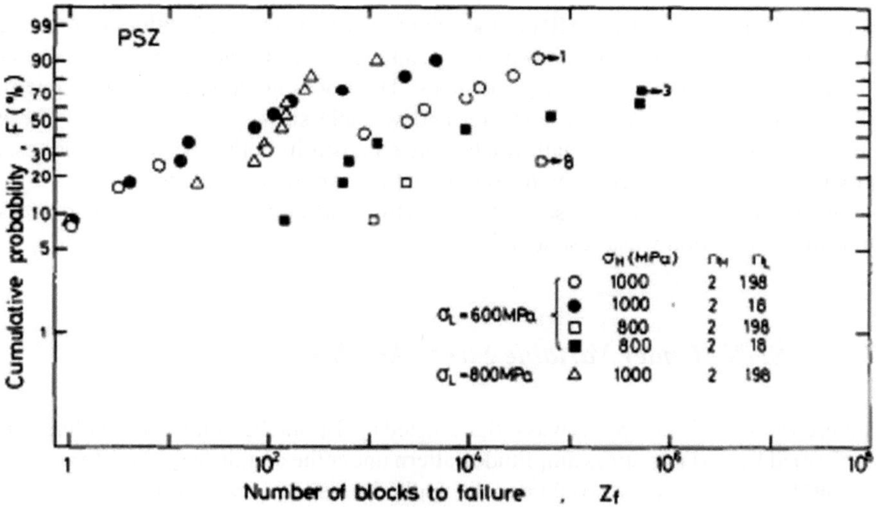

Fig. 7.30 Distribution of number of blocks to failure in zirconia, Hoshide et al. [6]. With kind permission of Elsevier

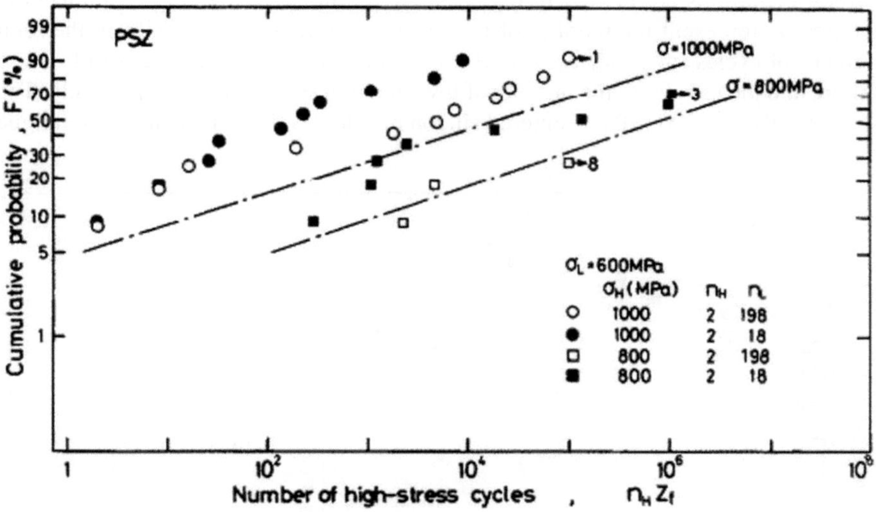

Fig. 7.31 Distribution of number of high-stress cycles in zirconia, Hoshide et al. [6]. With kind permission of Elsevier

$= 800$ MPa and $\sigma_H = 1000$ MPa in the figure may be associated with case A in the analysis above. Fractographic morphology in fatigue was found to be almost similar to that observed in constant stress rate tests. There was no significant difference in fracture morphology between constant and variable stress amplitude tests. These observations suggest that a main fracture mechanism in both types of fatigue tests should be the stable crack growth from a predominant inherent flaw.

In summary, the characteristics in fatigue life distribution under a variable stress amplitude condition was presented for PSZ.

7.6.2 Si_3N_4 Under Variable Stress Amplitude

Sintered silicon nitride (SSN) was fatigue tested by 3-point bending test. Loading was performed by variable stress amplitude pattern under the conditions of test frequency $F = 40$ Hz, stress ratio $R = 0.05$ and the applied stress wave was sinusoidal for both high stress and low stress loadings. The fatigue tests in a specified loading condition were carried out using 10–11 specimens.

The cumulative probabilities to failure, F of the number of blocks to failure under several conditions are plotted in Fig. 7.32. As in the case of PSZ, the numbers at the arrows represent the number of unbroken specimens at 10^7 cycles in the total number of cycles $(n_L + n_H)Z_f$. As mentioned earlier Z_f is the number of blocks to failure and the ns refer to the number of low and high stress cycles, respectively. The stress level $\sigma_L = 400$ MPa in sintered silicon nitride (SSN) is identical to the fatigue

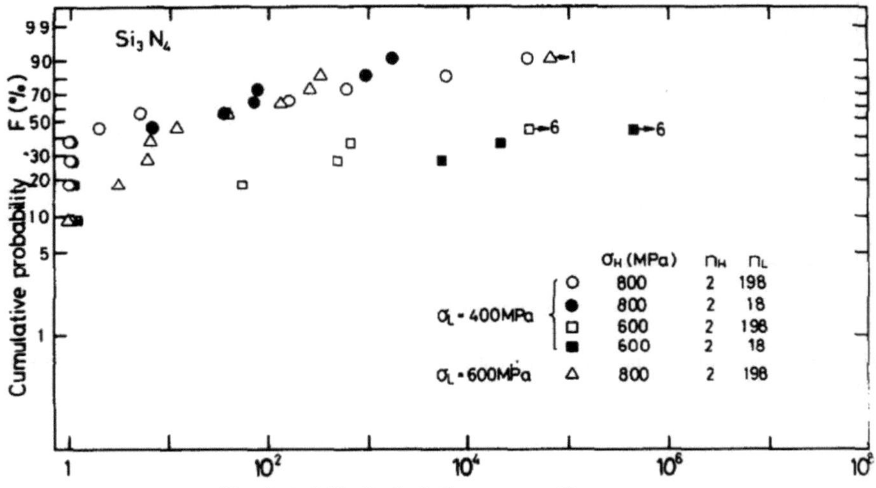

Fig. 7.32 Distribution of number of blocks to failure in silicon nitride, Hoshide et al. [6]. With kind permission of Elsevier

limit in the constant stress amplitude test. The summation of the high stress cycles to failure $n_H Z_f$, is indicated to investigate an effect of the low stress set to be the fatigue limit on the fatigue life. The dot-dash lines in Fig. 7.33 present the fatigue life distributions under constant stress amplitudes equal to those of the high stress levels.

Experimental and calculated results are compared in Figs. 7.34, 7.35 and 7.36, where the fatigue life is plotted as F versus the total number of cycles $(n + nH)Z_f$. The agreement is fairly good despite the small number of samples used in the experiments. Note that the analytical curves are deflected at the 3 points indicated by the circled numbers 1, 2, and 3. The first region up to point 1 indicates fracture during the first loading cycle due to low stress. In the region between 2 and 3 some specimens that survived in the first sequence of the low stress cycling, fail during the first loading cycle due to the subsequent high stress. The region above point 3 presents the ordinary life distribution.

It might be of interest to add that the fatigue resistance of some materials may be improved by understressing followed by a process of gradually increasing the amplitude of the alternating stress in small increments, a procedure called coaxing.

In summary, fatigue life was considered under variable stress amplitude loadings. It is assumed that fatigue life is identical to the period required for the crack growth from a predominant natural flaw present in a specimen. The test was performed under repeated two-step loadings. One should note that a fatigue threshold exists in fatigue deformation and therefore a threshold-based design approach is necessary.

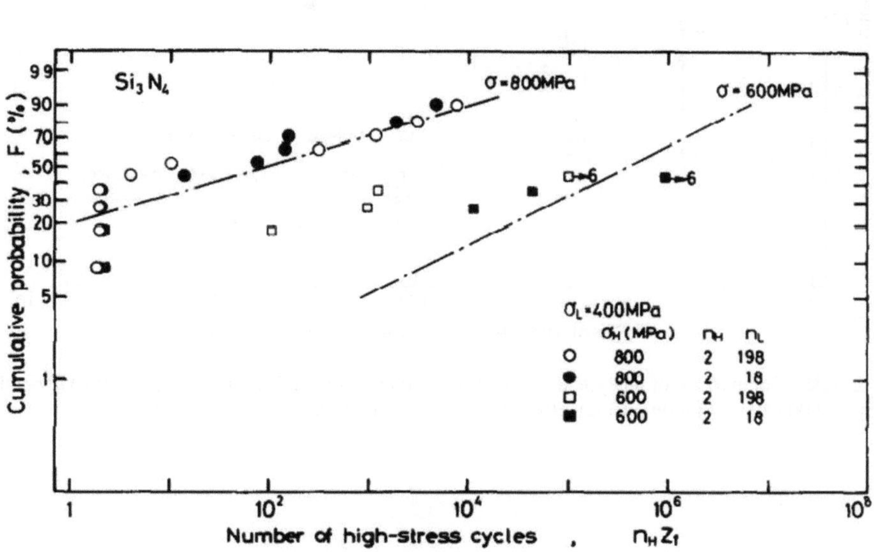

Fig. 7.33 Distribution of number of high-stress cycles in silicon nitride, Hoshide et al. [6]. With kind permission of Elsevier

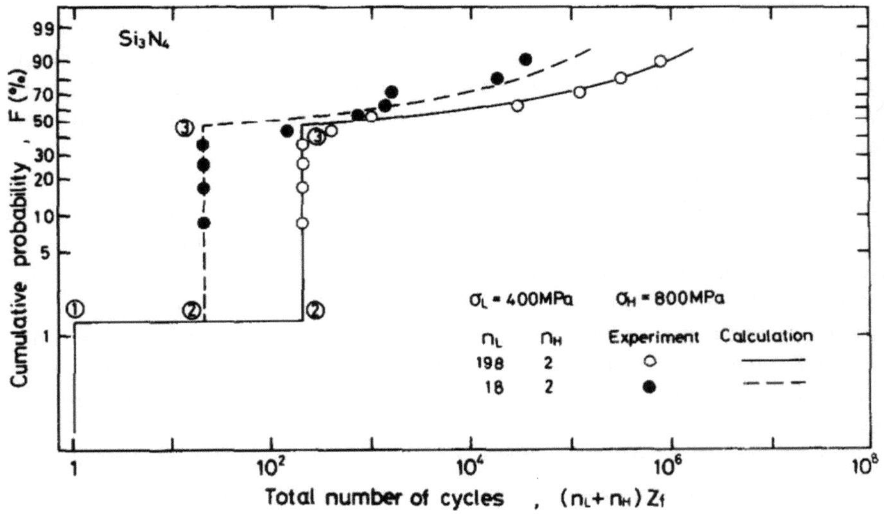

Fig. 7.34 Comparison of analytical distribution with experimental results for $\sigma_L = 400$ MPa and $\sigma_H = 800$ MPa in silicon nitride, Hoshide et al. [6]. With kind permission of Elsevier

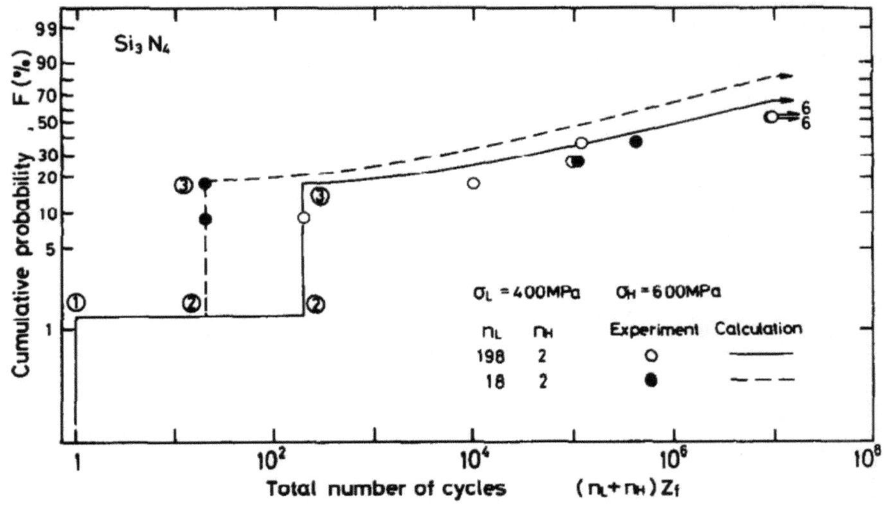

Fig. 7.35 Comparison of analytical distribution with experimental results for $\sigma_L = 400$ MPa and ($\sigma_H = 600$ MPa in silicon nitride), Hoshide et al. [6]. With kind permission of Elsevier

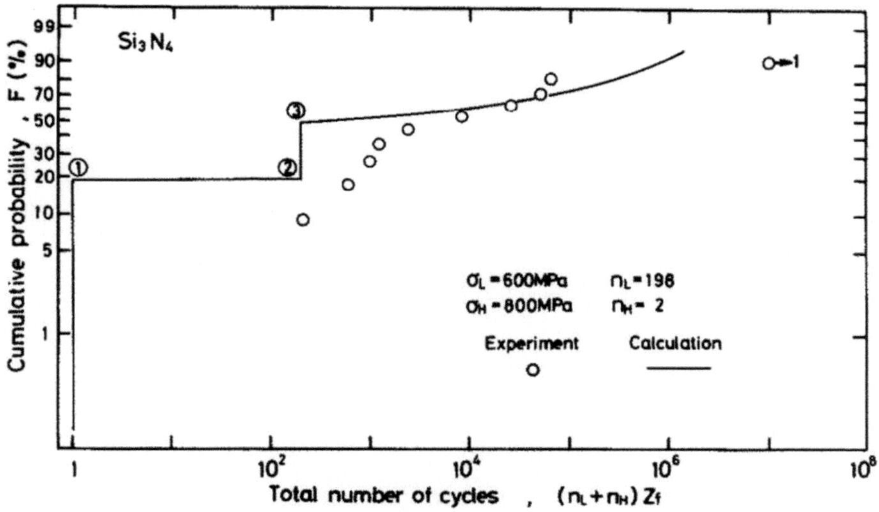

Fig. 7.36 Comparison of analytical distribution with experimental results for $\sigma_L = 600$ MPa and $\sigma_H = 800$ MPa in silicon nitride, Hoshide et al. [6]. With kind permission of Elsevier

References

1. R. Stephens, D. Chen, B. Hom, fatigue crack growth with negative stress ratio following single overloads in 2024-T3 and 7075-T6 aluminum alloys, in *Fatigue Crack Growth Under Spectrum Loads*, ed. by R. Wei, R. Stephens (ASTM International, West Conshohocken, 1976), p. 27–40
2. R.L. Carlson, G.A. Kardomateas, P.R. Bates, Int. J. Fatigue **13**, 453 (1991)
3. L.P. Borrego, J.M. Ferreira, J.M. Pinho da Cruz, J.M. Costa, Eng. Fract. Mech. **70**, 1379 (2003)
4. Z. Peng, H.-Z. Huang, J. Zhou, Y.-F. Li, Metals **8**, 456 (2018)
5. E. Takakura, S. Horibe, J. Mater. Sci. **27**, 6151 (1992)
6. T. Hoshide, T. Ohara, T. Yamada, Eng. Frac. Mech., **31**, 339 (1988)
7. W. Geary, Int. J. Fatigue **14**, 377 (1992)
8. D.S. Dawicke, Overload and Underload Effects on the Fatigue Crack Growth Behavior of the 2024-T3 Aluminum Alloy, NASA Contractor Report 201668
9. V. Zitounis, P.E. Irving, Int. J. Fatigue **29**, 108 (2007)
10. T.H. Topper, M.T. Yu, Int. J. Fatigue **7**, 159 (1985)
11. P.C. Paris, F. Erdogan, A critical analysis of crack propagation laws. J. Basic Eng. Trans ASME (Ser D) **85**, 528 (1963)
12. M.A. Miner, Cumulative damage in fatigue. J. Appl. Mech. **12**, 159 (1945)
13. A. Palmgren, Die Lebensdauer von Kugellagern. Z. Vereins Deut. Ingenieure **68**, 339 (1924)
14. P.C. Paris, H. Tada, J.K. Donald, Int. J. Fatigue **21**(Supplement), S35 (1999)
15. O.H. Basquin, *The Exponential Law of Endurance Tests* (American Society for Testing and Materials, Phildelphia PA USA, 1910)
16. M.A. Miner, J. Appl. Mech. **12**, 159 (1945)
17. Y. Zhang, B.R. Lawn, K.A. Malament, van P. Thompson, E. Dianne Rekow, The Inter. J. Prosthodontics **19**, 442 (2006)
18. V. Kostopoulos, Y.Z. Pappas, Y.P. Markopoulos, J. Eur. Cer. Soc., **19**, 207 (1999)

Chapter 8
Structural Observations in Fatigued Specimens

Abstract The resistance of structural ceramics, such as oxides, carbide or nitrides to exposure of cyclic loading determine their life. The cyclic deformation process starts with dislocation movements at the microscopic level forming persistent slip bands which become the nucleus of short cracks. The fatigue fracture occurs at three stages, namely (I) initiation, (II) propagation and (III) final rupture. The structural damage in components can occur in stress range far below the static material strength. The damage is not reversible and do not recover when rested (while not in service). Progression markings, known also as beach marks are observed on fractured surface. Striations represent the advance of the crack front by one load application in many ductile metals, whereas beach marks locate the position of the crack front when repetitive. Among the typical fracture surface markings for brittle materials are Wallner lines, which are striations from crack propagation along the crystal. Deformation by dislocations is the basic understanding in ambient temperature in alloys requiring some plasticity. High temperature deformation also of ceramics by dislocation explains plastic deformation. Thus, alumina at high temperature deforms by dislocation mechanism. Also, twin deformation is involved. MgO single crystal can even deform up to a few percent plastic strain due to dislocation processes which depend on parameters such as pressure, temperature and strain rate. Further, dislocations are involved in deformation of partially and fully stabilized zirconia, B_4C and Si_3N_4 as proven by TEM.

Keywords Beach marks · Striations · Dislocations · Alumina · Partially and fully stabilized zirconia · B_4C · Si_3N_4

8.1 Introduction

All material, among them structural ceramic components in service are often subjected to some sort of cyclic mechanical loading. The resistance of structural ceramics, such as oxides, carbide or nitrides to exposure of cyclic loading determine their life. When no visible macroscopic or microscopic discontinuities (or flaws) are observed in metals and alloys, the cyclic deformation process starts with dislocation movements at the microscopic level forming persistent slip bands which become the

J. Pelleg, *Cyclic Deformation in Oxides, Carbides and Nitrides*, Structural Integrity 22,
https://doi.org/10.1007/978-3-030-86118-6_8

nucleus of short cracks. Let us thus re-emphasize cyclic fatigue as the stress, strain and deformation induced in the material which after some specific loading cycles produce ultimately some sort of failure (rupture, fracture, disintegration, breakage etc.) in the material. The fatigue fracture occurs at three stages, namely (I) initiation, (II) propagation and (III) final rupture. Regarding these stages:

(1) During a large number of cycles damage develops on microscopic level and growth until a macroscopic crack is formed.
(2) Growth of the macroscopic crack until a critical size (length) is reached.
(3) Component breaks because it can no longer sustain the load.

Note, however, that in certain cases the second stage cannot be observed, and instead a microscopic crack grows rapidly inducing rapid failure. The structural damage—the most common failure in components—can occur when the stress range is far below the static material strength. The applied stresses may be tensile, compressive or torsional; crack initiation and propagation are due to the tensile component.

Proper understanding of fatigue and its characterization of structural ceramics is crucial for safe use of components in service due to their applications as gas turbine rotors or blades and diesel engine valves. Nevertheless, fatigue components designers have to be thoroughly aware that macroscopic and microscopic discontinuities and features, such as holes, slots, sharp changes of load direction, etc. are locations of stress concentrations where the fatigue process begins. Generally, the damage inflicted is not reversible and therefore—as observed in testing specimens—do not recover when rested. Not only the number of cycles endured by a component determines its life, but the stress range also is an important factor to be considered with the allowable limit of its level. Thus, clearly, the greater the applied stress range, the shorter the life. Further, fatigue is influenced by a variety of factors such—just to mention some—temperature, surface finish, microstructure, grain size, environment, etc.

In the next sections progression markings and dislocation will be discussed of the systems relevant to this book, clearly with the limited availability in the open literature.

8.2 Progression Markings, (Beach Marks) and Striations

Fatigue failures in metals display progression markings known as beach marks (sometimes called clamshell marks). These are macroscopic progression marks on a fatigue fracture surface that indicate successive positions of the advancing crack front. Striations represent the advance of the crack front by one load application in many ductile metals, whereas beach marks locate the position of the crack front when repetitive, or when fluctuating loading is stopped for a period of time. Striations are extremely small ridges, visible only with an electron microscope. Although it has

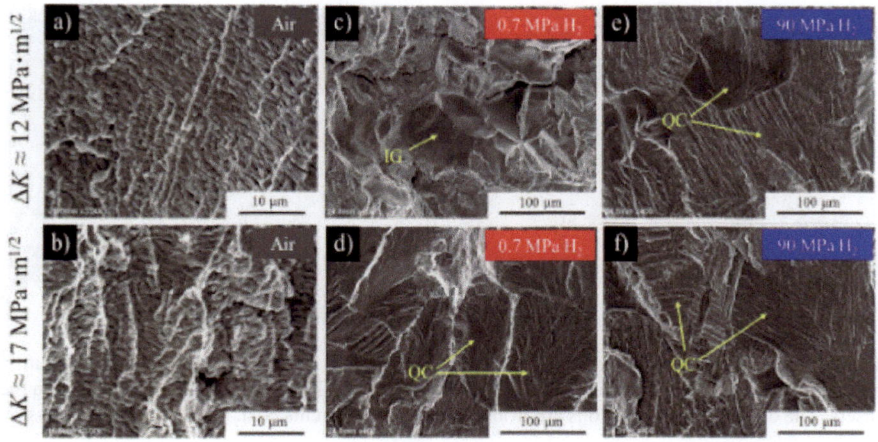

Fig. 8.1 SEM images of the fracture surfaces of the specimens subjected to the ΔP-constant tests in **a** and **b** air, **c** and **d** 0.7 MPa H_2, and **e** and **f** 90 MPa H_2 at different ΔK values. The crack growth directions are from bottom to top. Birenis et al. [1]. With kind permission of Elsevier

been suggested by some that many loading cycles are required to form a single striation, it is now generally thought that each striation is the result of a single loading cycle. In addition to the above difference, the major difference between striations and beach marks is in their size, as a matter-of-fact beach marks are much larger than striations. If they are present, they are normally visible by the naked eye. One can distinguish between fatigue striations and striation-like artifacts on a fractured surface in that the former, -namely true fatigue striations-, never cross or intersect one another.

Typical SEM fracture surface of specimens in air and hydrogen subjected to ΔP constant tests, showing ductile narrow striations are shown in Fig. 8.1. Clearly ΔK is the stress intensity stress intensity factor. Ductile transgranular features in steel with narrow ductile striations covered the entire fracture surface in the case of air as seen in (a) and (b). On the other hand, the fracture surfaces formed in hydrogen gas exhibited a combination of ductile transgranular and intergranular (IG) fracture features in the Stage I regime and quasi-cleavage (QC) features decorated with brittle-like striations in the Stage II regime seen in Fig. 8.2. Prior to failure the fatigue crack nucleates at stress concentrations. Generally, the fatigue fracture surface is perpendicular to the direction of the applied stress. Typical fatigue-fracture surface with progression markings (beach marks) looks like the one shown.

In Fig. 8.3. It is a SEM image of an Al alloy fractured surface. The fatigue crack nucleates at the stress concentration (an inclusion vicinity is also a site of stress concentration). Generally, the fatigue fracture surface is perpendicular to the direction of an applied stress.

A fatigue failure—as observed in metals such as steel—can be recognized from the appearance of the fracture surface, which shows a smooth and polished surface

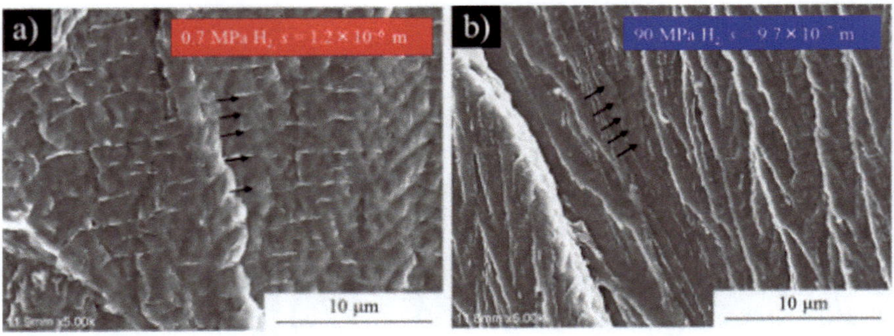

Fig. 8.2 SEM images of the brittle-like fatigue striations in **a** 0.7 MPa H_2 at $\Delta K = 17$ MPa $m^{1/2}$ and **b** 90 MPa at $\Delta K = 16$ MPa $m^{1/2}$ (Stage II), together with the average striation spacing, *s* in the insets. The crack growth directions are from bottom to top. Birenis et al. [1]. With kind permission of Elsevier

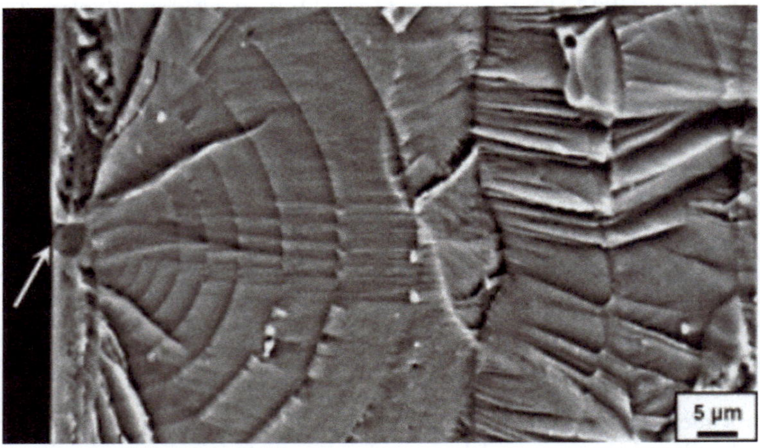

Fig. 8.3 SEM of fracture surface of an Al alloy showing pronounced progression markings (beach marks) produced by repeated spectrum-loading programs incorporating a single, much larger-than-average compressive load. Note that the markings radiate from a surface-breaking inclusion (arrowed) and become less pronounced after the crack has crossed a grain boundary into a differently oriented grain. Lynch [2]. With kind permission of Elsevier

that corresponds to the slow growth of crack, when the crack faces smoothen out by constant rubbing against each other and a rough/granular region corresponds to the stage of fast growth, after critical conditions is attained where member has failed in a ductile manner when cross section was no longer able to carry the applied load. The region of a fracture surface that formed during the crack propagation step may results in characteristic pattern of concentric rings spread over the smooth region of the fracture surface, known as beach marks or striations, radiating outward from the point of initiation of the failure. In Fig. 8.3 the arrow indicates an inclusion the

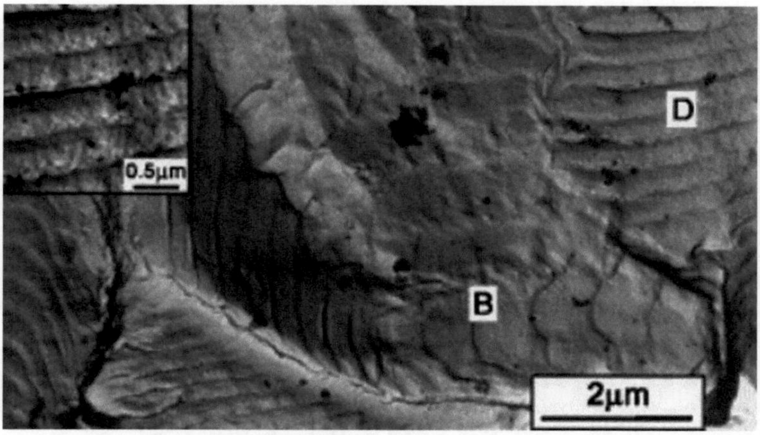

Fig. 8.4 TEM of direct-carbon replica of CA fatigue fracture surface of an age-hardened high-purity Al–Zn–Mg alloy showing striations. Adjacent areas exhibit 'ductile' (D) striations and 'brittle' striations (B) with widely differing spacings and orientations. The fine-scale roughness (inset) superimposed on striations may be associated with crack advance by a nano-void coalescence process. Lynch [2]. With kind permission of Elsevier. CA stands for constant amplitude

location from where the progression markings radiate out. Each beach mark band represents a period of time over which crack growth occurred. On the other hand,—as mentioned above and worth reemphasizing to understand more deeply the concept—fatigue striations are microscopic in size and subject to observation with the electron microscope, either TEM or SEM. The relatively widely spaced striations are caused by variations in the stress amplitude during the life of the component. On a much finer level, a large number of striations may be sometimes being seen. The width of each striation here is equal to the distance by which the crack grows during one cycle. Any point with stress concentration such as sharp corner or notch or metallurgical inclusion can act as point of initiation of fatigue crack. Figure 8.4 shows on a fracture surface region of a ductile and a brittle failure region of striations. It is an accepted concept that conventional fatigue process in ceramics ends by brittle fracture in the final stages of crack growth. Unstable crack initiation in this case is feasible due to the relatively brittle nature of ceramics.

8.2.1 Yttria Stabilized Zirconia (YSZ)

Microstructures of SEM images of beach marks are illustrated in Fig. 8.5a and b of YSZ on the fractured surface after loading. The beach marks show the progressive fatigue failure of the zirconia specimen. The apparent fracture line is visible in (b) and (c) alongside the beach marks which has occurred during the brittle failure of the material. The zirconia was fully dense after proper sintering as seen in (d) with

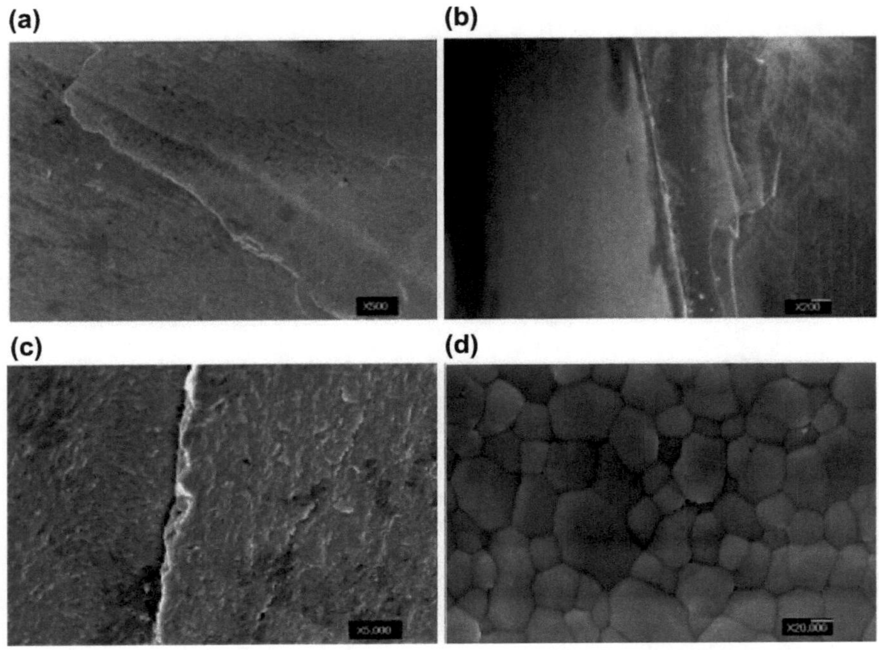

Fig. 8.5 Scanning electron microscopic images showing fracture condition of broken segments. **a**, **b**, beach marks show fracture segments from the specimen on SA25 (original magnification × 500) and SA50 (original magnification × 200). **c**apparent cracks observed on SA50 fracture segments (original magnification × 5000). **d**, intact surface of zirconia specimens (original magnification × 20,000). Lan et al. [3]. Open access

possible small pores present at grain boundaries. Since zirconia is use in dentistry, occlusal surface areas of implant abutment are indicated above by SA 25 and SA 50 (Recall that occlusion is defined as the way the teeth meet when the lower jaw (mandible) and upper jaw (maxilla) come together. It is how the teeth contact in any type of functional relationship). SA 50 implies that the total occlusal surface contact area is half of SA100. The fracture resistance of the implant abutment of the zirconia sintered at low temperatures for dental use was investigated by servo-hydraulic testing machine and SEM. The fracture resistance of monolithic zirconia specimens is primarily affected by their thickness. The minimum occlusal thickness of all specimens was 0.5 mm. Thicker zirconia specimens had higher fracture resistance.

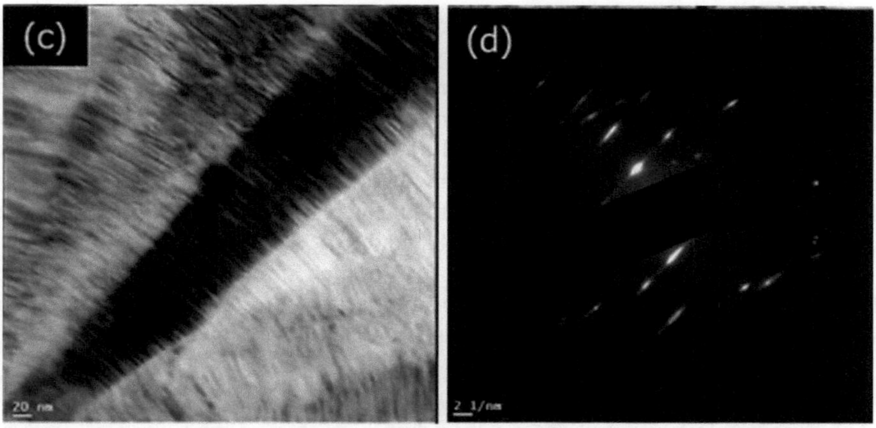

Fig. 8.6 **c** a single grain showing high contrast indicating different orientations and horizontal striations caused by stacking faults in the lattice which were confirmed in the diffraction pattern seen in **d**. J. Binner, UNIVERSITY OF BIRMINGHAM EDGBASTON BIRMINGHAM, B15 2TT GB, Final Report for FA9550-15-1-0465, 2015–2019, unlimited distribution

8.2.2 Striations in SiC/SiC Composite

As indicated earlier SiC is a reinforcing agent in various material including SiC matrix itself, and therefore most if not all information in the free literature relates to various type of composites. Here, such a SiC/SiC composite is presented as an example for striations through the grains and the corresponding diffraction pattern (Fig. 8.6).

8.2.3 Striations in Si_3N_4

Two types of silicon nitrides were subjected to fatigue tests. The APB silicon nitride doped with 6% yttria and 3% alumina had an amorphous grain boundary phase, while the CBP doped with 8% yttria and 1% alumina) had predominantly a crystalline grain boundary phase. The grain boundary phase of CBP refers to the phase that resides between pure silicon nitride, where all the dopants reside. The difference in fatigue lives, at high strain rates, between the two types of silicon nitride can be attributed to dislocation activity and subsequent cracking of the crystalline grain boundary phase in the CBP material. The crystalline nature of the grain-boundary phase promotes the propagation of a single crack, while, ABP samples showed no dislocation activity within the grain boundary phase, which enhances intergranular fracture around the amorphous boundary phase resulting in fragmentation. Surface flaws grow during each cycle—also in APB Si_3N_4—since the surface is the most favorable site for microcracks nucleation at contact regions between small and large grains. This process causes fatigue striations formation. During the applied load, the stresses in the contact region become very large, resulting in plastic deformation of the larger grain, which may lead to its transgranular failure. SEM micrographs in Fig. 8.7b illustrate striations in Si_3N_4. In Fig. 8.7a the fracture micromorphology consisted in APB both intergranular and transgranular fracture modes with grater propensity than in CPB for intergranular fracture. CPB Si_3N_4 generally failed by axial splitting. Figure 8.7b shows large fragments recovered from APB Si_3N_4 with areas where striations can be seen (Figs. 8.8 and 8.9).

In summary, the difference in fatigue lives at high strain rates between the two kinds of Si_3N_4 can be attributed to dislocation activity (to be discussed later) and subsequent cracking of the crystalline grain boundary phase in CPB material.

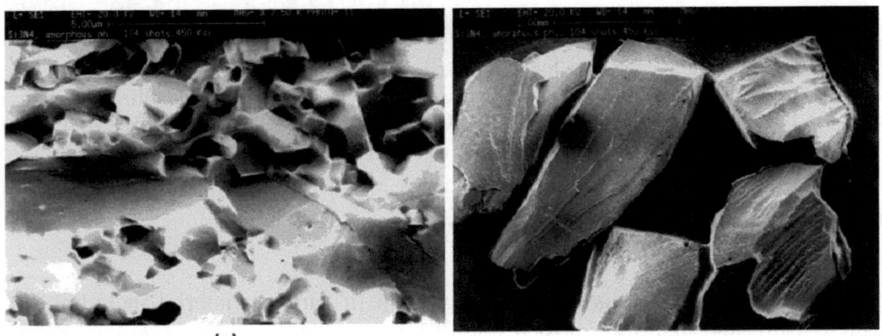

(a) (b)

Fig. 8.7 SEM micrographs showing **a** fracture surface in ABP silicon nitride showing mixed mode of fracture with a greater propensity for intergranular fracture, and **b** lower-magnification view of fatigue failure. Only the largest fragments are shown. The large number of fragments compared with failure of CBP silicon nitride (Fig. 8.10) indicates the prior presence of multiple cracks. Sharma et al. [4]. With kind permission of John Wiley and Sons

Fig. 8.8 Fracture surface of a piece of polycrystalline alumina. Wallner lines, twist hackle, and gull wings are seen on the fracture surface of the large grain in the middle of the image. Fracture moved through this grain (transgranular fracture). Fracture around grains (intergranular fracture) is seen at the top of the image. SEM; picture width ~ 50 μm. Varner [5]

Fig. 8.9 Fracture surface of a piece of polycrystalline alumina. The large grain in the center broke by transgranular fracture (note the fracture markings on its fracture surface). Most of the rest of the fracture shown in this image is intergranular. SEM; picture width ~ 50 μm. Varner [5]

8.3 Wallner Lines in Alumina

Among the typical fracture surface markings for brittle materials are Wallner lines, which are striations from crack propagation along the crystal. Fracture through grains is transgranular fracture. Examples are shown for alumina in Figs. 8.8 and 8.9. Transgranular fracture might occur when grain boundaries are strong in comparison with the strength of grains or when grain boundaries of cleavage planes match up with each other, namely when misorientations between them are small. In transgranular fracture the surfaces are relatively smooth as seen in Figs. 8.8 and 8.9. The fracture through grains leads to fracture markings called Wallner lines.

8.4 Dislocation Structure

8.4.1 Introduction

The concept of the role of dislocations in deformation making metals deform plastically, has been substantiated over almost two centuries (since 1934), as the solid base which makes them the ground breaking idea of explaining mechanical and physical properties. Thus, some ceramics which are hard and brittle are unlikely to deform plastically at ambient temperatures, which requires some sort of plasticity. Oxides fall into this category of ceramics, although exceptions exist. For example, MgO deforms by $\{1\bar{1}0\}\langle 110\rangle$ glide. However, in high temperature testing of ceramics, the ability to deform oxides—such as alumina for example—has been shown that the dislocation plays an important role in the process. Moreover, other, none oxide brittle ceramics, at high temperature deformation show some degree of plasticity and the role of dislocations emerges as the main reason for their plastic respond to applied stresses. Thus, high temperature plastic deformation of ceramics opened the field of dislocations theories in ceramics also to explain plastic deformation. The application of TEM to study dislocations in ceramics has been an important step in understanding their role in the deformation of ceramics.

In this section information on oxide, carbide and nitride ceramics are considered relevant to their dislocation presence and deformation activity.

8.4.2 Dislocations in Alumina

Alumina (α-Al_2O_3) is one of the important high-temperature structural materials and dislocations in alumina play an important role in its plastic deformation. As mentioned above it has been established that deformation in metals occurs by the generation and migration of dislocations in the lattice. Ceramics brittle in their nature, including alumina show a brittle-ductile transition temperature where plastic deformation can occur. The absence of ductility at room temperature does not mean that dislocations are absent in some brittle ceramics like alumina. They are there, but inactive in the deformation until some incentive-such as temperature rise during some mechanical operation-induces plasticity. Further, hydrostatic pressure also was observed to induce plastic deformation in sapphire (single crystal of Al_2O_3) at much lower temperatures than the brittle-ductile transition temperature, for prism plane slip at ~ 200 °C and for basal slip at ~ 400 °C. However, Hockey and Lawn reported that deformation was already induced at room temperature in alumina using a sharp indenter. The present author believes that during the indentation a temperature rise could have occurred, also an applied pressure has been involved inducing the observed plastic deformation in the tests of Hockey and Lawn during the indentation with the sharp indenter. Two different grain sizes of a polycrystalline alumina under different scratching conditions were performed following sintering in air in 1700 °C for 10 h. Thus, for the scratching experiments, two types of indenters were used, one a blunt conical indenter of about 100 μm tip radius and a sharp Vickers pyramid indenter with an included angle of 136°. Thus, different scratching conditions were generated. The normal load on the polished specimen surface was 6 N and the sliding speed of the indenter was 6 mm s^{-1}. Clearly the scratching was performed on the polished specimens. The scratching induced twins also in addition to the dislocations.

The structure of alumina before scratching is shown in Fig. 8.10 and as seen it is almost dislocation free. The topography of the groove formed by the sharp indenter is shown in Fig. 8.11. The grains are fractured during scratching (even dislodged) (a) and the damaged region is clearly seen by SEM in (b). It is characterized by a high density of cracks between the grains. TEM indicates in Fig. 8.12a, a severely deformed zone near the bottom of the groove where a large number of dislocation and twins are observed and a large number of intergranular and transgranular cracks are seen; it is a plastic zone. The zone spreads to a depth of about 20–40 μm from the surface of the groove. In order to resolve the individual dislocations, a region of low density dislocations were chosen for TEM identification. The groove generated by the blunt indenter was very smooth as illustrated in Fig. 8.13a. Basal dislocations in the plastic zone are illustrated in Fig. 8.14. In Fig. 8.14b, c and d the dark field images

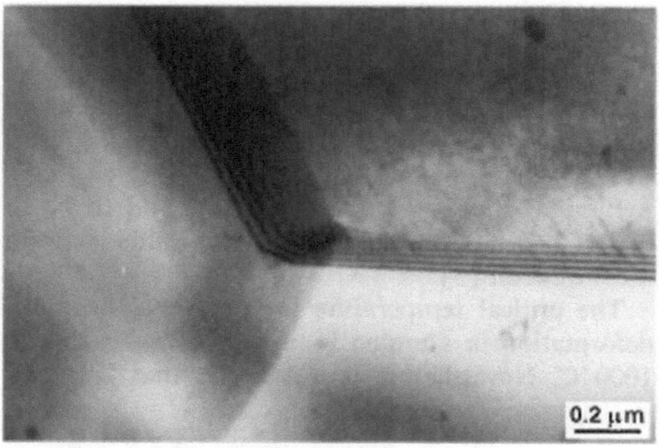

Fig. 8.10 The initial microstructure of alumina before scratching. Zarudi et al. [6]. With kind permission of Springer Nature

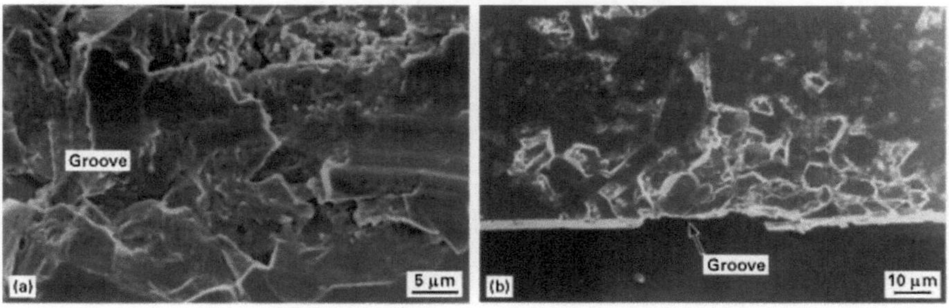

Fig. 8.11 Alumina after scratching with a sharp indenter; **a** topography of the groove; **b** cross-section view. Zarudi et al. [6]. With kind permission of Springer Nature

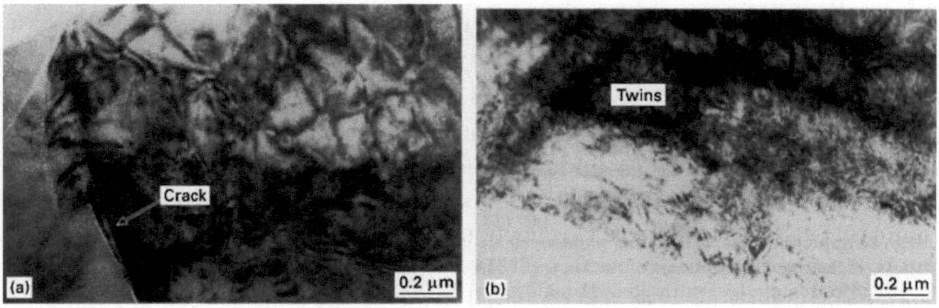

Fig. 8.12 Structure of the plastic zone after scratching with a sharp indenter; **a** highly deformed zone under the groove with cracks; **b** dislocations and twins in the plastic zone. Zarudi et al. [6]. With kind permission of Springer Nature

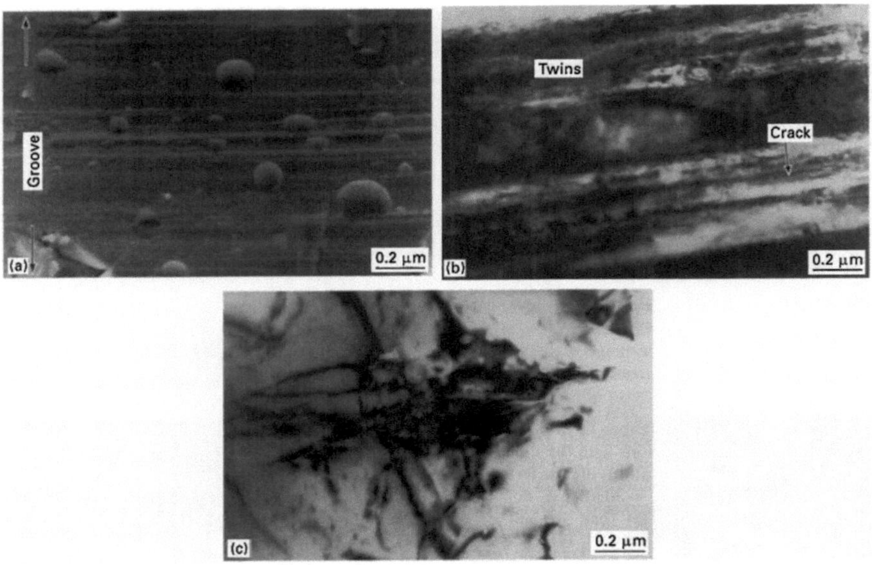

Fig. 8.13 Alumina after scratching with a blunt indenter: **a** topography of the groove; **b** twins in the plastic zone; **c** dislocation structure of the plastic zone. Zarudi et al. [6]. With kind permission of Springer Nature

of the same region were obtained by using the reflections $g = \overline{3}300$, $g = \overline{3}300$ and $g = 03\overline{3}0$, respectively. The direction of the Burgers vector of a type A-A dislocation is $\left[11\overline{2}0\right]$.

Since the dislocations lie in the (0001) basal plane, the dislocations are basal dislocations with $(0001)\frac{1}{3}\left[11\overline{2}0\right]$ slip system. The B-B dislocations are basal dislocations with with Burgers vector $\left[1\overline{2}10\right]$ and the slip system deduced to be $(0001)(0001)\frac{1}{3}\left[1\overline{2}10\right]$. Thus it seems that after scratching two different basal slip systems exist in alumina. (note that dislocations of type A-A are out of contrast); (c) dark-field image in reflection $30\overline{3}0$ (note that dislocations of type B-B are out of contrast); (d) dark-field image in reflection $03\overline{3}0$ (note that dislocations of type A-A and B-B are seen clearly).

Deformation twins were also observed in the plastic zone as seen in Figs. 8.12b and 8.13b. Diffraction pattern analysis proved that there are two types of twins. The first type twin is shown in the diffraction of Fig. 8.16b. The matrix zone axis is $\left[11\overline{2}0\right]$. It has been proven by diffraction pattern that the twin plane is (0001). Since the shear direction of the twin is perpendicular to the electron bean it must be $\left[\overline{1}100\right]$ a basal twin axis of matrix is $\overline{5}7\overline{2}3$ and the reflection of the matrix and twin in Fig. 8.16c are with a rotation of 180° about axis X_1 Y_1. This means that the direction of electron mean is parallel in the twin plane, namely, $K_1 = \left(10\overline{1}1\right)$. Thus, the twin is rhombohedral. Note that the cracks were initiated at the intersection of the basal and rhombohedral twins as seen in Fig. 8.16a there was a temperature rise in the zone undergoing plastic deformation by the scratch test. Together with the stress field in the vicinity of the cutting edge probably activated the dislocations and twins.

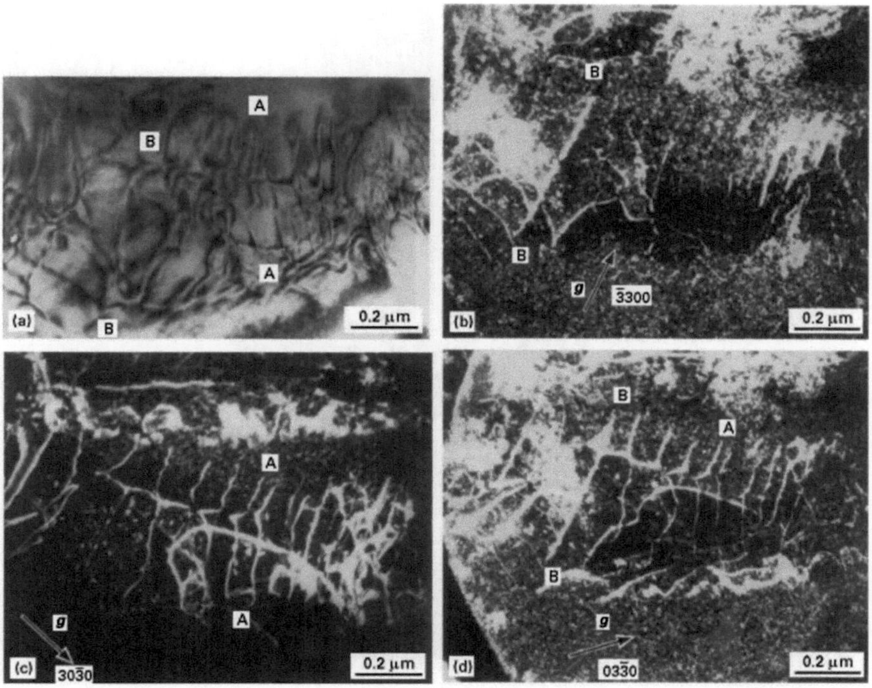

Fig. 8.14 Basal dislocations in the plastic zone: **a** bright-field image under multiple beam conditions (note dislocations of type A-A and B-B); **b** dark-field image of reflection $\bar{3}300$ (note that dislocations of type A-A are out of contrast); **c** dark field image in reflection $30\bar{3}0$ (note that dislocations of type B-B are out of contrast); **d** dark-field image in reflections $03\bar{3}0$ (note that dislocations of types A-A and B-B are seen clearly), Zarudi et al. [6]. With kind permission of Springer Nature

In summary, during single point scratching of alumina, slip and twin systems have been observed at room temperature. The five slip and twin systems are: basal slip systems (two), $(0001)\frac{1}{3}[11\bar{2}0]$ and $(0001)\frac{1}{3}[1\bar{2}10]$, one prism plane slip system, $\{11\bar{2}0\}\bar{1}100$, one basal twin with $K_1 = (0001)$ and shear direction $\eta_1 = [\bar{1}100]$, and one rhombohedral twin system with $K_1 = (10\bar{1}1)$ and $\eta_1 = [\bar{1}012]$. Macroscopic slip and twin systems enable room temperature deformation in alumina. Temperature rise during scratching and the resulting stress field in the vicinity of the cutting edge activated the dislocations and twis inducing plastic deformation. Note that K_1 is the twin composition plane.

8.4.3 Dislocations in Magnesia

MgO is a typical ionic ceramic with refractory properties. As a single crystal MgO can even deform up to a few percent plastic strain due to dislocation processes which

Fig. 8.15 Prism plane dislocations in the plastic zone: **a** bright-field image under multiple-beam conditions of types C–C and D-D; **b** dark-field image in reflection $\overline{2}300$; **c** dark-field image in reflection $11\overline{2}0$ (note that the dislocations of type C–C are out of contrast). Zarudi et al. [6]. With kind permission of Springer Nature

strongly depend on external parameters such as pressure, temperature, strain rate etc. MgO single crystals were deformed by rolling contact with a steel ball. The rolling-contact state of stress produces a high density of dislocations in a localized zone. Dislocation interaction in this zone produces cleavage-type cracks after a large number of rolling-contact cycles. The orientation of the crystal influences the character of dislocation accumulation. The triaxial state of stress produced in the contact region of rolling contact elements results in local plastic deformation and damage. The damage is highly localized and progressive with repeated stress applications, ultimately leading to spalling failures much like fatigue-initiated pitting damage in bearings and gears. The technological importance of rolling contact is in that it can terminate the service life of the bearings and gears before failure by wear. The rolling-contact stress was applied to the crystals with a hardened steel ball mounted on a shaft held between conical pivot bearings as illustrated schematically in Fig. 8.17.

Following an experiment, the specimens were cleaved to study subsurface deformation. The dislocation arrays were revealed by etching the crystal. This etchant produces microscopic pits where dislocations emerge at the surface of the crystal.

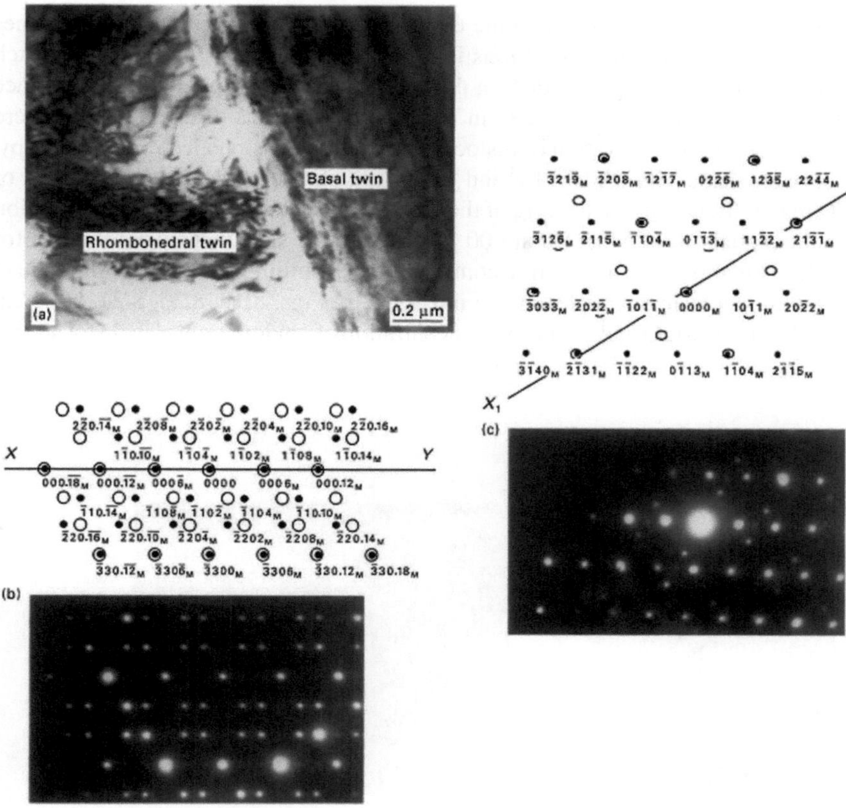

Fig. 8.16 Twins in the plastic zone after scratching: **a** bright field image; **b** diffraction pattern from the basal twin ((•), reflection of the matrix which are indexed; (○), reflections of twin); **c** diffraction pattern from the rhombohedral twin ((•), reflexion of the matrix which are indexed; (○), reflections of twin). Zarudi et al. [6]. With kind permission of Springer Nature

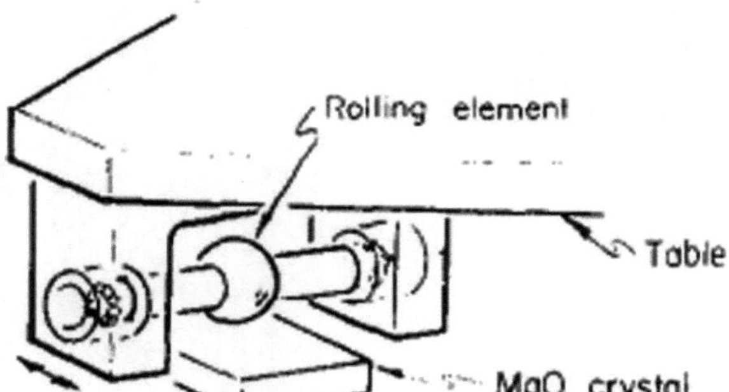

Fig. 8.17 Configuration for applying rolling-contact stress. Durfane and Glaeser [7]. With kind permission of Elsevier

The arrays of pits so produced define the surface intersection of easy glide planes where slip has occurred. Dislocations lying on these planes form the rows of etch pits normal to the rolling direction on the crystal surface and parallel to the surface on the crystal cross section as seen in Fig. 8.18. Thin films of crack regions were prepared for direct observation of dislocations by transmission electron microscopy. Tracks were made with 10, 10^2, 10^3 and 10^5 rolling-contact cycles at a ball velocity of 5.98 cm s^{-1}. To permit the locking of the dislocations, prior to specimen preparation the bulk crystals were annealed at 600 °C for 3 h. The specimens were thinned for TEM investigation. The dislocation configurations resulting from the rolling-contact stress with increasing cycle levels are illustrated in Figs. 8.19, 8.20, 8.21 and 8.22. The undeformed MgO had a very low concentration of random grown-in dislocations,

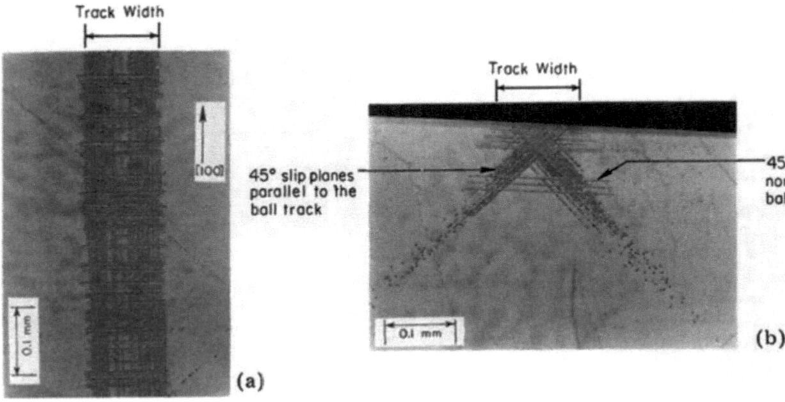

Fig. 8.18 Dislocation etch-pit patterns after one rolling-contact pass in the [1001] rolling direction at a 570 g load (rolling direction normal to the page): **a** surface; **b** cross section. Durfane and Glaeser [7]. With kind permission of Elsevier

Fig. 8.19 Dislocation configuration after 10^3 rolling-contact cycles with 570 g load. Durfane and Glaeser [7]. With kind permission of Elsevier

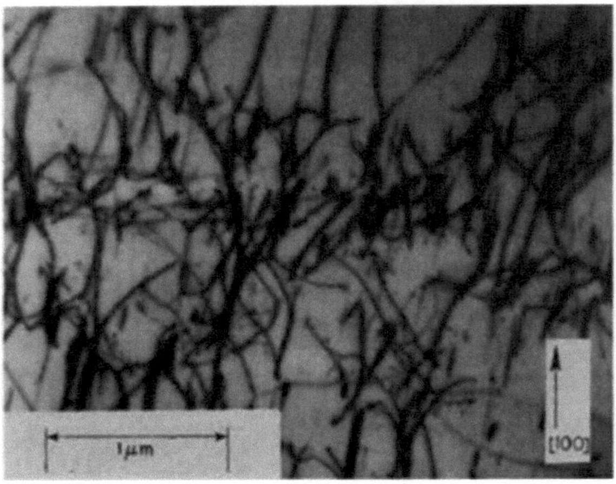

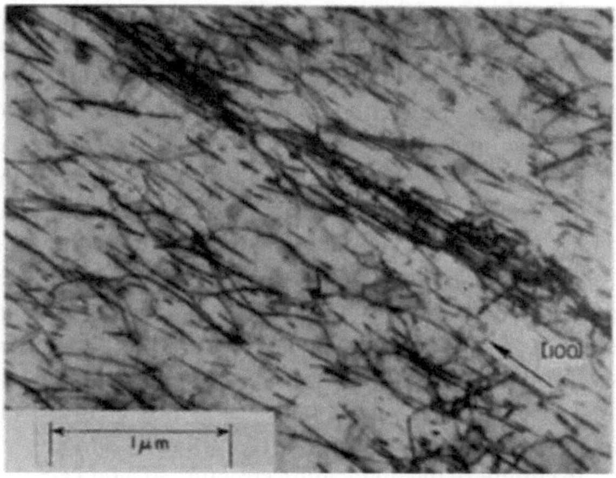

Fig. 8.20 Dislocation configuration after 10^2 rolling-contact cycles with 570 g load. Durfane and Glaeser [7]. With kind permission of Elsevier

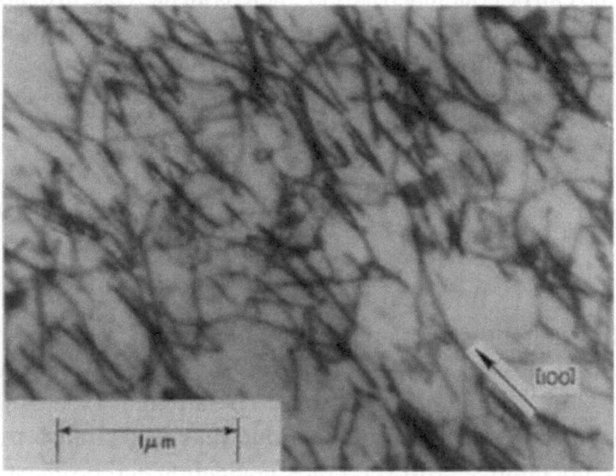

Fig. 8.21 Dislocation configuration after 10^3 rolling-contact cycles with 570 g load. Durfane and Glaeser [7]. With kind permission of Elsevier

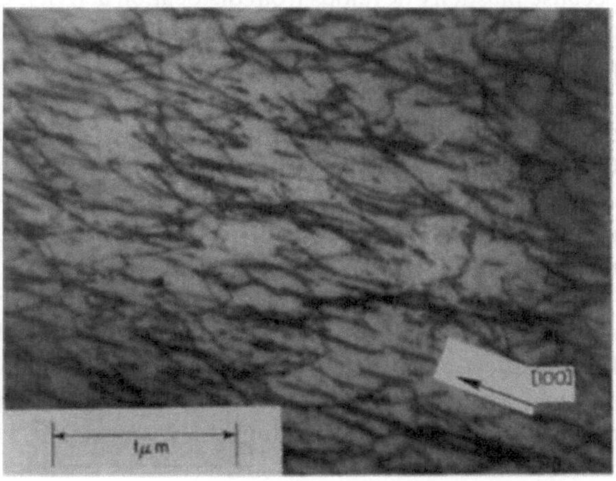

Fig. 8.22 Dislocation configuration after 10^5 rolling-contact cycles with 570 g load. Durfane and Glaeser [7]. With kind permission of Elsevier

but already after 10 rolling contact cycles extensive dislocation interactions became visible as seen in Fig. 8.19. Dislocation density and interactions increased tangles and dipole formation are indicated with increasing concentration (seen in Fig. 8.20). Little change in density of dislocation and interaction was observed between 10^2 and 10^3 rolling-contact as seen in Fig. 8.21. The dislocation configuration after 10^5 rolling-contact cycles seen in Fig. 8.22, showed an increase in dislocation density, in the density of dislocation tangles and a reduction in the size of dipoles when compared with Fig. 8.21 where the rolling-contact was 10^3 cycles.

The hardness and dislocation interactions as observed by transmission electron microscopy all increased in a consistent manner with increasing cycles. The hardness therefore also reflected a change in dislocation configuration. Variation of hardness with rolling-contact depends on rolling direction, as expected increases with cycles and the measured Knoop hardness is higher in the [110] direction compared with that of [100].

In summary the dislocation structure of MgO single crystals was presented above, and can be concluded as: (1) There is an increase in dislocation density with increasing rolling-contact stress cycles. (2) The nature of dislocations is orientation dependent. (3) Dislocation interactions increase with the number of cycles and (4) the hardness also increases with cycles and is orientation dependent.

8.4.4 Dislocations in Yttria Partially Stabilized Zirconia (YSZ)

In situ high temperature (up to 600 °C) compression tests by flash-sintered densification process of yttria stabilized zirconia (YSZ) is presented in this section. Below 400 °C a high ultimate compressive strength exceeding 3.5 GPa and a high inelastic strain of ~ 8% are obtained due mainly to transformation toughening. At higher temperatures, however, crack nucleation and propagation are significantly retarded, because plasticity develops from dislocation activity. Generally, the high dislocation density induced in flash-sintered ceramics may have implications for improving the plasticity of sintered ceramic materials.

Plan-view TEM samples of flash-sintered 3YSZ were prepared through the conventional approach, which includes manual grinding, polishing, dimpling, and final polishing in an ion milling system. Low energy ion polishing (2 kV) was used to minimize ion milling-induced damage. Microstructure characterization was performed by TEM and energy-dispersive X-ray spectroscopy (EDS) chemical mapping.

The 3YSZ was heated in a thermomechanical analyzer with platinum electrodes at a constant heating rate of 25 °C/min under an electrical field of 150 V/cm. The ultrafast densification process occurred at a furnace temperature of 1150–1200 °C in a few seconds, which is significantly lower compared to conventional sintering temperature of ~ 1900 °C to sinter 3YSZ in just a few seconds. The applied heat and electrical field were removed right after the onset of flash sintering to prevent grain growth. Even though the densification process of 3YSZ occurred at relatively low temperature, X-ray diffraction pattern shows a dominant tetragonal phase without evident monoclinic and cubic phases. SEM and TEM images are illustrated in Fig. 8.23. The cyclic loading and partial unloading were carried out at a strain rate of $5 \times 10^{-3}\ s^{-1}$ at room temperature and 400 °C as shown in Fig. 8.24. No crack was observed for the specimens tested at room temperature during the cyclic tests, and the pillar exhibited a significant amount of recoverable strain during unloading as seen in Fig. 8.24b. After cyclic tests, a monotonic compression test was conducted on the same micropillar up to 7% as seen in (e) of which 6% was recoverable strain and 1% irrecoverable strain without crack illustrated in (c). But in the same pillar catastrophic failure set in after a monotonic compression test as seen in (d) at a strain of 8% and shown by a green curve in (e). In comparison, YSZ subjected to cyclic loading tests at 400 °C as seen in Fig. 8.24i showed much less recoverable strain than that tested at room temperature. Also, small cracks formed as shown in Fig. 8.24g at a strain of 4% during the cyclic loading test. The subsequent monotonic microcompression test on the same pillar highlighted as a red curve in Fig. 8.24i shows that the pillar exhibited a UCS of 2 GPa, and a residual strength of 1.5 GPa at a strain of 9%, and multiple cracks formed after compression tests shown in Fig. 8.24h.

In conclusion, in situ study on mechanical behavior of flash-sintered YSZ was performed inside SEM at elevated temperatures up to 600 °C. Compared with bulk YSZ, micropillars sustain large strains up to 8%. However, the pillars fracture catastrophically at 400 °C after the nucleation of cracks. The brittle-to-ductile transition temperature is already observed at 400 °C in flashed-sintered YSZ much lower than the 800 C in conventional bulk YSZ. The enhanced plasticity at elevated temperatures arises from the transition from phase transformation toughening to dislocation creep, as the dominant inelastic deformation mechanism due to the existence of a high density of dislocations in flash-sintered YSZ and/or to early initiation of grain boundary sliding of ultra-fine grains.

8.4.5 Dislocations in Yttria Fully Stabilized Zirconia

Due to the lack of cyclic (fatigue) dislocations in fully stabilized zirconia, and because it was felt that it is important to get the taste and obtain some filling on dislocations in this material, the subject is considered here in terms of crystals deformed by compression.

Although 8% yttria is sufficient to fully stabilize zirconia, ~ 10 mol% yttria stabilized cubic zirconia single crystals is considered in this section. Recall that above 2400 °C pure zirconia is cubic (fluorite structure), it transforms below this temperature to the tetragonal phase and finally below about 1200 °C it is monoclinic. The high temperature cubic phase can be stabilized to lower temperatures by adding aliovalent cations. The most commonly known is the yttria stabilized zirconia (i. e. Y_2O_3-ZrO_2).

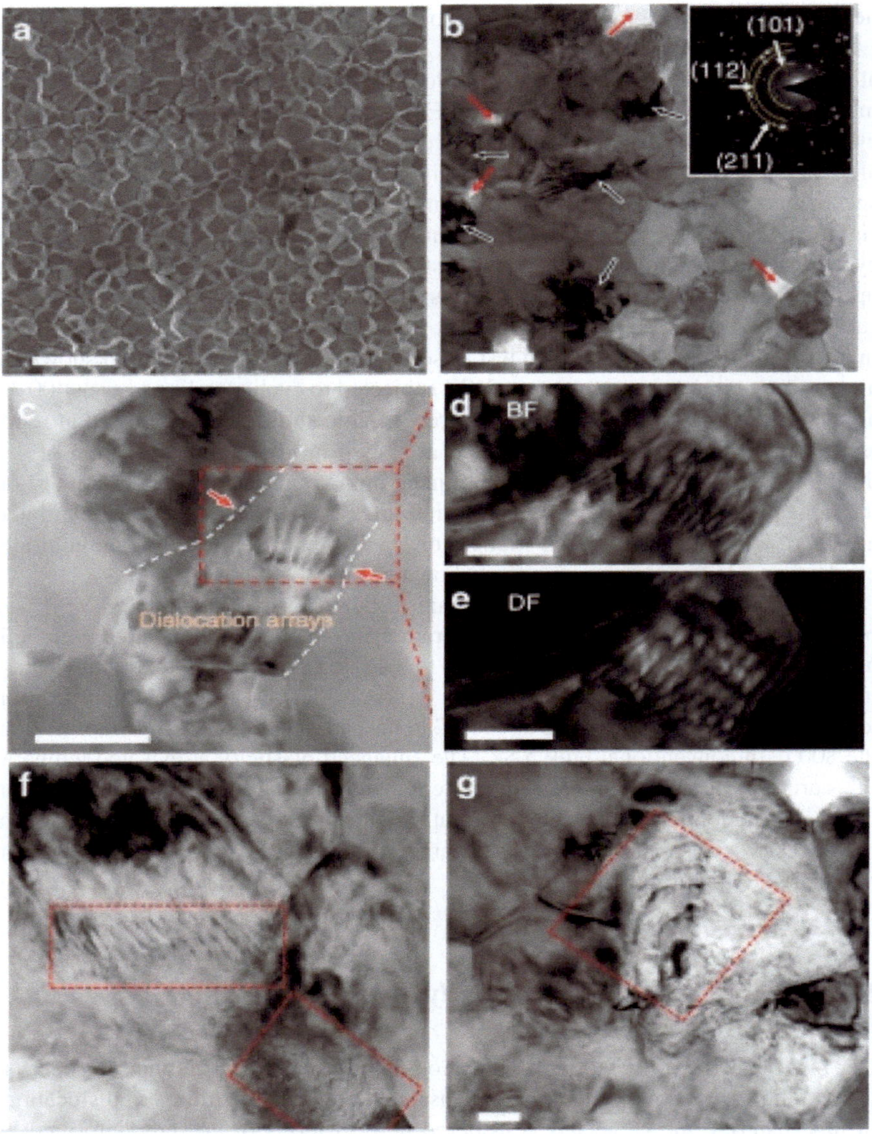

Fig. 8.23 Microstructure of flash-sintered 3YSZ. **a** SEM images of the unpolished flash-sintered 3YSZ. The average particle size is ~ 1 μm. Scale bar, 3 μm. **b** Bright-field TEM micrograph showing subgrains with grain boundaries and defects generated during flash sintering (labeled by black arrows), and nanopores (~1.4%) (indicated by red arrows). The inserted SAD pattern shows diffraction rings. Scale bar, 200 nm. **c** A STEM micrograph showing a dislocation array inside a grain in the flash-sintered 3YSZ. The dislocation array was generated in a bottleneck region of the grain as indicated by red arrows. Scale bar, 100 nm. **d, e** Bright-field (BF) and dark-field (DF) TEM micrographs showing the dislocation array in the boxed region in **c**. Scale bar, 50 nm. **f–i** Numerous BF TEM images of the flash-sintered 3YSZ showing the existence of dislocations and dislocation arrays in grains. Scale bar, 50 nm. Cho et al. [8]. Open access

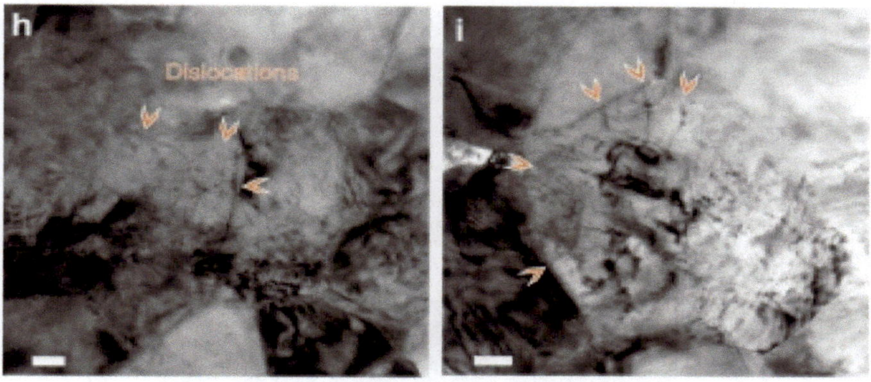

Fig. 8.23 (continued)

Fully stabilized zirconia single crystals oriented for compression along [100] with (010) and (001) side faces for 10 mol% Y_2O_3 and along [112] with (010) and (001) side faces for 15 mol% and 20 mol% Y_2O_3 were deformed at a constant strain rate of $\dot{\varepsilon} = 10^{-5}\,s^{-1}$. A few strain rate cycling experiments were carried out for comparison with the results of the relaxation tests (as seen in Fig. 8.25). In Fig. 8.25 the results of compression tests along the hard (100) orientation of 10 mol% yttria at a strain rate $\dot{\varepsilon} = 10^{-5}s^{-1}$ at different temperatures are illustrated. The critical flow stress, σ_y is plotted in Fig. 8.26 as a function of temperature. Stress–strain curves for 15 mol% yttria are plotted in Fig. 8.27. The yttria stabilized zirconia is brittle below ~ 700 °C even in the soft orientation. The deformation is stable at and below 800 °C and above 1400 °C but shows serrations at intermediate temperatures. High voltage transmission electron microscope HVEM was used to obtain all the micrographs from ZrO_2—10 mol% Y_2O_3 deformed along ⟨100⟩ were taken near a ⟨001⟩ pole perpendicular to a {001} side face. In some cases, the directions of the Burgers vectors were determined using the $\mathbf{g} \cdot \mathbf{b} = \mathbf{0}$ contrast extinction rule. The dislocation structure of ZrO_2—10 mol% Y_2O_3 deformed along the compression direction d = [100] at 700 °C to a strain of 1.3%, is shown in Fig. 8.28. Slip is inhomogeneous as seen in the upper part of the figure. Dislocations look quite straight running in the two ⟨110⟩ directions on the surface plane consistent with dislocation arrangement on {110}. In the lower part of the figure, dislocations are curved being arranged on {111} planes. Thus, slip systems with {110} and {111} planes coexist. Dislocation structures after deformation at 800 °C are displayed in Fig. 8.29. In Fig. 8.29a all dislocations at the (200) **g** vector are imaged with the non-zero orientation factors. Dislocations with 1/2[110] and 1/2[110] Burgers vectors within the image plane are extinguished and shown in (b) and (c) with (220) and (220) **g** vectors respectively. The dislocations have Burgers vectors in the image plane and are curved, therefore they should be arranged on two of the {111} planes. The dislocations have a large screw component. The fact that the dislocations are curved, indicates that they are pinned by obstacles being the reason for bowing out between them. The images

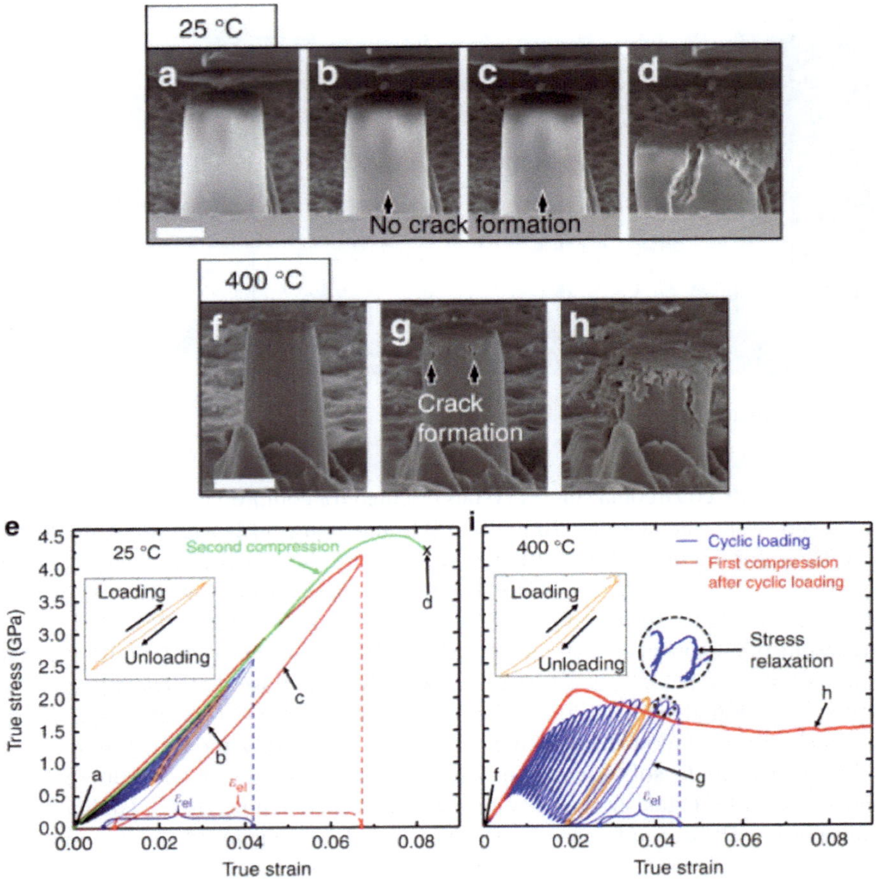

Fig. 8.24 Cyclic loading and subsequent monotonic compression tests at a strain rate of 5×10^{-3} s^{-1} at 25 and 400 °C and the corresponding stress–strain curves. (**a**), (**f**) SEM images of micropillars before cyclic loading tests. Scale bar, 2 μm. (**b**), (**g**) SEM micrographs of micropillars after 30 cyclic loading tests. The cyclic stress–strain curves are shown as blue curves in (**e**) and (**i**). (**c**), (**h**) After the first monotonic compression tests highlighted in red curves. (**d**) After the second monotonic compression test highlighted in green. The 27th loading–unloading curves at each temperature shown in orange color and enlarged in the inserted stress–strain curves clearly show the hysteresis loops. The upper portion of a loading–unloading curve is enlarged in a circle to illustrate the stress relaxation for 1 s of holding at 400 °C. Cho et al. [8]. Open access

shown at 700 °C (Fig. 8.2) and at 1050 °C in Fig. 8.30 have similar structures. Jogs are seen in Fig. 8.29b and are marked by J and their fraction with respect to the cusps is not high, and debris formed by them and marked by D is not high and should not influence the flow stress. Above 200 °C slip becomes homogeneous and thus the pinned and bowed out dislocations disappear as seen in Fig. 8.31a at 1300 °C. The many triple nodes indicate a three-dimensional network; similarly, at 1400 °C as

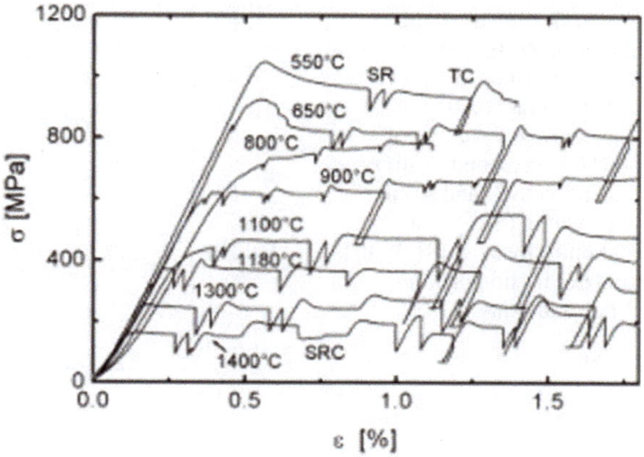

Fig. 8.25 Stress–strain curves of ZrO_2-10 mol% Y_2O_3 deformed along ⟨100⟩ at a strain rate of 10^{-5} s^{-1}, including stress relaxation (SR), strain rate cycling (SRC) and temperature change (TC) tests. Tikhonovsky et al. [9]. With kind permission of John Wiley and Sons

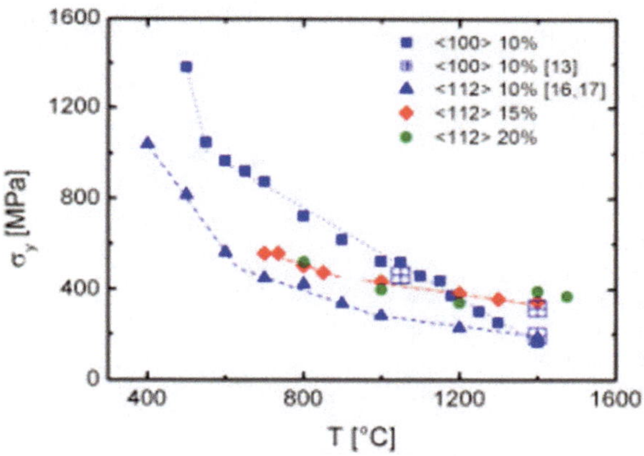

Fig. 8.26 Flow stresses σ_y versus temperature T at a strain rate of 10^{-5} s^{-1}. Solid squares: ZrO_2-10 mol% Y_2O_3, ⟨100⟩ orientation. Diamonds: ZrO_2—15 mol% Y_2O_3, ⟨112⟩ orientation. Circles: ZrO_2-20 mol% Y_2O_3, ⟨112⟩ orientation. Triangles: ZrO_2—10 mol% Y_2O_3, ⟨112⟩ orientation, reference data at a strain rate of 10^{-6} s^{-1}. Open squares plus crosses: ZrO_2—10 mol% Y_2O_3, ⟨100⟩ orientation, taken at 10^{-6} s^{-1} (1150 °C and 1400 °C) and 10^{-4} s^{-1} (1400 °C). Tikhonovsky et al. [9]. With kind permission of John Wiley and Sons.13, 16 and 17 refer to Baufeld and associates

demonstrated in (b) and (c) showing dislocations lying on two sets of {110} planes edge on.

The temperature dependence of the dislocation density is seen in Fig. 8.32. Thus dislocation glide is temperature dependent and its activation is expected to follow an Arrhenius type relation given as

$$\dot{\varepsilon} = \dot{\varepsilon}_0 \exp\left(-\frac{\Delta G}{kT}\right) \tag{8.1}$$

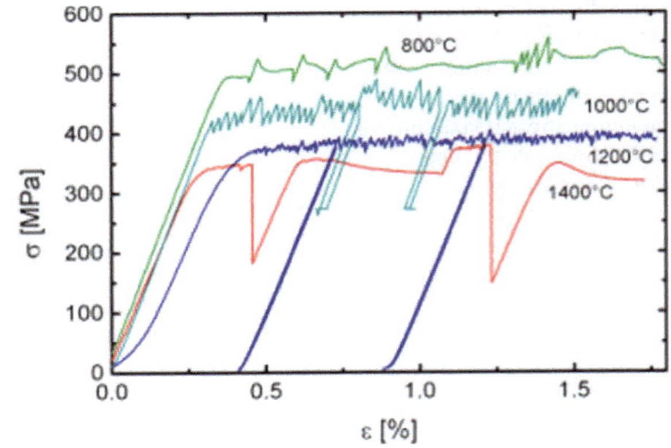

Fig. 8.27 Stress–strain curves of ZrO_2—15 mol% Y_2O_3 deformed along ⟨112⟩ at a strain rate of 10^{-5} s^{-1}. The cusps in the curve of 800 °C correspond to stress relaxation (SR) and strain rate cycling (SRC) tests. Tikhonovsky et al. [9]. With kind permission of John Wiley and Sons

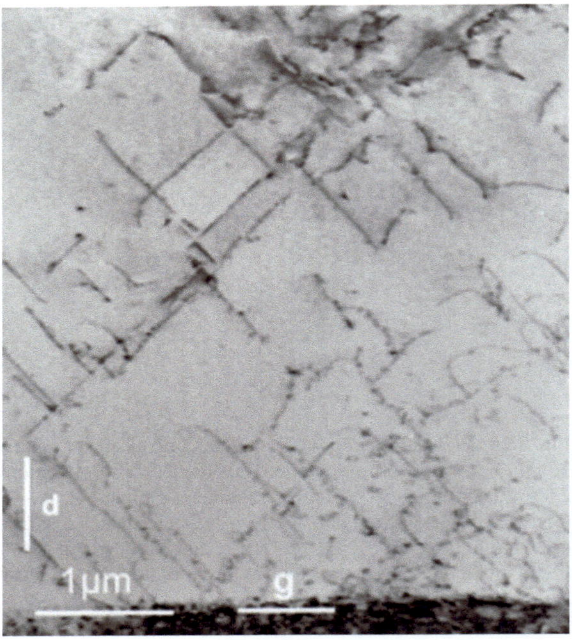

Fig. 8.28 Dislocation structure of ZrO_2—10 mol% Y_2O_3 deformed along the compression direction **d** = [100] at 700 °C to a strain of 1.3%. Imaging diffraction vector **g** = (020) near the [001] pole. With this **g** vector, all dislocations of slip systems with non-zero orientation factors are visible. Tikhonovsky et al. [9]. With kind permission of John Wiley and Sons

with

$$\Delta G(\tau^*) = \Delta F(\tau^*) - V(\tau^*)\tau^* \tag{8.2}$$

In the above relations, $\dot{\varepsilon}_0$ is the pre-exponential factor containing the mobile dislocation density and the distance for forward jumps. It is usually considered constant. ΔG is the Gibbs free energy for activation, ΔF is the free energy of activation, V is the activation volume and τ^* is the effective shear stress which is related to the total

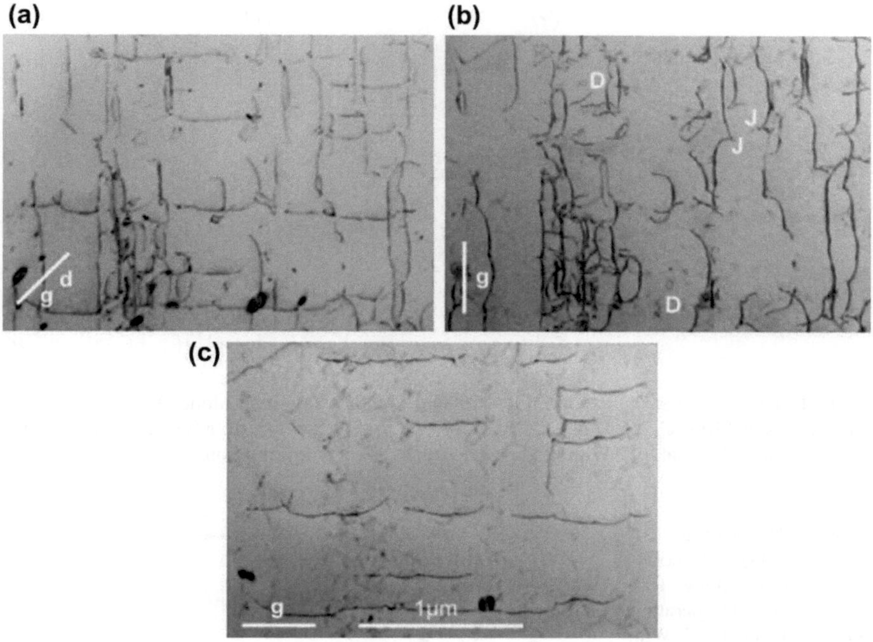

Fig. 8.29 Dislocation structure of ZrO_2—10 mol% Y_2O_3 deformed along **d** = [100] at 800 °C up to 1.1% imaged under different **g** vectors at the [001] pole. (**a**) **g** = (200). (**b**) **g** = (220). (**c**) **g** = (220). Tikhonovsky et al. [9]. With kind permission of John Wiley and Sons

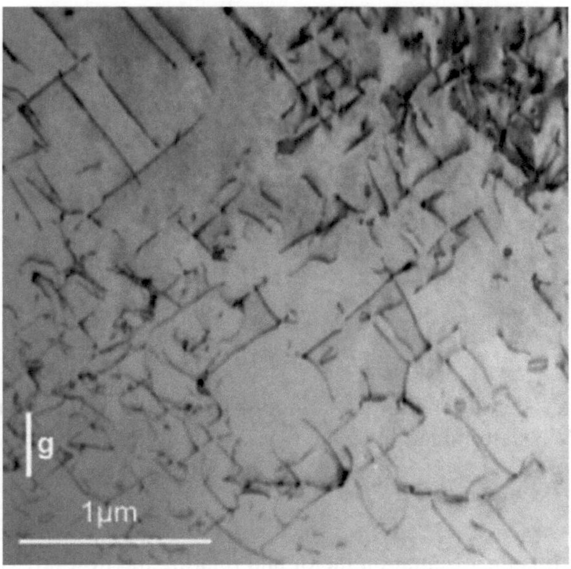

Fig. 8.30 Dislocation structure of ZrO_2—10 mol% Y_2O_3 deformed along **d** = [100] at 1050 °C up to 1.8%. **g** = (200). Tikhonovsky et al. [9]. With kind permission of John Wiley and Sons

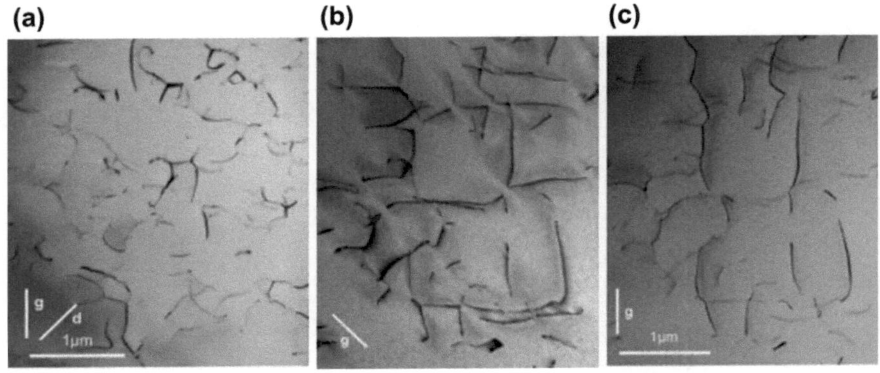

Fig. 8.31 Dislocation structure of ZrO_2—10 mol% Y_2O_3 deformed along **d** = [100] at high temperatures. **a** 1300 °C, 2.6%. **g** = (220). **b** 1400 °C, 2.6%, **g** = (020). **c** Same area as (**b**), **g** = (220). Tikhonovsky et al. [9]. With kind permission of John Wiley and Sons

Fig. 8.32 Dislocation density of ZrO_2—10 mol% Y_2O_3 deformed along ⟨100⟩ as a function of temperature. Tikhonovsky et al. [9]. With kind permission of John Wiley and Sons

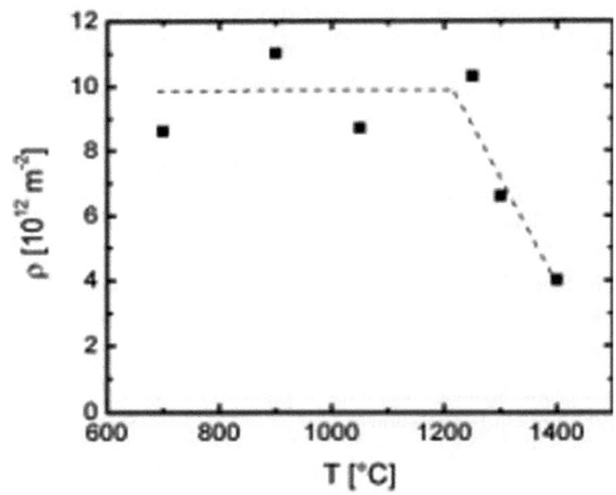

shear stress τ by

$$\tau = \tau_i + \tau^* \tag{8.3}$$

τ_i is the athermal stress component resulting from long-range dislocation interactions. V is determined from the rate sensitivity by

$$V = kT(m_s r) \tag{8.4}$$

m_s is the orientation factor and relating the flow stress, σ to the shear stress by

$$\tau_s = m_s \sigma \tag{8.5}$$

At high temperatures the strain rate sensitivity related to the stress exponent, is expressed as

$$m^* = \frac{d \ln \dot{\int}}{d \ln \sigma} = \frac{\sigma}{r} \tag{8.6}$$

The activation parameters of the plastic deformation evaluated from stress relaxation tests. The strain rate sensitivity, r of the flow stress can be calculated from the inverse slope of the relaxation curves by plotting ln(–σ) versus σ. Thus

$$r = \frac{d\sigma}{d \ln \dot{\varepsilon}} = \frac{d\sigma}{d \ln(-\sigma)} \tag{8.7}$$

In order to determine the activation energy of the plastic deformation, temperature change tests were performed. The specimens were partially unloaded before changing the temperature and the deformation was resumed at the new temperature after equilibrium was reached. The activation enthalpy was calculated according to

$$H = -kT^2 \left(\frac{\Delta\sigma}{\Delta T} \right) \frac{1}{r} \tag{8.8}$$

T being the average temperature $T = (T_1 T_2)^{1/2}$ and r the average value of the strain rate sensitivities from relaxation immediately before and after the temperature change $r = (r_1 + r_2)$.

In summary, the microstructures were investigated by high voltage electron microscopy. The results are considered in terms of long range interactions between dislocations. Zirconia stabilized by 10 mol% yttria was deformed along the hard < 100 > direction and specimens with higher mol% (15 and 20) along the soft < 112 > orientation between 1400 °C and the lowest possible temperature without brittle failure. Below 700 °C the failure is brittle. The dislocation structures illustrated in this section refer only to 10 mol% ittria stabilized zirconia.

8.4.6 *Dislocations in B_4C*

Although there is a lack of cyclic dislocations in fatigue deformation, it was felt that the illustrations in crept B_4C might be useful in illustrating the nature of dislocations in this material. The importance of including in this book—as far as available—B_4C is because its unique properties. This carbide is a promising ceramics that exhibits a combination of its high hardness, resistance to shock-impact and its high temperature respond (including creep) indicating an extensive deformation without failure in the total strain range of 13–18%.

Sintered coarse grained B_4C specimens were subjected to uniaxial compression creep testing. The applied stress was in the range 250–500 MPa at temperature in the range 1600–1800 °C under argon gas atmosphere to prevent oxidation of the samples. Polished surfaces of B_4C samples before and after creep testing were characterized by SEM and TEM. A series of TEM micrographs show in Fig. 8.33 show the dislocations in B_4C. Straight-line dislocations with Burger's vector $\mathbf{b} = <1\ 1\ 0>$ gliding on $\{1\ 1\ 1\}$ planes are seen in (A). Higher dislocation densities are seen in (B) and (C) with interactions between dislocations from multiple slip planes and with fine B_4C particles indicated by arrows. The obstacles to dislocation motion

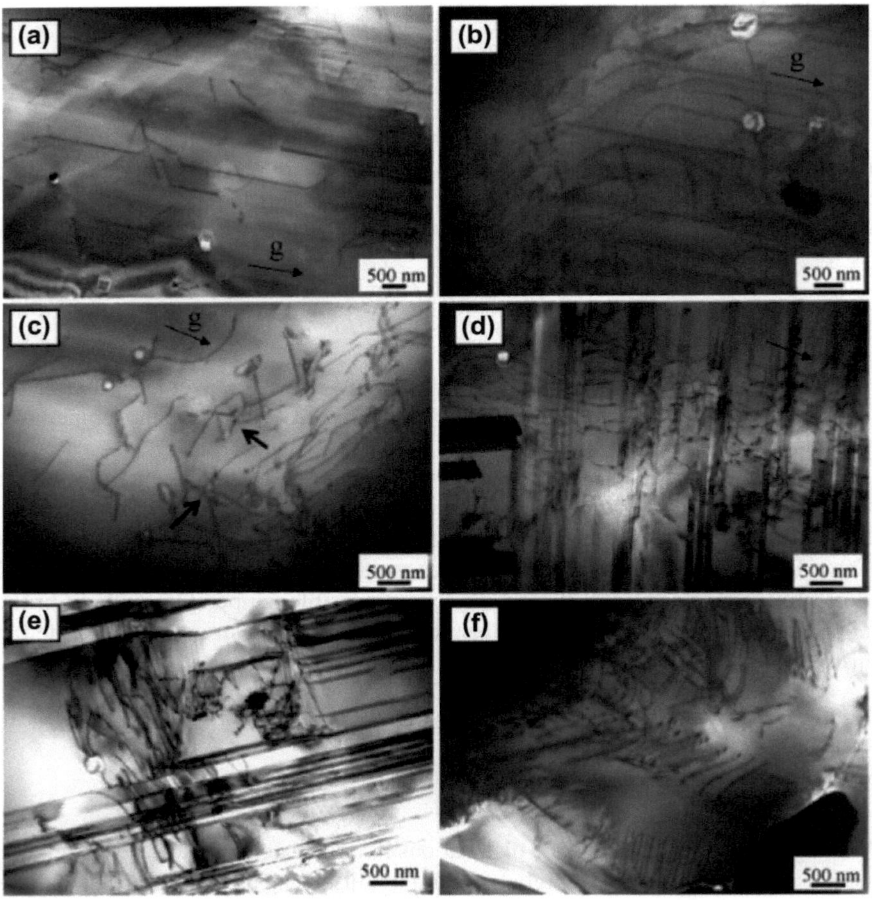

Fig. 8.33 TEM micrographs of creep-deformed B_4C specimen (~13% total deformation) showing: **a** straight-line dislocations with Burger's vector $b = <1\ 1\ 0>$ gliding on $\{1\ 1\ 1\}$ planes and long cross-slipping screw dislocation segments, **b, c** areas of higher dislocation densities and the dislocation debris, **d, e** interaction of dislocations and twins, and **f** dislocation pile-ups at the grain boundaries. Arrows indicates the presence of dislocation nodes. Moshtaghioun [10]

in addition to grain boundaries are twins and are observed in (D)–(F). A significant number pile-ups are observed at the grain boundaries in (F). Complex nodes in B_4C comprise perfect and partial dislocations which do not all lie in the same plane. The overall dislocation density in B_4C is ~ 3×10^{12} m^{-2}. This dislocation density is higher than that observed in the as-processed B_4C samples which is $\sim 10^8$ m^{-2}, which is typical of creep-deformed single crystals of several materials.

It might be of interest to note that the measured stress exponent is $n \sim 3$ and the activation energy for creep deformation is $Q \sim 632$ kJ/mole which indicate classical dislocation climb-glide power-law creep.

In summary, high-temperature creep studies of coarse-grained, dense B_4C performed and presented here, show that B_4C can undergo extensive deformation of 13–18% total strain without failure, which makes it an important ceramic for technological applications.

8.4.7 Dislocations in Si_3N_4

In a very early publication cyclic deformation in fatigue of hot-pressed Si_3N_4 was discussed. Some figures are reproduced here for the sole reason that it refers to cyclic fatigue (not readily available in the open literature). A flat specimen of Si_3N_4 was placed in a four-point bending fixture and subjected to a static load of 38,000 psi at 1200 °C for 4h. Most of the grains showed no evidence of an increased dislocation density over the original low density that is typical of the as-received material. Selected areas of highly localized deformation near the outer most-stressed surfaces of specimens tested at 1000 and 1200 °C is shown in Fig. 8.35b. A high density of tangled dislocations is shown in Fig. 8.35b. The helical dislocations and the tangles are not found in the as-received material and must have formed as a result of fatigue deformation. From a flat specimen, prepared from the HS-110 grade Si_3N_4, thin foils were prepared from both the tensile and the compressive surfaces, and typical microstructures from transmission photomicrographs are shown in Fig. 8.36a and b. The as received and tensile creep specimen are compared in these illustrations.

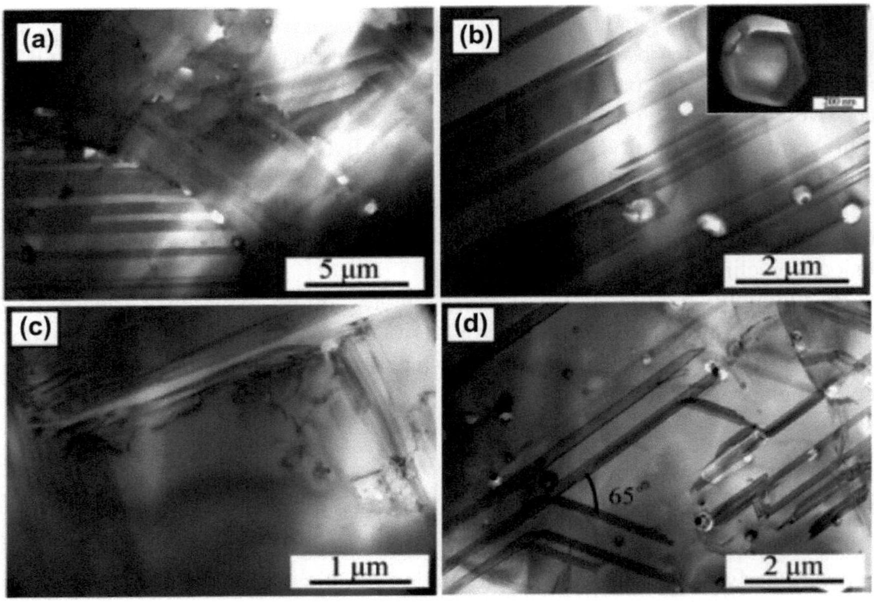

Fig. 8.34 Bright-field TEM images of B_4C before creep testing showing: **a** twins and trapped small pores at grain boundaries, **b** small pores within larger grains and trapped small grains within coarse grains (inset), **c** dislocation debris in some grains, and **d** twins, with crystallographic orientation forming variants at ~65° angle. Moshtaghioun [10]

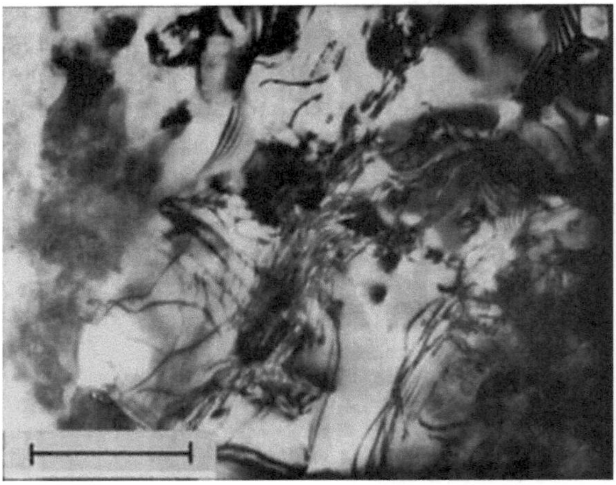

Fig. 8.35 Thin-foil transmission micrographs of fatigue specimens. **b** Localized deformation in a large grain from HS-110 specimen tested at 1200 °C. A high density of tangled dislocations is shown. Kossovsky [11]. With kind permission of John Wiley and Sons

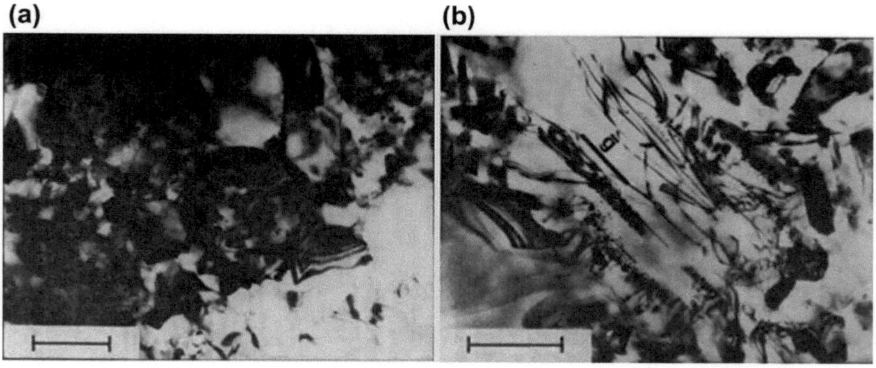

Fig. 8.36 Thin foil transmission photomicrographs of HS-110 creep specimen. **a** As-received, not-tested material for comparison. Large-headed arrows indicate cavitation at triple points. **b** Dislocation loops and helices occasionally observed in large grains. Tensile creep specimen tested at 1000 °C and 20,000 psi. Kossovsky [11]. With kind permission of John Wiley and Sons12435681112910

References

1. D. Birenis, Y. Ogawa, H. Matsunaga, O. Takakuwa, J. Yamabe, Ø. Prytz, A. Thøgersen, Acta Mater. **156**, 245 (2018)
2. S.P. Lynch, Mater. Sci. Eng. A **468**, 74 (2007)
3. T.-H. Lan, C.-Y. Pan, P.-H. Liu, M.M.C. Chou, Appl. Sci. **9**, 2585 (2019)
4. V. Sharma, S.N.Nasser, K.S. Vecchio, J. Am. Ceram. Soc. **81**, 129 (1998)
5. J.R. Varner, Fracture modes and appearances in ceramics. New York State College of Ceramics, Alfred University
6. I. Zarudi, L.C. Zhang, D. Cockayne, J. Mater. Sci. **33**, 1639 (1998)
7. K.F. Durfane, W.A. Glaeser, Wear **37**, 21 (1976)
8. J. Cho, Q. Li, H. Wang, Z. Fan, J. Li, S. Xue, K.S.N. Vikrant, H. Wang, T.B. Holland, A.K. Mukherjee, R.E. García, X. Zhang, Nat. Commun. **9**, 2063 (2018)
9. A. Tikhonovsky, M. Bartsch, U. Messerschmid, phys. stat. sol. (a) **201**, 26 (2004)
10. B. M. Moshtaghioun,https://www.sciencedirect.com/science/article/pii/S0955221914005871-! A.D. Rodríguez, N.P. Padture, J. Eur. Cer. Soc. **35**, 1423 (2015)
11. R. Kossovsky, J. Amer. Cer. Soc. **36**, 531 (1973)
12. B.J. Hockey, B.R. Lawn, J. Mater. Sci. **10**, 1275 (1975)
13. B. Baufeld, D. Baither, M. Bartsch, U. Messerschmidt, phys. stat. sol. (a) **166**, 127 (1998)
14. B. Baufeld, M. Bartsch, U. Messerschmidt, D. Baither, Acta Metall. Mater. **43**, 1925 (1995)
15. B. Baufeld, B.V. Petukhov, M. Bartsch, U. Messerschmidt, Acta Metall. Mater. **46**, 3077 (1998)

(a)

References

Chapter 9
Work Hardening (Softening)

Abstract At high temperatures where brittle materials show plasticity, work hardening may occur, but it does not necessarily occur, however obstacles such as grain boundaries or twin boundaries which hinder dislocation motion can harden the ceramic. The stress amplitude at which failure occurs for a given number of cycles is known as the fatigue strength, and the number of cycles required for a material to fail at a certain stress range is its fatigue life. The procedure for analyzing fatigue test for fatigue life is by the stress amplitude which was applied to alumina, stabilized zirconia, silicon carbide and silicon nitride. The effect of grain size is one of the most important factors that influences the mechanical properties of materials. The Hall–Petch relation predicts that yield strength increases with decreasing grain size. The effect of grain size is broadly discussed in alumina, in stabilized zirconia, SiC and Si_3N_4. Generally, the fatigue properties depend on temperature. Of the various thermal processes, deformation by dislocations is of major importance. In general, the temperature of components in service varies between a high operating temperature and a low non-operational ambient temperature, which means that thermal stresses can be introduced. High-temperature fatigue conditions imply that the fatigue load and temperature vary both as a function of time. These effects are considered in discussing oxides, carbides and nitrides and are presented in this chapter limited only by the avalibility of such information in the literature.

Keywords Work hardening · Stress amplitude · Strain amplitude · Grain size · Temperature · Alumina · Zirconia · SiC/SiC · Si_3N_4

9.1 Introduction

Generally, work hardening also known as strain hardening is a feature that characterizes ductile materials such as metals but also ductile ceramics. Ceramics are mostly brittle at room temperature except a few exemptions. However, at high temperatures where brittle materials show plasticity, strain hardening may occur but it does not necessarily occur. However, hardening clearly can occur by obstacles such as grain boundaries or twin boundaries which act as obstacles to dislocation motion, and thus harden the ceramic. Further, other mechanisms can strengthen ceramics.

J. Pelleg, *Cyclic Deformation in Oxides, Carbides and Nitrides*, Structural Integrity 22,
https://doi.org/10.1007/978-3-030-86118-6_9

Several mechanisms that strengthen ceramics in general are listed below: (i) Work (strain) hardening in ductile ceramics; (ii) solid solution strengthening by substitutional or intestinal atoms by pinning dislocations; (iii) second phase hardening, which clearly includes precipitates resulting from the precipitation process; (iv) transformation hardening (such as occurs in zirconia for example) and (v) boundaries as already mentioned.

The reason that some brittle ceramics who became ductile ceramics at elevated temperatures does not work harden is the simultaneous recovery process that may occur, a process which depends on test conditions and composition. Such ceramics such as MgO show superplasticity rather than strain harden due to the recovery process. Note the paradox of strain hardening, which occurs in ductile material capable of deformation, whereas hardening is defined as the resistance to plastic deformation. So, ductility and plastic deformation go hand by hand with strain hardening and thus the importance of strain hardening is because the possibility for deformation. In this chapter oxide, carbide and nitride ceramics will be presented, limited only by the availability of experimental and research.

9.2 The Effect of Stress Amplitude

In the S–N relations the stress amplitude at which failure occurs for a given number of cycles is known as the fatigue strength, where N is the number of cycles required for a material to fail at a certain stress range in its fatigue life. The stress range is

$$\Delta\sigma = \sigma_{max} - \sigma_{min} \tag{9.1}$$

Figure 9.1 and Eqs. (9.1)–(9.3) are reproduced from Chap. 1.

$$\sigma_a = \frac{\sigma_{max} - \sigma_{min}}{2} \tag{9.2}$$

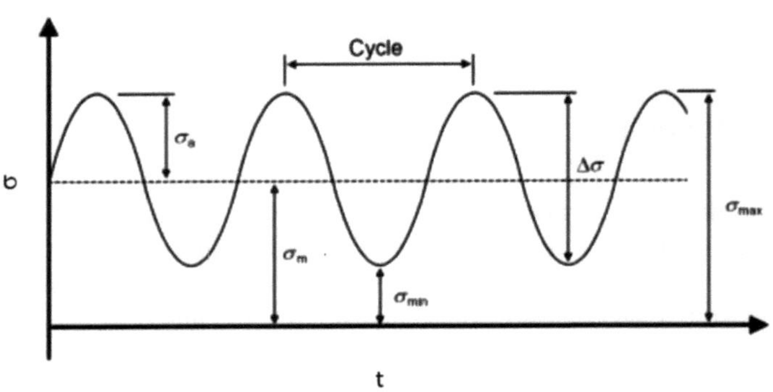

Fig. 9.1 Several cycles in a fatigue stress cycle

$$\sigma_m = \frac{\sigma_{\max} + \sigma_{\min}}{2} \tag{9.3}$$

The stress amplitude can be constant or variable. In materials showing S–N relation with a knee region like steel, the loading in constant amplitude test, the knee level region the fatigue lives become very long and only small changes in stress range are observed. For conservative design the knee point should occur at 1×10^7 cycles, but often the concept suggested is that a fatigue limit below the knee point, fatigue life should be infinite. During variable amplitude loading the problem is more complex and the fatigue life is affected by various factors that might induce damage. There are a number of special phenomena that are not observed for constant amplitude loading. Tensile overloads may interact with other mechanisms which might operate, such as creep, yielding or fracture while compressive overloads may influence local buckling or produce secondary bending stresses. Further and of more serious nature are the local stresses that occur in the crack tip region which often accelerate fatigue crack growth rate, but under some circumstances, decelerate the growth of a fatigue crack. The majority of design data is still based on constant amplitude testing. Note, that the local and structural geometric stress concentrations that greatly influence the fatigue strength will have similar influence for both constant and variable amplitude loading.

Not all materials show a knee in the S–N curves. Details on S–N relations and concepts are discussed in Chap. 1. In this chapter the stress amplitude in some oxides, carbides and nitrides are presented, limited only by the availability of in formation in literature.

The fatigue life, N_f in terms of stress amplitude is defined under a stress ratio $R = -1$ as

$$N_f = S_f(\sigma_a - \sigma_{ac})^{-2} \tag{9.4}$$

where N_f is defined as the number of cycles to failure, σ_a is the stress amplitude, and S_f and σ_{ac} are respectively the fatigue resistance coefficient and fatigue limit, below which N_f tends to be infinite. Differentiate the logarithm of Eq. (9.4) to obtain (according to Zhen and Lu)

$$\frac{\Delta N_f}{N_f} = \frac{\Delta S_f}{S_f} - \frac{2\Delta\sigma_a}{(\sigma_a - \sigma_{ac})} + \frac{2\Delta\sigma_{ac}}{(\sigma_a - \sigma_{ac})} \tag{9.5}$$

It is seen that when the stress amplitude, σ_a approaches σ_{ac}, N_f becomes infinite. The two constants S_f and σ_{ac} of Eq. 9.5 should be determined by experiments for its use by substituting the experimental values in the equation and then the fatigue life can be drawn (see below in alumina). As usually done for solving Eq. 9.5, two test data of the fatigue life determined at two different stress amplitudes is substituted to Eq. (9.4) to solve for the fatigue life for obtaining S_f and σ_{ac}. Thus, the fatigue test data can be analyzed by the following procedures:

(1) rearranging the test data according to the increasing order of fatigue life,

(2) substituting two values of S_f and σ_{ac} in Eq. (9.4) taken from test data in sequence from two groups of stress amplitudes,
(3) Calculate the values of S_f and σ_{ac} after checking if they follow the log-normal or normal distribution,
(4) Substitute the value of S_f and σ_{ac} into Eq. (9.4) to obtain the expression of the fatigue life and drawing a curve with the given survivability, namely the so called P-S-N curves. (See below for alumina).

9.2.1 *The Effect of Stress Amplitude in Alumina*

A simplified procedure for analyzing fatigue test results of alumina (applicable to other ceramics) for fatigue life is presented below: For a given survivability of fatigue life the so called expression of P-S-N is suggested. The slope of P-S-N curves decreases with increasing survivability and tends to zero when it is equal or higher than 99.%Sv_f. Sv clearly stands for survivability.

Using the procedures proposed above, the values of the fatigue resistance coefficient, S_f, and the theoretical fatigue limit, σ_{ac}, can be calculated and the results so obtained are listed in Table 9.1.

Two groups of the fatigue test results determined at $\sigma_a = 169$ and 148 MPa are chosen to calculate the values of S_f and σ_{ac}, the constants in Eq. (9.4). Using this equation of the fatigue life of alumina with 50% survivability, S_v can be written as

$$N_f = 2.14 \times 10^7 (s_a - 138.1)^{-2} \quad (9.6)$$

and for Fig. 9.2.

The expressions of the fatigue life of ceramics with 97.7% S_v and 99.9% S_v, respectively, can be obtained as follows

$$\text{for } 97.7\%\ S_v \quad N_f = 1.44 \times 10^4 (s_a - 125.3)^{-2} \quad (9.7)$$

$$\text{for } 99.9\%\ S_v \quad N_f = 2.68 \times 10^2 (s_a - 118.4)^{-2} \quad (9.8)$$

The fatigue resistance coefficient S_f follows the log normal distribution and σ_{ac}, the theoretical fatigue limit follows the normal distribution as seen in Fig. 9.3.

The expressions for fatigue life with 2.3% S_v and 0.1% S_v are

$$\text{For } 2.3\%\ S_v \quad N_f = 3.17 \times 10^{10} (s_a - 150.9)^{-2} \quad (9.9)$$

$$\text{For } 0.1\%\ S_v \quad N_f = 1.70 \times 10^{12} (s_a - 157.8) \quad (9.10)$$

Table 9.1 The fatigue test results of alumina ceramics and the calculated results of the constants in fatigue formula, Eq. (9.4). Zheng and Lu [1]. With kind permission of Elsevier

No.	Log N_f $\sigma_a = 169$ MPa	Log N_f $\sigma_a = 148$ MPa	Log A	σ_{ac} MPa	σ_{bb} MPa
1	1.4314	2.7340	4.2923	142.0	238
2	1.9090	2.7427	4.9713	135.0	242
3	2.0294	2.8639	5.0920	135.0	249
4	2.6365	3.4814	5.6951	135.2	249
5	2.7235	3.6684	5.9430	137.3	263
6	3.1139	4.5441	5.9700	143.0	265
7	4.0455	4.6010	7.1711	124.6	266
8	4.1553	4.8915	7.2348	132.2	266
9	4.3636	4.9020	7.2880	123.5	269
10	4.3874	5.2068	7.3398	134.6	278
11	4.4362	6.0086	7.4607	143.9	293
12	4.4518	6.6232	7.6800	146.1	296
13	5.0969	6.6284	7.8993	143.7	
14	5.4362	6.6609	8.3251	141.2	
15	5.5670	7.0000	8.3874	143.0	
16	5.6749	7.0414	8.5218	142.5	
17	5.9159	7.4031	8.7305	143.4	
18	5.9952	7.5.65	8.8106	143.5	
19	6.5705	7.7251	9.4850	140.4	
20	7.1644	7.9165	10.2810	132.8	
$\bar{x}$†	4.3552	5.5075	7.3294	138.1	264.5
s†	1.6239	1.7677	1.5860	6.4	14.0

†$\bar{x}$ and s are, respectively, the mean value and the standard deviation.

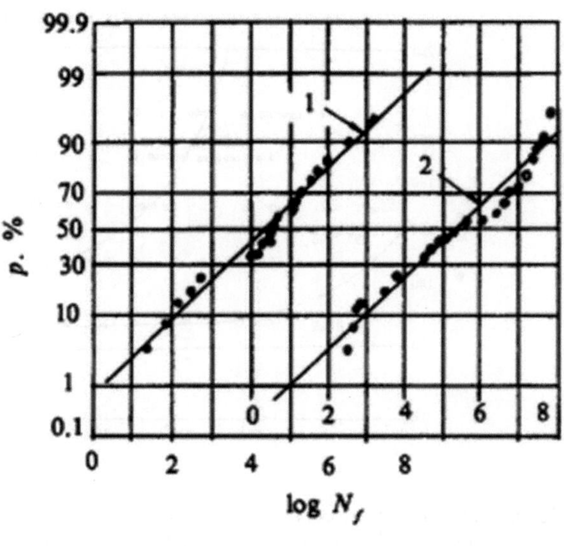

Fig. 9.2 The probability distribution of the test results of the fatigue life of alumina ceramics (1) $\sigma_a = 169$ MPa, (2) σ_a, = 148 MPa. Zheng and Lu [1]. With kind permission of Elsevier

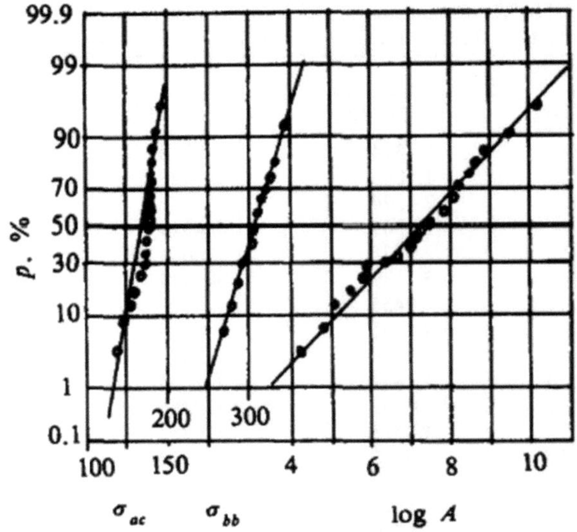

Fig. 9.3 Distribution of the calculated results of the theoretical fatigue limit σ_{ac} and the fatigue resistance coefficient S_f, and the flexible strength σ_{bb}. Zheng and Lu [1]. With kind permission of Elsevier

The curves of fatigue life of alumina are presented according to Eqs. 9.6–9.9 in Fig. 9.4.

Details on relations 9.6–9.10 are presented in the paper of Zheng and Lu. Note the dispersion about six orders in the fatigue resistance, S_f is unusually high compared to what is observed in metals (Table 9.1).

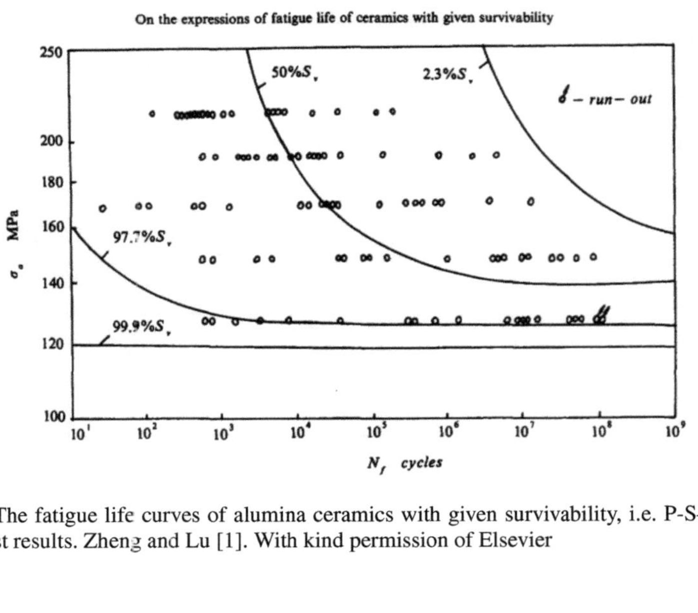

Fig. 9.4 The fatigue life curves of alumina ceramics with given survivability, i.e. P-S–N curves and the test results. Zheng and Lu [1]. With kind permission of Elsevier

The slope of P-S–N curves (i.e. the so called P. S. N curves are expressions of fatigue life of ceramics with given survivability) decreases with increasing survivability, and tends to be zero when the survivability is higher tha 99.9% S_v. Therefore, for high survivability, the stress amplitude applied to the ceramic elements should not be over the theoretical fatigue limit. Otherwise the fatigue life will be very short.

In summary, the fatigue resistance coefficient follows a log-normal distribution. A great dispersion in fatigue life (Fig. 9.4) is obtained, probably because the nature of fatigue experiments (customary to make at least 5 tests in each case). The test results and the observed dispersion is associated with the dispersion of the fatigue resistance coefficient. The slope of the fatigue life curve greatly decreases with increasing the survivability which tends to zero when the survivability is in the range of 99.9%. Therefore, the stress amplitude applied to the ceramic elements should not be over the theoretical fatigue as indicated earlier.

9.2.2 The Effect of Stress Amplitude in Zirconia

Thermal barrier coatings (TBC) of plasma sprayed 8% yttria stabilized zirconia considered for protection in gas turbine engines is a promising process to produce porous strain tolerant coatings. However, the highly porous structure is vulnerable to crack propagation, especially near the topcoat-bondcoat interface where high stresses are generated due to thermal cycling. Therefore, a topcoat layer with high toughness near the topcoat-bondcoat interface could be beneficial to enhance thermal cyclic lifetime of suspension plasma sprayed (SPS) TBCs. The fully stabilized zirconia could be considered a good protective coating and the denser this layer is the better it will function as a protective TBC.

Coating lifetime depends on the stress amplitude and during shock testing the stresses in the coating fluctuate from tensile to compressive or vice-versa due to cyclic thermal loading resulting from the cyclic temperature variation. This fluctuation of the stresses will result in thermo-mechanical fatigue in the coating which will influence the lifetime of the coating. The coating's lifetime is dependent on $\Delta\sigma$. A higher $\Delta\sigma$ in the coating would imply a lower lifetime and vice-versa. Therefore, $\Delta\sigma$ could be a helpful parameter to understand the lifetime of the coatings and specially to compare different coatings. The stress amplitude is a key factor in the TBC failure, since the mean stresses do not show significant difference on whether they operate under tension or compression. The highest stresses typically occur near the topcoat (TC)-bondcoat (BC) interface during thermal cycling, thus causing failure of the coating near the TC-BC interface. $\Delta\sigma$ in the current presentation refer to the difference between maximum and minimum stresses during the cycle with respect to the Y direction distance from the TC-BC interface. Figure 9.5 shows $\Delta\sigma$ above a peak location in the Y direction distance from the peak at the TC-BC interface.

The single layer TBC is illustrated in Fig. 9.6 and the bilayer configuration is shown in Fig. 9.7.

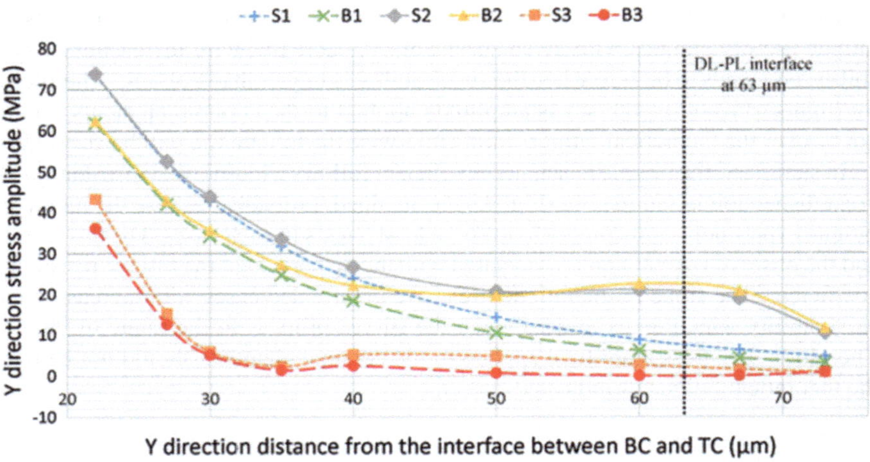

Fig. 9.5 Y direction stress amplitudes ($\Delta\sigma$) in all models against the Y direction distance from a peak of the TC-BC interface. Gupta et al. [2]. Open access. The Ss refer to single layers, and the Bs to bilayers, respectively

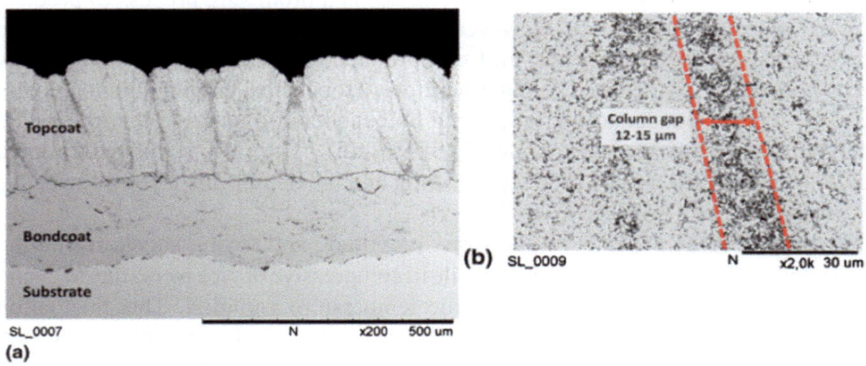

Fig. 9.6 Single-layer TBC SEM microstructure images in as-sprayed condition **a** an overview, and **b** column gap at a higher magnification. Gupta et al. [2]. Open access

Failure or crack initiation can happen near the TC-BC interface where $\Delta\sigma$ is hiher. $\Delta\sigma$ decreased at a decreasing rate on moving away from the interface and remained constant at the porous topcoat layer (PL)—dense topcoat layer (DL) interface. (Fig. 9.4; TC refers to the top coat). In the present experiments, two TBC systems were sprayed—a single layer topcoat TBC and a bilayer topcoat TBC. The results indicate that the bilayer topcoat microstructure is beneficial for increasing the thermal cyclic lifetime of SPS coatings.

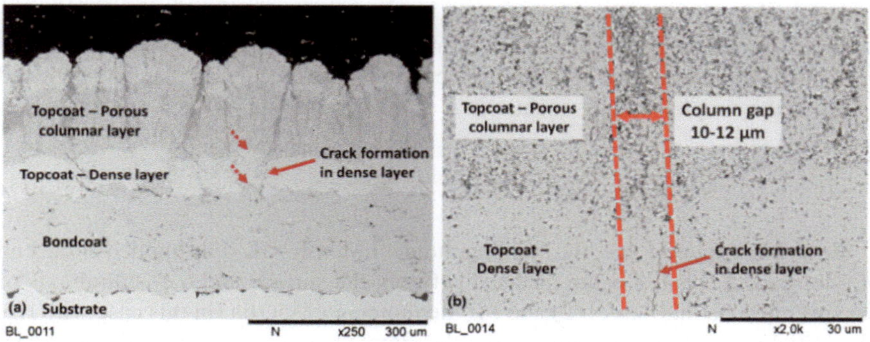

Fig. 9.7 Bilayer TBC microstructure in as-sprayed condition **a** an overview, and **b** the DL-PL interface at a higher magnification. Gupta et al. [2]. Open access

In summary, the life time under thermal shock testing of a bilayer topcoat with dense layer close to the TC-BC interface, followed by a porous layer fabricated by SPS is compared to the single-layer porous columnar SPS topcoat in TBCs. The bilayer TBC exhibited significantly higher thermal shock lifetime as compared to the single-layer TBC. Failure in the single layer TBC occurred in the topcoat near the TC-BC interface compared to failure by cracking in the topcoat near the TC-BC interface and close to the PL-DL interface in the bilayer (see Figs. 9.6 and 9.7). It is seen that the dense layer and vertical cracs/column gaps are beneficial to enhance the TBC lifetime since they reduce the thermal shock induced stresses.

9.2.3 The Effect of Stress Amplitude in Partially Stabilized Zirconia

The cyclic fatigue behavior of 3 mol percent yttria tetragonal zirconia (3Y-TZP), which is known as partially stabilized zirconia is presented below, which is used as dental bridges for extended lifetime. The problem with such and other ceramic materials is their susceptibility to fatigue, which reduces over time the strength with the consequent lifetime reduction. The strength reduction due to fatigue is caused by natural cracks initially present in the microstructure. Crack propagation under subcritical conditions, such as $\sigma < \sigma_c$ and $K_I < K_{IC}$ is a consequence of stress assisted reaction of water which is promoted when cyclic stresses are applied. Clearly dental restorations are exposed to aqueous environment and cyclic loading. Recall that K_I is the stress intensity factor and thus K_{IC} is the critical stress intensity factor.

The strength and lifetime data were obtained using a specially designed bending machine under simple and oscillatory loading conditions, respectively. A bending testing device was used to determine the initial strength (simple bending) and the lifetime (oscillatory bending) of the zirconia specimens. The tensile stress, σ developed on the sample surface upon the application of the force, F can then be calculated based on the simple bending equation:

$$\sigma = \frac{Mz}{I} = \frac{FLz}{I} \tag{9.11}$$

Here, F is applied force, the bending moment $M = F \times L$ (F is applied at the end of a long cantilever), z is the distance between the surface and the neutral axis of the specimen (in this case, h/2) and I is the moment of inertia (in this case $wh^3/12$). The initial strength data obtained from the simple bending tests were described by Weibull statistics, using the equation

$$\ln\ln\left(\frac{1}{1-F}\right) = m\ln\sigma_c - m\ln\sigma_{c,0} \tag{9.12}$$

where F is the probability of failure, σ_c the inert strength, $\sigma_{c,0}$ the characteristic inert strength (at 63% failure probability) and m is the strength Weibull modulus. The failure probability was calculated assuming the relation: $F_i = (i - 0.5/p)$, where F_i is the failure probability of the ith specimen (in a strength-increasing rank) and p is the total number of specimens tested (for the resulting Weibull plot see the original work).

The lifetime of the TZP specimens was also described by a Weibull distribution, according to the equation

$$\ln\ln\left(\frac{1}{1-F}\right) = m^*\ln N_f - m^*\ln N_{f,0} \tag{9.13}$$

where N_f is the number of cycles until failure, $N_{f,0}$ the characteristic number of cycles at 63% failure probability and m* is the lifetime Weibull modulus. The lifetime Weibull plots measured in air and water are seen in Fig. 9.8. The characteristic number of cycles $N_{f,0}$ is the slope intercept with the x axis of the fitting curves and the lifetime modulus m^*. For the subcritical crack growth under cyclic loading the empirical law of Paris can be applied, which correlates the velocity of the propagating crack, υ with the stress intensity factor by

$$\upsilon = A\Delta K_I^n = A(1-R)^n K_{I,\max}^n = A^* K_{I,\max}^n \tag{9.14}$$

where ΔK_I is the amplitude of the stress intensity factor applied ($K_{I,max} - K_{I,min}$), $K_{I,max}$ and $K_{I,min}$ the maximum and minimum applied stress intensity factors, R the stress ratio $K_{I,min}/K_{I,max}$, and A and n are the subcritical crack growth parameters

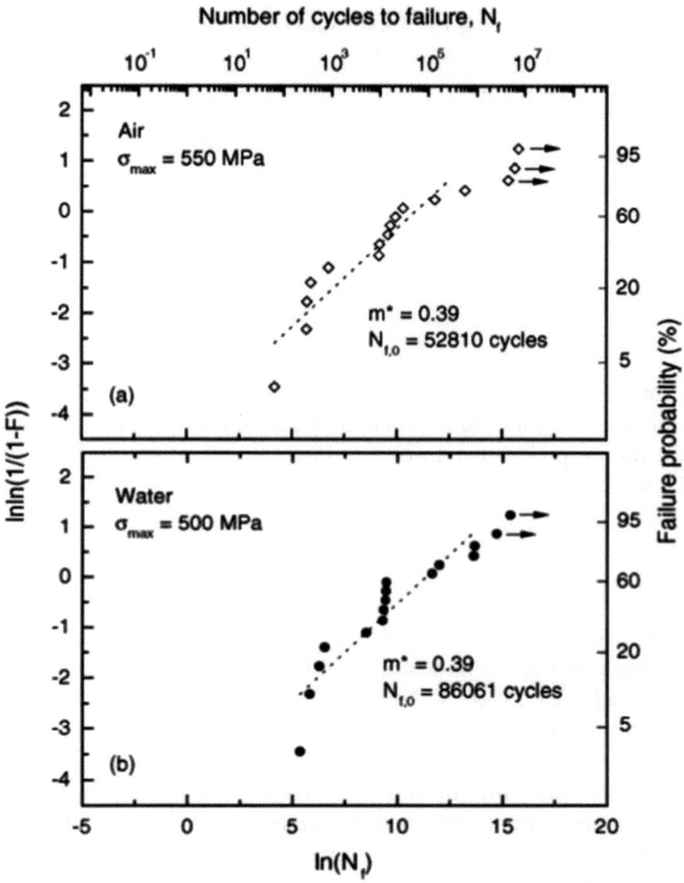

Fig. 9.8 Weibull distribution for the lifetime of TZP specimens measured in **a** air ($\sigma_{max} = 550$ MPa) and **b** water ($\sigma_{max} = 500$ MPa). Studart et al. [3]. With kind permission of Elsevier

for cyclic fatigue. For cyclic loading the crack velocity, v is given by the increment in crack size (da) per number of cycles applied (dN), i.e. da/dN. $K_{I,max}$ and K_I are directly related to the maximum stress σ_{max} and to the stress amplitude $\Delta\sigma$, respectively, through Griffith's law as seen below

$$K_{I,\max} = \Delta\sigma Y\sqrt{a} \tag{9.15}$$

$$\Delta K_I = \Delta\sigma Y\sqrt{a} \tag{9.16}$$

where Y is a geometrical constant that depends on crack location and shape (= 1.3 for surface cracks). For the alternation cyclic stress (R = −1), the stress amplitude

$\Delta\sigma$ is equal $2\sigma_{max}$. Introducing Eq. (9.16) into (9.14) and integration in the limits a_i (the cycles N = 0) and a_c (when $N = N_f$), the results allow the determination of the number of cycles to failure as a function of the applied stress amplitude as

$$N_{f,0} = \frac{2K_{IC}^{2-n}\sigma_{c,0}^{n-2}}{AY^2(n-2)(\Delta\sigma)^2} = \frac{B\sigma_{c,0}^{n-2}}{(\Delta\sigma)^n} \tag{9.17}$$

with

$$B = \frac{2K_{IC}^{2-n}}{AY^2(n-2))} \tag{9.18}$$

For the steps to obtain Eq. (9.17) from Eqs. (9.16) and (9.14) see Munz and Fett.

Assuming that the same flaw population determines the strength and life time (namely the failure probability F_i (σ_c) = F_i (N_i)) and combining Eqs. 9.12, 9.13, 9.17–9.18 one obtains for m^*

$$m^* = \frac{m}{n-2} \tag{9.19}$$

In Eqs. (9.18) and (9.19) direct correlation between Weibull parameters of m^* and $N_{f,0}$, the strength Weibull parameters m and $\sigma_{c,0}$ and the subcritical crack growth parameters A and n are given. These equations were used to assess the subcritical crack growth parameters A and n for the TZP under cyclic loading in air and in water (10 Hz, R = −1). It was found (see Table 9.2) that the crack velocity in water is higher than in air by about one order of magnitude (A is higher). Further, no difference in the n value in air and water is observed as seen in Table 9.2, indicating that the same cyclic degradation of TZP, namely the same operative mechanism is acting in air and water. The subcritical crack growth behavior is illustrated in Fig. 9.9 as plots of crack velocity against the maximum stress intensity factor $K_{I,max}$. As seen the stress intensity factors ($K_{I,max}$) significantly lower than the critical stress intensity factor K_{IC} (namely at 46 and 52% of the critical K_{IC} value in water and in air, respectively).

Table 9.2 Subcritical crack growth parameters A, B and n obtained in this work for the TZP specimens submitted to cyclic loading (10 Hz, R = −1). Studart et al. [3]. With kind permission of Elsevier

3Y-TZP	A	n	A*	B
Air	5.73×10^{-34}	29.00	3.07×10^{-25}	4.83×10^{11}
Water	6.07×10^{-33}	28.95	3.15×10^{-24}	4.97×10^{10}

The values of A, A* and B were obtained using stress intensity factor and velocity data in MPa m$^{1/2}$ and m/cycle, respectively.

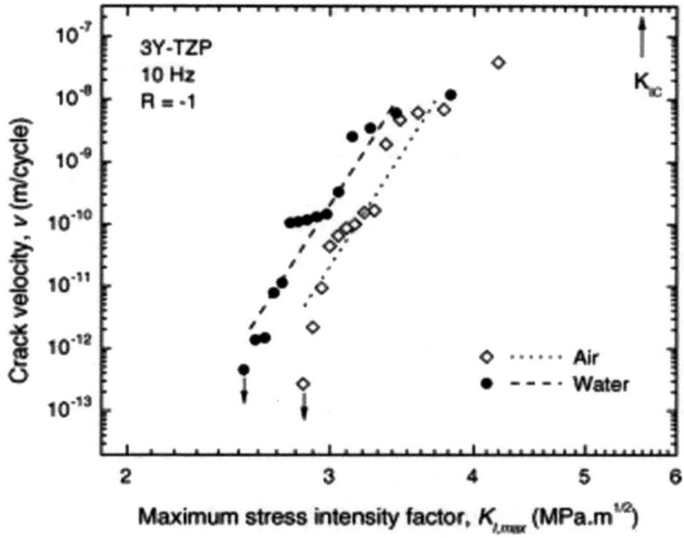

Fig. 9.9 Crack velocity as a function of the maximum stress intensity factor applied to TZP specimens tested under cyclic loading (10 Hz, R = −1) in air and in water. Studart et al. [3]. With kind permission of Elsevier

The arrows in Fig. 9.9 indicates that below it no subcritical crack growth takes place. With the crack growth parameters, A and n, the lifetime of TSZ is evaluated as a Wöhler plot in Fig. 9.10 at any loading conditions with an initial strength of 820 MPa.

In summary, the crack growth parameters of 3 T-TZP in air and in water was evaluated (as mentioned earlier 3 T-TZP is a dental material). 3 T-TZP exhibits significant subcritical crack propagation at stress levels much lower than the critical stress intensity factor. TZP can be used (lifetime more than 20 years) as dental bridges in spite of its susceptibility to cyclic fatigue and water-assisted crack growth as long as their initial mechanical strength is sufficiently high.

9.2.4 The Effect of Stress Amplitude in Si_3N_4

Cyclic fatigue tests of sintered Si_3N_4 were carried out by alternating cantilever bending fatigue tests. The room temperature cyclic fatigue depends mainly on the number of cycles rather than on time. As have been mentioned earlier the interest in Si_3N_4 as a leading candidate for various applications among them as gas turbine rotors, is because of its high mechanical strength and high heat resistance. In addition to the sintered Si_3N_4 room temperature tests were also performed. Thus the tests were performed at room temperature and high temperature for the sake of

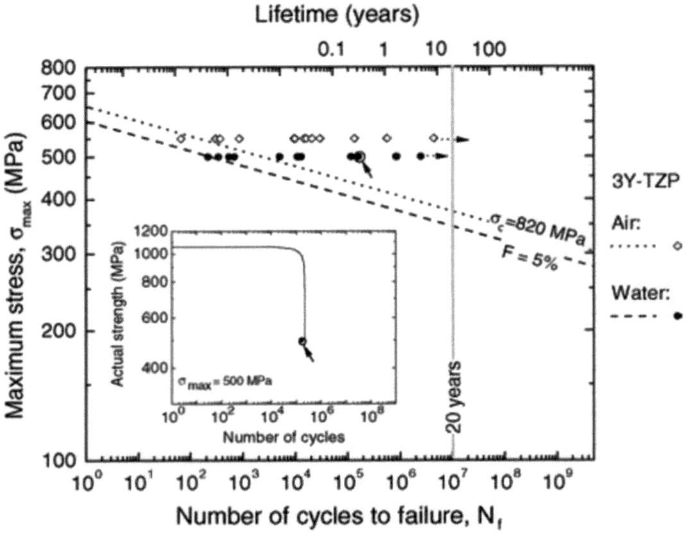

Fig. 9.10 Wohler diagram displaying the number of cycles to failure as a function of the maximum applied stress for a TZP component with initial mechanical strength of 820 MPa. The number of cycles to failure obtained for the TZP samples tested in air and in water are also included in the graph, as well as the lifetime in years assuming a mastication frequency of 1400 cycles/day. The inset graph shows the evolution of actual strength as a function of the number of cycles for the TZP specimen indicated with a circle and a full arrow

comparison of their behavior. The designations of the two kinds sintered specimens are SSN-A and SSN-B. The stress amplitude vs. the number of cycles at frequencies from 0.03 Hz to 3 kHz is illustrated in Fig. 9.11 (tapered beam specimens). The use of high frequencies reduces the time required to test more than 10^{10} cycles. Figure 9.12 shows the mean stress/stress amplitude diagram of SSN-A1 at room temperature, σ_1 on the horizontal axis represents the mean tensile strength. Open circles represent specimens surviving at 10^7 cycles, and closed circles represent the failed specimens. The dotted line represents tensile strength. The solid line from σ_e to σ_t is the modified Goodman line, given by Eq. (9.20), which is usually used for design purposes.

$$\sigma_a = \sigma_e\left[1 - \left(\frac{\sigma_m}{\sigma_t}\right)\right] \tag{9.20}$$

where σ_a is the stress amplitude, σ_m is the mean stress, σ_t is the tensile stress and σ_e is the alternative fatigue strength at a stress ratio $R = -1$. The effect of volume on cyclic fatigue tests under alternative load was examined on two specimens with different volumes and is illustrated in Fig. 9.13, where the stress amplitude is plotted against the number of cycles. The tests carried out one a tensile specimens and a cantilever beam specimens of SSN-B. The fatigue strength at the same number of

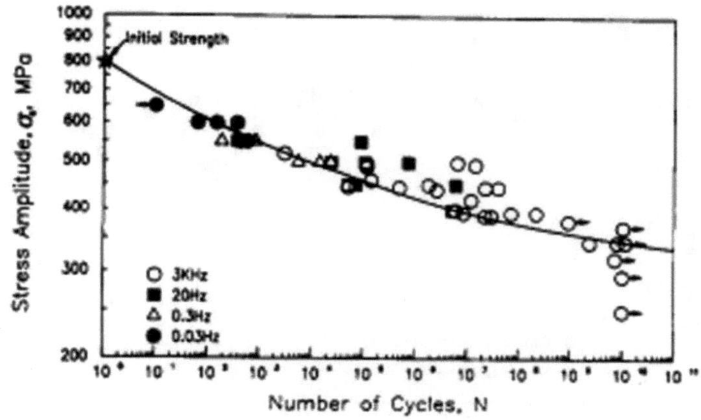

Fig. 9.11 S–N curve of cantilever beam specimens of SSN-A2 under alternating load, at frequencies from 0.03 Hz to 3 kHz. Masuda et al. [4]. With kind permission of Elsevier

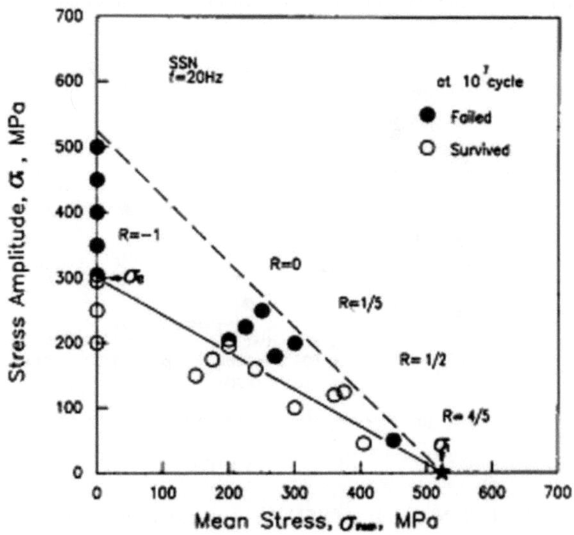

Fig. 9.12 Mean stress versus stress amplitude for fatigue failure in SSN-A1 at 107 cycles measured by high cycle fatigue test at room temperature. The solid line represents tensile strength. Masuda et al. [4]. With kind permission of Elsevier

cycles decreased with increasing effective volume. However, the slope of the S–N curve of the cantilever beam specimens, which have a small effective volume, is larger than that of tensile specimens, which have a large volume. The threshold stress of cyclic fatigue may be obtained by extrapolating for the point at which two curves intersect.

The fatigue behavior at high temperatures of 800 and 1000 °C is compared in Fig. 9.14 that for the sintered Si_3N_4 with that at room temperature. The plot is a

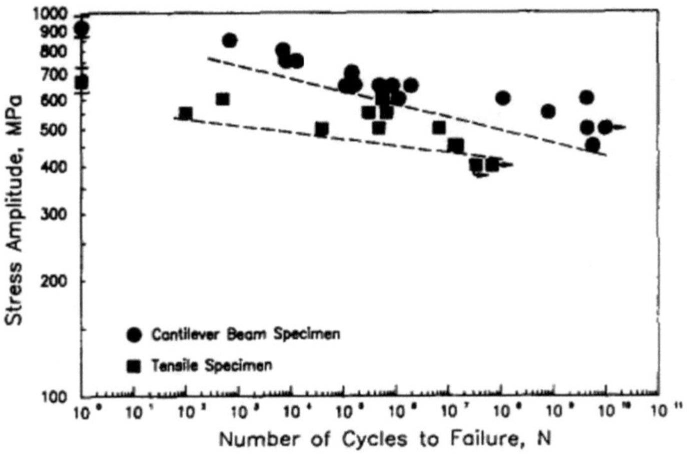

Fig. 9.13 The effect of volume on cyclic fatigue at room temperature. Cantilever beam specimens and the tensile specimens of SSN-B were used; the former have a small volume and the latter have a large volume. Masuda et al. [4]. With kind permission of Elsevier

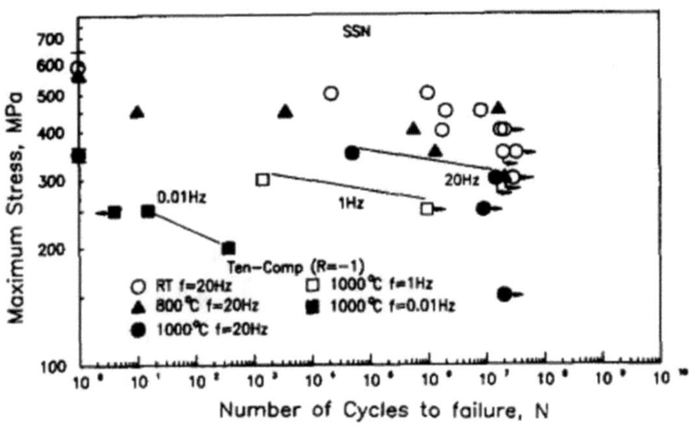

Fig. 9.14 S–N diagram of SSN-A3 using axial tension–compression load control at room temperature, 800 and 1000 °C. The symbols on the vertical axis represent the tensile strength of the materials used. Masuda et al. [4]. With kind permission of Elsevier

log–log plot of stress vs. the number of cycles to failure. The mean tensile strengths were 585, 560 and 360 MPa, respectively. Arrows pointing to the left indicate specimen fracture immediately on loading and in those pointing to the right tests were terminated at the cycles indicated without failure. In spite of the great scatter it can be determined that the fatigue life greatly increased as the stress amplitude decreased.

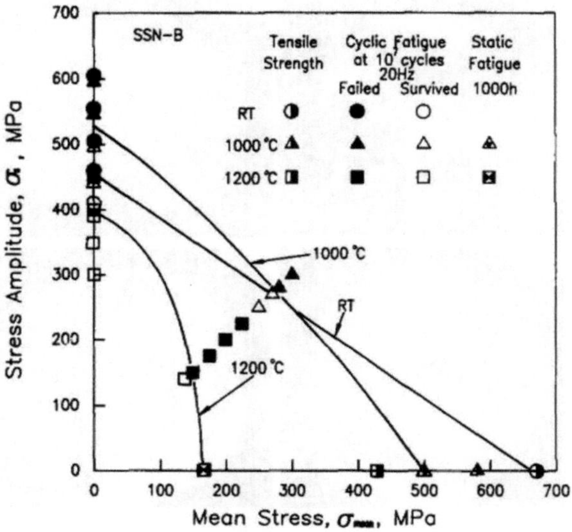

Fig. 9.15 Tensile mean stress effects on cyclic fatigue of SSN-B at room temperature, 1000 and 1200 °C. Masuda et al. [4]. With kind permission of Elsevier

The effects of the mean stress and static fatigue on cyclic fatigue at RT, 1000 and 1200 °C are seen in Fig. 9.15. The mean stress effect is expressed by Eq. (9.21) as

$$\left(\frac{\sigma_a}{\sigma_e}\right)^n + \left(\frac{\sigma_m}{\sigma_s}\right)^n = 1 \tag{9.21}$$

Recall the several empirical relationships expressed in Eqs. (1.11) Chap. 1 that are similar to Eq. (9.21). Here, σ_a is stress amplitude, σ_m mean stress, σ_e alternating fatigue strength and σ_s static fatigue strength. The n value increased with increasing temperature. The vertical axis represents a fully reversed fatigue condition and the horizontal axis represents static fatigue strength and ultimate tensile strength. With increasing temperature, the static fatigue strength decreased.

The fracture origins and failure mode observed by optical microscope are illustrated as fractured surfaces in Fig. 9.16. Pores and inclusions are preferential sites for crack nucleation and propagation. Microcracks at grain boundaries were observed in specimens that failed under cyclic load, as shown in Fig. 9.17.

In summary, Microcracks is caused by stress concentration around an initial flaw or heterogeneous portion in the microstructure. Cyclic fatigue failure occurs as the initial flaw grows to its critical size during testing (or in use). Crack propagation is faster under cyclic load than under static load. Strength degradation by cyclic fatigue at room temperature depends rather on the number of cycles than on time. At high temperatures frequency is the dominant factor determining the resistance to fatigue;

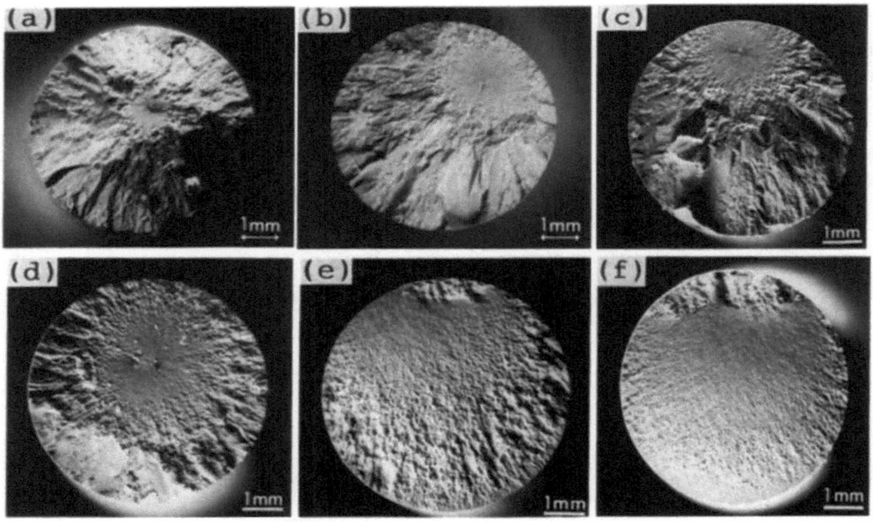

Fig. 9.16 Typical fracture surfaces of fatigue specimens of SSN-A3. **a** RT, 20 Hz, ± 500 MPa, 2.2 × 104 cycles; **b** RT, 20 Hz, + 400 MPa, 1.8 × 107 cycles; **c** 800 °C, 20 Hz, ± 350 MPa, 1.3 × 106 cycles; **d** 1000 °C, 20 Hz, ± 350 MPa, 5.0 × 104 cycles; **e** 1000 °C, 1 × 10^2 Hz, ± 200 MPa, 3.7 × 10^2 cycles; (f) 1000 °C, 150 MPa, 204 h. Masuda et al. [4]. With kind permission of Elsevier

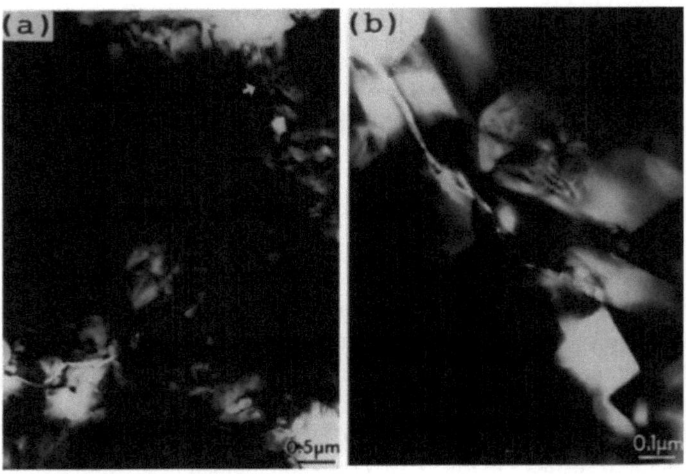

Fig. 9.17 Typical microstructures of SSN-A3 after fatigue test at room temperature. **a** ±450 MPa, 8.2 × 10^6 cycles; **b** enlarged view of area marked with arrow in (a). Masuda et al. [4]. With kind permission of Elsevier

the lower the frequency the less resistant is Si_3N_4 to fatigue. It has been shown that the mean stress and stress amplitude effect the fatigue strength (Fig. 9.15). Cyclic fatigue progresses by formation of microcracks at grain boundaries and at high temperatures in the slow crack growth creep deformation might be involved.

9.3 The Effect of Strain Amplitude

9.3.1 Introduction

The strain amplitude, the half-width of the stress–strain hysteresis loop at zero stress is an important parameter in low cycle fatigue (LCF) tests because it is directly connected to the fatigue life of structural components by the Coffin-Manson relationship.

In the strain–life approach seen in Fig. 9.18 the relationship between the total strain amplitude and the number of load reversals to failure is given as

$$\frac{\Delta\varepsilon}{2} = \frac{\sigma_f^{'}}{E}(2N_f)^b + \varepsilon_f^{'}(2N_f)^c \tag{9.22}$$

where $\varepsilon_f^{'}$ is the fatigue ductility, c is the fatigue ductility exponent, $\sigma_f^{'}$ is the fatigue strength coefficient, $\Delta\varepsilon$ is the strain amplitude range, $2N_f$ is the number of load reversals to failure, and b is the Basquin exponent. The plot is expresses in log–log coordinates.

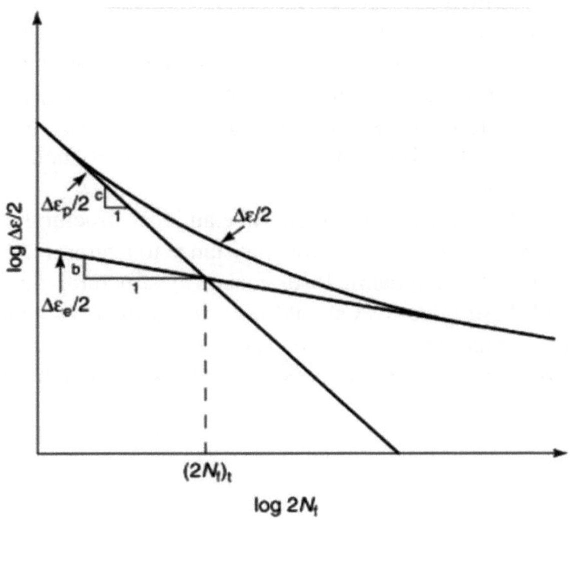

Fig. 9.18 Schematics of the plot of elastic and plastic strain amplitude vs. life showing total strain amplitude. McBagonluri and Soboyejo [5]. Open access

In Fig. 9.18 the representation of the upper curve is by separate power law fits to stress amplitude vs. life, and a plastic strain amplitude vs. life, which are marked as $\frac{\Delta\varepsilon_e}{2}$ and $\frac{\Delta\varepsilon_p}{2}$, respectively for the elastic and plastic contributions to the total strain amplitude $\frac{\Delta\varepsilon}{2}$. Often the total strain amplitude, $\frac{\Delta\varepsilon}{2}$ is denoted as ε_a. In the figure the elastic and plastic denotations are marked in Eq. (9.22) as σ'_f and ε'_f. The four constants of Eq. (9.22), namely σ'_f, b, ε'_f an c can be determined from Fig. 9.18. The intercepts at $N_f = ½$ are σ'_f and ε'_f, while the exponents b and c give the slopes on the log–log plot; these values are negative due to the downward slopes.

9.3.2 *Effect of Strain Amplitude in SiC_f/SiC_m*

Unfortunately, no data on the effect of the strain amplitude of pure SiC is available in the open literature. The main reason might be associated with its use as reinforcing agent. As a mater matter of fact limited or no data exist on the other oxides which are the subject of this book, despite the abundant information for metals, such as Al, C, steel etc.

The silicon carbide fiber, SiC_f and silicon carbide matrix, SiC_m composite, SiC_f/SiC_m was fabricated in an argon atmosphere at constant strain amplitude in the frequency range of 1–50 Hz. The strain amplitude ranged from 1 to 11 μm at the constant frequency of 10 Hz. The influence of the microstructure on the dynamic mechanical properties, among them the internal friction and storage modulus of this composite after high-temperature treatment (800 °C and 1400 °C) in the air was investigated by three-point bending vibration test. The as-prepared samples have maximum storage modulus and internal friction, while the composites treated at 800 °C in the air have the minimum storage modulus due to a large number of defects produced within the composite structure, and the composites treated at 1400 °C have the minimum internal friction due to the formation of α-cristobalite. in the interface between the matrix and fibers, resulting in stronger interface bonding. Further, the storage modulus is sensitive to amplitude but not frequency, but the internal friction is sensitive to both, the frequency because of anelasticity and the amplitude due to the static hysteresis.

SiC_f/SiC_m is considered to be an ideal structural material in turbine engines due to its light weight and large resistance to aggressive environments. Turbine engines suffer severe dynamic loads caused by structural vibration and noise during service, and these vibrations should not exceed their resonance limit, so as not to damage the engine. Internal friction tests can be used as a non-destructive testing method to assess the damping process of materials through the evolution of the microstructure.

The as-prepared SiC_f/SiC_m composite is shown in Fig. 9.19, while Fig. 9.20 the surface morphology of the heat treated materials in air at 800 °C and 1400 °C for 1 h are shown.

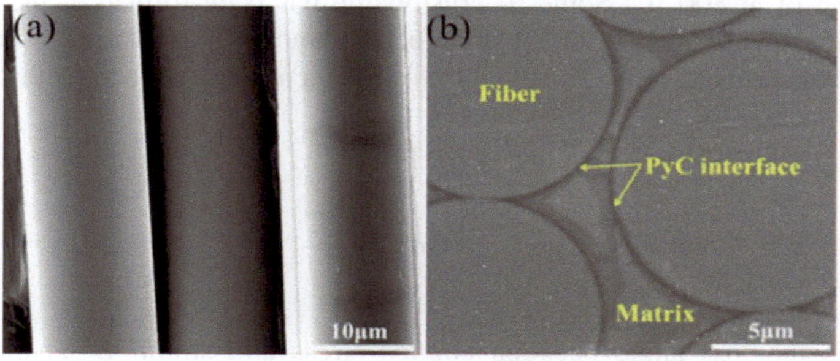

Fig. 9.19 Microstructure morphologies of the as-prepared SiC_f/SiC_m composite. **a** warp direction and **b** weft direction. Yang et al. [6]. Open access. Pyrocarbon (PyC) is interfacial phase between fiber and matrix. The single PyC is used as a coating of the fiber bundels

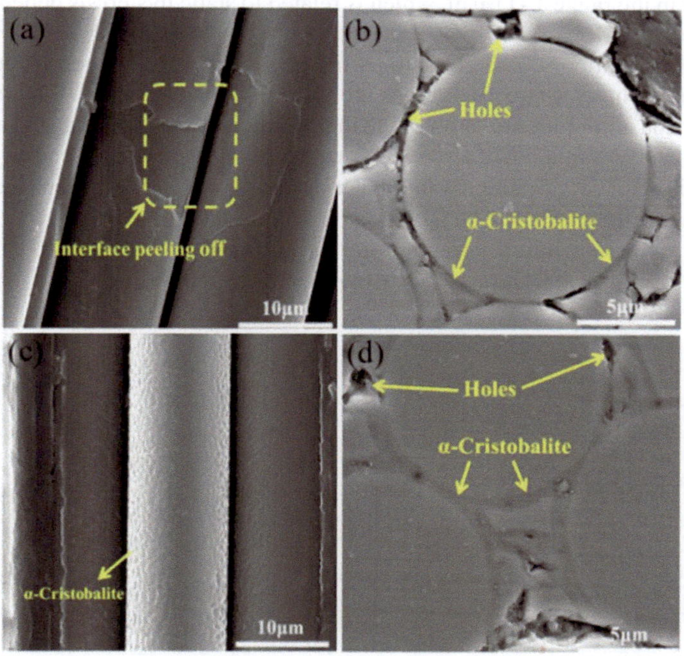

Fig. 9.20 Microstructure morphologies of the 2.5D SiC_f/SiC_m composites under different treatment conditions. (**a, b**) A-800, (**c, d**) A-1400. Yang et al. [6]. Open access

The PyC layers partially peeled off from the fiber surface following heat treatment at 800 °C as seen in Fig. 9.19a and some holes also developed at the interface as shown in (b).

It seems that the interface PyC layers oxidation occurred resulting in debonding and peeling of at 800 °C. Some α- crystobalite has been formed at the interface between fibers and matrix. The oxidation became more severe at 1400 °C as shown in Fig. 9.20c and with more crystobalite formation at the interface, with holes still present as seen in (d).

The internal friction, Q^{-1} is calculated by

$$Q^{-1} = \tan\delta = \frac{E''}{E'} \quad (9.23)$$

where δ is the loss angle between applied stress and strain, E' is the storage modulus and E'' is the loss modulus. Experimental results show that the as-prepared composite, A-P has the highest internal friction values, while the A-1400 composite has the lowest values with the change in frequency and amplitude as seen in Fig. 9.21. As seen in Fig. 9.21 (b), the internal friction increases with the increase of strain amplitude in the range 1 -11 μm, which is related to static hysteresis dependence on amplitude.

In summary, after 800 °C for 1 h, partial peeling off of the interface and a few α-cristobalite phases were observed on the surface of the composite. The increase in frequency and amplitude are observed to increase the internal friction, and a too high frequency and amplitude will also cause resonance of the material, leading to cyclic fatigue failure. The vibration of the material has to be reduced to ensure the safe use of the turbine engines. The heat treated samples at 1400 °C (A-1400 samples) possessed minimum internal friction due to the formation of α-cristobalite at the interface. The internal friction is sensitive not only to frequency but also to the strain amplitude.

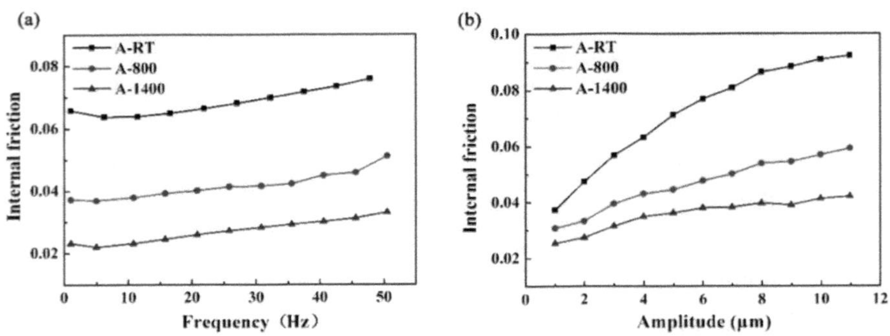

Fig. 9.21 Plots of the internal friction of the composites versus frequency (**a**) and amplitude (**b**). Yang et al. [6]. Open access

9.3.3 *Effect of Strain Amplitude in Si_3N_4*

The change in the cyclic stress–strain at the early stage of strain cycling in the monolithic-Si_3N_4(A) is controlled by strain amplitude at $\Delta\varepsilon_t/2 = 1.04 \times 10^{-3}$ and at the strain rate $\dot{\varepsilon}_t = 5.0 \times 10^{-6}\ s^{-1}$. This is shown in Fig. 9.22. The increase in the stress and decrease in the inelastic strain are found in the cyclic straining process, which occur mainly on the tension stress and strain sides. Consequently, the hysteresis loop shifted to the positive strain- and the negative stress-directions at the first strain cycle returns gradually to a symmetric position with respect to the origin.

Push–pull low cycle fatigue tests were performed in monolithic-Si_3N_4 and composite-Si_3N_4/SiC_w materials at 1573 K (SiC_w stands for whiskers). Two kinds of monolithic specimens (in addition to the composite) were used and are designated by Si_3N_4 (A) and (B). A similar plot to the one shown in Fig. 9.22 is illustrated in Fig. 9.23 also. The subscripts of this curves indicate the different conditions of the tests. The variation of the stress and the inelastic strain ranges in the strain controlled cyclic straining process are shown in Fig. 9.24 for the monolithic-Si_3N_4 (A) at $\dot{\varepsilon}_t = 5.0 \times 10^{-6}\ s^{-1}$ at 1573 K. The gradual increase in the stress range and decrease in the inelastic range are clearly shown with the increasing strain cycles, exhibiting the increasing resistance to cyclic deformation during strain cycling. Cyclic straining at $\dot{\varepsilon}_t = 5.0 \times 10^{-6}\ s^{-1}$ at 1573 K and strain range of $\Delta\varepsilon_t = 2.8 \times 10^{-3}$ at 1573 in the monolithic-Si_3N_4 (B) and the composite-Si_3N_4/SiC_w are shown, respectively in Fig. 9.25. It seems when comparing Figs. 9.25a and Fig. 9.23 for example, most likely due the processing difference (hot-pressing in the former and sintering in the latter). The strain range- and strain rate-controlled low cycle fatigue tests for the monolithic-Si_3N_4(A) and (B), and the composite-Si_3N_4/SiC_w to failure at 1573 K is

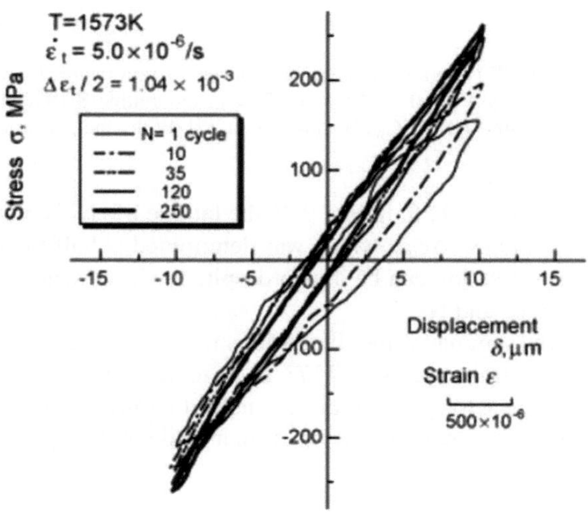

Fig. 9.22 Cyclic stress–strain response at early stage in cyclic triangular strain wave loading process at $\Delta\varepsilon_t/2 = 1.04 \times 10^{-3}$ and $\dot{\varepsilon}_t = 5.0 \times 10^{-6}\ s^{-1}$ at 1573 K in monolithic-Si_3N_4(A): **a** stress range $\Delta\sigma$ vs. N and **b** inelastic strain range $\Delta\varepsilon_{ins}$ vs. N. Hatanaka et al. [7]. With kind permission of Elsevier

Fig. 9.23 Cyclic stress–strain response at early stage in cyclic triangular wave loading process at $\Delta\sigma/2 = 150$ MPa and $\dot{\sigma} = 0.2$ MPa/s at 1573 K in monolithic-Si_3N_4(A). Hatanaka et al. [7]. With kind permission of Elsevier

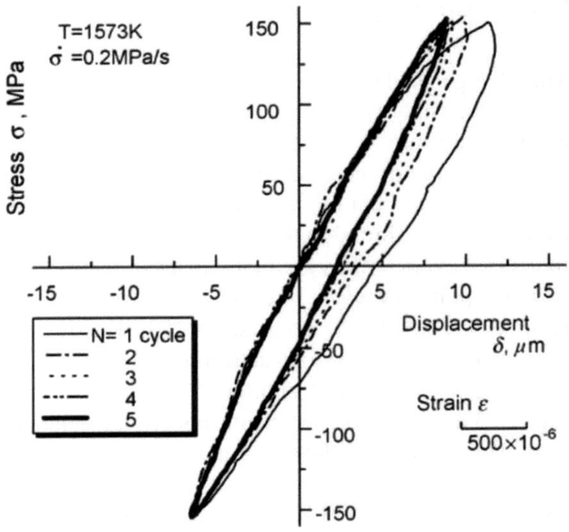

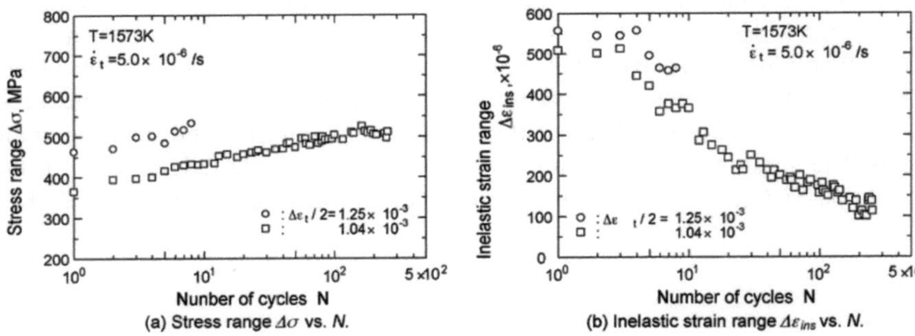

Fig. 9.24 Variations in stress and inelastic strain ranges in cyclic triangular strain wave loading at $\dot{\varepsilon}_t = 5.0 \times 10^{-6}\ s^{-1}$ at 1573 K in monolithic-Si_3N_4(A): (a) monolithic-Si_3N_4(B) and (b) composite-Si_3N_4/SiC_w. Hatanaka et al. [7]. With kind permission of Elsevier

seen in Fig. 9.26. In Fig. 9.26 the fatigue life N_f is plotted against the inelastic strain strain range $\Delta\varepsilon_{ins}$, which was determined at half the number of cycles to failure. A linear relation can be drawn despite the large scatter in the double logarithmic plot of $\Delta\varepsilon_{ins}$ and N_f.

In summary, the inelastic strain decreases with increasing the number of stress/strain cycles at 1573 K in monotonic-Si_3N_4, which indicates an increased resistance to cyclic deformation. Furthermore, the resistance to the inelastic strain is much less under tensile stress than under compressive stress.

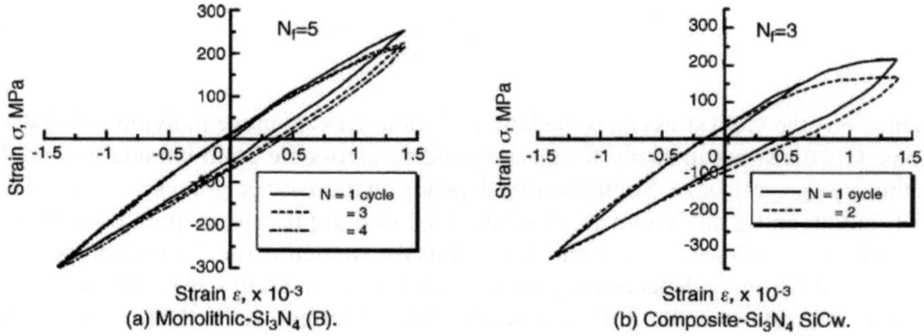

Fig. 9.25 Cyclic stress–strain response under cyclic straining controlled at strain rate $\dot{\varepsilon}_t = 5.0 \times 10^{-6}\ \mathrm{s}^{-1}$ and strain range $\Delta\varepsilon_t/2 = 2.8 \times 10^{-3}$ at 1573 K in monolithic-Si_3N_4(B) and composite-Si_3N_4/SiC_w: **a** monolithic-Si_3N_4(B) and **b** composite-Si_3N_4/SiC_w. Hatanaka et al. [7]. With kind permission of Elsevier

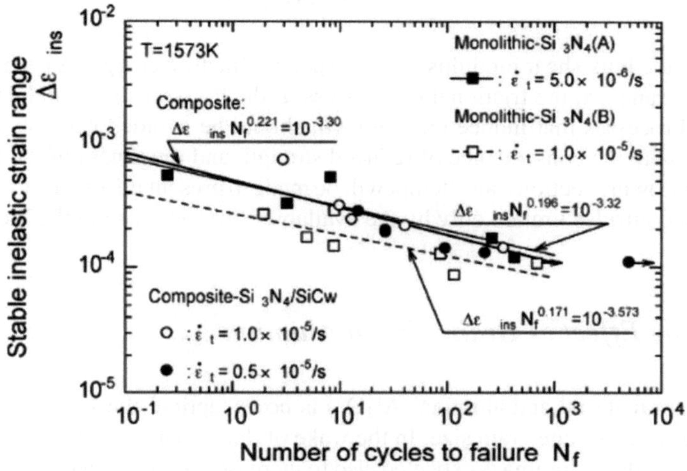

Fig. 9.26 Low cycle fatigue life plotted against inelastic strain range at 1573 K in monolithic-Si_3N_4 (A) and (B), and composite-Si_3N_4/SiC_w: **a** monolithic-Si_3N_4 (B) and **b** composite-Si_3N_4/SiC_w. Hatanaka et al. [7]. With kind permission of Elsevier

9.4 The Effect of Grain Size

Grain size is one of the most important factors that influences the mechanical properties of materials. The Hall–Petch relation predicts that yield strength increases with decreasing grain size, a fact that has been proven experimentally. The Hall–Petch relationship is expressed as

$$\sigma = \sigma_0 + \frac{k}{\sqrt{d}} \tag{9.24}$$

Where σ is the yield stress σ_0 is the lattice friction stress to move individual dislocations, k is the stress intensity for plastic yielding across the grain boundaries and d is the average grain size. Strengthening depends on the average grain size, grain size distribution and grain boundary structure. Dislocation pile ups at grain boundaries provide the resistance to deformation and thus the strengthening. Fatigue properties of materials strongly depends on grain size, an increase in grain size results in reduction of the endurance limit. It is expected that the fatigue strength increases with decreasing inverse grain size compared to the large grain size when tested under the same number of cycles. Tanaka and Mura relates the number of cycles for crack nucleation to the inverse grain size by

$$N_n = \frac{4GW_s}{\pi(1-\upsilon)d(\Delta\tau - 2\tau_i)^2} \tag{9.25}$$

where G is the bulk shear modulus, W_s the specific fracture energy, Δτ the resolved shear stress range, τ_i the frictional shear stress, υ the Poisson's ratio and d the grain diameter. Processes that induce grain growth should be avoided because softening may set in with the consequence of reduced strength and toughness of the material.

In the following sections, an attempt will be made to present information on oxides, carbides and nitrides limited only by the availability of such data in the literature.

9.4.1 The Effect of Grain Size in Alumina

Generally, in material and alumina (Al_2O_3) is no exception, the strength increases with decreasing average grain size. In the wake of this experimental observation the fracture strength of alumina is expected also to increase. One can state based on these experimental facts that generally the mechanical properties of alumina are grain size dependent and the fine-grained structure have superior properties over the coarse-grained counterpart. However, the strength of ceramic is very often-almost invariably-limited by brittle failure. Therefore, it is very important to design appropriately the microstructure of ceramics for improving the fracture properties. Grain size is one of the methods to design a ceramic for optimal performance in service. Of the many fracture properties grain size change can affect the variation of toughness, hardness, density etc., and alumina is one of the primary representatives of ceramics influenced by these qualities. Due to these attributes alumina is of technical and commercial interest for many engineering requirements.

At a certain grain size, the alumina similarly to other ceramics exhibit a maximum as illustrated in Fig. 9.27 where a plot of the fracture energy vs. the grain size

Fig. 9.27 Measured variation of toughness with grain size for alumina. Bower and Ortiz [8]. With kind permission of ASME

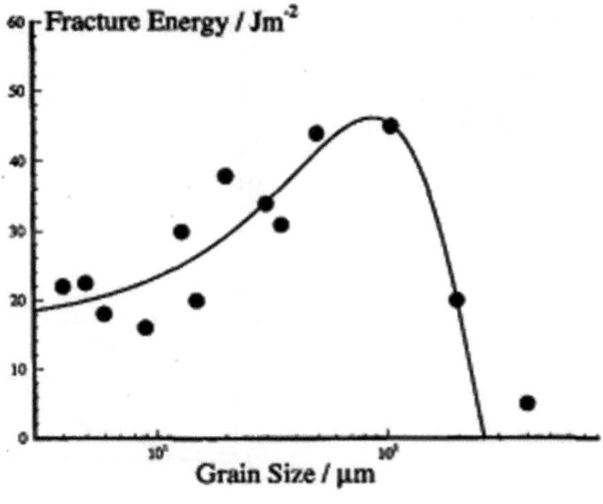

variation is illustrated. It has been indicated that hardness and density also are grain size dependent. (Fig. 9.28).

During the use (or test) before fracture cracks resulting from microcrack coalescence propagate through the ceramics. The microcrack density variation with grain size is illustrated in Fig. 9.29. The residual stress is also displayed. The variation of toughness with grain size for alumina is shown in Fig. 9.29. The relations used to obtain the predicted illustrations and their deviation is fully described in the work of professor Bower and Ortiz, and the interested readers should consult their publication.

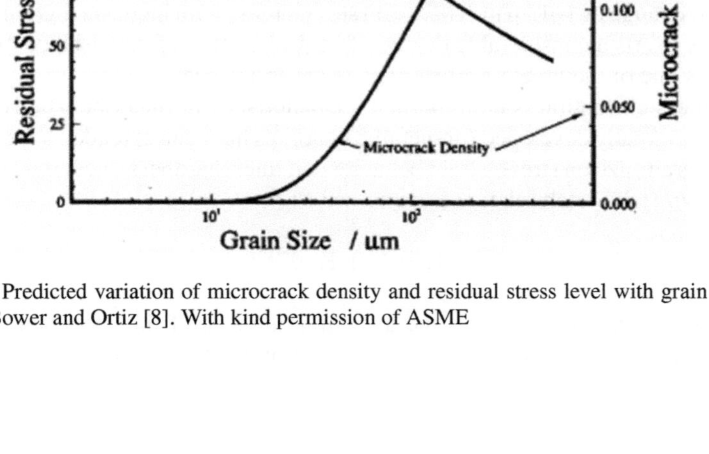

Fig. 9.28 Predicted variation of microcrack density and residual stress level with grain size. for alumina. Bower and Ortiz [8]. With kind permission of ASME

Fig. 9.29 Predicted variation of toughness with grain size for alumina. Bower and Ortiz [8]. With kind permission of ASME

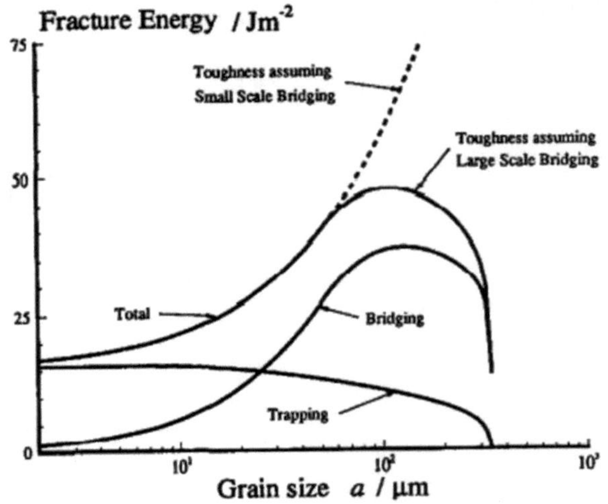

One application of alumina involves protection against abrasion and erosion wear. The wear properties of a ceramic can be improved not only by controlling its material characteristics but also by controlling the fabrication process, which defines the material's microstructure. Of the various factors that influence the wear properties of the alumina (such as dopant diffusivity at grain boundaries, the fabrication route, the sintering mechanism) the grain size is the subject of this section and of current interest. Just to recall that the wear of ceramic materials is an issue of great industrial importance because wear leads to the frequent replacement of components. The proper selection of wear-resistant materials increases the useful life of equipment and considerably reduces maintenance costs. Alumina is the most commonly used material in applications requiring wear resistance because its good fracture toughness, high hardness, high erosion resistance, appropriate for high temperature applications and last but not least its relatively low cost and availability. The strong dependence of the wear rate on the alumina grain size is shown in Fig. 9.30. In this figure the behavior of alumina is shown when undergoing erosive wear, with grain sizes in the range of 1–15 μm. The wear rates increase with increasing grain size. The grain size exerts a considerable degree of control on the wear resistance of alumina and it has one of the greatest effect on its wear resistance.

Hardness evaluation is an important characteristics of material. The Hall–Petch relation-originally developed for metallic materials has been extended to describe the effect of grain size on stress–strain behavior and microhardness of ceramic materials. The known Hall–Petch relation is expressed as.

$$\mathrm{H} = \mathrm{H} + \mathrm{K_H d^{-1/2}} \tag{9.26}$$

H is the microhardness, d is the average grain size of the specimen, H_0 is the intrinsic grain microhardness and is dependent on the frictional lattice resistance to

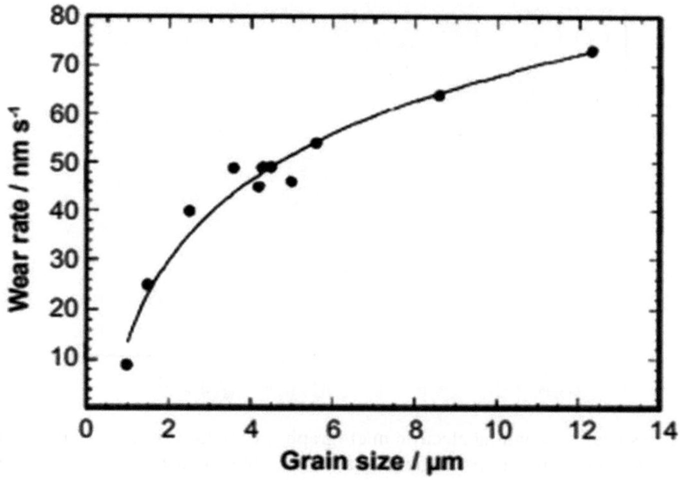

Fig. 9.30 Wear rate-grain size dependence in the wet erosive wear of pure polycrystalline alumina. Milak et al. [9]. Open access

dislocation motion, and K_H is the Hall–Petch constant, which is a material-specific strengthening coefficient. Particularly, it has been shown that the plastic deformation of Al_2O_3 due to room temperature abrasion of the surface occurs by basal twinning and pyramidal slip defined by $\frac{1}{3}\langle 11\bar{2}0\rangle\{\bar{1}101\}$. Further, it was found that in low angle grain boundaries in alumina bicrystals that edge and screw dislocations with Burgers vector $\frac{1}{3}1\bar{1}20$ are present.

Two different aluminas were used in the experiments. The alumina powders were sintered using spark plasma sintering (SPS) technique. The sintering temperatures, the resulting grain sizes and the relative densities are listed in Table 9.3. Micrographs

Table 9.3 Sintering temperature, grain size and relative densities for Al_2O_3, of sintered samples. All specimens were prepared at a heating rate of 10 K/min. Novitskaya et al. [10]. With kind permission of Elsevier

Specimen	Sintering temperature (K)	Average grain size (nm)	Relative density (%)
Al_2O_3 (E)	1233	164 ± 33	96.0 ± 0.8
	1273	259 ± 64	98.5 ± 0.2
	1323	339 ± 115	99.1 ± 0.6
	1423	1461 ± 361	99.3 ± 0.2
	1473	3689 ± 1293	99.1 ± 0.6
	1573	10572 ± 2329	98.5 ± 0.8
Al_2O_3 (C)	1353	437 ± 195	99.5 ± 0.9
	1423	3419 ±1353	99.3 ± 0.7
	1473	4650 ± 1501	97.8 ± 0.5
	1573	15991 ± 4105	99.1 ± 0.4

Fig. 9.31 Representative scanning electron micrographs of fracture surfaces of the sintered specimens demonstrating the grain morphologies for (a) Al_2O_3 sintered at T = 1423 K. Novitskaya et al. [10]. With kind permission of Elsevier

of the cross sections of the fractured samples were obtained by scanning electron microscopy (samples were coated with iridium to prevent charging) and a representative sample of the fracture surface is illustrated in Fig. 9.31. For the grain size analysis, more than 300 grains for each sample were analyzed. Vickers indentation was performed on prepared, polished and clean surfaces. The Hall–Petch correlation between microhardness and grain size for alumina sample is shown in Fig. 9.31. The plot is linear only for the current experimental data.

It has been shown that the Hall–Petch correlation between microhardness values and grain sizes applies also to alumina sintered by SPS. Vickers indentations was applied to evaluate the mechanical properties. Alumina has high hardness and toughness values of 20 GPa and 5.8 MPa $m^{1/2}$, respectively. The number of slip systems could be the main reason for the improved mechanical performance of the alumina which consists of only17 slip systems.

9.4.2 *The Effect of Grain Size in Y-TZP*

The grain size effect in bending tests of 2 mol% yttria tetragonal zirconia polycrystals (2Y-TZP) is presemtened in this section. The grain size is in the range of 0.2–1.6 μm. To obtain different grain sizes, the powder (containing the desired proportions of the components) after compacting and isostatic pressing (at 400 MPa) was sintered at temperatures between 1400 °C and 1600 °C for 0.5 to 10 h. The surface residual stress induced by machining the specimens was eliminated by a heat treatment at 1350 °C for 2 h in air. Static and cyclic fatigue tests at stress ratios R = 0.2 and −1, respectively were performed. Table 9.4 shows the average grain size, achieved densities and mechanical properties such as fracture toughness and Young's modulus

Table 9.4 Sintering conditions, average grain sizes, porosities and mechanical properties of the 2Y-TZP ceramics. Matt et al. [11]. With kind permission of Elsevier

Materials	Sintering conditions	Average grain size (μm)	Density (% theor. density)	Fracture toughness K_{Ic} (MPa$\sqrt{m}$)	Young's modulus (GPa)
2Y-TZP-I	1400°C, 0.5 h	0.2	99.7	5.1	205
2Y-TZP-II	1450°C, 2.0 h	0.3	99.6	5.2	205
2Y-TZP-III	1500°C, 2.5 h	0.6	99.4	5.3	200
2Y-TZP-IV	1600°C, 2.5 h	0.9	98.3	4.4	188
2Y-TZP-V	1600°C, 10 h	1.6	96.4	4.6	174

of the investigated ceramics as produced under different sintering conditions. As seen in Table 9.4, 5 materials were used in the experiments. Table 9.5 lists the experimental fatigue data obtained in the tests (Fig. 9.32).

In the cyclic fatigue test the maximum number of cycles was fixed at $N_L = 2 \times 10^6$. Under cyclic loading with the stress ratio R = −1 most specimens failed within this limit even when the maximum stress applied was lower compared to the tests with R = 0.2 (see Table 9.5). Figure 9.33 shows the grain size dependence of

Table 9.5 Compilation of experimental fatigue data. Number of specimens (i) failed during the first loading cycle, (f) failed during fatigue test, (s) not failed after the limit number or lifetime of cycles. Matt et al. [11]. With kind permission of Elsevier

Materials	Loading conditions	Stress (MPa)	Number of specimens (i)	(f)	(s)
2Y-TZP-I	Static	$\sigma_s = 690$	1	7	15
	$R = 0.2$	$\sigma_{max} = 600$	1	8	15
	$R = -1$	$\sigma_{max} = 600$	2	21	0
2Y-TZP-II	Static	$\sigma_s = 690$	2	4	18
	R = 0.2	$\sigma_{max} = 600$	0	3	21
2Y-TZP-III	Static	$\sigma_s = 710$	1	8	15
	$R = 0.2$	$\sigma_{max} = 620$	0	9	15
	$R = -1$	$\sigma_{max} = 550$	0	18	6
2Y-TZP-IV	Static	$\sigma_s = 640$	0	16	8
	$R = 0.2$	$\sigma_{max} = 560$	0	15	9
	$R = -1$	$\sigma_{max} = 500$	2	22	0
2Y-TZP-V	Static	$\sigma_s = 450$	0	7	17
	$R = 0.2$	$\sigma_{max} = 400$	3	5	16
	$R = -1$	$\sigma_{max} = 350$	0	18	6

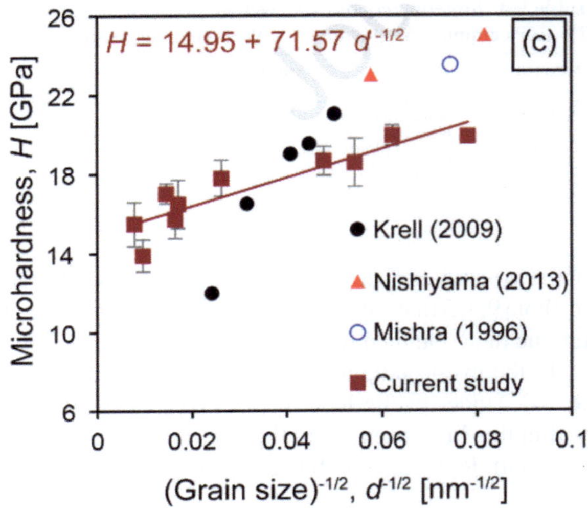

Fig. 9.32 Hall–Petch plot with data from previous studies, but the linear fit only to data for the current data. Novitskaya et al. [10]. With kind permission of Elsevier

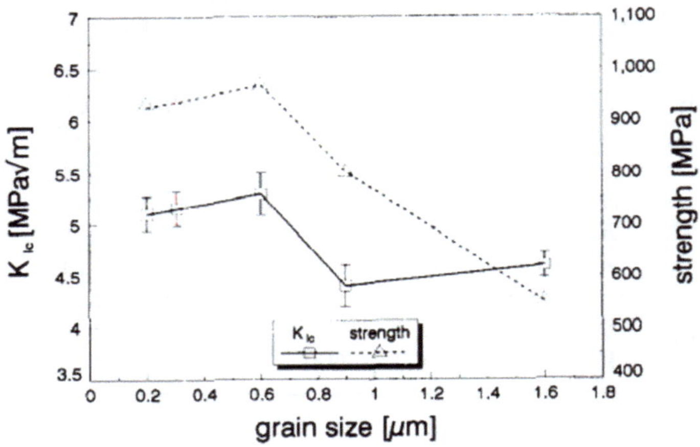

Fig. 9.33 Flexural strength and fracture toughness as a function of grain size. Matt et al. [11]. With kind permission of Elsevier

the flexural strength and fracture toughness. Due to the higher porosity and larger grain size, the flexural strength of the materials 2Y-TZP-IV and -V is considerably lower. In Fig. 9.34 the amount of the monoclinc phase on the fractured surfaces in as-sintered and as-ground state of static and cyclic fatigue tests as a function of grain size is shown. In the figure m, c and t stand for monoclinic, cubic and tetragonal phases within 2Y-TZP, and the monoclinic phase with respect the total phases is illustrated on the ordinate. The amount of the stress induced t-m transformation at the ground surface showed a nearly linear increase with increasing grain size, while the increase in the as-sinterd case is much less significant (Fig. 9.34).

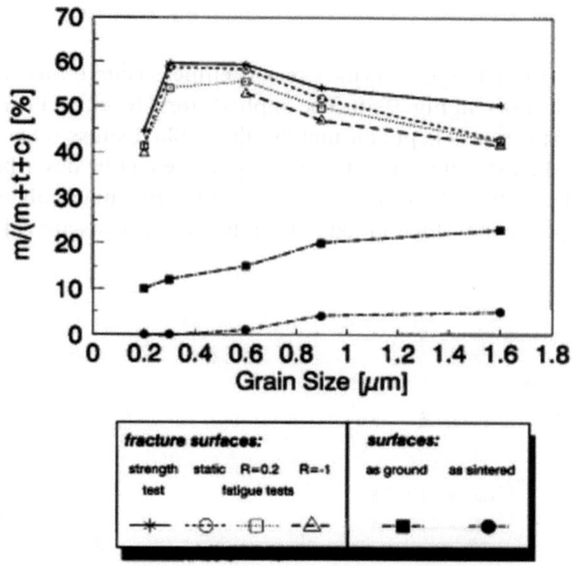

Fig. 9.34 Monoclinic ZrO_2 phase content in as-sintered and as-ground states, at fracture surfaces after initial strength measurements and after static and cyclic fatigue tests as a function of grain size. Matt et al. [11]. With kind permission of Elsevier

Figure 9.35 shows the correlations of the static and cyclic fatigue thresholds K_{10} as a function of grain size and porosity. As a matter of fact, the ratio K_{10}/K_{Ic} is plotted on the ordinate. The fatigue thresholds are determined by the fact that some specimens withstand fatigue loading without failure (survivor specimens). The threshold ratio K_{10}/K_{Ic} can then be calculated, with the ratio of the applied stress, σ_a and the expected (initial) strength of the first survivor specimen $\sigma_{i(n+1)}$, taken from the reference samples (Fig. 4 of Mat et al.).

The expression for the rato is given in Eq. (9.27).

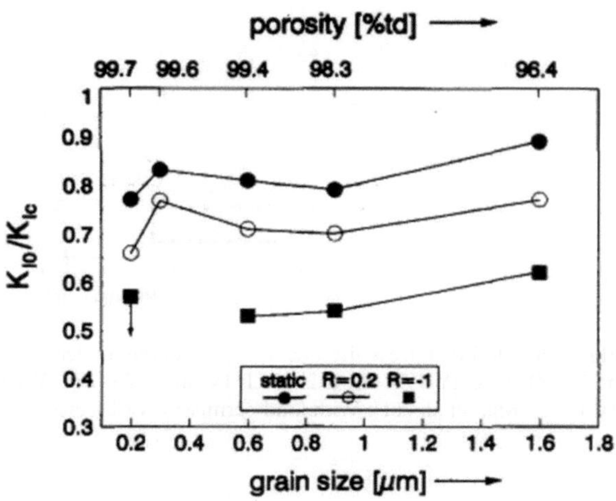

Fig. 9.35 Fatigue thresholds of the five 2Y-TZP ceramics tested. Matt et al. [11]. With kind permission of Elsevier

$$K_{10}/K_{Ic} = \sigma_a/\sigma_{i(n+1)} \tag{9.27}$$

From fatigue data the stress-lifetime, s/t curve was evaluated for the five specimens as shown in Fig. 9.36. The applied stress level, σ_a in each fatigue test of a particular test series is kept constant but the applied stress intensity is different for each individual specimen and the time to failure can be described as σ_a/σ_i (see Fig. 9.36). In these stress-lifetime (s/t) curves of the five materials -shown in Fig. 9.36-, only the specimens which failed during the fatigue tests are considered (see Table 9.5).

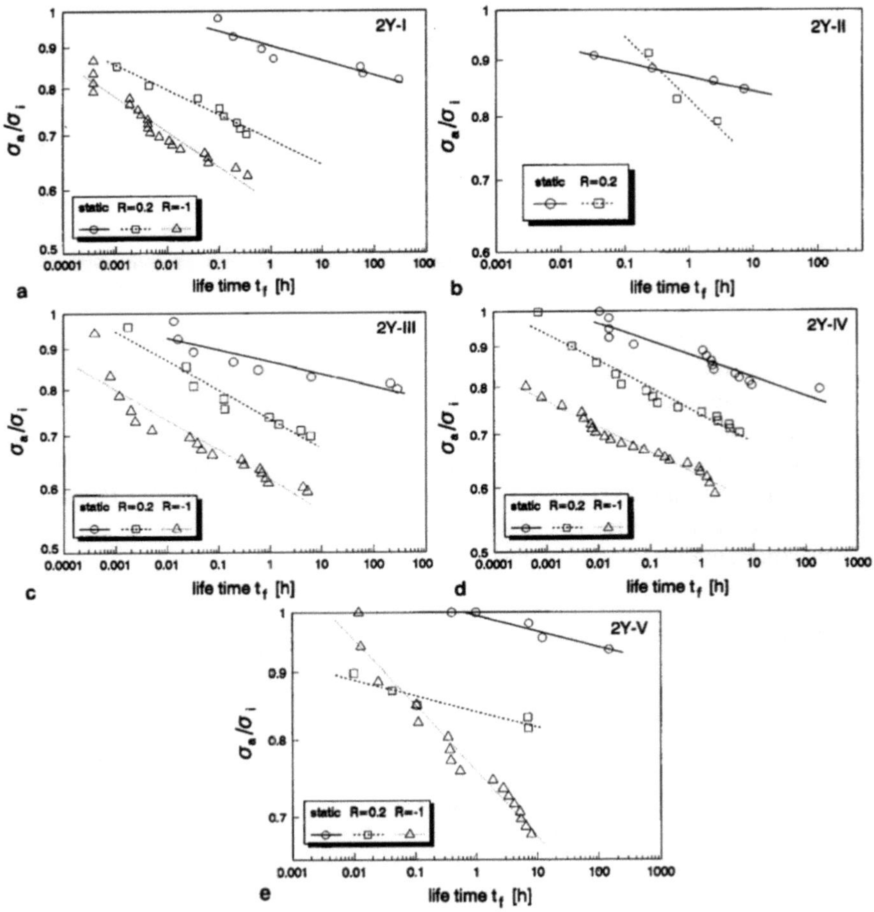

Fig. 9.36 Modified stress-lifetime (σ_a/σ_i - t_f) curves under static and cyclic loading for **a** 2Y-TZP-I, **b** 2Y-TZP-II, **c** 2Y-TZP-III, **d** 2Y-TZP-IV and **e** 2Y-TZP-V ceramics. For fatigue conditions see Table 9.5. Matt et al. [11]. With kind permission of Elsevier

In summary, the stress induced tetragonal-monoclinic transformation at ground surface of the 2Y-TZP ceramics increase almost linearly with grain size (Fig. 9.34). Clearly it increases also with higher applied stresses. Under cyclic loading 2Y-TZP showed shorter lifetime compared to static loading. The highest fatigue resistance under static and cyclic loading of the high-strength materials is found for the material with the intermediate grain size. The results of cyclic fatigue tests in air are influenced by subcritical crack growth, but this effect is negligible at high failure probabilities.

9.4.3 The Effect of Grain Size in 8 mo% Y-TZP (Fully Stabilized)

Generally, addition of 8 mol % (13.75 wt %) Y_2O_3 into zirconia structures is needed to form a fully stabilized zirconia. Its structure becomes cubic solid solution, which has no phase transformation from room temp up to 2500 °C. As good ceramic ion conducting materials, fully yitria stabilized zirconia (YSZ) has been used in oxygen sensor and solid oxide fuel cell (SOFC) applications among its other technological uses. For example, the use of fully stabilized zirconia (FSZ), which has improved translucency due to the increase in cubic phase concentration is used for optical purposes and to a large extent for clinical (dental) applications. Let us re-emphasize cubic phase stabilization is achieved by adding yttria in concentrations of ~8 mol% or higher, while in partially stabilized zirconia (PSZ), tetragonal phase is stabilized with 2–5 mol% yttria content (stabilization, partial or fully can be achieved also by addition of other appropriate oxides). However, a reduction in mechanical resistance is expected, once it has lost the toughening effect resultant from the tetragonal-to-monoclinic phase transformation. Reducing the grain size and/or increasing the concentration of stabilizing oxide reduces the transformation rate, and certain surface treatments may enhance the resistance of Y-TZP to low temperature degradation (LTD).

In the following the grain sizes indicated in Table 9.6 are related to various properties of the ceramic. TEM micrograph of the as-milled 8YSZ nanopowder is displayed

Table 9.6 Mechanical properties of near full dense specimens produced via different sintering regimes. Mazaheri et al. [12]. With kind permission of Elsevier

Sintering method	Average grain size (μm)	Fractional density	Hardness (GPa)	Fracture toughness (MPa m$^{1/2}$)
CS	2.14	97.5	12.87 ± 0.12	1.61 ± 0.07
LMS	2.35	98	12.97 ± 0.19	1.9 ± 0.09
HMS	0.9	98	13.72 ± 0.22	3.17 ± 0.10
TSS	0.29	97.6	13.51 ± 0.17	3.16 ± 0.06

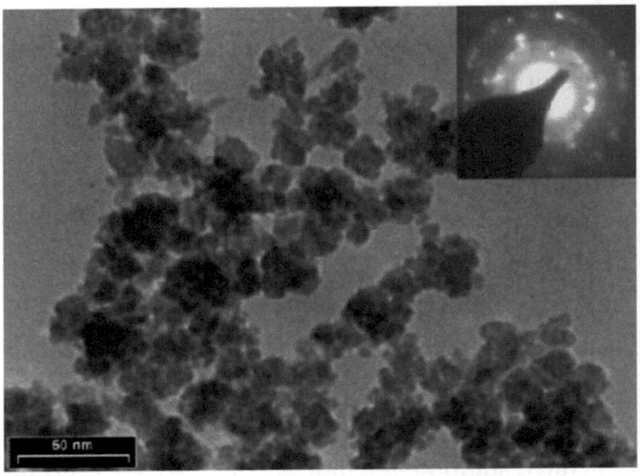

Fig. 9.37 Transmission electron microscopy image and diffraction pattern of nanocrystalline 8YSZ powder synthesized via glycine–nitrate method after milling for 24 h. Mazaheri et al. [12]. With kind permission of Elsevier

in Fig. 9.37. Selected area diffraction pattern of the sample embedded in the figure, confirms the nanocrystallinity of the powder. The fractional density as a function of temperature for various grain sizes is shown in Fig. 9.38. Instead, specific indication of the grain size the designation of the specimen types is listed. Thus, in Table

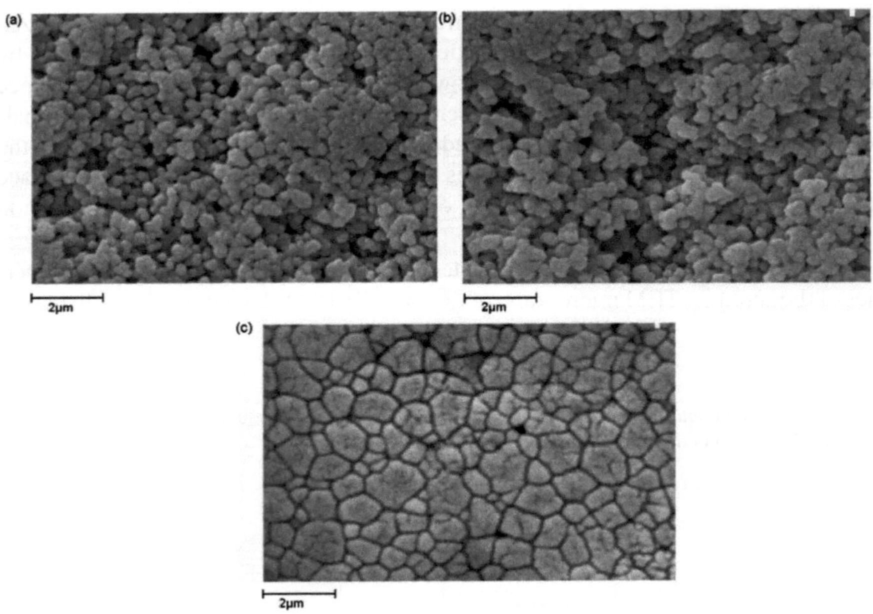

Fig. 9.38 Scanning electron microscopy (SEM) micrographs of 8YSZ specimens sintered by **a** CS, **b** HMS and **c** LMS regimes at 1250 °C. Mazaheri et al. [12]. With kind permission of Elsevier

9.6, the sinterings to achieve the respective average grain sizes are for: conventional sintering (CS) 2.14 μm, low rate microwave sintering (LMS) 2.35 μm, high rate microwave sintering (HMS) 0.9 μm and two step sintering (TSS) 0.29 mm. The images of the sintered specimens obtained by SEM are illustrate in Fig. 9.38. The effect of temperature on the fractional density of the sintered specimens indicated above is illustrated in Fig. 9.39. All lines representing various grains show sigmoidal shape. Similar shape curves -i. e. sigmoidal- are obtained for the fractional density as a function of the sintering time as seen in Fig. 9.40. The variation of grain size vs. temperature and fractional density are illustrated in Figs. 9.41 and 9.42, respectively.

The microstructures of the sintered full-dense-specimens are illustrated in Fig. 9.43. It is apparent that the two-step sintering process seems to be the best approach, since grain growth is suppressed. Further, the HMS approach has smaller grain size than the CS and LMS. Finally, grain size distribution is plotted in Fig. 9.44 as frequency vis. density for the sintered HMS and TSS specimens.

In summary, 8 mol% YSZ in various sintering methods has been presented above. The TSS (two step processing) technique to reach full density is the mos significant approach providing the smallest grain size by suppressing grain growh. Further, the HMS (high microwave sintering rate) provides better grain structure than the CS (conventional sintering) and the LMS (low microwave sintering rate). The heating rates for the LMS and the HMS procedures were 5 and 50 °C min^{-1}, respectively. Further LMS provides higher densities at lower temperatures, but at this temperature this method is not capable of providing a final fractional density better than ~ 97% of the theoretical density. The drawback of TSS -despite its best sintering qualities- is the long total time which might testrict its application. It is important to re-emphasize that

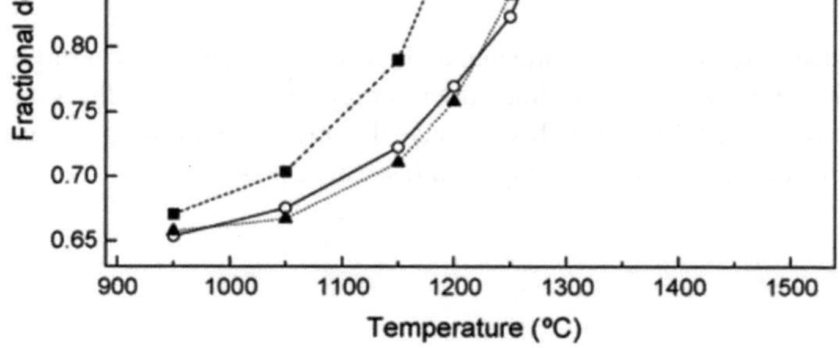

Fig. 9.39 The fractional density variations of the specimens sintered by CS, LMS and HMS as a function of sintering temperature. Mazaheri et al. [12]. With kind permission of Elsevier

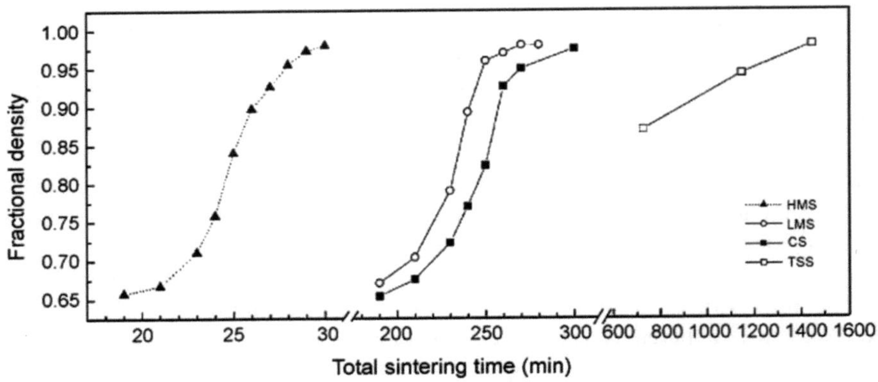

Fig. 9.40 Fractional density versus total sintering time for the specimens sintered via CS, LMS and HMS routes. Mazaheri et al. [12]. With kind permission of Elsevier

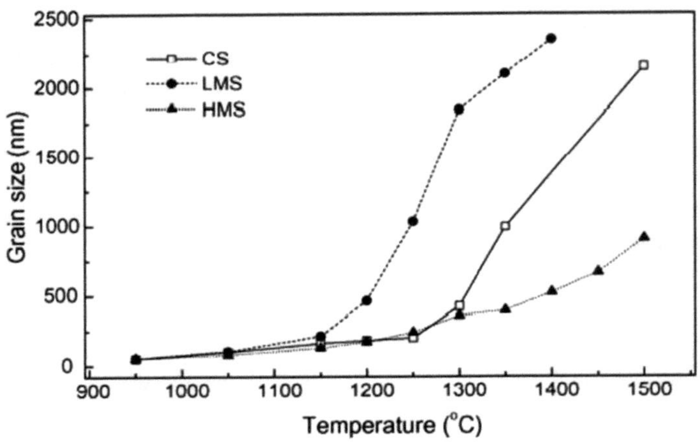

Fig. 9.41 The grain size variations of CS, LMS and HMS samples as a function of temperature. Mazaheri et al. [12]. With kind permission of Elsevier

fine grain strucrure improves the mechanical properties of ceramics. It might be worth noting that the TSS having the finer microstructure, gave about the same fracture toughness as that of HMS, probably because the microstructural homogeneity in the latter case is a consequence of applying a fast-firing process.

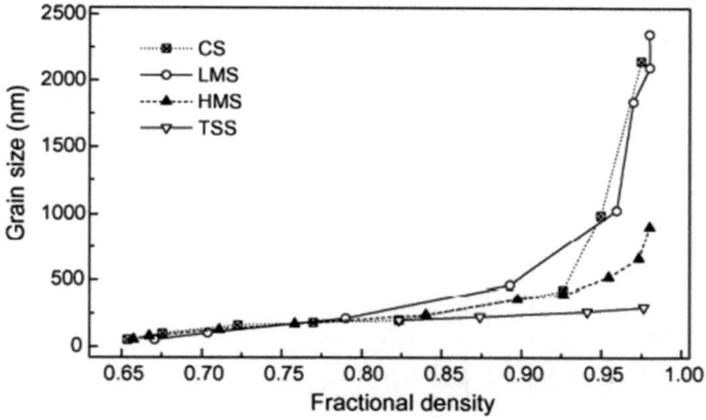

Fig. 9.42 Grain size/fractional density trajectory obtained by CS, LMS, HMS and TS. Mazaheri et al. [12]. With kind permission of Elsevier

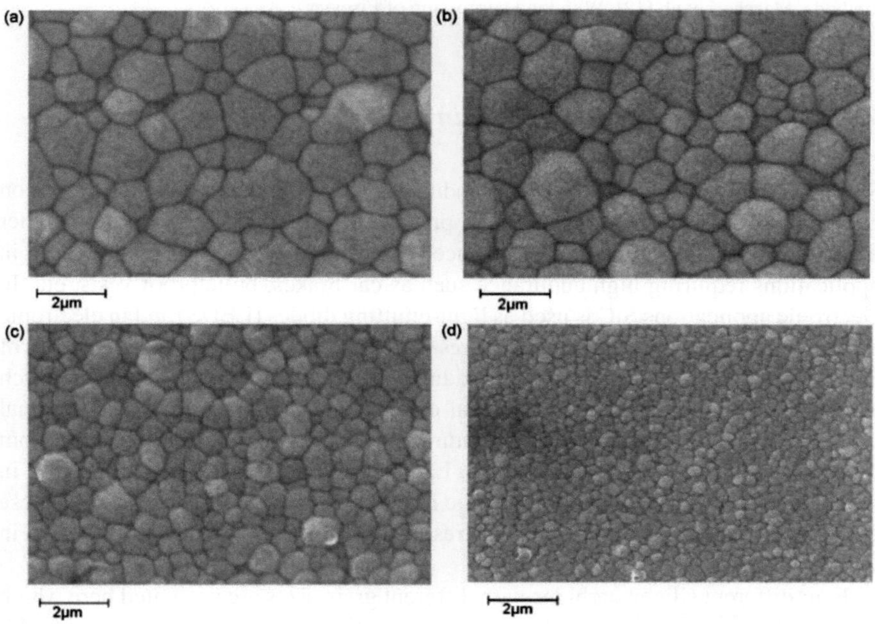

Fig. 9.43 The SEM micrographs of near full dense specimens sintered by **a** CS, **b** LMS, **c** HMS and **d** TSS methods. Mazaheri et al. [12]. With kind permission of Elsevier

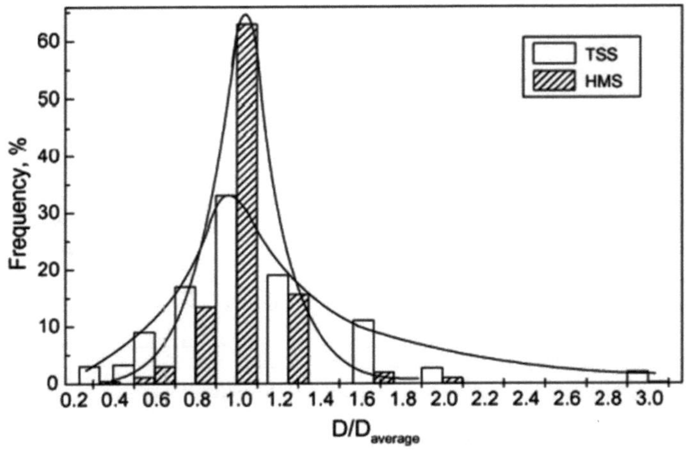

Fig. 9.44 The grain size distribution of near full dense specimens sintered by HMS and TSS methods. Mazaheri et al. [12]. With kind permission of Elsevier

9.4.4 *The Effect of Grain Size in SiC*

Silicon carbide, also known as carborundum, is a semiconductor containing silicon and carbon that occurs in nature, but is processed technology by bonding together by sintering silicon and carbon to produce a very hard ceramic. It is widely used in applications requiring high endurance such as car brakes, bulletproof vests, etc. In electronic applications SiC is used as light-emitting diodes (LEDS) and in electronic devices that operate at high temperatures or high voltages, or both. Fabrication of SiC is done with the aid of sintering additives. Further, the properties of SiC such as a high specific stiffness, low thermal expansion, low density and high thermal conductivity, make it an important structural material for many applications at room temperature and for space applications. Last but not least SiC is of primary use in biomedical applications, i.e., orthopaedic implants. It is thus expected that for these applications, the components will be stressed for prolonged time in dry but also in wet environments.

Four different silicon carbides with different grain sizes are presented here which were produced by using liquid phase sintering (LPS) with alumina and yttria and solid state sintering (SSS) using boron and carbon. For each, a fine grained and a coarse grained SiC was obtained. The grain size of the samples having equiaxed grains and were determined by the linear intercept method. Fine grained and a coarse grained SiC were obtained in each case of of the production method. The SiC fine grains were in the range of 1 ± 0.1 μm and the coarse grained specimen in the range of 4 ± 0.4 μm. Figures 9.45, 9.46, 9.47 and 9.48 illustrate the cyclic fatigue curves for the various SiCs and the environments of the test. Here, the load vs. displacement under an applied stress intensity K_{Ic} ($K_{min}/K_{max} = 0.1$) at 5 Hz is presented. The number

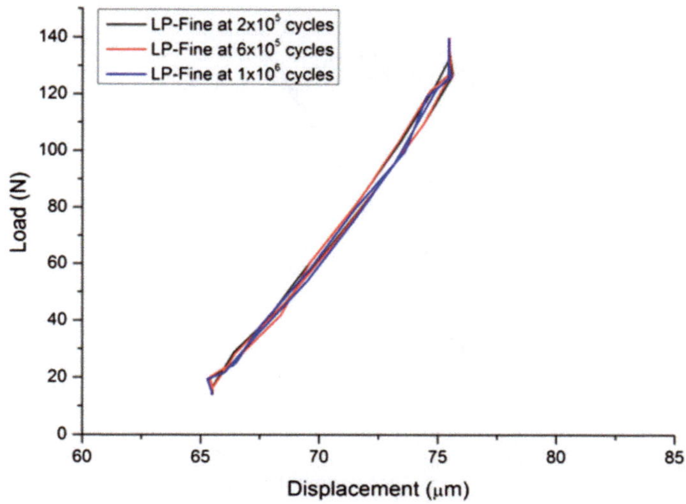

Fig. 9.45 Load versus displacement of LP-Fine during cyclic loading at 5 Hz for: (black) after 2×10^5 cycles, (red) after 6×10^5 cycles and (blue) after 1×10^6 cycles. The test was conducted at $K_I = 3.99$ MPa $m^{1/2}$ and R $(K_{min}/K_{max}) = 0.1$. Nasrin Al Nasiri, Thesis, Imperial College London Department of Mechanical Engineering, 2014. Free access

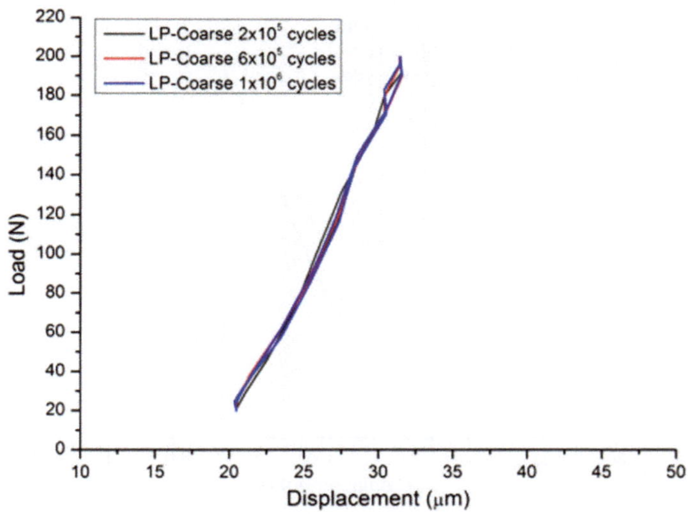

Fig. 9.46 Load versus displacement of LP-Coarse during cyclic loading at 5 Hz for: (black) after 2×10^5 cycles, (red) after 6×10^5 cycles and (blue) after 1×10^6 cycles. The test was conducted at $K_I = 5.85$ MPa $m^{1/2}$ and R $(K_{min}/K_{max}) = 0.1$. Nasrin Al Nasiri, Thesis, Imperial College London Department of Mechanical Engineering, 2014. Free access

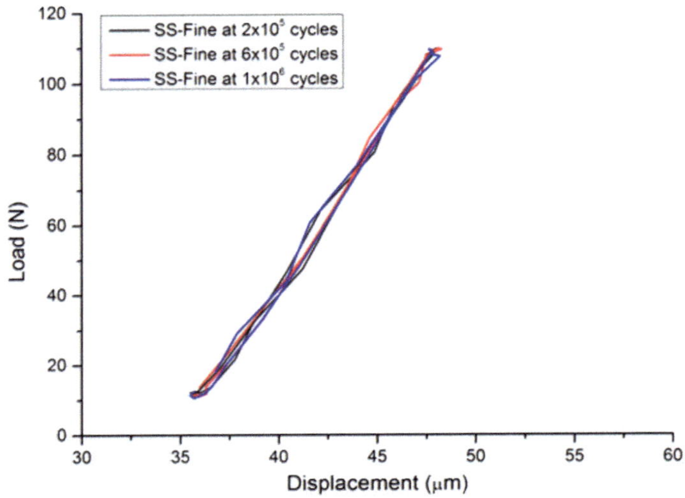

Fig. 9.47 Load versus displacement of SS-Fine during cyclic loading at 5 Hz for: (black) after 2 $\times 10^5$ cycles, (red) after 6 $\times 10^5$ cycles and (blue) after 1 $\times 10^6$ cycles. The test was conducted at $K_I = 3.0$ MPa $m^{1/2}$ and R $(K_{min}/K_{max}) = 0.1$. Nasrin Al Nasiri, Thesis, Imperial College London Department of Mechanical Engineering, 2014. Free access

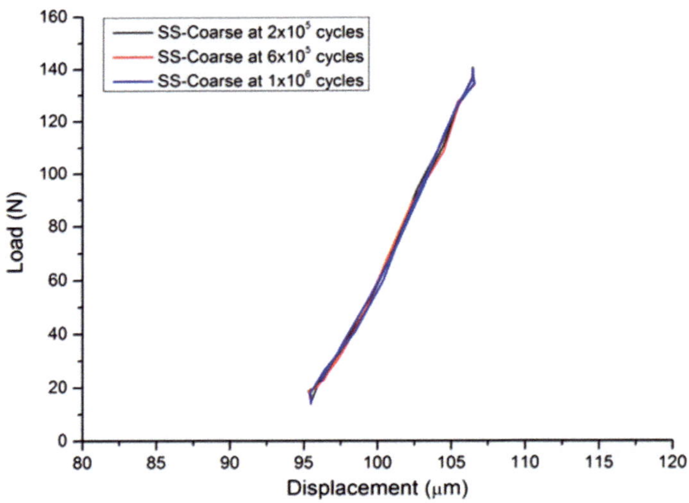

Fig. 9.48 Load versus displacement of SS-Fine during cyclic loading at 5 Hz for: (black) after 2 $\times 10^5$ cycles, (red) after 6 $\times 10^5$ cycles and (blue) after 1 $\times 10^6$ cycles. The test was conducted at KI = 3.5 MPa $m^{1/2}$ and R $(K_{min}/K_{max}) = 0.1$. Nasrin Al Nasiri, Thesis, Imperial College London Department of Mechanical Engineering, 2014. Free access

Table 9.7 Powder weight compositions and hot pressing schedule. Nasrin Al Nasiri, Thesis, Imperial College London Department of Mechanical Engineering, 2014. Free access

Microstructure	SiC type	Sintering additives wt.%	Temperature °C	Time Hour
LP-Fine	α-SiC	$6\ Al_2O_3 + 4\ Y_2O_3$	1950	0.5
LP-Coarse	α-SiC	$6\ Al_2O_3 + 4\ Y_2O_3$	1950	3
LP-Coarse	α-SiC	$6\ Al_2O_3 + 4\ Y_2O_3$	2050	3
LP-Coarse	β-SiC	$6\ Al_2O_3 + 4\ Y_2O_3$	1950	3
LP-Coarse	β-SiC	$6\ Al_2O_3 + 4\ Y_2O_3$	2050	3
SS-Fine	α-SiC	0.1 B + 3 C	2050	0.5
SS-Fine	α-SiC	0.2 B + 3 C	2050	0.5
SS-Coarse	β -SiC	0.3 B + 3 C	2050	3
SS-Coarse	β-SiC	0.4 B + 3 C	2050	3
SS-Coarse	β-SiC	0.5 B + 3 C	2050	3
SS-Coarse	β-SiC	0.5 B + 3 C	2150	3

of cycles applied are indicated in the illustrations. Recall that K_{Ic} is defined as the plane strain fracture toughness. It is a measure of the resistance to crack extension under predominantly linear-elastic conditions with little to no plastic deformation occurring at the crack tip, namely, at the low toughness conditions.

As seen in the figures, there is no change in slope which suggests that the crack was not advancing under cyclic loading. Optical observation of the crack confirmed no further crack growth occurred and the crack remained at its pre-crack position prior testing. It migh be of interest to compare the static fatigue figures of these four SiCs with those of the cyc lic fatigue shown in Figs. 9.45, 9.46, 9.47 and 9.48. Table 9.7 lists the relevant data of the SiCs processed in the experiments (Fig. 9.49).

9.4.5 The Effect of Grain Size in Si_3N_4

Very limired information is available where the combined effect of fatigue and grain size coexist in a single application. The reason might be that Si_3N_4 like SiC are mainly used as composites rather as pure material. As a matter of fact, thsese ceramics are often constitute a composite.

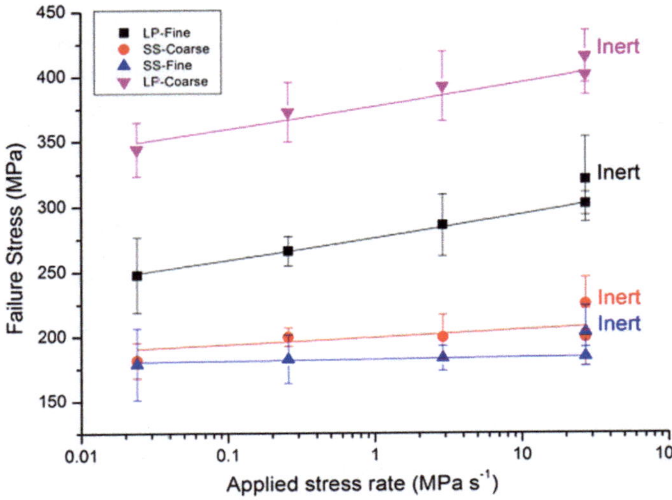

Fig. 9.49 Failure stress versus stress rate of the four SiC materials in water with their corresponding inert strength measured in oil. Each symbol represents an average value of 5 samples. Note the strength degradation of LP-SiC with decreasing loading rate in contrast with the more constant strength for SS-SiC. Nasrin Al Nasiri, Thesis, Imperial College London Department of Mechanical Engineering, 2014. Free access

3-point bending fatigue test was subjected under a sinusoidal cyclic stress at 250 Hz and stress ratio R (minimum stress/maximum stress) of 0.1 (span 30 mm, piezo electric bimorph actuator). The 3-point bending strength test results of the silicon nitride sintered in the current experiments and the usual silicon nitride sintered to up to 1200 °C are shown in Fig. 9.50. The bending strength of the former is about 2.0 GPa at room temperature. This value is higher than the values observed in the usual Si_3N_4 being 1.8 GPa at 800 °C, 1.0 GPa at 1200 °C and 1.0 GPa at 1200 °C. Other properties are listed in Table 9.8.

Fig. 9.50 Relationship between 3-point bending strength and temperature. Yoshimura et al. [13]. Free access

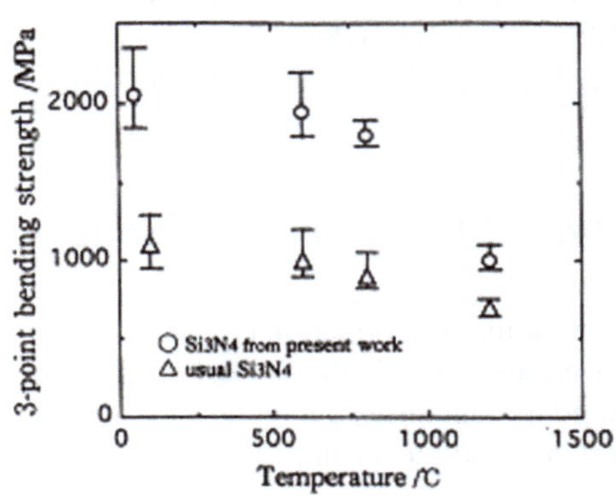

Table 9.8 Mechanical Properties of the Silicon Nitride Ceramics. Yoshimura et al. [13]. Free access

Density/10^3kg·m^{-3}	3.24
Fracture toughness /MPa·m$^{1/2}$	5.8
Charpy impact value /kJ·m^{-2}	20
3-point bending fatigue limit /GPa※	1.3

※Result after 10^7cycles-stress-loading

The green bodies were sintered at 1600 °C in 10 MPa nitrogen atmosphere and sintering was determined for grain size control. SEM micrographs of the sintered Si_3N_4 are illustrated in Fig. 9.51. The high strength silicon nitrides were developed by reducing the grain size which as expected results in the observed high strength. The finer the grain size of the microstructure is, the stronger the resulting mechanical properties.

Density/10^3kg·m^{-3}	3.24
Fracture toughness /MPa·m$^{1/2}$	5.8
Charpy impact value /kJ·m^{-2}	20
3-point bending fatigue limit /GPa※	1.3

※Result after 10^7cycles-stress-loading

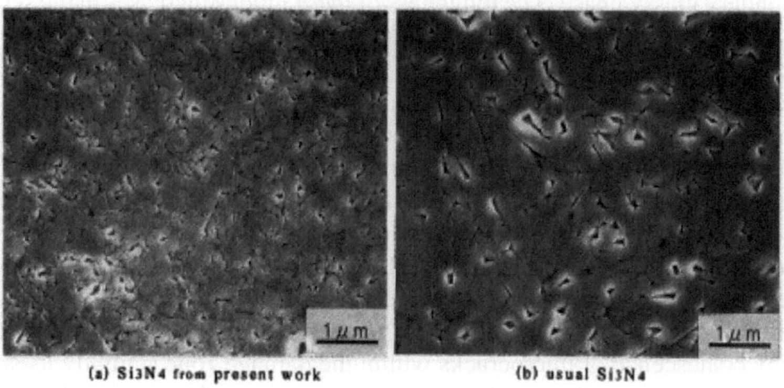

Fig. 9.51 SEM micrographs of the etched surfaces. **a** Si_3N_4 from present work, **b** usual Si_3N_4. Yoshimura et al. [13]. Free access

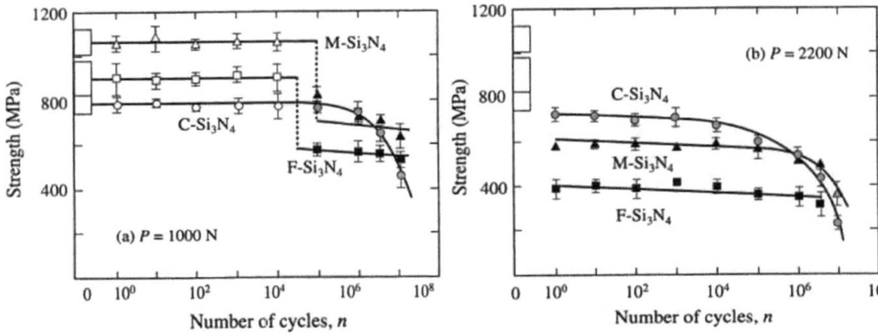

Fig. 9.52 Strength as a function of number of contact cycles F-Si_3N_4, M- Si_3N_4, C- Si_3N_4, demonstrating role of microstructure. Indentation with WC spheres, r = 1.98 mm, in air, at loads P indicated. Black symbols indiocate failure from cone cracks, grey synbols from quasi-plastic zones, open symbols from natural flaws. Boxes at left axis indicate "laboratory" strength (unindented specimens). Lawn et al. [14]. Public Full-text

Further interest in Si_3N_4 is its microsrtructural influence on Hertzian contact fatigue (for example in the bearing applications) and clearly -as mentioned earlier-in dental applications. In both cases, in bearing and dental applications, the contact stresses are critical and the question is how the microstructure affects the fatigue strength after repeated contacts. For this purpose three silicon nitrides were used as detailed: (i) fine silicon nitride (F-Si_3N_4), a bimodal microstructure with mainly small equiaxed α grains (<1 μm) and some elongated β grains; (ii) medium nitride (M-Si_3N_4), a bimodal microstructure with a larger fraction of elongated β grains of (length $\approx$ 4 μm) and (iii) coarse material (C-Si3N4), a microstructure with predominantly elongated β grains (length $\approx$ 9 μm). Each material contains ~ 10 vol % interboundary glassy phase. The toughness of these nitides increases in the sequence from fine to coarse and between them the (M-Si_3N_4). The results of tests shown in Fig. 9.52 at loads of P = 1000 (a) and P = 2200 (b) indicate the severity of the fracture tests, which were performed in air and are plotted as strength against the number of cycles. The damaged region centered on the tensile face. In Fig. 9.52 (a) the load P = 1000 is lower than the required to initiate damage in a single cycle. Note, however, that after ~ 10^5 cycles the strength drops abruptly in F- Si_3N_4 and M-Si_3N_4, due to initiation of cone cracks, while in C- Si_3N_4 the drop is more gradual from progressive accumulation of quasi-plastic damage.

Also note that at P = 2200 damage already occurred after a single cycle and the strength falls off progressively thereafter. In the tougher C- Si_3N_4 the degradation occurs by coalescence of microcracks within the damage zone (initially its strength degradation is the less). Staic and cyclic fatigue contacts of F- Si_3N_4, M- Si_3N_4 and C- Si_3N_4 are compared in Fig. 9.53 at P = 2200 in air. In this plots the contact time is plotted in the upper axis, while the number of cycles in the lower axis. The static and cyclic fatigue in F- Si_3N_4 and M- Si_3N_4 data coincide, however in the case of the C- Si_3N_4 the cyclic fatigue data falls significantly at ~ 10^4 cycles. The observations

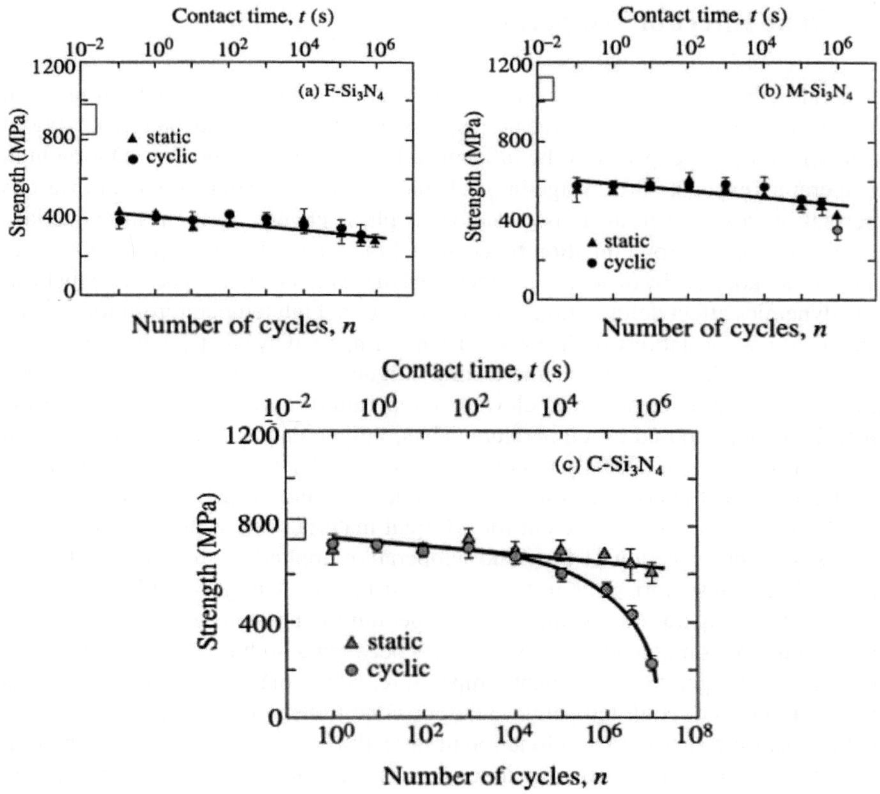

Fig. 9.53 Strength as a function of contact time for **a** F-Si_3N_4, **b** M- Si_3N_4, **c** C- Si_3N_4, comparing damage accumulation at static and cyclic contacts. Indentation with WC spheres, r = 1.98 mm. P 2200 N, in air. Black symbols indicate failure from cone cracks, grey symbols from quasi plastic zones. Boxes at left axis indicate "laboratory" strengths. Lawn et al. [14]. Public Full-text

of the experiments show that there is a mechanical component contribution in the quasi-plastic process.

Summarizing the essential points of the effect of the grain size of Si_3N_4 are: (i) Si_3N_4 like other materials regardless to the level of their toughness, are susceptible to fatigue. (ii) The finer the grain size of the microstructure is, the stronger the resulting mechanical properties. (iii) Contact stresses are acting in bearing and dental application of Si_3N_4; comparing static and fatigue data indicate that in both cases their strength level and failure is grain size dependent. Strength degradation is a result of crack evolution and finally failure sets in as a consequebce of crack coalescence.

9.5 The Effect of Temperature

Generally, the properties such as the tensile strength, and yield strength of material depend on temperature and fatigue properties are no exception, and materials exposed to temperature are expected to be influenced by its level and duration. During high temperature exposure the change in mechanical properties can be related to various thermal processes such as transformation (if phase change exists in the material), diffusion processes, precipitation processes (if the material is not a pure substance) and last but not least by dislocation effects. All these factors, but in particular dislocation dynamics affect deformation which can occur at high temperatures more easily. The steadiness or stability of the material can be impaired by creep phenomena under sustained load. Microcracks characterizing fatigue are prone to grow to macrocrack level during the deformation at elevated temperature more readily than when the material is not exposed to temperature. Thus, fatigue damage accumulation might be enhanced. In general, fatigue is not an intergranular but a transgranular failure mechanism and the combined action of cyclic load and an increased temperature should be expected to be different for different materials and different temperature ranges. Combined action of time and temperature applied to structures at service are likely to cause variations in the specific structure. A major problem is that at the elevated temperature operation the temperature is not constant. In general, the temperature of components in service varies between a high operating temperature and a low non-operational ambient temperature, which means that thermal stresses can be introduced. High-temperature fatigue conditions imply that the fatigue load and temperature vary both as a function of time, thus cyclic load, time and temperature are variables to be considered in fatigue. In the following, oxides, carbides and nitrides are presented limited by the avalibility of such information in the literature.

9.5.1 The Effect of Temperature in Alumina

Fatigue and fracture test are performed at high temperature up to 1000 °C are presented in this section. The static and cyclic fatigue tests performed in vacuum and air were intended to compare the strength behavior of alumina under these test conditions by using four-point bending tests. The fundamental physical and mechanical properties of alumina are listed in Table 9.8 and in Table 9.9 the statistical evaluation of the four-point bending fracture strength at room and higher temperatures are listed. Cyclic fatigue tests were conducted under three types of load patterns shown in Fig. 9.54 with the stress ratio R of $(\sigma_{min}/\sigma_{max}) = 0$. One sinusoidal (5 Hz) and two trapezoidal stress patterns with different maximum stress hold times. The results of the tests (ten specimens at each condition) are seen in Fig. 9.55. The center of the standard deviation of the data shown is the mean strength while the upper and lower ends of the error bars indicate the deviation range.

Table 9.9 Physical and Mechanical Properties of the Material. Nakayama and Ohira [15]. Open access SEPB stands for Single-Edge Precracked Beam

Density (Mg/m^3) *1	3.92
Fracture toughness ($MPa \cdot m^{1/2}$) *2	4.4
Hardness (HV 10)	1670

*1 Archimedes method
*2 SEPB method (JIS R1607)

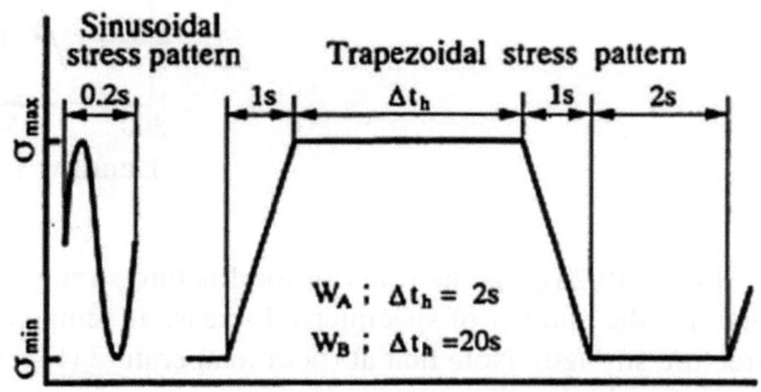

Fig. 9.54 Load patterns for cyclic fatigue tests. Nakayama and Ohira [15]. Open access

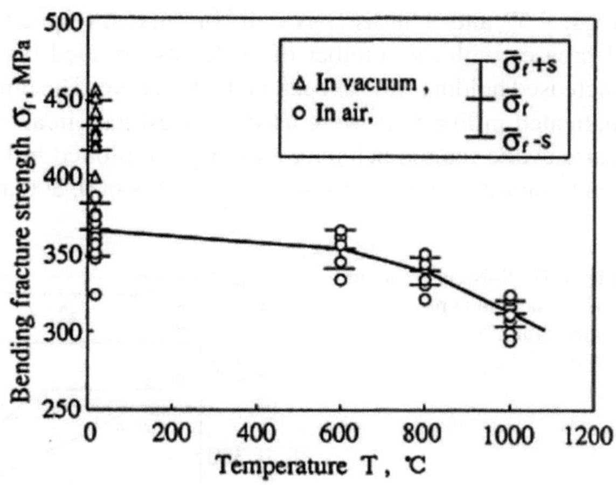

Fig. 9.55 Temperature dependence of bending fracture strength. Nakayama and Ohira [15]. Open access

The distributions of the fracture strength at room temperature in air and vacuum are plotted in Fig. 9.56 on probability paper. Here the cumulative probability distribution of the fracture strength is shown. The cumulative probability, P_f on the ordinate is determined by

$$P_f = \frac{i - 0.5}{n} \tag{9.28}$$

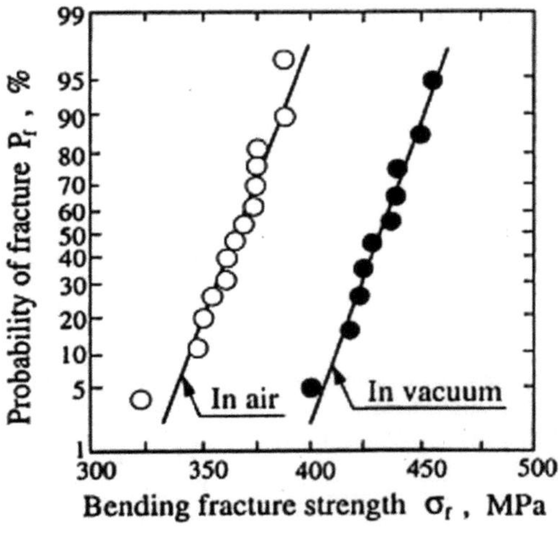

Fig. 9.56 Distribution of bending fracture strength of alumina in air and in vacuum plotted on normal probability paper. Nakayama and Ohira [15]. Open access

In Eq. (9.28), i is the order of the fracture strength from the minimum extreme, and n is the number of specimens. Increase in temperature lowers continuously the fracture strength. Note that at room temperature (Fig. 9.55) the fracture strength in vacuum is higher than that in air.

The cyclic fatigue tests at room temperature in air and vacuum are shown in Figs. 9.57 and 9.58 as S–N and S-t curves, respectively. The maximum strength decreases with the number of cycles as revealed by the S–N relation or with the increased holding time as seen in the S-t curve. The three types of loading conditions indicated in Fig. 9.54 were used to investigate in air and vacuum the high temperature cyclic fatigue behavior and they are plotted in Figs. 9.59, 9.60 and 9.61. The temperatures are indicated in Table 9.10 as 600, 800 and 1000 °C.

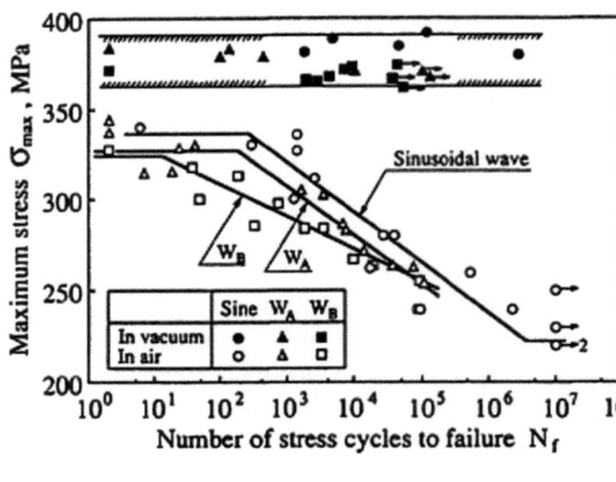

Fig. 9.57 S–N curves in air and in vacuum at room temperature. Nakayama and Ohira [15]. Open access

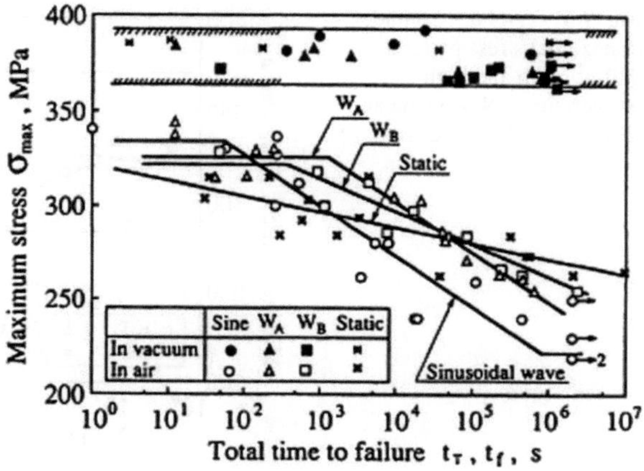

Fig. 9.58 S-t curves in air and in vacuum at room temperature. Nakayama and Ohira [15]. Open access

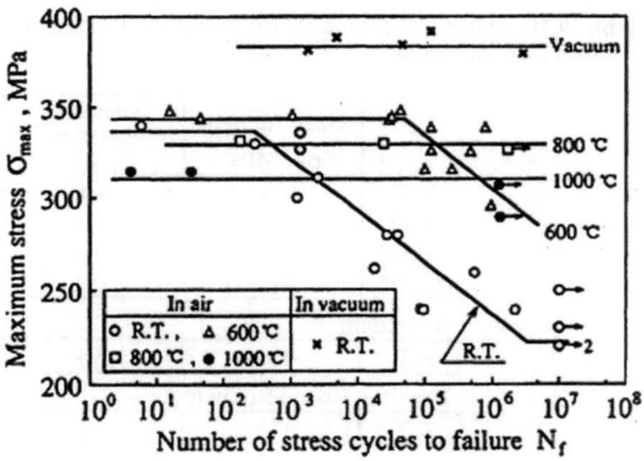

Fig. 9.59 S–N curves under sinusoidal stress pattern at room and high temperatures. Nakayama and Ohira [15]. Open access

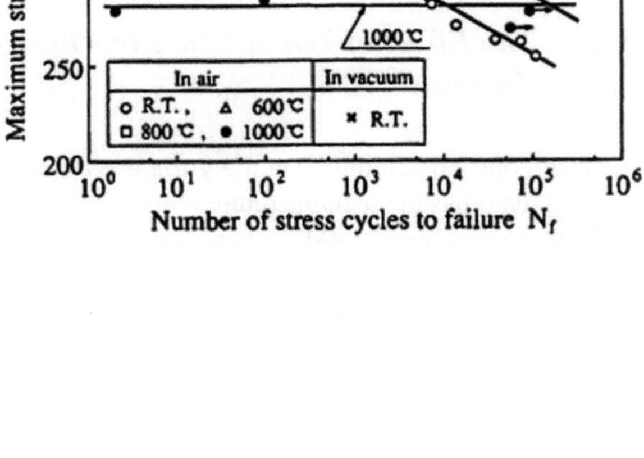

Fig. 9.60 S–N curves under trapezoidal stress pattern WA at room and high temperatures. Nakayama and Ohira [15]. Open access

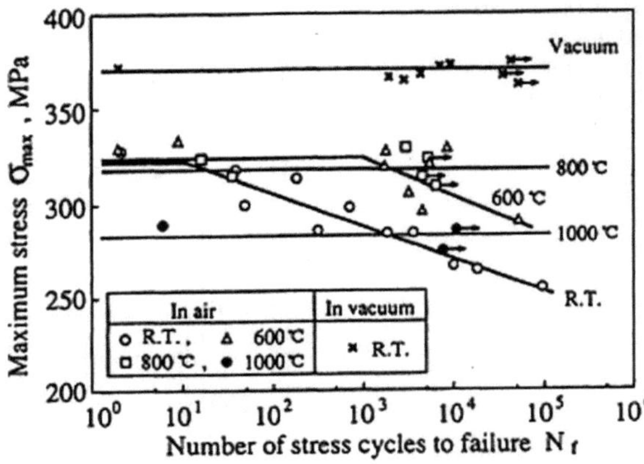

Fig. 9.61 S–N curves under trapezoidal stress pattern WB at room and high temperatures. Nakayama and Ohira [15]. Open access

Table 9.10 Statistical evaluation of four-point bending fracture strength at room and high temperatures. Nakayama and Ohira [15]. Open access

	Mean strength σ_f, MPa	Standard deviation s, MPa	Coefficient of variation c.v.
In air			
R.T.	364	16.8	0.046
600°C	352	13.1	0.037
800°C	338	9.2	0.027
1000°C	311	9.2	0.029
In vacuum			
R.T.	431	15.9	0.037
Result in Round Robin test			
R.T.	373	23.1	0.062

In summary, fracture tests and fatigue tests were carried out on alumina under four-point bending loads. Results of static and cyclic fatigue tests at room temperature in air indicate that cyclic fatigue degradation is time-dependent and cycle-dependent degradations which are promoted by water vapor (air contains water).

9.5.2 *The Effect of Temperature in Yttria Fully Stabilized Zirconia*

Cubic zirconia containing 10 mol% yttria was deformed along the hard ⟨100⟩ direction and materials in the temperature range 550–1400 °C. Recall-as indicated in this book- that above about 2400 °C, pure zirconia has the cubic fluorite structure. Below that temperature, it transforms to a tetragonal phase and finally below about 1200 °C to a monoclinic one. However, the cubic high-temperature phase can be stabilized to lower temperatures by adding aliovalent cations. The most often used system is the zirconia- yttria (ZrO_2–Y_2O_3) system. Addition of about 8 mol% yttria or more, stabilizes the cubic phase (c-ZrO2) down to room temperature. Brittleness in cubic zircona below ~ 000 °C can be modified to plastic deformation in the soft < 112 > orientation down to ~ 400 °C and by applying hydrostatic pressure down even to 250 °C. Stress strain curves are shown in Fig. 9.62 for the temperatres indicated. Below 500 °C the specimens broke without significant plastic deformation and at orientation factors in b) and c) with (220) and (220) **g** vectors always one set of dislocations at the low temperatures a yield drop can be observed.

Stress–strain curves of 15 mol % yttria are illustrated in Fig. 9.63. Preparation of specimens for TEM below 700 °C failed because of the brittleness of the deformed material. At this temperature slip is inhomogeneous. In the upper part of the dislocation structure illustrated in Fig. 9.64 the dislocations look quite straight running

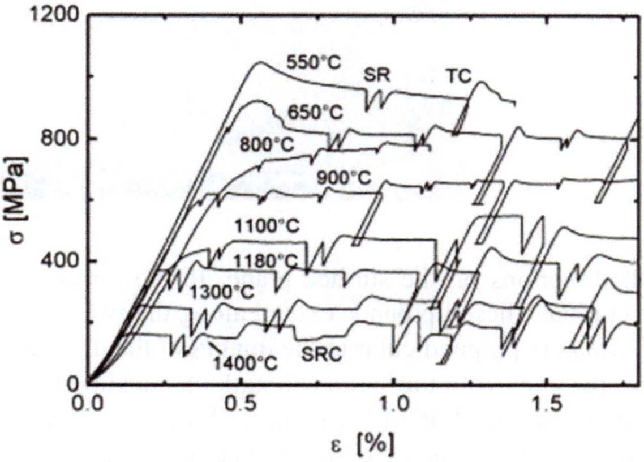

Fig. 9.62 Stress–strain curves of ZrO_2-10 mol% Y_2O_3 deformed along ⟨100⟩ at a strain rate of 10^{-5} s^{-1}, including stress relaxation (SR), strain rate cycling (SRC) and temperature change (TC) tests. A. Tikhonovsky, M. Bartsch, and U. Messerschmidt, phys. stat. sol. (a) 201, 26 (2004). With kind permission of John Wiley and Sons

Fig. 9.63 Stress–strain curves of ZrO_2- 15 mol% Y_2O_3 deformed along ⟨112⟩ at a strain rate of 10^{-5} s^{-1}. The cusps in the curve of 800 °C correspond to SR and SRC tests. Tikhonovsky et al. [16]. With kind permission of John Wiley and Sons

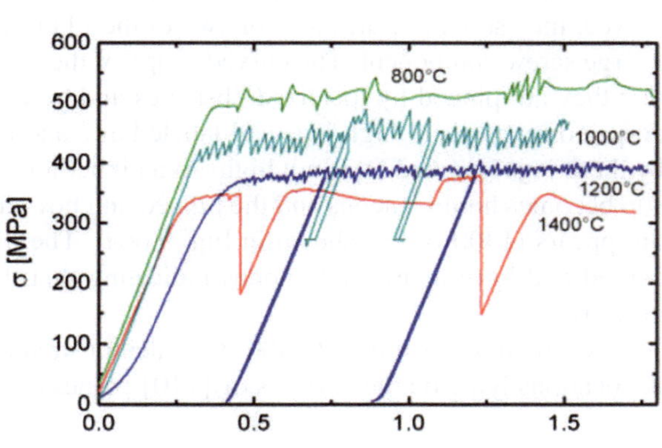

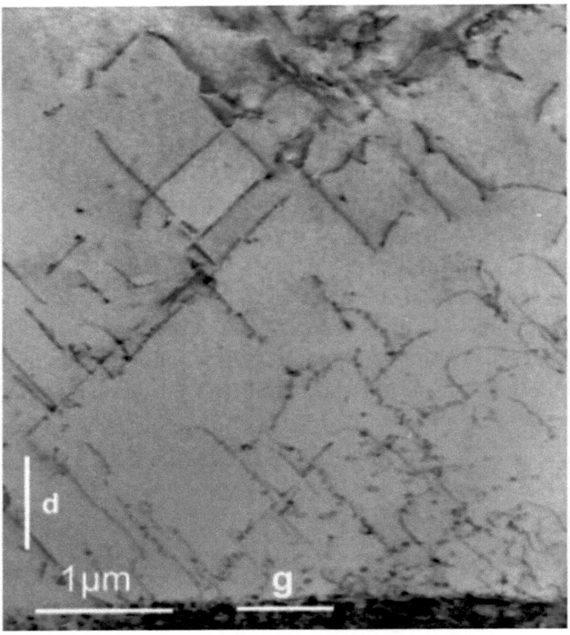

Fig. 9.64 Dislocation structure of ZrO_2-10 mol% Y_2O_3 deformed along the compression direction **d** = [100] at 700 °C to a strain of 1.3%. Imaging diffraction vector **g** = (020) near the [001] pole. With this **g** vector, all dislocations of slip systems with non-zero orientation factors are visible. Tikhonovsky et al. [16]. With kind permission of John Wiley and Sons

in the two ⟨110⟩ directions on the surface plane. In the lower part of the figure, dislocations are curved. These lip bands extend along the two ⟨110⟩ directions (the directions approximately perpendicular to the images of the dislocation lines which, mostly, cross the specimen with the shortest length on the slip plane). This points at the dislocations being arranged on {111} planes. Thus, slip systems with {110} and {111} planes coexist. This is characteristic for other temperatures, too. The dislocation structure following 800 °C deformation with two sets of dislocations, indicating multiple slip, is illustrated in Fig. 9.65. At the (200) **g** vector in Fig. 9.65a), all dislocations are imaged with non-zero orientation factors in b) and c) with (220) and (220) with 1/2[110] and 1/2[110] Burgers vectors within the image plane are extinguished (since the dislocations have Burgers vectors in the image plane and are curved, they should be arranged on two of the {111} planes). The dislocations have a large screw component. The curved shape of the dislocations with cusps indicates that they are pinned by localized obstacles and bow out between them. Part of the cusps corresponds to high jogs and labeled as J and the density of debris produced by these jogs (marked D) is not high. At a higher temperature above about 1200 °C, slip becomes homogeneous and the pinned and bowed-out shape of the dislocations disappears (1300 °C) as shown in Fig. 9.66a). The dislocations are now smoothly curved and have many triple nodes indicating that they form a three-dimensional network.

The situation is similar at 1400 °C as demonstrated in Figs. 9.66b, c which show dislocations lying on the two sets of {110} planes oriented edge on.

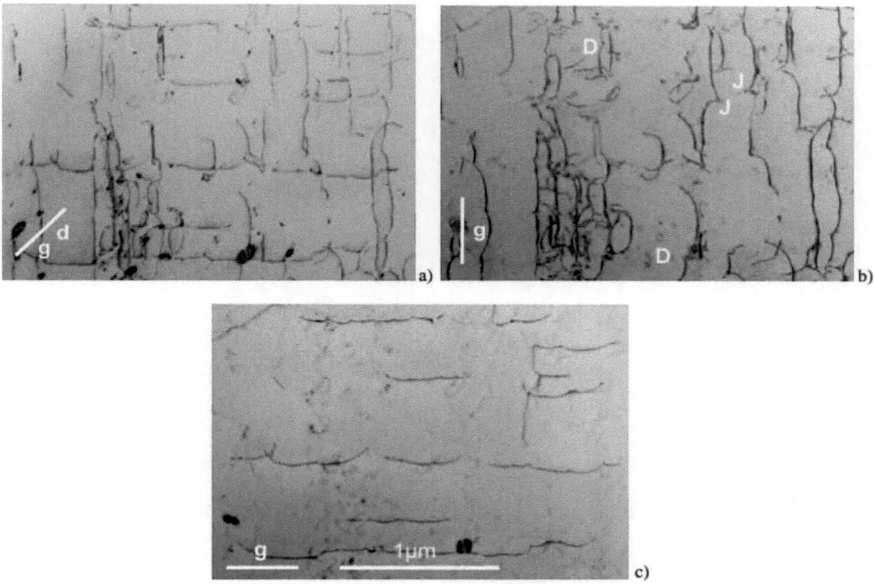

Fig. 9.65 Dislocation structure of ZrO_2-10 mol% Y_2O_3 deformed along d = [100] at 800 °C up to 1.1% imaged under different **g** vectors at the [001] pole. **a g** = (200). **b g** = (220). **c g** = (220). Tikhonovsky et al. [16]. With kind permission of John Wiley and Sons

In summary, 10 mol% yttria was deformed in the temperature range 550–1400 °C. Stress strain curves at this range were obtained for 10% and 15% yttria addition to stabilize the cubic structure down to room temperature.

9.5.3 The Effect of Temperature in Partially Stabilized Zirconia

As mentioned earlier, several oxides (yttria, calcia, magnesia etc.) can stabilize zirconia partially or fully. In this section partially stabilized zirconia by magnesia is presented. Fatigue crack growth through magnesia stabilized zirconia at 20, 450 and 650 °C has been observed dynamically in a high temperature loading stage for the SEM (scanning electron microscope). A backscattered electron image of magnesia (MgO) partially stabilized zirconia is shown in Fig. 9.67. The material properties such as typical modulus and tensile strength values are listed in Table 9.11. Crack growth in ambient, 450 and 650 °C is shown in Fig. 9.68. The fatigue growth in these regions is stable at 20 °C, stable to quasi-stable at 450 °C and unstable at 650 °C. If straight lines passed through the data, the use of the Paris relation is possible according to

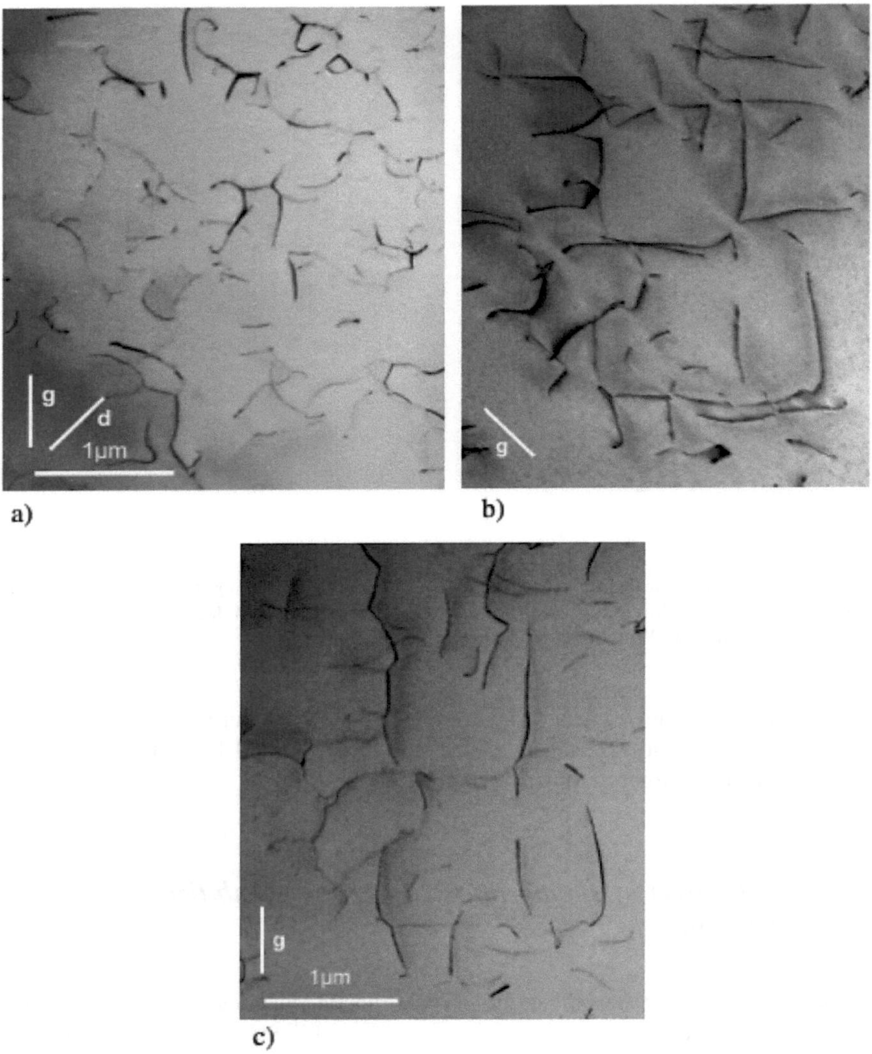

Fig. 9.66 Dislocation structure of ZrO_2-10 mol% Y_2O_3 deformed along **d** = [100] at high temperatures. **a** 1300 °C, 2.6%. **g** = (220). **b** 1400 °C, 2.6%, **g** = (020). **c** Same area as (b), **g** = (220). Tikhonovsky et al. [16]. With kind permission of John Wiley and Sons

$$\frac{da}{dN} = B\Delta K^2 \tag{9.29}$$

where B and s are constants derived from the data (the straight line through the data) and their values ate listed in Table 9.12. Sequential crack growth at 20 °C is shown in

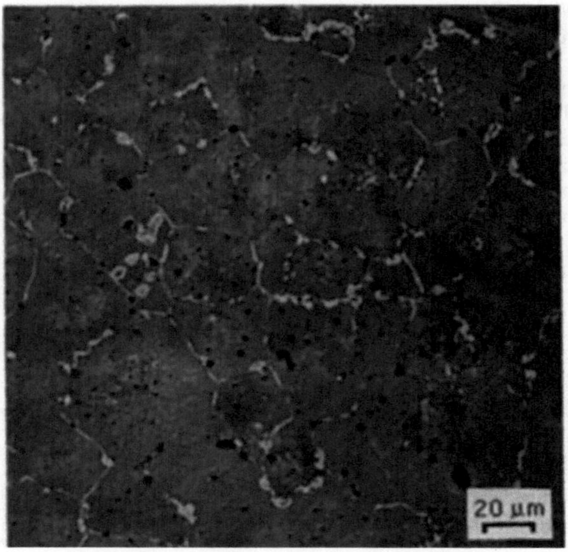

Fig. 9.67 Backscattered electron image of ion-etched magnesia partially stabilized zirconia. Average size of the grains is about 32 µm. The white material within some of the grain boundaries is monoclinic phase zirconia. The remainder of the material within the grains is tetragonal zirconia. Davidson et al. [17]. With kind permission of Elsevier

Table 9.11 Material properties. Davidson et al. [17]. With kind permission of Elsevier

Temperature (°C)	Yield stress (MPa)	Modulus (GPa)	Yield strain
20	330	205	0.0016
450	221	172	0.0014
650	200	165	0.0012
Tensile yield curves: $\sigma = \sigma_0 \epsilon_t^n$	Temperature (°C)	σ_0(MPa)	n
	20	2400	0.305
ϵ_t = total strain	450	1720	0.230
	650	1215	0.092

Fig. 9.69. The sequence of crack growth as a function of the number of cycles is seen. From the almost difficult to see crack in the very first photograph, the crack advances with the number of cycles, two microns after 50 cycles with further 0.4 µm between 50 and 150 cycless. At about 120 cycles no advance of the crack is observed but a slip line appeared at the crack tip earlier and lengthened to 120 cycles. Subsequently, the crack slowly grew down the slip line until 360 cycle is reached. At this point another slip line formed (photograph j). The crack growth sequence at 450 °C with much faster growth rate is shown in Fig. 9.70. The crack grew an average 1.9 µm per cycle, but there was no indication of slip line formation and breakdown. It seems that the microcracks did not alter the crack growth process, e. g., they were not observed to link-up with the main crack. Crack tip parameters are listed in Table 9.13. CTOD in the list stands for crack tip opening displacement. Crack opening displacement

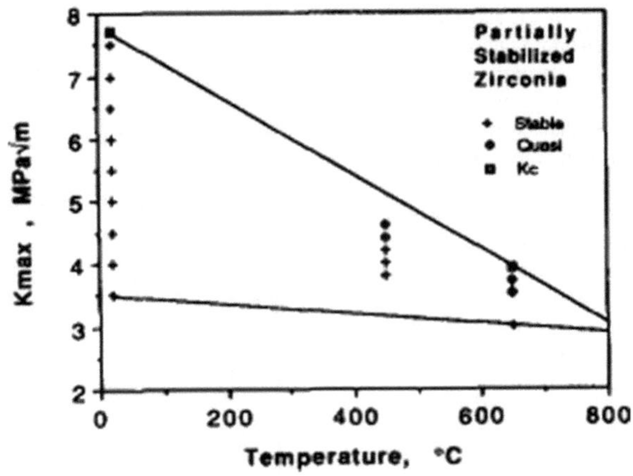

Fig. 9.68 A diagram of the region of stability for fatigue crack growth found for this partially stabilized zirconia. The lines indicate the envelope in which fatigue crack growth was found to be the dominant failure mechanism. Davidson et al. [17]. With kind permission of Elsevier

Table 9.12 Fatigue crack growth and fracture parameters. Davidson et al. [17]. With kind permission of Elsevier

T (°C)	Frequency (Hz)	ΔK_{th} (MPa$\sqrt{m}$)	B (m/cy)	s	K_c (MPa$\sqrt{m}$)	Reference
20	50	3.0	1.4×10^{-23}	28	—	DYR [2]
20	10	3.5	6.2×10^{-25}	25	7.7	
450	10	3.2	7.3×10^{-33}	46	5.2	
650	10	3.0	3.2×10^{-44}	70	3.9	

(COD) has been measured at 20, 450 and 650 °C at several levels of ΔK. In each case, the dependence of COD on distance behind the tip (d) fit the function

$$COD = C_0\sqrt{d} \tag{9.30}$$

The constant C_0 in Eq. (9.30) is the same as COD for d = 1 mm (d is the distance behind the crack tip). A comparison between COD elastic and COD measured is shown in Fig. 9.71. The largest COD elastic values correspond to the sequence of photographs shown in Fig. 9.69.

In summary, fatigue crack growth in partially stabilized zirconia by magnesia at ambient temperature was presented in this section. Fatigue cracks at ambient temperature do not grow at a uniform rate, rather, a slow rate of growth occurs as a slip line forms at the crack tip, and then the crack grows faster as it advances down the slip line. The levels of fracture toughness measured at all temperatures is attributed -based on a strain distribution model- to deformation within the plastic zone of the crack tip. Fatigue crack growth at 450 and 650 °C exhibited a threshold stress intensity factor and crack closure. The region of stable crack growth was found

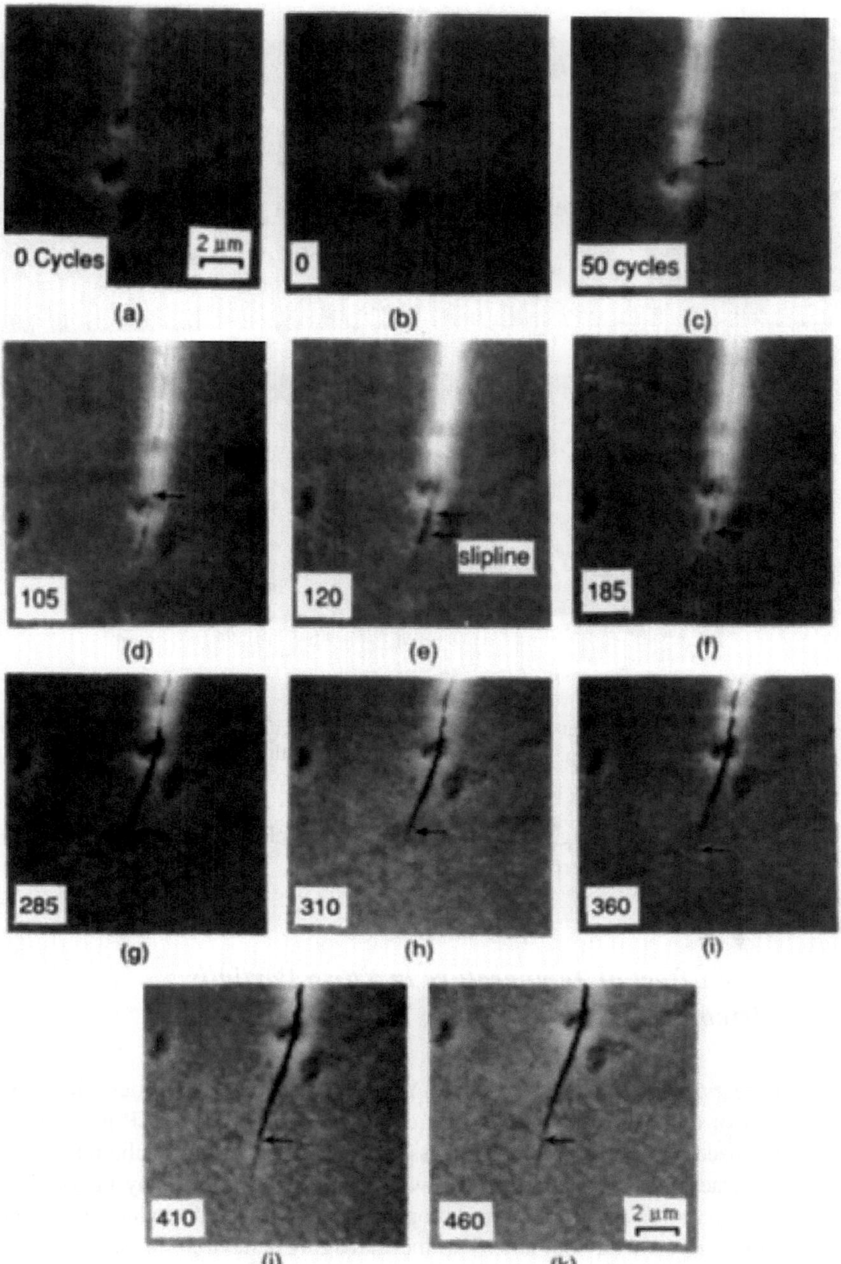

Fig. 9.69 Fatigue crack advance at ambient temperature, ΔK = 5.6 MPax/m, R = 0.1. This sequence begins with the crack at minimum load in (a), then (b) at maximum load. The arrow indicates the location of the crack tip. By (d) a slip line has formed ahead of the crack tip; subsequent cycling breaks down this slip line and allows the crack to advance. By cycle 360, the crack has advanced completely down the slip line, and in (j) another slip line has formed. Davidson et al. [17]. With kind permission of Elsevier

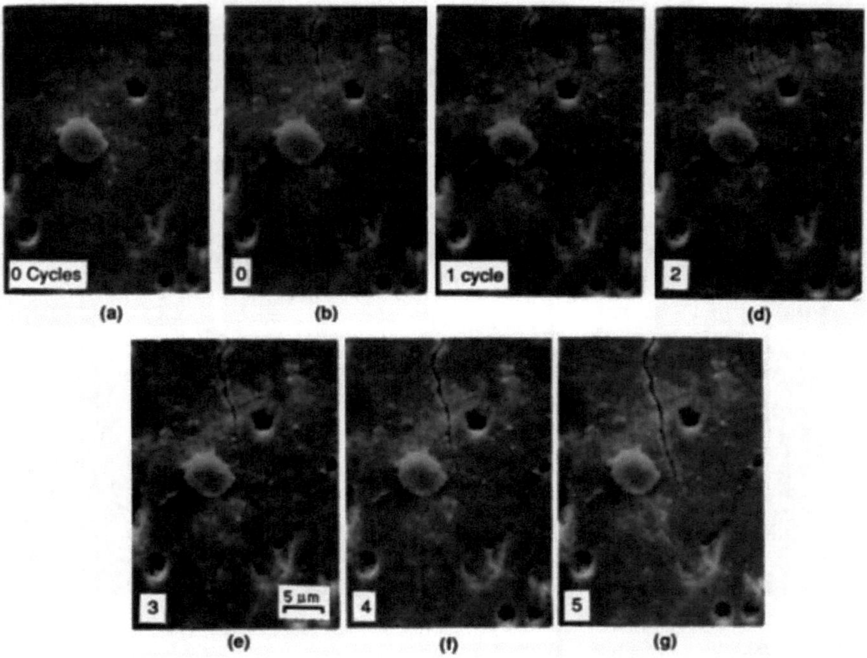

Fig. 9.70 Fatigue crack growth at 450 °C, ΔK = 3.74 MPa√m. This sequence shows crack advance on each cycle. Compare with Fig. 9.69. Davidson et al. [17]. With kind permission of Elsevier

to be very narrow, decreasing with temperature. The crack growth sequence at 450 °C occurs with a much faster growth rate.

9.5.4 The Effect of Temperature in Yttria Partially Stabilized Zirconia

In recent years, yttria-stabilized zirconia Y-TZP has attracted considerable interest in the field of dental restoration. In particular, full-strength 3Y-TZP grade that is commonly used in clinical practice. An essential requirement for cyclic fatigue lifetime is the fracture resistance of Y-TZP which is a function of many factors influencing its preparation and the acquired properties of the finished product. The effects of long-term fatigue behaviour and hydrothermal aging behaviour on the service lifetime of zirconia dental devices has to be known for practical applications. As known hydrothermal aging is also considered a low temperature degradation spontaneous phase change of zirconia under water or humid environment. Many factors such as different grain size, residual or applied stresses, density, content of Y_2O_3 in tetragonal phase, proportion of cubic phase, doping treatments, all contribute towards the

Table 9.13 Crack tip parameters. Davidson et al. [17]. With kind permission of Elsevier

ΔK (MPa$\sqrt{m}$)	CTOD (elastic) (μm)	CTOD (meas.) (μm)	Crack tip strain	Plastic zone size (1) (μm)
20°C				
4.6	0.068	0.05	0.011	(2)
4.9	0.072	0.08	0.037	19
5.1	0.075	0.08	0.013	16
5.3	0.078	0.082	0.035	14
5.6	0.082	0.11	0.037	(2)
5.6	0.082	0.094	0.037	(2)
5.6	0.082	0.066	0.056	19
5.6	0.082	0.075	0.050	19
5.6	0.082	—	0.047	12
5.6	0.082	0.035	0.41	12
450°C				
4.3	0.076	0.095	0.020	26
650°C				
2.0	0.024	0.020	0.017	(2)
3.2	0.057	0.036	0.012	20
3.2	0.057	0.040	0.011	20

Notes: (1) PZS is in the direction of loading; computed from the strain distribution with $\epsilon_y = \sigma_y/E$.
(2) No strain distribution determined.

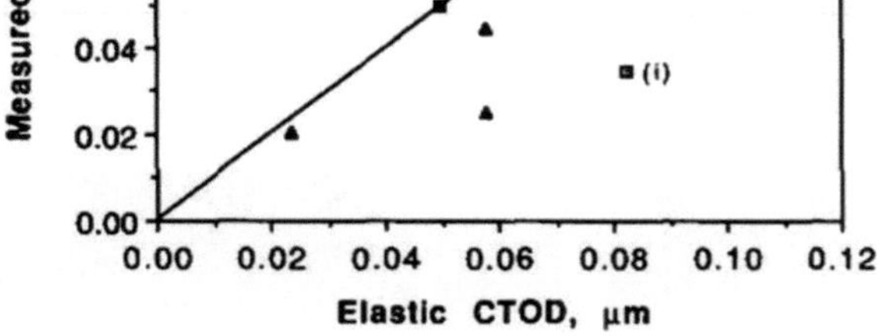

Fig. 9.71 Comparison between CTOD measured and calculated for an elastic crack. Shown is the wide variation in CTOD corresponding to various steps in the crack growth process illustrated in Fig. 9.69. Davidson et al. [17]. With kind permission of Elsevier

aging process. Some of these factors-which are acting in dental use may trigger detrimental consequences and shorten the lifetime of the dental zirconia ceramics. Table 9.14 and Fig. 9.72 indcate all the partially stabilized zirconia consisdered in the low temperature aging treatment. The designations ST, MT and TT refer to three kind specimens as super-translucence, medium-translucence and tooth translucence, respectively. The temperatures and the time schedule are included in the Table. The microstructural features are listed in Table 9.15. The initial mechanical properties such as the flexural strength and the fatigue lifetime are listed in Table 9.16. The surface phase transformation curves for the ST and MT samples as a function of the hydrothermal ageing time are shown in Fig. 9.73. As seen in the illustrations (a) and (b) the lower is the temperature the higher is the aging time. The volume fraction of the monoclinic samples for ST and MT is shown in Fig. 9.74. The illustration represents the aging kinetics at 37 and 134 °C

In summary, partially stabilized zirconia samples (3 mol%) designated as ST, MT and TT were considered at low temperature aging treatment. The ST and MT samples have higher initial mechanical strength than the TT (5 mol%) samples. The results indicate that in cyclic fatigue, lifetime of ST and MT samples is much longer than 20 years regarding the number of cycles to failure in use under human conditions without aging treatment, while the cyclic fatigue lifetime for TT is about 10 years.

Table 9.14 Aging procedure for all materials. Wei et al. With kind permission of Elsevier

Schedule1: ST		Schedule 2: MT		Schedule3:TT	
T_i (°C)	Δt_i (h)	T_i (°C)	Δt_i(h)	T_i (°C)	t_i(h)
134	2 + 2	134	2 + 2	134	200
121	6 + 6	121	6 + 6+6	90	1720
111	10 + 5	111	10 + 20 + 6		
100	22 + 6+8	100	60 + 22 + 20		
90	50	134	2 + 8+3		
		90	100		

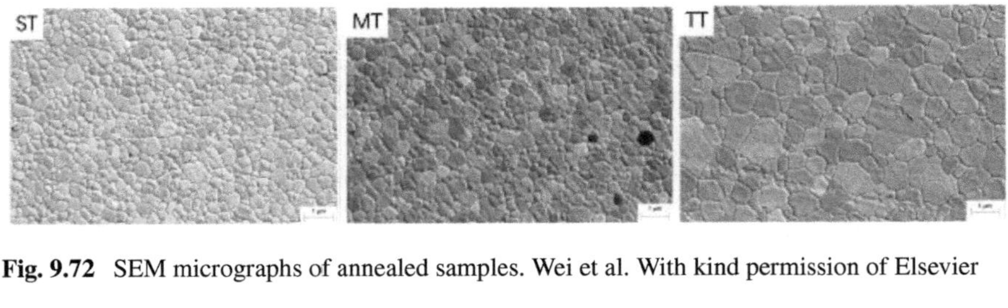

Fig. 9.72 SEM micrographs of annealed samples. Wei et al. With kind permission of Elsevier

Table 9.15 Microstructural features of zirconia materials before aging. Wei et al. With kind permission of Elsevier

Material	Y_2O_3 content (mol%)	wt% cubic ±0.5	wt% tetragonal ±0.5	mol% Y_2O_3 in c ±0.05	mol% Y_2O_3 in t ±0.05	Grain size(nm) ±10
ST	3	22.3	77.6	6.64	2.62	420
MT	3	20.9	79.1	6.42	2.5	490
TT	5	58.6	41.4	6.00	3.64	930

Table 9.16 Statistical parameters obtained from the Weibull distributions of the initial mechanical strength and fatigue lifetime. Wei et al. With kind permission of Elsevier

Material	Bending tests			Cyclic fatigue tests		
	m	σ_0	$\sigma_{0.05}$	m*	N_{f0}	σ_{max} (MPa)
ST	8.81	1207	864	0.45	30,710	686
MT	4.49	1263	657	0.51	64,150	691
TT	6.49	568	360	0.45	69,862	318

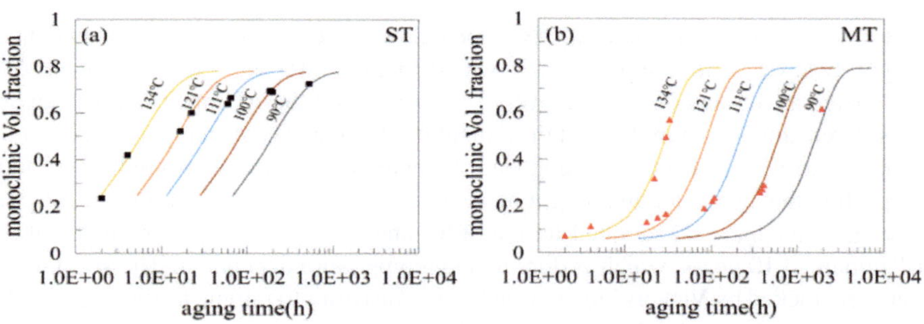

Fig. 9.73 Ageing kinetics between 90 and 134 °C for ST and MT samples. Wei et al. With kind permission of Elsevier

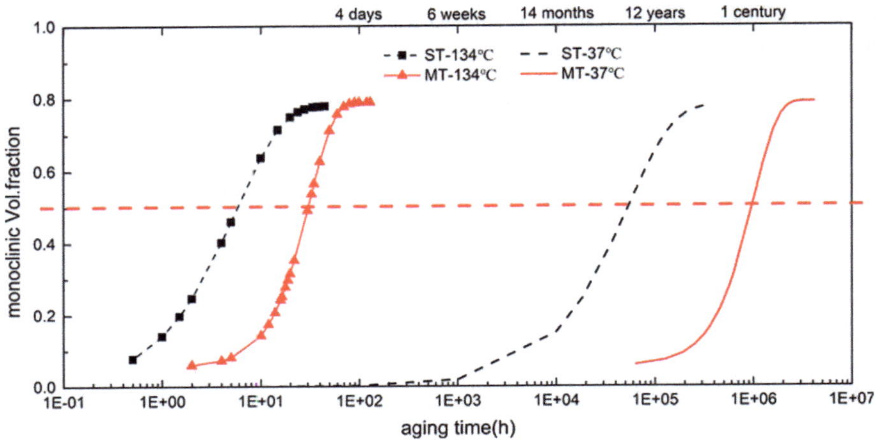

Fig. 9.74 Ageing kinetics at 37 and 134 °C for ST and MT samples. Wei et al.. With kind permission of Elsevier

Aging reduces the observed lifetime the zirconia dental ceramics in the human oral environment for anterior teeth restoration; the use of TT provides a longer service, while ST are used as dental restorative material. MT material can be used as a prosthesis material for the posterior teeth.

9.6 The Effect of Temperature in SiC

Two kinds of SiC specimens are considered below to see the temperature dependence fatigue behavior, the composition of which include 0.03 wt% and 0.39 wt% Al, designated as S-1 and S-2, respectively. Thus, even S-2 the Al doped sample can be considered almost as pure SiC. At the temperature range of room temperature (RT) – 1500 °C, no effects of cyclic loading were observed in both specimens indicating and suggesting that cyclic loading might be insignificant. However, there is static fatigue damage in the material service lifetime of ST and MT by about 6 years, and that of TT remained 10 years, which is doping and temperature dependent. The specimens were precracked by Vickers indentation (or Knoop) prior to fatigue testing. The loads used were 49 N and 98 N at room and high temperature, respectively. Dynamic, static and cyclic fatigue experiments were performed on the specimens by four-point bending test, using a hydraulic fatigue testing machine. Test temperatures were RT, 770 and 1500 °C. Testing at RT was carried out in air and at high temperatures in pure argon. In cyclic fatigue a sine-wave loading was used at a stress ratio R = 0.2 and a frequency range of 1 to 3 Hz in the RT and 770 °C, whereas a trapezoidal wave (time of holding at the maximum stress = 5 s) was used at 1500 °C (under creep conditions). The tendency for crack propagation under static loading is shown in Fig. 9.75, while in cyclic fatigue it is exhibited in Fig. 9.76. The effect of Al on the σ-ε relation is illustrated in Fig. 9.77. Without Al the precracked sample (S-1) at 1500 °C shows a higher stress than the one containing Al. The surface microcrack introduced by a Knoop indener at a load of 9.8 N is shown in the cyclically loaded sample at high temperature (Fig. 9.79). The crack tends to propagate slowly in the S-2 (0.39 wt % Al) and its tip is blunted, but this crack behavior was not observed

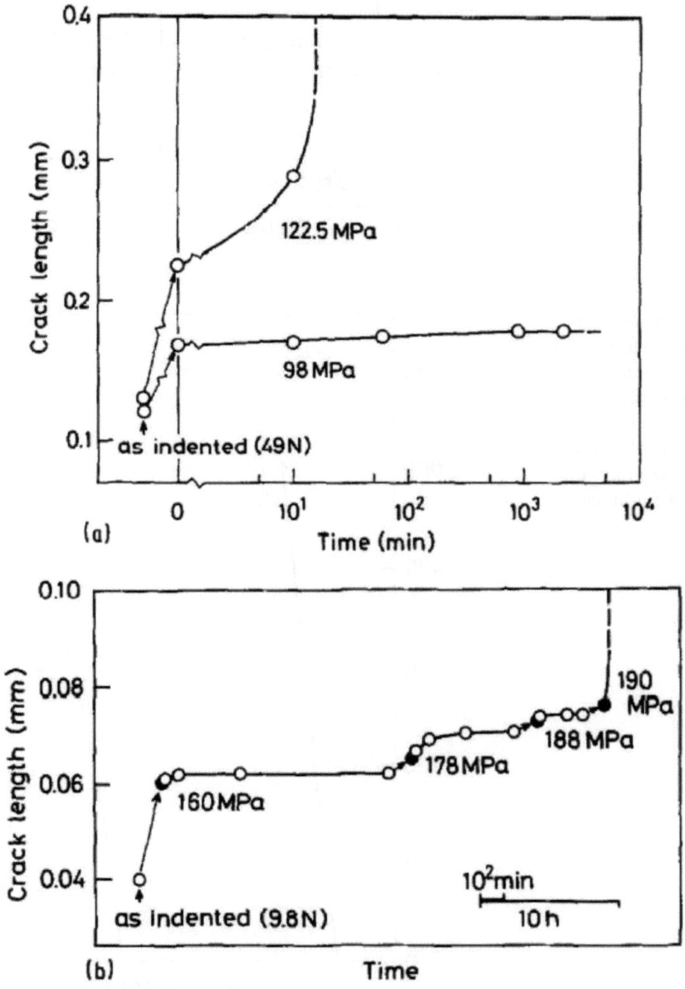

Fig. 9.75 Crack propagation in S-1 during static loading in air at room temperature indentation of **a** 49 N, **b** 9.8 N loading. Horibe and Sumita [18]. With kind permission of Springer Nature

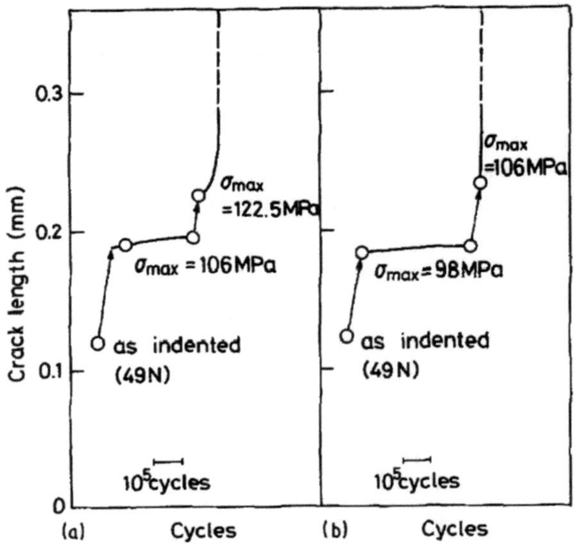

Fig. 9.76 Crack propagtion observed during cyclic loading in air at room temperature: **a** S-I, **b** S-2. S. Horibe and Sumita [18]. With kind permission of Springer Nature

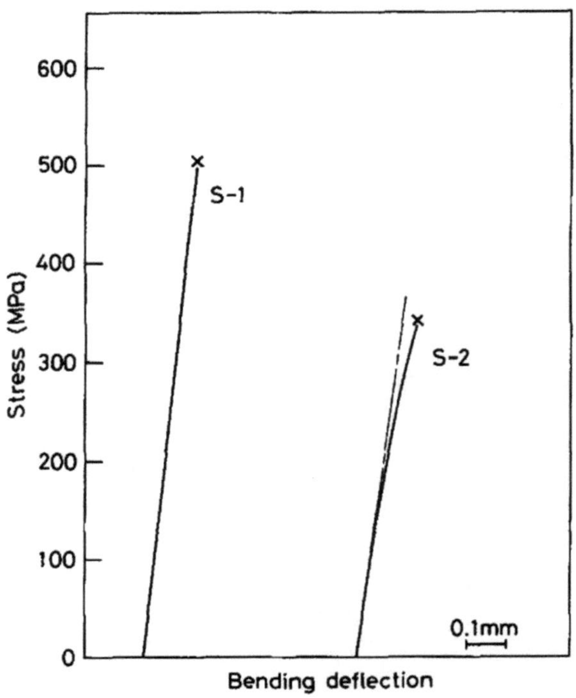

Fig. 9.77 Relationship between stress and bending deflection in argon gas at 1500 °C (stress rate 8.2 × 10^{-1} MPa sec^{-1}, indentation load 9.8 N). Horibe and Sumita [18]. With kind permission of Springer Nature

in S-1 (Al free). The Al free SiC fails at earlier times as seen in Fig. 9.78, has a higher strength, but the samples containing Al have a higher ductility and crack-tip blunting effect at high temperatures. SEM images of the fractured surface of the S-1 and S-2 specimens broken at room temperature are shown in Figs. 9.80 and 9.81, respectively. The temperature dependence of the fracture stress is shown in Fig. 9.82.

In summary, the effect of Al doping on precracked sintered β-SiC and the emphasis on its temperature dependence in the RT-1500 °C range was presented in this section. The fact that no loading effects in both SiC samples were observed over the whole temperature range, suggests that cyclic fatigue might be insignificant in SiC. However, the static fatigue behavior is different and depends on temperature and Al doping. Further, the static fatigue observed is environmentally assisted cracking. At intermediate temperatures significant static fatigue was not observed, but in S-2

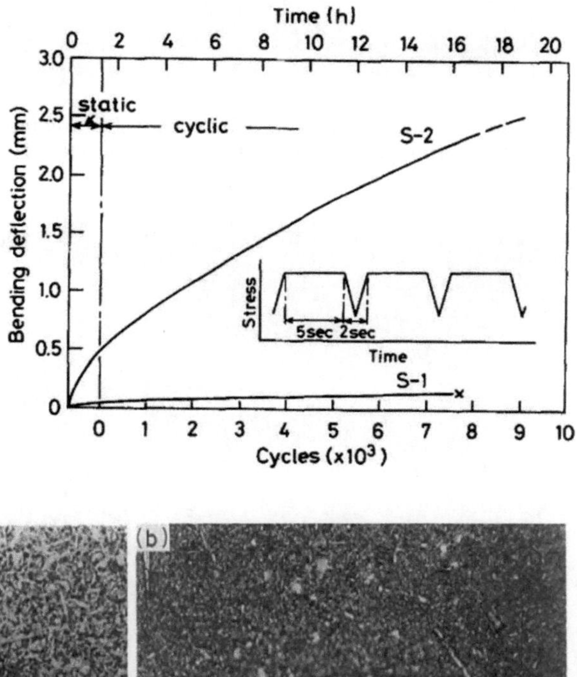

Fig. 9.78 Variation of bending deflection during cyclic loading in argon gas at 1500 °C ($\sigma_{max} = 0.89\sigma_f$, R = 0.28, indentation 9.8 N). Horibe and Sumita [18]. With kind permission of Springer Nature

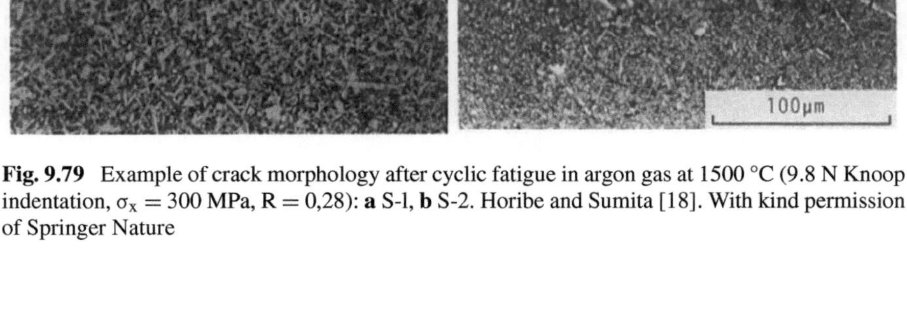

Fig. 9.79 Example of crack morphology after cyclic fatigue in argon gas at 1500 °C (9.8 N Knoop indentation, $\sigma_x = 300$ MPa, R = 0,28): **a** S-l, **b** S-2. Horibe and Sumita [18]. With kind permission of Springer Nature

Fig. 9.80 Fracture surface of the S-2 specimen shown in Fig. 9.79b which was broken at room temperature: **a** low magnification, **b** high magnification. Horibe and Sumita [18]. With kind permission of Springer Nature

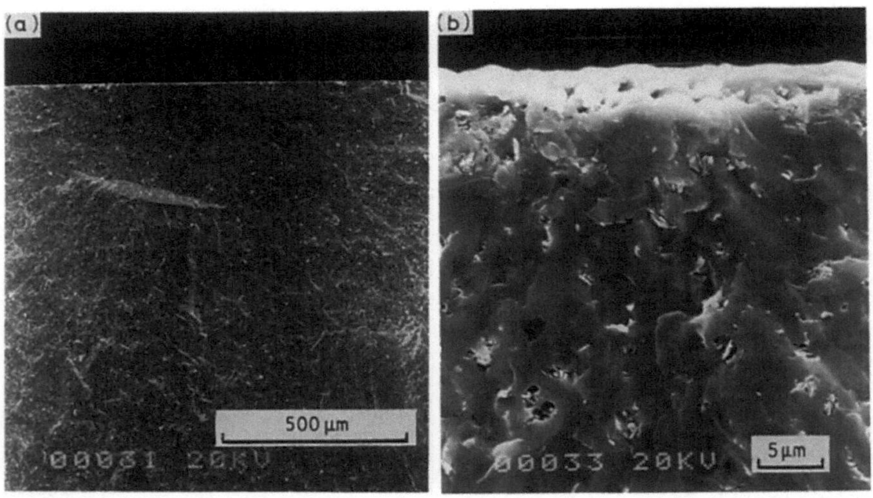

Fig. 9.81 Fracture surface of the S-1 specimen shown in Fig. 9.79a which was broken at room temperature: **a** low magnification, **b** high magnification. Horibe and Sumita [18]. With kind permission of Springer Nature

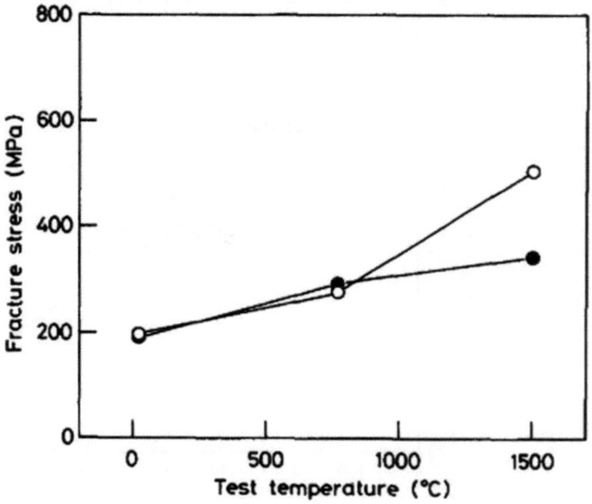

Fig. 9.82 Temperature dependence of the fracture stress of SiC materials (stress rate 8.2×10^{-1} MPa sec^{-1}, indentation load 9.8 N): (○) S-I, (●) S-2. Horibe and Sumita [18]. With kind permission of Springer Nature

a sligh damage (crack growth) occurred for yet unknown reason. Finally at higher temperatures of 1500 °C static fatigue seems to be controlled by creep. In samples containing Al (S-2) considerable plastic deformation was observed (mainly in grain boundaries) and intergranular cracking with crack-tip blunting took place, while in S-1 creep deformation and healing of the initial flaws were seen at the same temperature.

9.6.1 The Effect of Temperature in SiC/SiC

As indicated earlier in this book SiC is used as a reinforcimg agent in various forms, as particulates, whiskers fibers etc. But is also used as a matrix with other materials aimed to attain some physical or mechanical property. Often SiC matrix is used with its own whisker or fiber strengthener. In all these cases we are basically considering SiC composites. In the following damage and fracture of fiber-reinforced SiC composites are predicted by using the Budiansky-Hutchinson-Evans shear-lag model to determine the microstress field of the thermal damaged composites. Further, experimental and predicted data based on the model are compared. The thermal cyclic fatigue loading at elevated temperatures in oxidizing atmosphere are considered. The matrix/fiber interface shear stress is determined as a function of temperature and cycles. Cracking, fiber/matrix interface debonding/sliding and fibers fracture are included in this section. The decrease of cyclic dependent fiber/matrix interface shear stress, τ_0 from 10 to 5 MPa due to fatigue hysteresis dissipation energy, is associated with an increase in the fiber/matrix interface debonding and sliding length. This shear stress decrease and interface debonded length increase is seen in Fig. 9.83. The cyclic temperature dependent thermal fatigue damage at the SiC/SiC interface shear

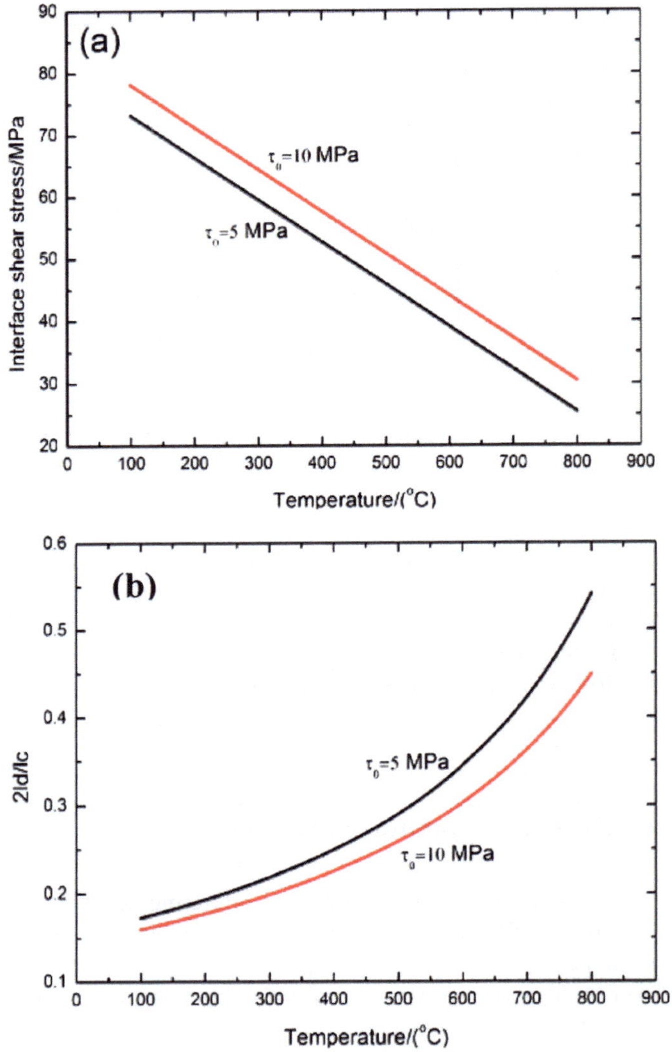

Fig. 9.83 **a** The interface shear stress versus temperature curves; **b** the interface debonded length versus temperature curves of SiC/SiC composite subjected to thermal cycle under constant loading for different interface shear stress of $\tau_0 = 5$ and 10 MPa. Longbiao [19]. Free access

stress (a) and the frictional coefficient, $\mu = 0.1$ and 0.5 (b) are shown in Fig. 9.84. The temperature range is 100 – 800 °C and the constant stress is $\sigma_{\tau ax} = 200$ MPa. When the fiber/matrix interface frictional coefficient increases from $\tau = 0.1$ to 0.5, the interface debonded length decreases. The effect of the fiber radius ($r_f = 6$ and

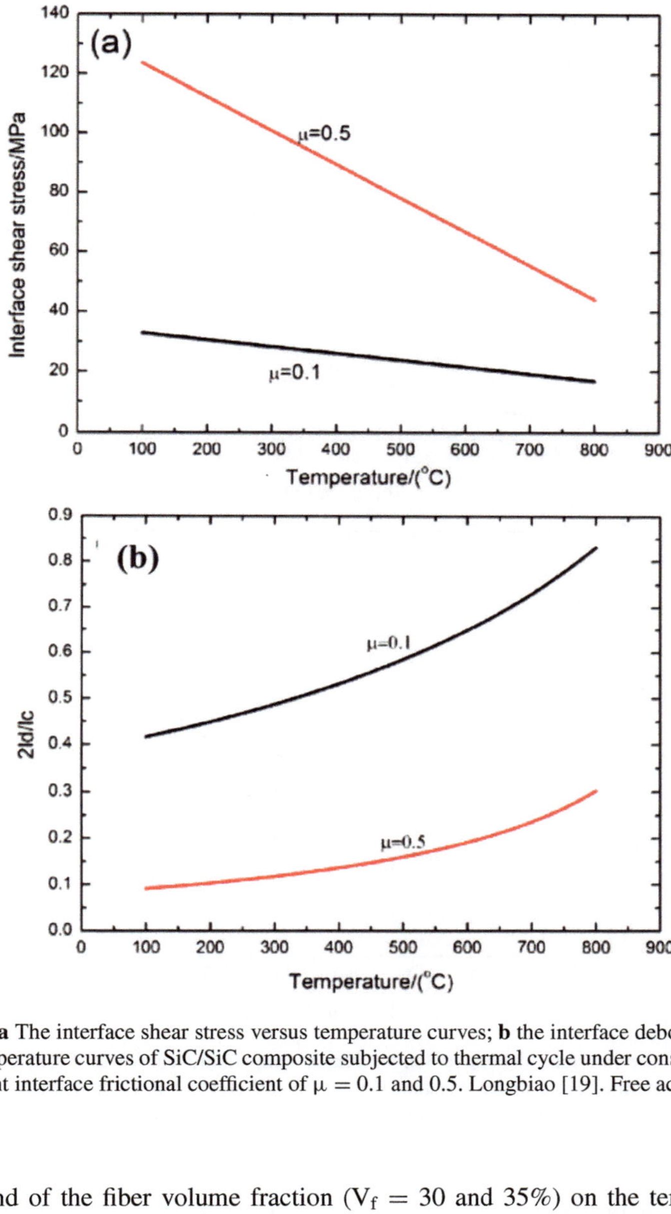

Fig. 9.84 **a** The interface shear stress versus temperature curves; **b** the interface debonded length versus temperature curves of SiC/SiC composite subjected to thermal cycle under constant loading for different interface frictional coefficient of $\mu = 0.1$ and 0.5. Longbiao [19]. Free access

9 μm) and of the fiber volume fraction (V_f = 30 and 35%) on the temperature-dependent thermal fatigue damage evolution of the fiber/matrix interface debonding ratio (2ld/lc) versus cyclic temperature curves under $\sigma_{max} = 300$ MPa at the temperature range 100–800 °C are illustrated in Fig. 9.85, (a) and (b), respectively. When the fiber radius increases from $r_f = 6$ to 9 μm, the fiber/matrix interface debonding and sliding length increase (Fig. 9.85a), while when the volume fraction increases from 30 to 35% the fiber/matrix interface debonding and sliding length decrease (Fig. 9.85b).

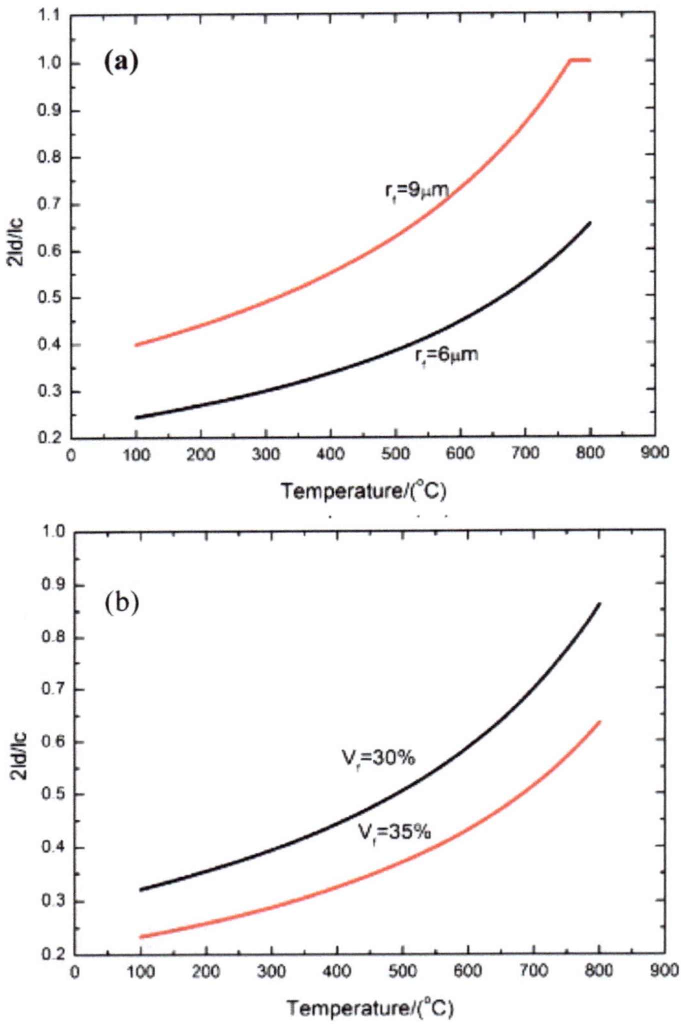

Fig. 9.85 **a** the interface debonded length versus temperature curves of SiC/SiC composite subjected to thermal cycle under constant loading for different fiber volume fraction of $r_f = 6$ and 9 μm; **b** for different fiber volume fraction of $V_f = 30$ and 35%. Longbiao [19]. Free access

The matrix crack spacing ($l_c = 200$ and 300 μm) as a function of temperature is illustrated in Fig. 9.86a, and that of the peak stress ($\sigma_{max} = 200$ and 250 MPa) in (b). An increase in the matrix crack spacing from $l_c = 200$ to 300 μm, results in a decrease in the fiber/matrix interface debonding and sliding range, and when the fatigue peak stress increases from $\sigma_{max} = 200$ to 250 MPa, the fiber/matrix interface debonding and sliding length increases. Of the several experimental/predicted figures

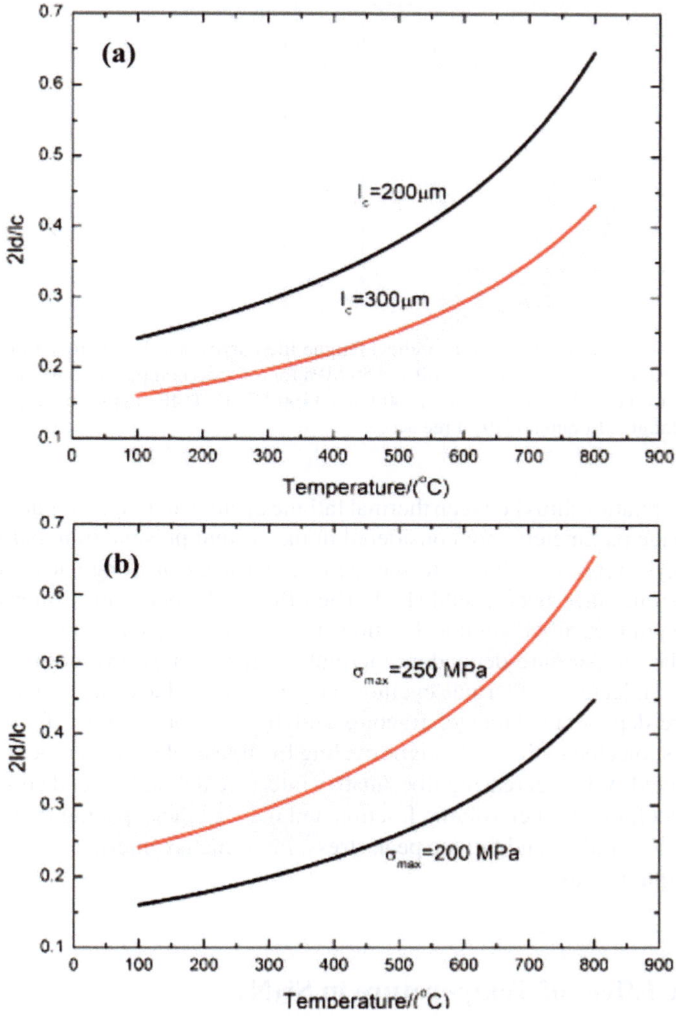

Fig. 9.86 **a** the interface debonded length versus temperature curves of SiC/SiC composite subjected to thermal cycle under constant loading for different fiber volume fraction of $l_c = 200$ and 300 μm; **b** $\sigma_{max} = 200$ and 250 MPa. Longbiao [19]. Free access

in the original publication two will be presented here. In Fig. 9.87 the experimental and the predicted lines are seen where the variation of the stress vs the number of cycles is illustrated.

In summary, the damage and fracture of fiber-reinforced ceramic matrix composite (CMC) subjected to thermal cycling at elevated temperatures in oxidizing atmosphere has been considered. The CMC is SiC/SiC, namely SiC fiber einforced SiC

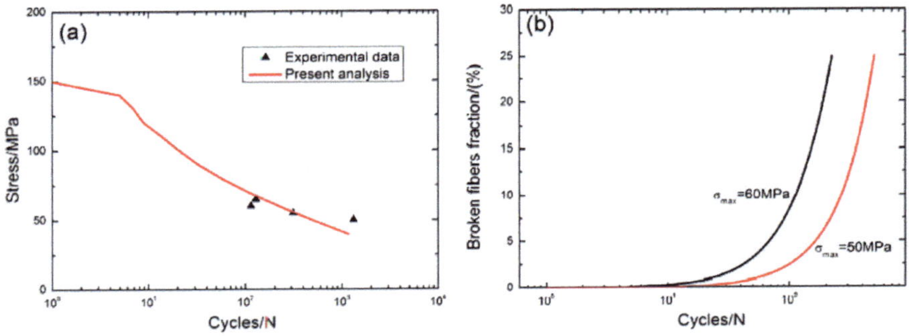

Fig. 9.87 **a** The experimental and theoretical fatigue life curves; and **b** the broken fibers fraction versus cycle number curves at $\sigma_{max} = 60$ and 50 MPa for SiC/SiC composite subjected to IF-TMF with respect to thermal cycling between 600 and 1100 °C. IF-TMF stands for in-phase thermo mechanical fatigue. Longbiao [19]. Free access

matrix. The relationships between thermal fatigue cycling, damage fatigue hysteresis-based damage parameters (not considered in the current presentation, but interested parties in hysteresis are advised to see the original work of Longbiao) and thermal fatigue lifetime have been established. The effects of fiber/matrix interface properties, fiber radius, fiber volume fraction, matrix crack spacing and fatigue peak stress on the temperature-dependent thermal fatigue damage evolution of SiC/SiC composite subjected to thermal cycling fatigue loading have been analyzed. The temperature-dependent damage, fracture and lifetime prediction of 2D SiC/SiC composite subjected to thermal fatigue cycling loading at elevated temperatures have been predicted. With decreasing fiber/matrix interface debonding and sliding range, frictional coefficient, fiber volume fraction and matrix crack spacing increase. With increasing fiber radius and fatigue peak stress, fiber/ matrix interface debonding and sliding length increase.

9.7 The Effect of Temperature in Si_3N_4

The use of Si_3N_4 as a structural material is widely spread because of its unique properties for high temperature applications. In particular, it has been identified as one of the few monolithic ceramic materials capable of surviving the severe thermal shock and thermal gradients generated in hydrogen/oxygen rocket engines. The list of major applications of Si_3N_4 illustrates its wide use in various fields, namely, automotive industry, bearings, rockets, medical as orthopedic applications, cutting tools, electronics, etc. (Fig. 9.88)

Two kinds of Si_3N_4 specimens designated as HS-110 and HS-130 were used in the experiments. The cyclic fatigue experiments at the temperatures indicated are shown in Fig. 9.89 for the two grades Si_3N_4 ceramics. Typical fracture surfaces fatigue fractured at 250 and 1200 °C are shown in Figs. 9.90 (A) and (B). The dark lines on

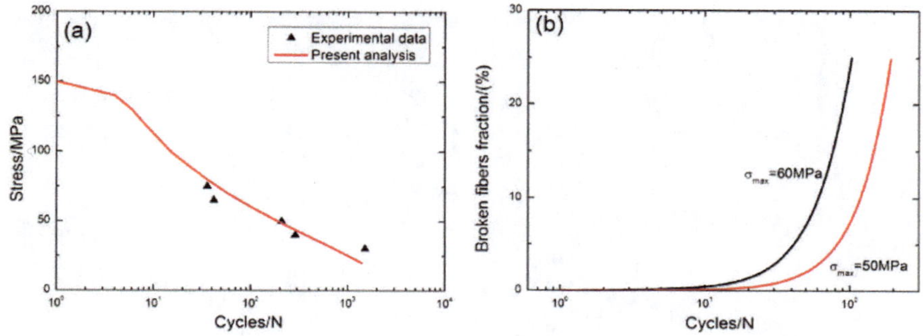

Fig. 9.88 **a** The experimental and theoretical fatigue life curves; and **b** the broken fibers fraction versus cycle number curves at $\sigma_{\max} = 60$ and 50 MPa for SiC/SiC composite subjected to OP TMF with respect to thermal cycling between 600 and 1100 °C. OP TMF stands for out of phase thermo mechanical fatigue. Longbiao [19]. Free access

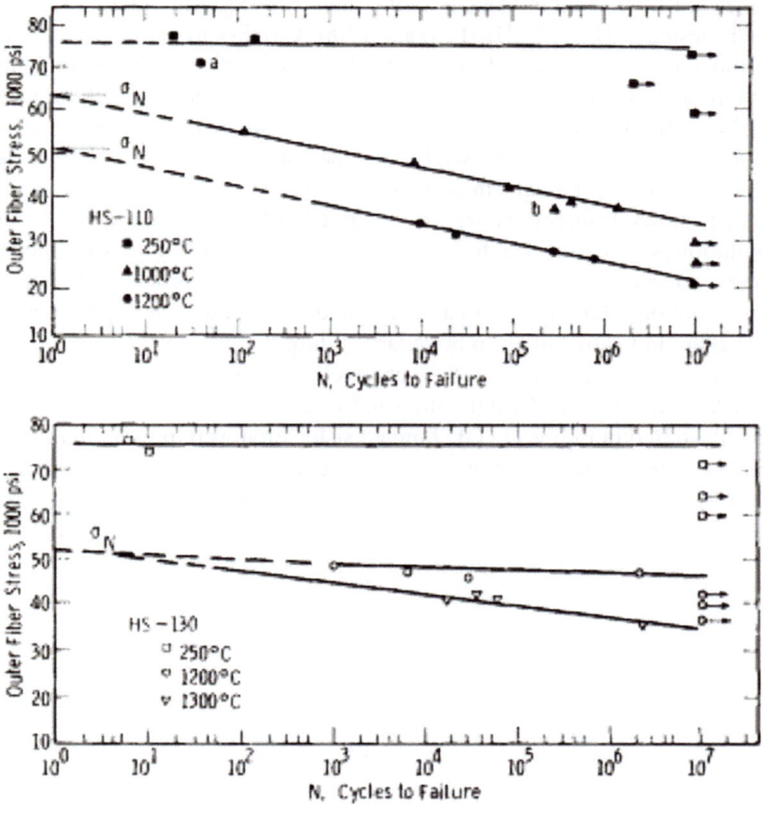

Fig. 9.89 Cyclic fatigue behavior for HS-110 and HS-130 hot-pressed Si_3N_4 at 1800 cpm. Ram Kossowsky et al. [20]. With kind permission of John Wiley and Sons

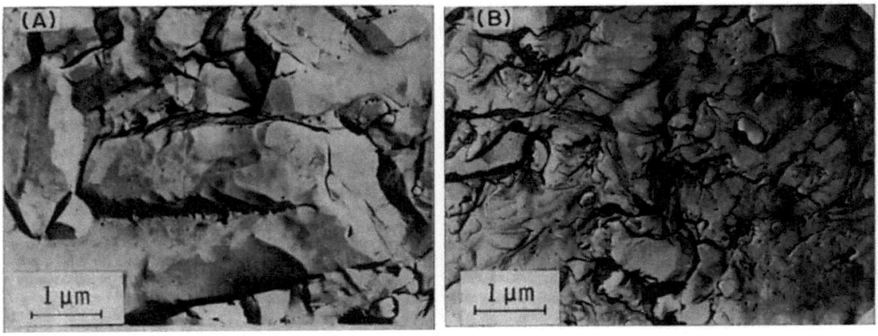

Fig. 9.90 Fracture surfaces from fatigue specimen of HS110 material at **a** 250 °C and **b** 1200 °C. Ram Kossowsky et al. [20]. With kind permission of John Wiley and Sons

the replica are cracks that indicate intergranulae cracking. The fractured surfaces of specimens tested at 1000 °C (HS-110) and 1300 °C (HS-130) show similar features. At the most stressed outer surfaces where localized deformation occurs, samples tested at 1000 and 1200 °C are shown in Fig. 9.91. Helices and stacking faults are seen inFig. 9.91 (A) and a high density of tangled dislocation are seen in (B) which were not observed in as received Si_3N_4, thus they must have been formed during the fatigue deformation. The deformation at 1300 °C, observed in micrographs of the foils tested, shows different features as illustrated in (C). Grain boundary separation and cavitation is observed. The foil was cut from the interior of the specimen where large stress concentration near a large inclusion (D) existed (probably the location of fracture initiation). Tripple point cavitation and grain boundary separation (shown by arrows) are indicative of grain boundary sliding.

In summary, two grades of Si_3N_4 were cyclic fatigue tested at the temperatures of 250, 1250 and 1300 °C. Fatigue life is influenced by the impurity level or inclusions as illustrated in Fig. 9.91 (D). Tangled sislocation were present in the fatigued specimens due to fatigue deformation. At the highest temperature of the fatigue tests, namely 1300 °C grain boundary separation and cavitation were observed.

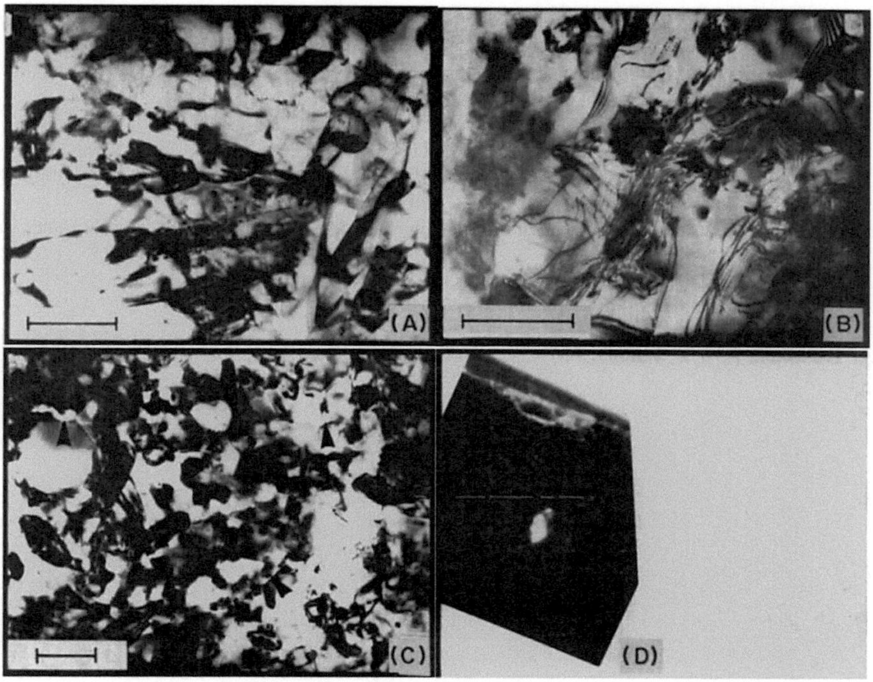

Fig. 9.91 Thin-foil transmission micrographs of fatigue specimens. **a** Helices and stacking faults in HS-110 specimen tested at 1000 °C and 20,000 psi. **b** Localized deformation in a large grain from HS-110 specimen tested at 1200 °C and 25,000 psi. **c** Grain boundary separations and cavitation (shown by large arrow heads) in specimen tested at 1300 °C and 40,000 psi; failure originated at the inclusion shown in **d** (1 μm scale bars). Ram Kossowsky et al. [20]. With kind permission of John Wiley and Sons

References

1. X.L. Zheng, B.T. Lu, Int. J. Fatigue **9**, 169 (1987)
2. M. Gupta, C. Kumara, P. Nylén, J. Therm. Spray Tech. **26**, 1038 (2017)
3. A.R. Studart, F. Filser, P. Kocher, L.J. Gauckler, Dental Mater. **23**, 106 (2007)
4. M. Masuda, T. Soma, M. Matsui, J. Eur. Cer. Soc. **6**, 253 (1990)
5. F. McBagonluri, W. O. Soboyejo, Mechanical Properties: Fatigue, in *Encyclopedia of Condensed Matter Physics*, (Elsevier, 2005) p 291
6. C. Yang, J. Wu, A. Ditta, S. Wu, Z. Zhao, Materials **14**, 145 (2021)
7. K. Hatanaka, Y. Ishiga, R. Kawazoe, M. Hasui, J. Ohgi, H. Ogawa, Mater. Sci. Eng. A **468**, 273 (2007)
8. A. F. Bower M. Ortiz, Trans. ASME, 115. 228 (1993)
9. P.C. Milak, F.D. Minatto, A. De Noni Jr, O.R.K. Montedo, Cerâmica **61**, 88 (2015)
10. E. Novitskaya, K. Karandikar, K. Cummings, M. Mecartney and O. A. Graeve, J. Mater. Res. Tech., Available online 23 January 2021
11. R. Matt, T. Liu, G. Grathwohl, Mater. Des. **14**, 159 (1993)
12. M. Mazaheri, A.M. Zahedi, M.M. Hejazi, Mater. Sci. Eng. A **492**, 261 (2008)
13. M. Yoshimura, T. Nishioka, A. Yamakawa, M. Miyake, J. Cer. Soc. Japan **103**, 407 (1995)

14. B.R. Lawn, D.K. Kim, S.K. Lee, Y.-G. Jung, I. M. Peterson, in *The Seventh International Fatigue Conference*, Fatigue 99 Beijing, China, Vol. 1, p. 69 (1999)
15. H. Nakayama and K. Ohira, Mater. Sci. Res. Inter., 4, 212 (1998), J. Soc. Mater. Sci., Japan
16. A. Tikhonovsky, M. Bartsch, and U. Messerschmidt, phys. stat. sol. (a) 201, 26 (2004)
17. D.L. Davidson, J.B. Cambell, J. Lankford, Acta metal. mater. **39**, 1319 (1991)
18. S. Horibe, M. Sumita, J. Mater. Sci. **23**, 3305 (1988)
19. L. Longbiao, J. Cer. Soc. Japan **127**, 67–80 (2019)
20. J. Ram Kossowsky, Amer. Cer. Soc. **56**, 531 (1973)
21. D. Munz, T. Fett, *Ceramics: Mechanical Properties, Failure Behaviour, Materials Selection* (Springer-Verlag, Berlin, 1999)
22. E.O. Hall, Proc. Phys. Soc. Lond. **64**, 747 (1951)
23. N.J. Petch, J. Iron Steel Inst. London. **173**, 25–28 (1953)
24. K. Tanaka, T. Mura, J. Appl. Mech. **48**, 97 (1981)
25. B. Budiansky, J.W. Hutchinson, A.G. Evans, J. Mech. Phys. Solids **34**, 67 (1986)

Chapter 10
Fatigue Fracture

Abstract Fatigue fracture is one of the primary damage mechanisms of structural components. Prior to final failure cracks or microcracks are induced. The fatigue cracking results from cyclic stresses that are below the ultimate tensile stress, or even the yield stress of the material. Surface flaws small are serving as stress risers where cracks can initiate, leading to fracture under repeated stresses. Porosity is detrimental to fatigue resistance and addition of small amounts a metal such as Ni for example, can improve the fatigue resistance by increasing the density. Fracture process can be divided into three stages crack formation, propagation and rapid fracture stage. Alumina, magnesia, zirconia, SiC and Si_3N_4 subjected to fatigue fracture are presented and the effects promoting or resisting their progress are considered.

Keywords Fracture · Alumina · Partially stabilized zirconia · Fully stabilized zirconia · Silicon carbide · Si_3N_4 · Temperature · Fatigue · Resistance · Flexural strength · Bondin

10.1 Introduction

Fatigue fracture is one of the primary damage mechanisms of structural components. Prior to final failure cracks or microcracks are induced. The fatigue cracking results from cyclic stresses that are below the ultimate tensile stress, or even the yield stress of the material. Surface flaws small as they may be serving as stress risers where tiny cracks can initiate, and as mentioned leading to fracture under repeated (or fluctuating) stresses and as mentioned having maximum value less than the UTS of the material. As the stress level increases, the crack growth rate increases at a given number of cycles determining the fatigue lifetime until failure sets in. The cyclic load induced imperfections or flaws grow to macroscopic cracks causing the failure. One has to note that fatigue is usually associared with tensile stress, but fatigue cracks-thus failure-have been also observed by applying compressive loads. Although most of the ceramics are brittle, and fatigue failure is brittle-like with relatively litttle plastic deformation even in ductile mateials (when tested by conventional static methods). It is immaterial what type of load applied is causing fatigue; the sudden and catastrophic applied stress causing fatigue may be axial, flexural or torsional. The advance of the

J. Pelleg, *Cyclic Deformation in Oxides, Carbides and Nitrides*, Structural Integrity 22,
https://doi.org/10.1007/978-3-030-86118-6_10

failure from an initial crack can be seen on the surface by beach marks, and when a critical crack size is reached it will propagate rapidly to complete fracture.

There are three stages of fatigue fracture: (i) initiation, (ii) propagation, and (iii) final fracture or rupture. The failure occurs due to the cyclic nature of the load which causes microscopic material imperfections (flaws) to grow into a macroscopic crack (initiation phase). Dislocations play a major role in the fatigue crack initiation phase. Experiments indicate that after a large number of cycles dislocations pile up and form structures called persistent slip bands (PSB). Further, classification of fatigue is low cycle fattigue (LCF) and high cycle fatigue (HCF). LCF is characterized by repeated plastic deformation, namely in each cycle, whereas HCF is characterized by elastic deformation. Clearly the terms indicated are a consequence of the number of cycles to failure, in LCF it is low and in HCF it is high. The LCF is characterized by high cyclic stress levels in excess of the endurance limit of the material and is in the range of 10^4–10^5 cycles. Note that in non-ferrous metals (also Ti alloys), no endurance limit in the form of a knee (horizontal line) exists. Thus, one is talking in fatigue about the number of stress cycles a material can endure before failing.

About 80% of material fail by fatigue, thus it is clear that many factors affect the prospensity of a material to fatigue fracture: fluctuating stress levels, inconsistencies (pores, impurity inclusions), component features induced by machining or other fabrications (sharp corners, or bolt holes for example) etc. The main point to remember is that local stress risers induced and created by these and other factors, which can cause catastrophic failure. Generally, fatigue occurs by tensile stress, but fatigue cracks were observed also by compressive loads. It is obvious that the greater is the applied stress, the shorter the fatigue-life. About 5–10 specimens are required for fatigue tests because of the great scatter in the test results and thus obtaining a meaningful average value. Most of the factors affecting fatigue life are: (i) Type and nature of loading, (ii) size of the component exposed to fatigue and the distribution of the stress or the strain, (iii) surface finish, (iv) stress or strain concentration, (v) the mean stress or strain, (vi) environment, (vii) material and structural properties, (viii) strain rate and (ix) frequency.

As mentioned above fatigue occurs in ceramics. The fatigue in cermic as in metals depend on microstructure and on some of the the criteria indicated earlier. Thus, for example the fatigue behavior in alumina, or zirconia has already been reported in the nineties of the previous century performed by direct push–pull testing that generated both static and cyclic fatigue, which indicated that under cyclic loading the time to failure is considerably shorter. In the following the fatigue behavior of some oxides, carbides and nitrides will be considered, clearly limited by the availability of such data in the open literature.

10.1.1 Fracture in Alumina

One hardly finds fatigue fracture in monolithic alumina, although abundant information on aluminas is available, in various composites, as reinforcing agents, or reinforced by themselves with other constituents.

Fine-grained alumina prepared by cold isostatic pressing and sintering process is considered here for its fatique behaviour under repeated impact loading. The fatigue resistance was found to be 17.12 while the endurance limit was around 270 MPa. Fatigue cracks are transgranular near the crack initiation region, while in other regions it is intergranular. The fracture of alumina is of considerable interest because—as mentioned in various places of this book—it has unique properties and it is being applied in a wide range of engineering applications from electronic substrates to cutting tools. Since alumina is exposed during its actual engineering applications to cyclic loading, therefore the interest in its fatigue behavior is related to various loading systems such as three or four-point bending tests, tensile loading, tension–compression loading, various types of bending (cantilever bending, plane bending, rotary bending) etc. Further, alumina in many cases encounter impacts (for example cutting tools encounter continuous vibrations and repetitive shock or impact, electrical circuitry is exposed to repetitive stresses on switching on and off), therefore experimental observations are necessary in impact fatigue fracture behavior. In the following, impact fatigue of fine grained isostatically pressed, sintrered alumina is presented in terms of fatigue resistance and fatigue endurance limit. In Fig. 10.1 the points with arrows indicate samples whose testing was terminated. The correlation between single impact strength and fatigue strength is presented in Fig. 10.2 as a plot of the impact strength versus N. The relation is linear (on a double logarithmic graph) and can be expressed as

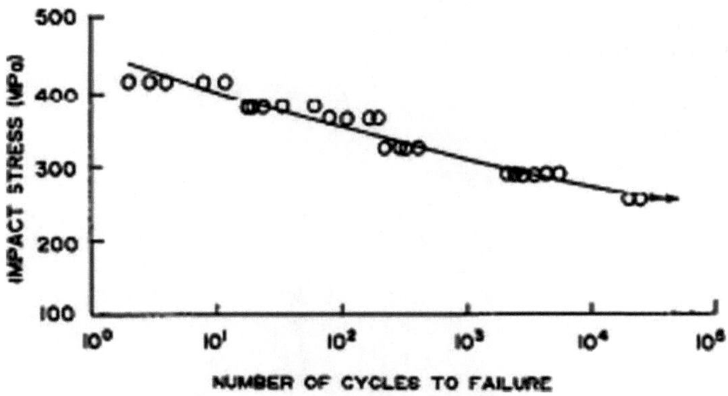

Fig. 10.1 Stress-endurance plot of sintered fine grained alumina. Points with arrows indicate unbroken samples for which test were terminated. S. Maity, D. Vasu and B. K. Sarkar, Mater. Let., 21, 63(1994). With kind permission of Elsevier

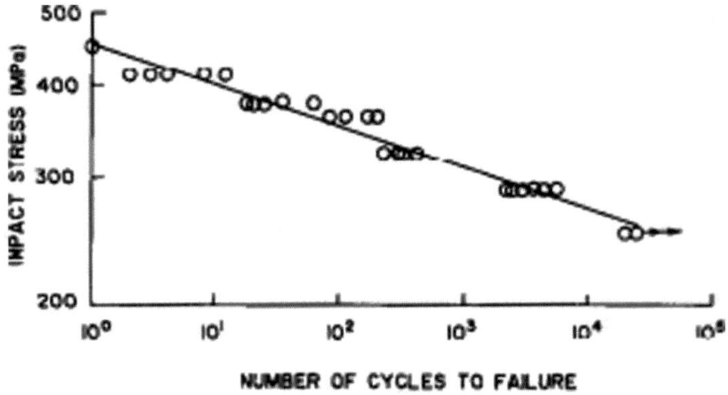

Fig. 10.2 S–N curve including single impact strength. S. Maity, D. Vasu and B. K. Sarkar, Mater. Let., 21, 63(1994). With kind permission of Elsevier

$$\sigma^n N = \text{constant} \tag{10.1}$$

Here, σ is the impact stress and N is the number of impacts to failure. From the slope of the line the exponent indicating fatigue resistance of the material can be evaluated. The fine grained microstructure used in the experiments is presented in Fig. 10.3. The grain size is ~3–4 μm which clearly improved the mechanical properties. However, impact produces a shock wave which induces greater damage than the slow cyclic stressing to the same level. The micrograph of the fractured surface after fatigue is illustrated in Fig. 10.4. Magnified views of regions (B) and (C) are shown in

Fig. 10.3 Typical microstructure of fine-grained alumina used in these experiments. S. Maity, D. Vasu and B. K. Sarkar, Mater. Let., 21, 63(1994). With kind permission of Elsevier

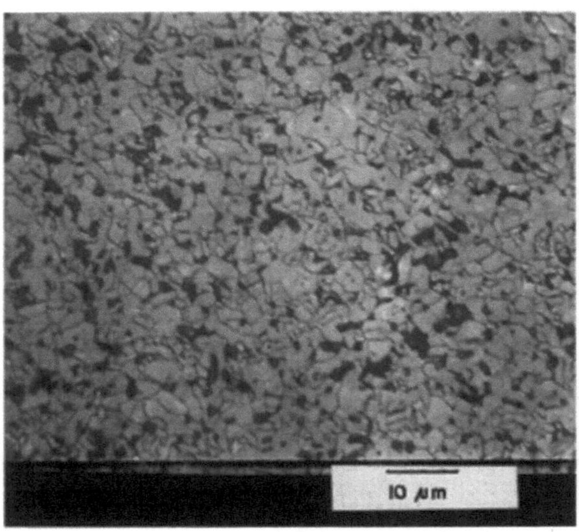

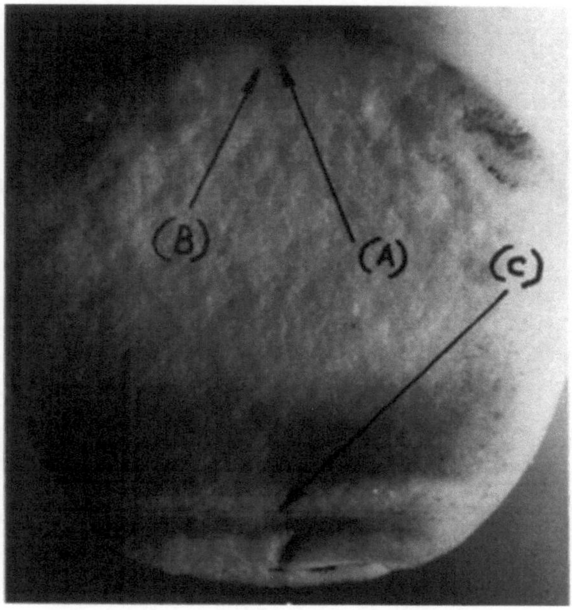

Fig. 10.4 Microphotograph of fractured surface after fatigue test $\sigma = 288.65$ MPa, N = 2864 impacts. (A) = point of crack initiation, (B) = mirror region, (C) = region of lip formation. S. Maity, D. Vasu and B. K. Sarkar, Mater. Let., 21, 63(1994). With kind permission of Elsevier

Figs. 10.5 and 10.6. One can observe in the fractured surface at (A) of Fig. 10.4 the fracture intitiation; in Fig. 10.5 at (B) of Fig. 10.4, the mirror region is seen which is characteristic to fatigue failure and in Fig. 10.6 at point (C) of Fig. 10.4, the crack

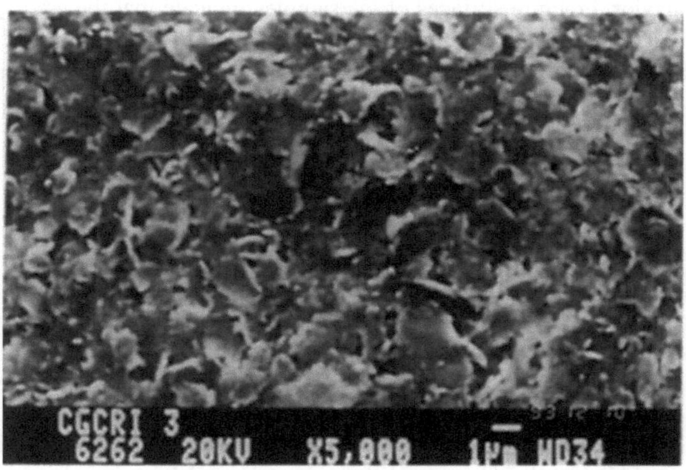

Fig. 10.5 Magnified view of the mirror region B (Fig. 10.4), indicating a severe transgranular fracture pattern. S. Maity, D. Vasu and B. K. Sarkar, Mater. Let., 21, 63(1994). With kind permission of Elsevier

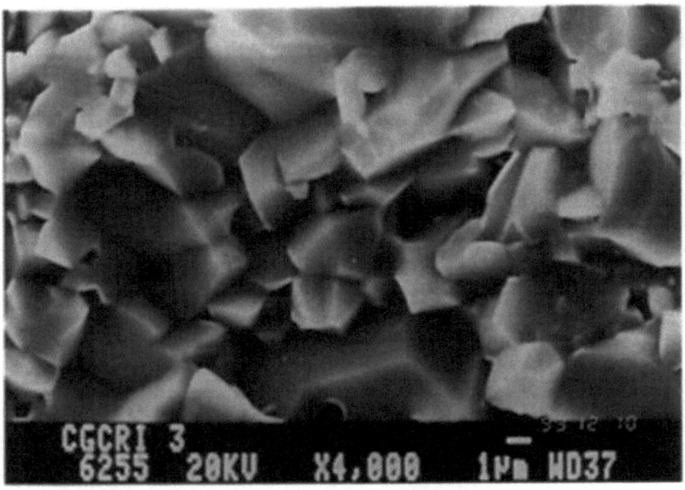

Fig. 10.6 Magnified view of region C (Fig. 10.4) showing mainly intergranular fatigue crack propagation path. S. Maity, D. Vasu and B. K. Sarkar, Mater. Let., 21, 63(1994). With kind permission of Elsevier

path follows an intergranular pattern. Region B is associated with slow crack growth and the applied stress energy is dissipated transgranularly, while in region (C) fast crack growth occurs and its pass follows an intergaranular pattern proceeding along the low energy grain boundaries where the residual stress can be accommodated.

In summary, under repeated impact loading fine grained alumina shows typical fatigue behavior. Fracture surfaces of the samples indicate that the crack growth is transgranular at the fast growth region, while it is intergranular in the slow crack growth region.

10.1.2 Fracture in Magnesia

Refractory materials (ceramics) are expected to be shock resistant for safe and prolonged use in service. A key factor among other parameters for industrial applications of ceramics is the resistance of crack propagation, and its characterization is essential to remedy such problems if encountered in service. For many industrial applications, refractories with reduced brittleness, which correspond to pronounced deviations from pure linear elastic fracture mechanics (LEFM), are favourable. The characteristic length l_{ch} was used to describe the material brittleness. The characteristic length following Hillerborg an Harmuth is given by

$$l_{ch} = \frac{G_F E}{\sigma_t^2} \tag{10.2}$$

where G_F is the specific fracture energy for forming two fracture surfaces, σ_t is the tensile strength and E is the Young's modulus. Generally, maintaining G_F constant and reducing σ_t^2/E, the brittleness of a ceramic decreases while the shock resistance increases. The reduction in the refractory brittleness is often achieved by introducing micro-cracks during the production process. Nonlinear fracture parameters of refractory ceramic such as the resistance for crack growth can be expressed by the relation between the specific fracture energy, R and the crack length "a" by

$$R_c = \frac{\Delta U_{ir}}{B\Delta a} \tag{10.3}$$

Here, ΔU_{ir} is the irreversible energy dissipation during the crack extension of Δa, and B is the thickness of the specimen. The wedge splitting test (WST) has been used to identify the fracture mechanical properties of brittle or quasibrittle materials and applied for magnesia. The experimental set-up is illustrated in Fig. 10.7. A video extensometer camera at the rear side of the specimen was used to measure the horizontal displacement at the load points. From the measured load–displacement curve, the fracture parameters including the specific fracture energy, G'_f and the nominal notch tensile strength, σ_{NT} can be determined. It is expressed as

$$G'_f = \frac{1}{A}\int_0^{\delta} F_H d\delta_H \tag{10.4}$$

$$\sigma_{NT} = \frac{F_{H,\max}}{bh}\left(1 + \frac{6y}{h}\right) \tag{10.5}$$

Fig. 10.7 Configuration of cyclic wedge splitting test (WST) and image acquisition set-up. Y. Dai, D. Gruber and H. Harmuth, J. Eur. Cer. Soc., 37, 5035 (2017). With kind permission of Elsevier

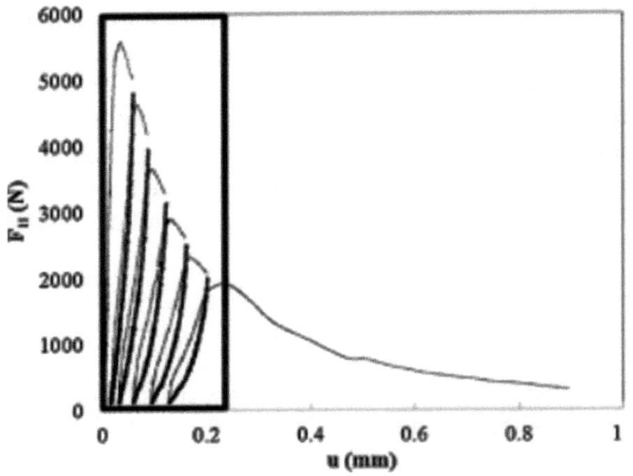

Fig. 10.8 Load–displacement diagrams for the magnesia material. Y. Dai, D. Gruber and H. Harmuth, J. Eur. Cer. Soc., 37, 5035(2017). With kind permission of Elsevier

Here, δ_H is the horizontal displacement at 15% of the maximum load. $F_{H,max}$ is the maximum horizontal load, band h are the width and the height of the ligament, respectively, and y is the vertical distance from the load vector to the centre of the ligament.

Three specimens of the magnesia were tested. Figure 10.8 shows a typical load–displacement curve for a magnesia sample. The horizontal displacement "u" was measured at the load points. For each cycle, the integrated area between the loading and unloading paths equals to the dissipated energy for the crack extension during this cycle. The enframed areas of Fig. 10.8 are displayed in more detail in Fig. 10.9. The mechanical results determined from the WST are exhibited in Table 10.1. The specific fracture energy G'_f and the nominal notch tensile strength σ_{NT} are calculated using Eqs. (10.4) and (10.5), respectively. The ratio of the specific fracture energy to the notch tensile strength was frequently applied to indicate the brittleness. A higher G'_f/σ_{NT} ratio means a lower brittleness. An interesting peculiarity is observed, namely, the slopes of the unloading curves merge at a common point, a focal point (Fig. 10.9). The focal point is used to approximate the stiffness at any point on the envelope of the load–displacement curve. Each data point, shown in Fig. 10.10 represents the stiffness of a corresponding loading–unloading cycle. Magnesia spinel is included for comparison (magnesia spinel has a higher normalized stiffness at each value of u (horizontal displacement)). The digital image correlation (DIC) was used to visualize the fracture process. The strain, calculated from the displacement field was used as an indicator for determining the fracture process. The evolution of the crack mouth open displacement (CMOD) over the loading–unloading cycle is seen in Fig. 10.11. During the unloading, decrease in the CMOD coincides with the release of elastic strain from the samples. The dissipated 'energy at a certain crack extension or CMOD is the irreversibly consumed energy. The relationships between the dissipated energy and CMOD are shown in Fig. 10.12. Each data point represents the measured data for one cycle and the last data point is determined from the end of the load–displacement curve as shown in Fig. 10.8 (15% of F_{max}). As seen in

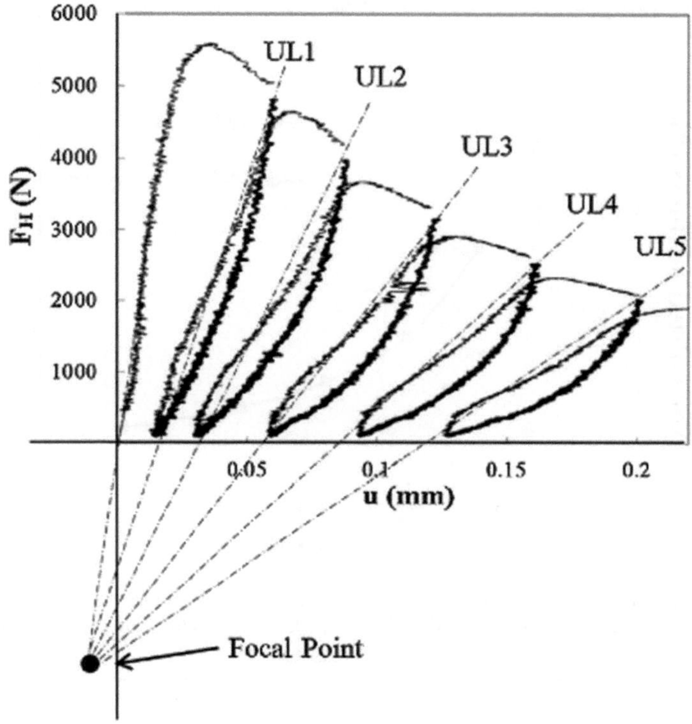

Fig. 10.9 Illustration of focal point determination for magnesia. Y. Dai, D. Gruber and H. Harmuth, J. Eur. Cer. Soc., 37, 5035(2017). With kind permission of Elsevier

Table 10.1 Mechanical properties for the magnesia material (M). Y. Dai, D. Gruber and H. Harmuth, J. Eur. Cer. Soc., 37, 5035(2017). With kind permission of Elsevier

	Magnesia			
	M-a	M-b	M-c	Mean value
$F_{H,max}$ (N)	5718	5594	4476	5263
σ_{NT} (MPa)	7.77	7.74	6.17	7.23
G_f' (N/m)	118	220	118	152
G_f'/σ_{NT} (mm)	15.2	28.5	19.2	20.9

Fig. 10.12, the slope of the dissipated energy with respect to the CMOD is reducing with the increase in the CMOD.

In summary, the wedge splitting test method along with the digital image correlation was successfully applied to investigate the fracture behavior in magnesia (and magnesia spinel, not discussed here). The fracture process zone (FPZ) facilitates

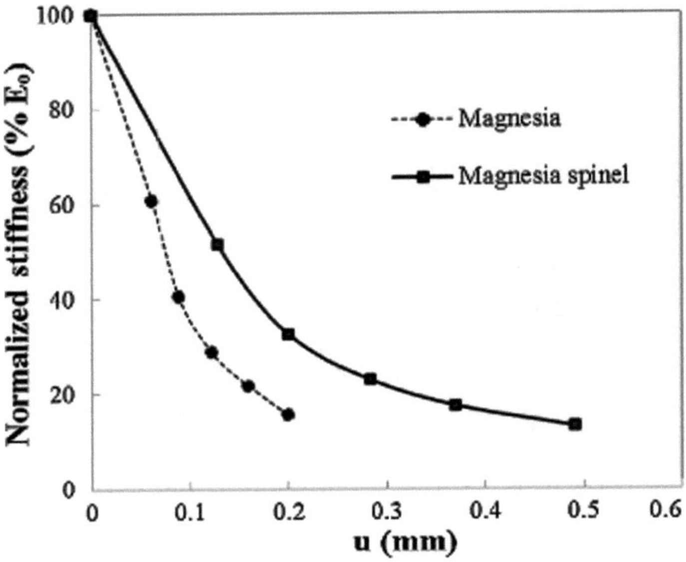

Fig. 10.10 The stiffness degradation for both material the magnesia and the magnesia spinel. Y. Dai, D. Gruber and H. Harmuth, J. Eur. Cer. Soc., 37, 5035(2017). With kind permission of Elsevier

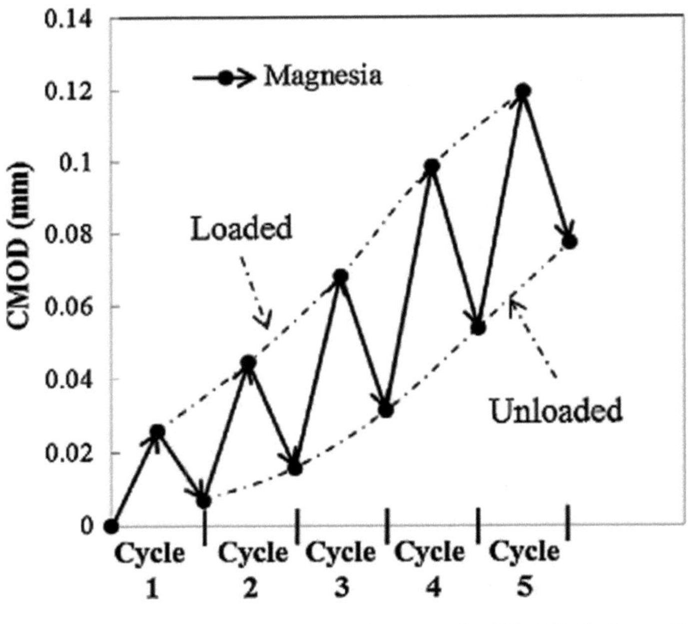

Fig. 10.11 Evolution of the CMOD over the cycles for magnesia. Y. Dai, D. Gruber and H. Harmuth, J. Eur. Cer. Soc., 37, 5035(2017). With kind permission of Elsevier

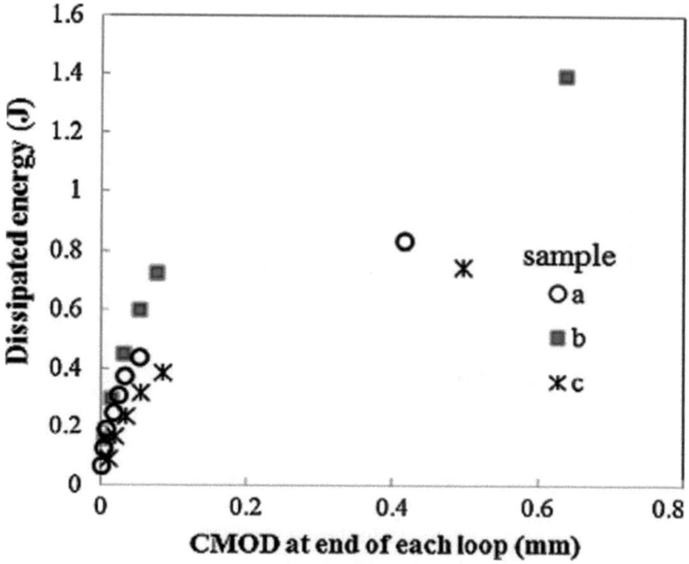

Fig. 10.12 Increase of dissipated energy with respect to CMOD for magnesia. Y. Dai, D. Gruber and H. Harmuth, J. Eur. Cer. Soc., 37, 5035(2017). With kind permission of Elsevier

stable crack propagation, high crack mouth opening displacement (CMOD) tolerance and tesults in a material with lower btittleness. The loading–unloading WST is a convenient method to measure the dissipated energy with a single specimen. By using the DIC, the FPZ development characterization and in-situ crack length measurement was realized. The fracture process comprises the following two main stages: (i) formation and extension of the fracture process zone and (ii) the subsequent development of the macro-crack.

10.1.3 Fracture in Zirconia

In a relatively recent publication on ceramic composites based on ZrO_2, pure monolithic zirconia fatigue behavior was aslo included for comparison in order to evaluate the strengthening effect of the additives. Dense (>98%) and homogeneous composites were obtained by spark plasma sintering (SPS) using ZrO_2 powder. Indentation load versus flexural strength of indented samples is shown in Fig. 10.13. Each data point is representing the mean value of ~12 specimens tested at a given load. Linear fitting was applied. The mean biaxual flexural strength is indicated in the figure, being for ZrO_2 1217 ± 10 MPa (for the composites these are 970 ± 18 MPa and 850 ± 20 MPa, respectively). Note that the pure ZrO_2 line of the flexural strength is below the lines of the composites at all loads. Densities and mechanical properties of

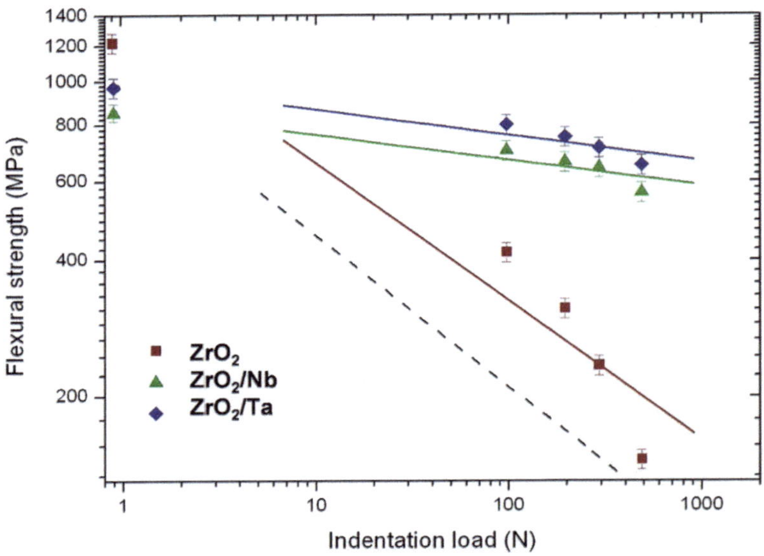

Fig. 10.13 Indentation load versus strength plots of 3Y-TZP/Ta, 3Y-TZP/Nb composites and zirconia ceramic. The indentation-strength data to the P − 1/3 strength response is shown by the diagonal dashed line. J. I. Beltrán, T. Rodriguez-Suarez, C. Pecharromán, M. C. Muñoz, J. S. Moya and J. F. Bartolomé, Scientific Reports, **7**, Article number: 44922 (2017). Open access

ZrO_2 are assembled in Table 10.2. Understanding fatigue is very important for design since a structure may fail at a small percentage of its fracture strength. The results for cyclic fatigue life for the specimens tested (with a disc-geometry) are presented in Fig. 10.14 as peak stress versus N the cycles to failure. The tests were interrupted at N = 107 cycles, for the unfailed samples, which are marked with an arrow. Three maximum stress levels, σ_{max} were selected for the initial strength obtained under static tests already indicated in Fig. 10.2 and also shown in Table 10.3. For the interpretation of the fatigue data of ZrO_2, the exponential model was used. The estimation of the model parameters was based on linear regression analysis. It was found that the fatigue limit for ZrO_2 is 1200 ± 15 MPa (see Table 10.3). Note in Fig. 10.14 the line of ZrO_2 is above those of the composites. This is a direct consequence of the smaller

Table 10.2 Densities and mechanical properties of all the studied specimens as well as volume fractions of tetragonal "t" and monoclinic "m" zirconia in polished and fractured surfaces and the resulting transformabilities of tetragonal zirconia. J. I. Beltrán, T. Rodriguez-Suarez, C. Pecharromán, M. C. Muñoz, J. S. Moya and J. F. Bartolomé, Scientific Reports, **7**, Article number: 44922 (2017). Open access

Specimen	Density [% φ_{th}]	Elastic modulus [GPa]	Flexural strength σ_f [MPa]	Hardness HV [GPa]	Fracture toughness K_{Ic} [MPa·m$^{1/2}$]	Volume fractions of t- and m-ZrO_2 [vol%]				Transformability of t-ZrO_2 V_{trans}
						Polished		Fractured		
						t	m	t	m	
ZrO_2	99	198 ± 5	1217 ± 10	13 ± 0.3	6 ± 0.3	99	1	97	3	2

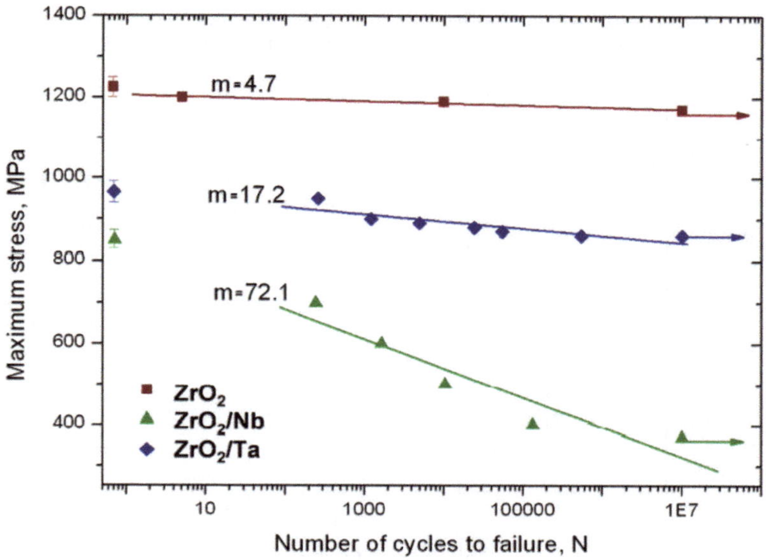

Fig. 10.14 Fatigue resistance S–N curves with values of slopes for un-indented polished specimens. J. I. Beltrán, T. Rodriguez-Suarez, C. Pecharromán, M. C. Muñoz, J. S. Moya and J. F. Bartolomé, Scientific Reports, **7**, Article number: 44922 (2017). Open access

Table 10.3 Fatigue ratio of tested ZrO_2. J. I. Beltrán, T. Rodriguez-Suarez, C. Pecharromán, M. C. Muñoz, J. S. Moya and J. F. Bartolomé, Scientific Reports, **7**, Article number: 44922 (2017). Open access

Composition	Flexural strength σ_f, [MPa]	Fatigue limit [MPa]	Fatigue ratio
Specimens without artificial flaws			
ZrO_2	1217 ± 10	1200 ± 15	0.98
Artificially-flawed samples			
ZrO_2	312 ± 7	130 ± 40	0.41

critical grain size in ZrO_2 and thus the bending strength of the monolithic ceramic is higher than the strength corresponding to both zirconia/metal composites. The figure in the presence of an indentation-flaw (crack) shown in Fig. 10.15 looks differently than that of Fig. 10.14. The indented specimens were tested under the same cyclic conditions as the un-indented ones. The average fracture toughness of ZrO_2 is 6 $\pm$ 0.3 MPa·m$^{1/2}$. The initial flaw size is 335 μm. Any flaw present in monolithic zirconia under very low subcritical loading can lead to catastrophic failure. Therefore, the fatigue ratio of zirconia is extremely sensitive to surface condition and the presence of defects that act as stress raisers and, consequently, the chances of failure of monolithic ceramic is high (higher than in ceramic/metal composites). It is important that the growth of defects should be limited before reaching critical size

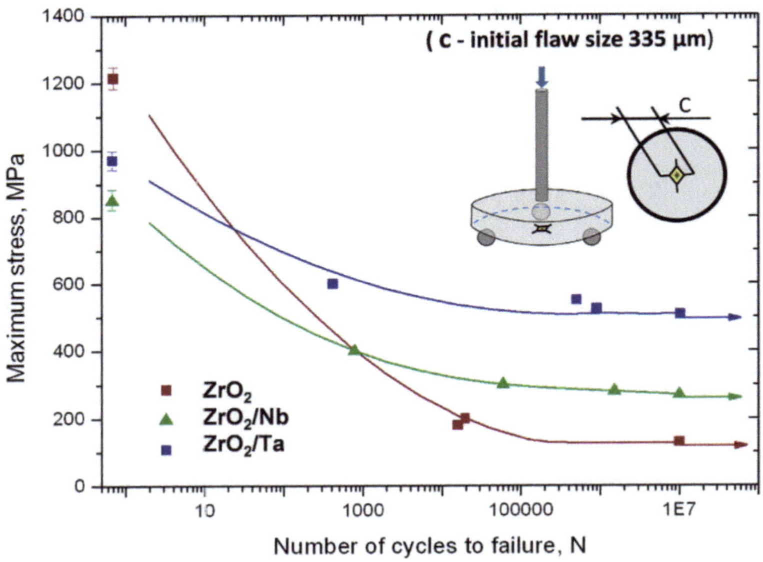

Fig. 10.15 S–N plot of the cycles to failure (N) of indented ZrO_2, ZrO_2/Ta and ZrO_2/Nb composites. Ultimate flexural strength of un-indented specimens is plotted as well. J. I. Beltrán, T. Rodriguez-Suarez, C. Pecharromán, M. C. Muñoz, J. S. Moya and J. F. Bartolomé, Scientific Reports, **7**, Article number: 44922 (2017). Open access

and prevent sudden fracture. In order to observe crack propagation, the experiment was stopped at various stages of cycling tests. At each stop the test specimen was dismounted from the jig for SEM analysis. The combination of this information with the test load is shown in the graphs of crack lengths as a function of the number of cycles in Fig. 10.16. The vertical black arrows correspond to the number of cycles of stress to cause complete fracture of the specimen. It has been found that zirconia with artificially induced flaws shows fatigue crack growth initiation when the applied load is only ~7% (80 MPa) of the flexural strength value of un-indented specimens. In other words, any flaw present in monolithic zirconia under very low subcritical loading can lead to catastrophic failure. Table 10.3 shows the flexural strength, σ_f, the fatigue limit and the fatigue ratio (ratio of the fatigue limit to the static strength) for specimens without and with artificial flaws. A significant decrease in the ratio of the monolithic ZrO_2 between un-indented and intended specimen is observed (Table 10.4). The presence of defects (and rough surface) act as stress raisers with higher probability to failure.

In summary, mechanical tests were performed immediately after indentation to avoid any subcritical crack growth in the secimens. Fatigue life of specimens was determined for the monolithic zirconia (and for composites; not discussed since it is outside the scope of this book). Biaxial flexural strength tests were subjected to a fatigue test. The stress ratio was 0.1 and the fatigue cycle was limited at 10^7 cycles.

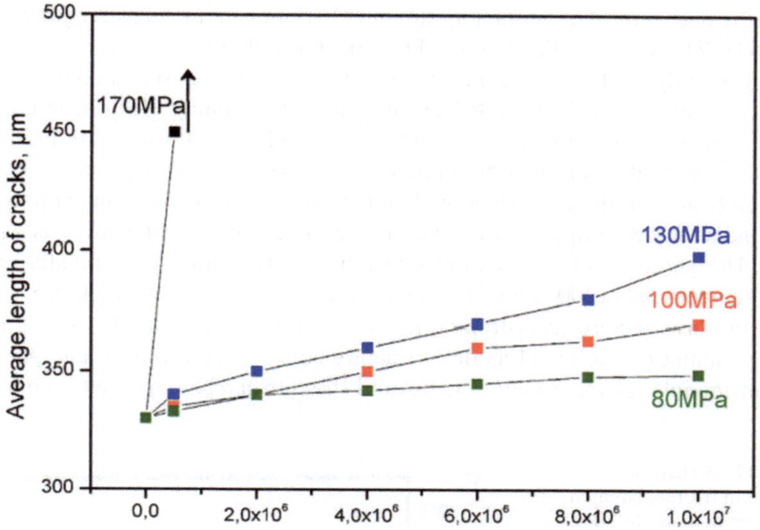

Fig. 10.16 Number of cycles and crack length as a function of the applied load for ZrO_2. J. I. Beltrán, T. Rodriguez-Suarez, C. Pecharromán, M. C. Muñoz, J. S. Moya and J. F. Bartolomé, Scientific Reports, **7**, Article number: 44922 (2017). Open access

Table 10.4 Fatigue ratio of tested zirconias. J. I. Beltrán, T. Rodriguez-Suarez, C. Pecharromán, M. C. Muñoz, J. S. Moya and J. F. Bartolomé, Scientific Reports, **7**, Article number: 44922 (2017). Open access

Composition	Flexural strength σ_f [MPa]	Fatigue limit [MPa]	Fatigue ratio
Specimens without artificial flaws			
ZrO_2	1217 ± 10	1200 ± 15	0.98
Artificially-flawed samples			
ZrO_2	312 ± 7	130 ± 40	0.41

10.1.4 Fracture in Partial Stabilized Zirconia (MgO)

A commercial grade MgO partially stabilized zircomia (PSZ) is presented below. MgO is one of the stabilizers of zirconia and the designated grade of the commercial material is TS grade contains 9 mol% MgO. In the as machined test samples tetragonal and monoclinic phases were present. Flexural tests were used to evaluate the fracture properties. A detaile experimental procedure, specimen size, shape and preparation for the test are described in detail by Govila in his work on strength characterization in a partially stabilized zirconia by MgO. The flexural strength was evaluated in air environment (in the temperature range of 20–1300 °C) as a function of temperature, applied stress and time. At room temperature 10 specimens were tested in four-point bending to determine the fast fracture strength. Variation in fracture strength, σ_F in the range 641–716 MPa is illllustrated in Fig. 10.17 as a Weibull plot. As known fatige failure generally starts at the surface where often processing comtaminants or other impurities usually reside. Detailed examination by scanning electron micrograph indeed showed at the initiation site oxides of Al and Si as illustrated in Fig. 10.18. The presence of SiO_2 densifies the ceramic via liquid phase sintering (see 10.18b) and therefore beneficial in the production, but detrimental for the mechanical properties at RT and at higher temperatures due to its presence in grain boundaries as a glassy phase. The variation of the flexural strength as a function of temperature at the higher temperatures (200–1300 °C) is shown in Fig. 10.19. The crack propagation during the fast fracture (catastrophic failure) appeas to be primarily intergranular, although sometimes a mixed mode of fracture appears consisting of intergaranular and transgranular crack growth. The flexural strength can be improved even in large

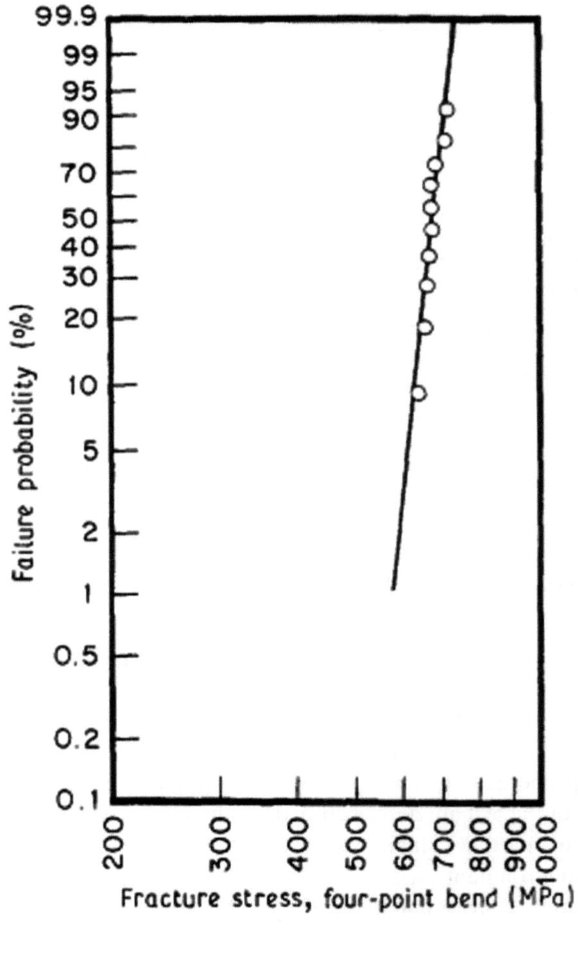

Fig. 10.17 Statistical variation in fracture strength for polycrystalline MgO-PSZ (TS grade), at 20 °C, σ_{ave} = 677 MPa, S. D. 32 MPa, m = 26. R. K. Govila, J. Mater. Sci. 26, 1545 (1991). With kind permission of Springer Nature. S.D. stands for standard deviation

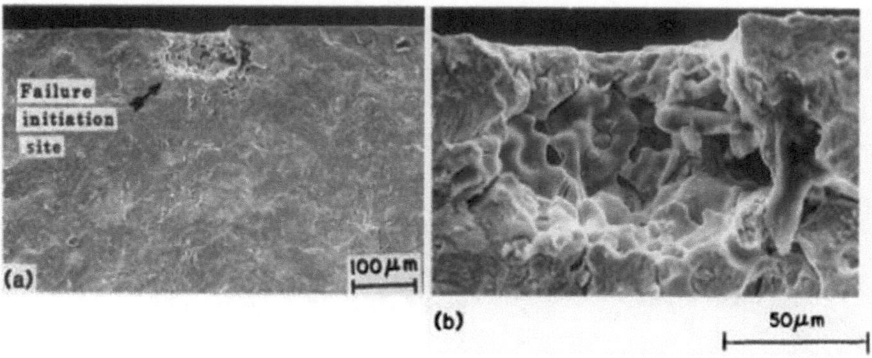

Fig. 10.18 **a** Scanning electron micrographs showing failure initiation occurring at a surface initiated flaw in MgO-PSZ (TS grade) specimen tested at 20 °C in a fast fracture mode (MCS = 0.5 mm min^{-1}). **b** Note the occurrence of melted or glassy phase material inside the failure region. Energy-dispersive X-ray spectra taken inside the failure region revealed the presence of aluminium and silicon. R. K. Govila, J. Mater. Sci. 26, 1545 (1991). With kind permission of Springer Nature

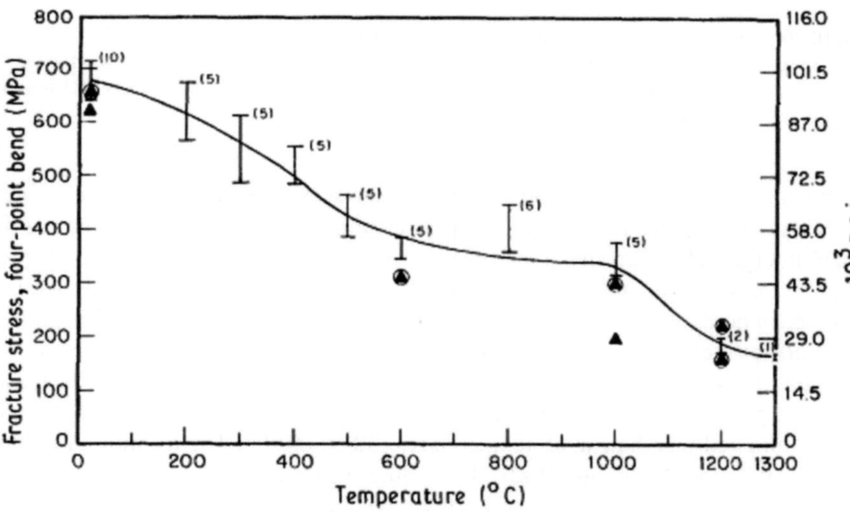

Fig. 10.19 Variation in fast fracture strength of MgO-PSZ as a function of temperature. All specimens (●) were tested at a MCS = 0.5 mm min^{-1}. (▲) Specimens with 10 kg Vicker's indentation; did not fail at the indentation site. R. K. Govila, J. Mater. Sci. 26, 1545 (1991). With kind permission of Springer Nature

components fabricated minimizing the presence of oxide inclusions and by improving the sintering conditions. Further, the decrease in σ_F with increasing temperature seen in Fig. 10.19 is due to increased stability of the tetragonal phase and decreased extent of the stress-induced martensitic phase transformation of the tetragonal phase to the monoclinic phase which increase with increasing temperatures. The identification of

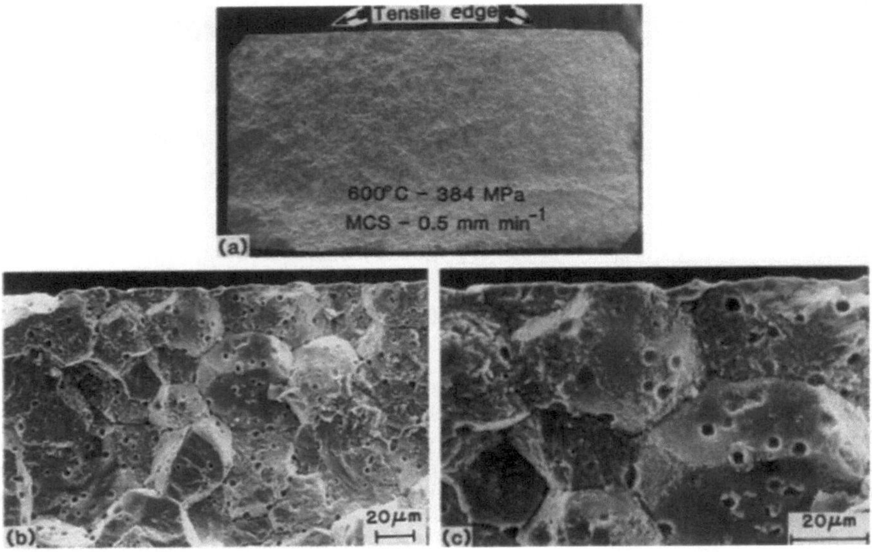

Fig. 10.20 Typical fracture surface as seen in SEM. **a** Overall view of the fracture surface. **b** Localized group of cubic grains showing intergranular cracking and possibly the failure origin. **c** Grain boundary cracking and opening. R. K. Govila, J. Mater. Sci. 26, 1545 (1991). With kind permission of Springer Nature

the fracture origin is difficult, and the fracture surfaces are similar to those observed at ~RT as seen in Fig. 10.20. Fractographic examination showed a localized region where intergranular cracking along several cubic grains is visible in Figs. 10.20b, c. Fracture initiation did not directly seem to be promoted by the pores. However, the presence of pores along the grain boundaries would weaken the bond between grains. Also, no failure sites associated with the porous region were observed. Fracture surfaces of soecimens tested in fast fracture mode are shown in Figs. 10.21 and 10.22, respectively. The fracture surface at 1200 °C (Fig. 10.21) is an example of surface initiated type flaw at all temperatures (30–1300 °C). The location of failure initiation is delineated in Fig. 10.21b. The failure site resulting in catastrophic failure. Evidence of transgranular cleavage is clearly visible, Fig. 10.21c. Failure initiation site might appear at the subsurface as seen at 1300 °C in Fig. 10.22a; back-scatterd mode clearly confirms this observation in Fig. 10.22b. No evidence of the presence of oxide inclusions of Al, Si and Fe is observed, but intergranular cracking along several cubic grains is seen in Fig. 10.22c and significant decomposition product along the grain boundaries is noticed in Fig. 10.22d. Further, no signs of plastic deformation is observed except that the cubic grains showed signs of cleavage fracture.

In summary, the flexural test of partially MgO stabilized zirconia was evaluated as a function of temperature, applied stress and time in the temperature range of 20–1300 °C. A decrease in fracture strength with increasing temperature was presented. Failure occurred in a brittle manner and the crack propagation during fast fracture

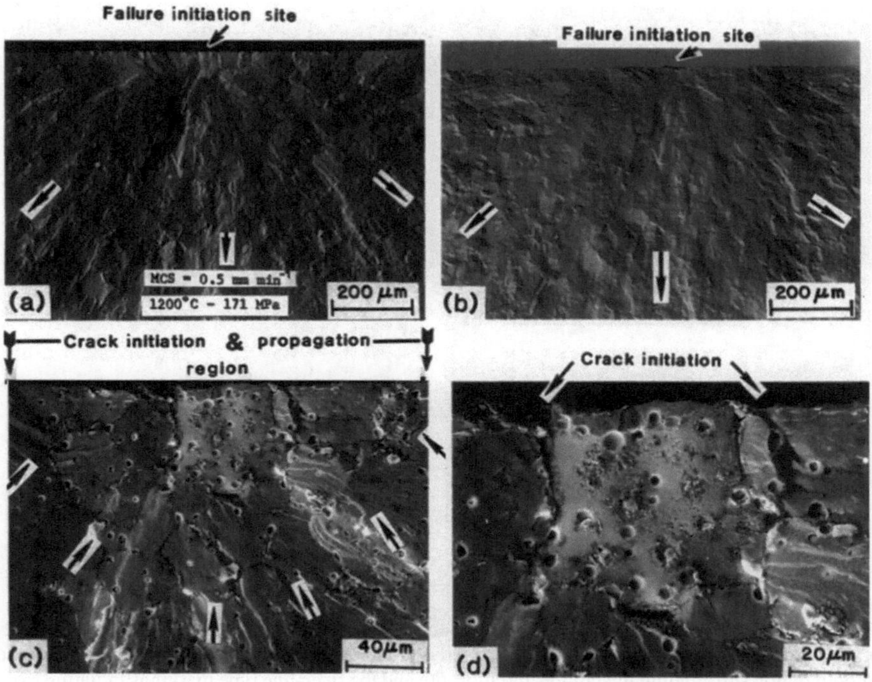

Fig. 10.21 Typical example of a surface initiated failure as seen in SEM. **a** Fracture surface. **b** Same area as seen in (a) but fractograph taken in a back-scattered mode indicating the location of failure-initiation site. Arrows indicate the direction of crack propagation. **c** Higher-magnification view of the failure-initiation site showing crack initiation and propagation region. Arrows outline approximately the boundary of the effective flaw resulting in catastrophic failure. **d** Higher magnification view showing crack initiation site and the associated grain-boundary sliding. R. K. Govila, J. Mater. Sci. 26, 1545 (1991). With kind permission of Springer Nature

was a mixture of transgranular and intergranular grack growth. The sudden decrease of the flexural stress rupture testing at 500–800 °C in air indicates the material's susceptibility to time-dependent failure, and outlines safe applied stress levels for a given temperature (see Fig. 10.19). Stress rupture tests at ~1000 °C reveals the onset of plastic deformation in the glassy phase with the consequent degradation of MgO.

10.1.5 Fracture in Yttria Partial Stabilized Zirconia

Sintered 3 mol% yttria stabilized tetragonal zirconia (3Y-TZP) with a grain size of 0.38 μm is presented in this section. Direct measurement of the fatigue crack growth rate in this polycrystalline material is considered where controlled surface flaws are introduced. The present 3Y-TZP is characterized by limited transformation plasticity,

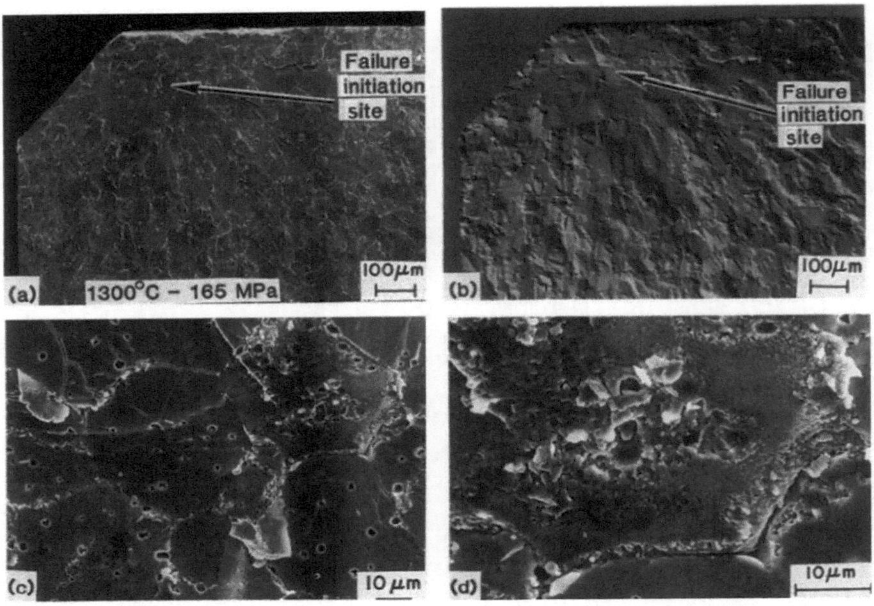

Fig. 10.22 SEM fractographs showing failure occurring at a subsurface site. **a** Fracture surface. **b** Same area as seen in (a) but in a back-scattered mode to reveal the surface structure. Note the smooth region surrounding the failure-initiation site and the river-line type markings radiating from it clearly identify the location of failure origin. **c** Higher magnification view of the failure site showing intergranular cracking associated with multiple cubic grains. Transgranular cleavage is also visible. **d** Higher magnification view showing eutectoid decomposition. R. K. Govila, J. Mater. Sci. 26, 1545 (1991). With kind permission of Springer Nature

not very large fracture toughness and high strength. Very little residual stress in this material is present having almost cubic single-phase homogeneous microstructure with very fine grain <0.5 μm. The specimen used contain an indentation crack and the residual stress due to indentation was removed by annealing. The specimen was then loaded to cause crack propagation and their size was measured. The loading was repeated after crack length measurement and resumed to a higher value to cause further crack growth. This procedure was repeated until the specimen failed. The indent was also introduced to a four-point-bend specimen. Cyclic cantilever bending fatigue tests were conducted using an MTS machine, operated under load control, with a sinusoidal wave form of 3 Hz. For each set of experiment, a set of constant maximum and minimum loads was chosen. The crack length was measured periodically by a replication technique. The crack growth rate as a function of the stress intensity factor, ΔK is plotted in Fig. 10.23. Here the bending fatigue crack growth data under constant maximum load and R ratio are shown in Fig. 10.23. Each set of data is from a single specimen. It is assumed that the lifetime of a smooth bar specimen under an uniaxial tension–compression cycling is governed by crack growth only according to the following relation

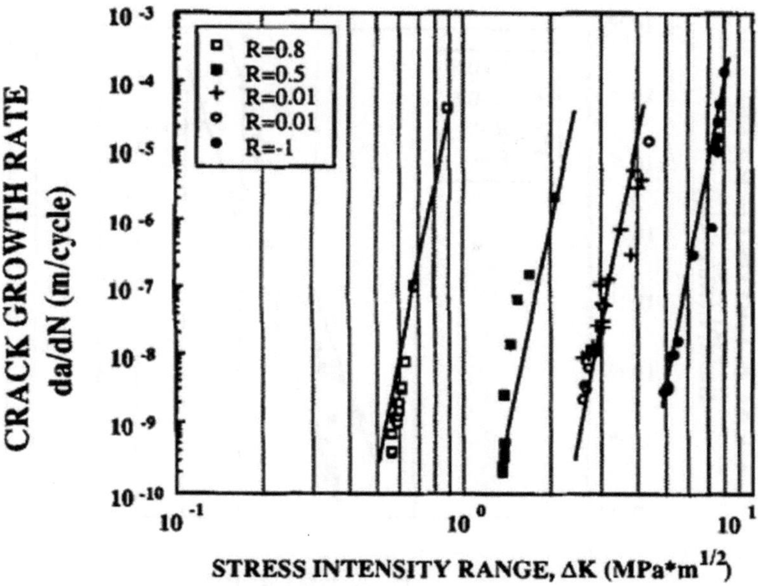

Fig. 10.23 Fatigue crack growth rate vs stress intensity range. S.-Y. Liu and I.-W. Chen, J. Am. Ceram. Soc., 74, 1206 (1991). With kind permission of John Wiley and Sons

$$\frac{da}{dN} = A(\Delta K)^m \tag{10.6}$$

where A is a constant, ΔK stress intensity range raised to the mth power. The initial crack size estimated from fractography is 25 μm. In Fig. 10.24 the crack growth rate aginst the maximum intensity factor is plotted, while in Fig. 10.25 the normalized growth rate versus K_{max} is shown. Observation of striations on fatigue fractures surfaces is a well-known feature in metals, which marks cycle-to-cycle advance of the crack before final fracture. Such fatigue striation were observed in ceramics also, an example of which is shown in Fig. 10.26 for the 3Y-TZP. In Fig. 10.26a K_{max} = 4.3 and 5 MPa $m^{1/2}$ ar R = 0.8. The striations were found for the 10 last cycles. Similar striations were observed for R = 0.8, 0.5, and 0.01 illustrated in Fig. 10.26b, c when the peak stress intensity factors were high enough, i.e., during the last scores of cycles before catastrophic failure. The maximum stress intensity factor (computed for each striation) is plotted against the crack length in Fig. 10.27. They are shown in Fig. 10.27 by open symbols. Fatigue striations are observed only when the peak stress intensity factor exceeds the crack resistance, K_r on the R-curve. It is the conclusion that the fatigue striations in the current tests are the result of combined fatigue crack growth and R-curve-limited (overload) crack growth.

Three stages of crack growth can be defined as illustrated in Fig. 10.28. In stage I, the maximum stress intensity factor, K_{max} exceeds the crack growth resistance, K_r. In stage II, $K_{max} < K_r$ so that crack growth cannot take place without stress cycling.

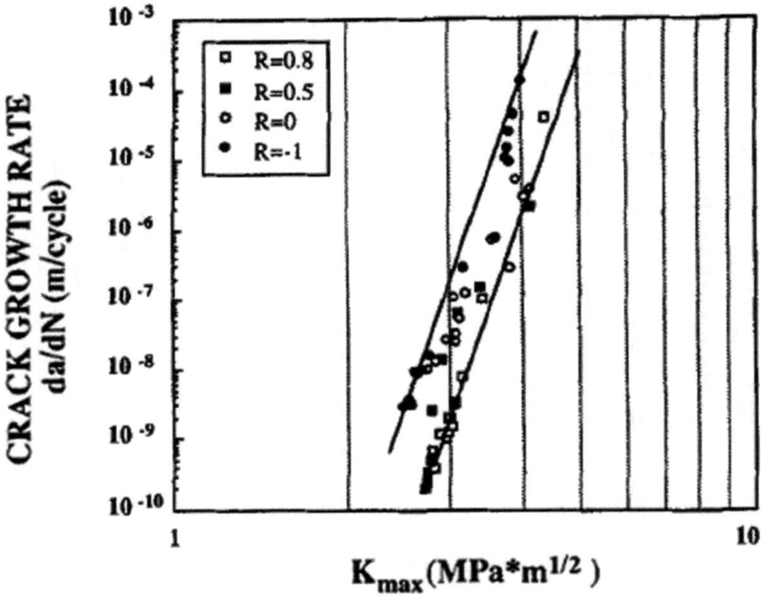

Fig. 10.24 Fatigue crack growth rate vs maximum stress intensity factor. Crack growth rate is slightly higher at lower R ratios. S.-Y. Liu and I.-W. Chen, J. Am. Ceram. Soc., 74, 1206 (1991). With kind permission of John Wiley and Sons

In this stage, crack propagation is solely due to fatigue. In stage III, K_{max} once again exceeds K_r. The slopes of both are relatively flat (see Fig. 10.28). Whether fatigue striation may be found for the same R ratio, depends on the relative values of K_{max} and K_r. Since the crack growth rate in stage III has a contribution from a nonfatigue mechanism, it is expected to be faster.

In summary, the fatigue lifetime of 3Y-TZP is controlled by the growth of preexisting flaws. Knowing the size and the geometry of the preexisting flaws and the crack growth law, the fatigue lifetime can be estimated. For the experimental details and the calculations the reader should consult the work of Liu and Chen. Fatigue striations were observed in 3Y-TZP and they form when overload fracture and fatigue fracture alternate during cyclic loading. Their formation is R ratio dependent, and very sensitive to the loading conditions and crack geometry. Fatigue crack in ceramics is suggested to follow a microplasticity mechanism. The dependence on the maximum stress intensity factor seem to be an inherent feature of brittle crack growth.

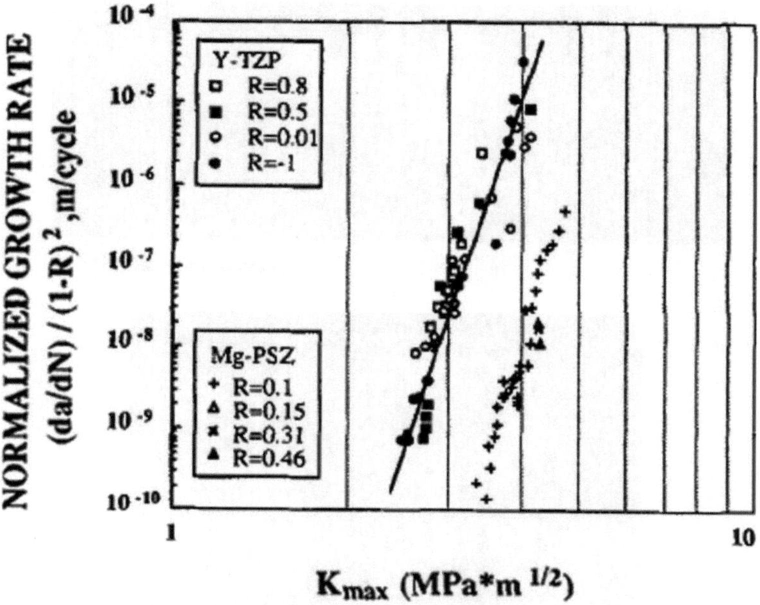

Fig. 10.25 Normalized fatigue crack growth rate vs maximum stress intensity factor for Y-TZP. No systematic dependence on R ratio is evident. S.-Y. Liu and I.-W. Chen, J. Am. Ceram. Soc., 74, 1206 (1991). With kind permission of John Wiley and Sons

10.1.6 Fracture in SiC

SiC is mainly used as a reinforcing agent for materials or being by itself reinforced. SiC whiskers, fibers or particulates are often used to further strengthen SiC matrix, and consequently limited information if at all can be found in the literature, in particular regarding fatigue cyclic fracture. Therefore, a composite based on SiC matrix strengthened by Nicalon fiber is presented in this section (Recal that Nicalon is a silicon carbide continuous tow that possesses increased strength and heat and corrosion resistance and mainly used in the aerospace industry as composite reinforcement). Among the many applications, SiC/SiC ceramic matrix composite is of high interest for potential use as a structural and barrier material in fusion systems (reactors). It possesses reasonable fracture toughness over a range of temperatures and, due to the low atomic number of its constituents, is appealing for many technologigal applications, for exampla aviation, etc.

It is of interest to get information on this composite material for evaluating its use as resisting crack growth. The fatigue strength tests were performed on an MTS machine with sinusoidal loading frequency of 10–20 Hz, in air at RT and at the stress ratio ($\sigma_{min}/\sigma_{max}$) of 0.1. For the high temperature tests Ar or vacuum were used to protect from oxidation. Further in situ measurements in air with sinusoidal loading frequency of 2–10 Hz were performed at RT. Fatige fractured surfaces in

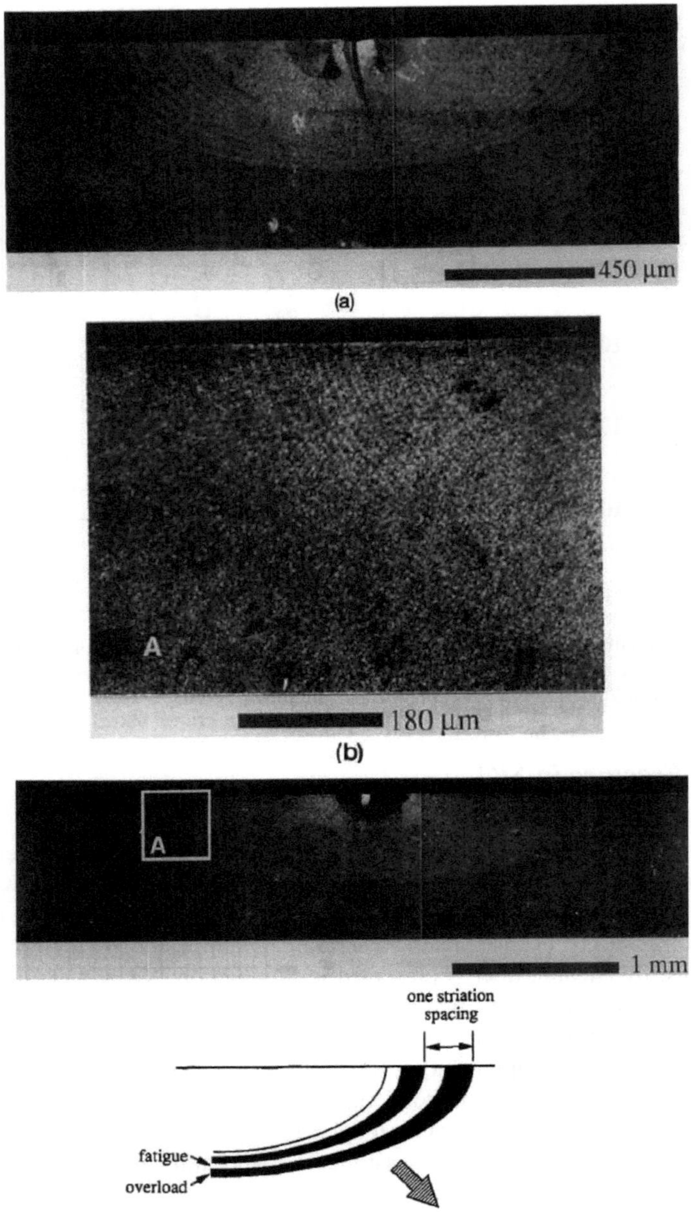

Fig. 10.26 Fatigue fracture surface of 3Y-TZP for **a** $R = 0.8$ and **b**, **c** $R = 0.01$. In both cases, fatigue striations appear during the last scores of cycles before catastrophic failure. **d** Schematic of fatigue striations in terms of two growth steps. S.-Y. Liu and I.-W. Chen, J. Am. Ceram. Soc., 74, 1206 (1991). With kind permission of John Wiley and Sons

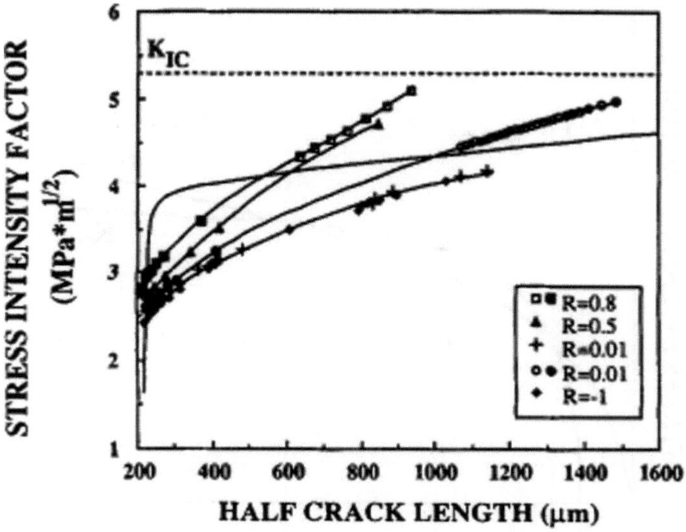

Fig. 10.27 Maximum stress intensity factor versus half crack length. The two sets of data at R = 0.01 were obtained using two different stress levels. Open symbols correspond to observations of striations. S.-Y. Liu and I.-W. Chen, J. Am. Ceram. Soc., 74, 1206 (1991). With kind permission of John Wiley and Sons

SiC/SiC composites were examined by SEM and optical microscopy. Plane woven microstructures of the composite have large pores between 0° and 90° fiber bundels as shown in Fig. 10.29a. Cracks usually intiate at sharp corners of large pores after loading as seen 10.29a. As seen the two cracks formed propagate into 0° and 90° bundels, respectively. Fibers inhibit crack propagation, but after some cycles the fibers loose this effect until the maximum stress is increased to make crack propagation reoccur again, after which crack propagation rate again decreases. This seems to be a pattern of cyclic crack propagation in SiC/ SiC-fiber composite. Crack length versus number of cycles is illustrated in Fig. 10.30. It was found that the interface sliding resistance was decreased by cyclic fatigue in SiC/SiC. The interface sliding and degradation lead to the breakage of bridge fibers and therefore the fibers loose the inhibiting crack propagation effect. As a result, the bridging fibers fracture promotes crack propagation. Fatigue at 1300 °C is almost the same as that of Enhanced SiC/SiC or as the Standard SiC/SiC at 1300 °C as shown in Fig. 10.32 under the maximum stress of 150 PMa in air at 2.5 $\times$ 10^4 cycles (Fig. 10.31).

In conclusion, fracture cracks initiate at the corners of large pores during loading of smooth specimens. The balance between the fiber bridging in the wake of propagating crack tip and the breakage of bridged fibers by the degradadation of the interfaces maintains a steady state cyclic crack propagation. The crack propagation rate decreases with time after the maximum load is applied. The fatigue resistance of the Hi-Nicalon^{TM}/SiC is similar to the Enhanced SiC/SiC but much better than the Standard SiC/SiC at 1300 °C in air.

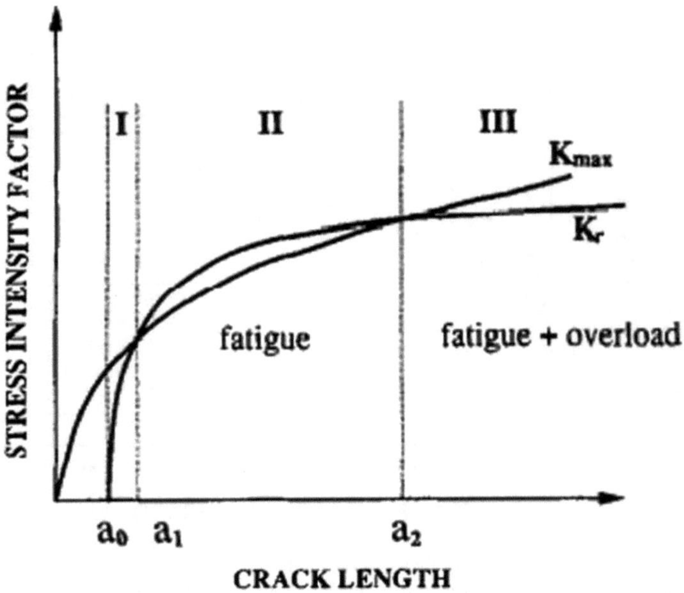

Fig. 10.28 Maximum stress intensity factor (K_{max}) and crack resistance (Kr) vs crack length: (I) growth due to overload, (II) growth due to fatigue, (III) growth due to overload and fatigue. S.-Y. Liu and I.-W. Chen, J. Am. Ceram. Soc., 74, 1206 (1991). With kind permission of John Wiley and Sons

10.1.7 Fracture in Si_3N_4

Due to the wide use of Si_3N_4 as structural components in varios applications characteized by high mechanical strength, it seems to be important to devote this section to its resistance to fracture which is associated with cyclic failure by crack initiation and popagation. Sintered Si_3N_4 wer fatigue tested by the resonant bending method to gradually collect fatigue data up to 10^{10} cycles. This way the effect of frequency on cyclic fatigue could be obtained at room temperature. The objective was to compare cyclic fatigue at room temperature and high temperature. The designation of the samples and their mechanical properties are listed in Table 10.5. The stress amplitude, σ_a versus the number of cycles at alternating loads is illustrated in Fig. 10.33 at frequencies in the range 0.03 Hz–3 kHz.

The stress amplitude variation with the mean stress for failure at room temperature is shown in Fig. 10.34. In this figure open circles represent specimens surviving at 10^7 cycles, and closed circles represent the failed specimens. The dotted line represents tensile strength. The solid line can be expressed as

$$\sigma_a = \sigma_e(1 - (\sigma_m/\sigma_t)) \tag{10.7}$$

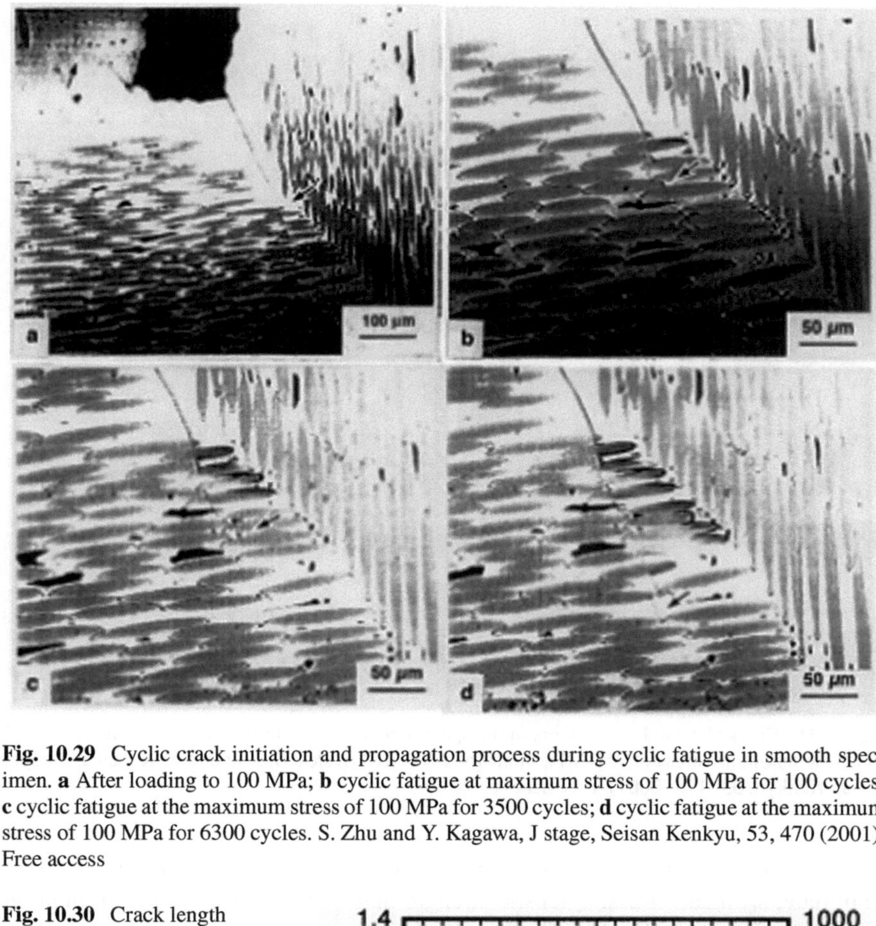

Fig. 10.29 Cyclic crack initiation and propagation process during cyclic fatigue in smooth specimen. **a** After loading to 100 MPa; **b** cyclic fatigue at maximum stress of 100 MPa for 100 cycles; **c** cyclic fatigue at the maximum stress of 100 MPa for 3500 cycles; **d** cyclic fatigue at the maximum stress of 100 MPa for 6300 cycles. S. Zhu and Y. Kagawa, J stage, Seisan Kenkyu, 53, 470 (2001). Free access

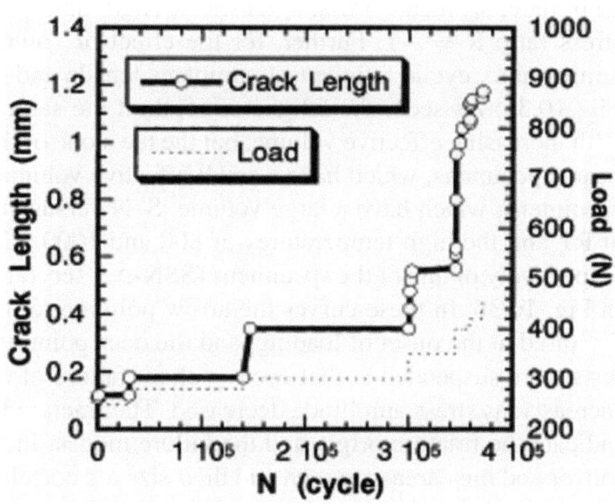

Fig. 10.30 Crack length versus number of cycle at different maximum stresses in the specimen cyclic fatigue fracture at the maximum load of 450 N. S. Zhu and Y. Kagawa, J stage, Seisan Kenkyu, 53, 470 (2001). Free access

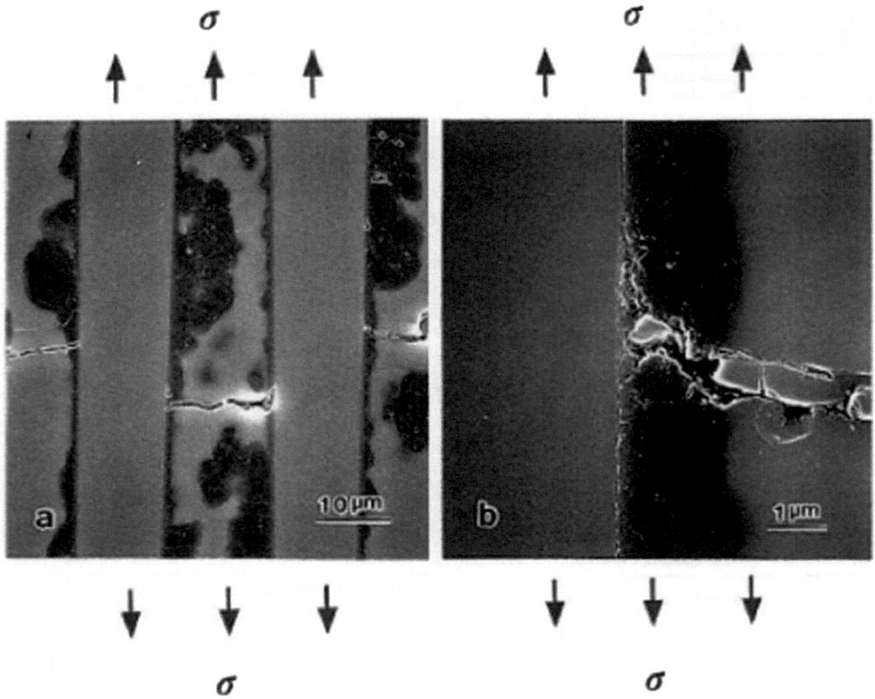

Fig. 10.31 Cracks bridged by 0° fibers in the specimen of Hi-Nicalon™/SiC fatigued at 1300 °C under the naximum stress of 150 MPa in air for 2.5 × 10^4 cycles. S. Zhu and Y. Kagawa, J stage, Seisan Kenkyu, 53, 470 (2001). Free access

which is expresses basically the modified Goodman diagram. Here, s_a, is stress amplitude, s_m mean stress, s_t tensile strength and s_e alternative fatigue strength at stress ratio $R = -1$. Further, for the effect of volume on cyclic fatigue at room tempertaure, cyclic fatigue test compares tensile and cantilever beam specimens in Fig. 10.35. As seen, the fatigue strength at the same number of cycles decreased with increasing effective volume, but the the slope of the S–N curve of the cantilever beam specimens, which have a small effective volume, is larger than that of tensile specimens, which have a large volume. S–N versus the number of cycles to failure at RT and the high temperatures of 800 and 1000 °C and 585, 560 and 340 MPa, respectively of one of the specimens (SSN-A3, see Table 10.5 (Fig. 10.37)) is shown in Fig. 10.36. In these curves the arrow pointing to the left indicates the specimen fractured at the onset of loading, and the ones pointing to the right indicate that the tests were suspended at that cycle without failure of the specimen. The fatigue life increased as stress amplitude decreased. The fractured surfaces shown in Fig. 10.37 indicate the fracture origin and the failure modes. In the specimen fractured at RT, mirror and mist areas are seen and their size are correlated with stress amplitude and the number of cycles up to failre (see legend). Microcracks at grain boundaries were

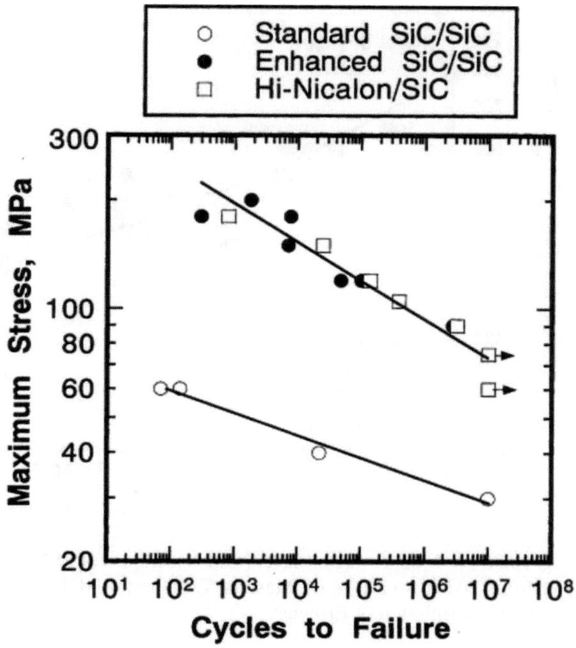

Fig. 10.32 The maximum stress versus cycles to failure for fatigue in Hi-Nicalon™/SiC, Enhance SiC/SiC, and Standard SiC/SiC in air at 1300 °C. S. Zhu and Y. Kagawa, J stage, Seisan Kenkyu, 53, 470 (2001). Free access

Table 10.5 Mechanical properties of sintered Si_3N_4. M. Masuda, T. Soma and M. Matsui, J. Eur. Ceram. Soc., 6, 253 (1990). With kind permission of Elsevier

Materials	*Density* (g/cm^3)	*Four-point bending strength*[a] (*MPa*)	*Fracture toughness*[b] ($MN/m^{3/2}$)	*Young's modulus* (*GPa*)
SSN-A1	3·21	740	6·2	265
SSN-A2	3·24	820	6·3	270
SSN-A3	3·28	860	6·2	275
SSN-B	3·25	970	6·1	300

[a] According to JIS R1601.[11]
[b] Chevron notch method.

observed in specimens that failed under cyclic load. It is suggested that cyclic fatigue mechanisms possibly involve a crack nucleation model and crack propagation model. Specifically, microcracks form as a result of stress concentration around an initial flaw or heterogeneous region in the microstructure. If the crack density reaches a critical value fatigue occurs. Crack propagation is faster under cyclic load than under static load. At the high temperature in the Si_3N_4 with a glassy phase pores and inclusions are the major defects which are the preferential crack nucleation and propagation sites.

In summary, the degradation in the strength of sintered Si_3N_4 by cyclic fatigue at RT depends on the number of cycles, however, at high temperatures it is frequency

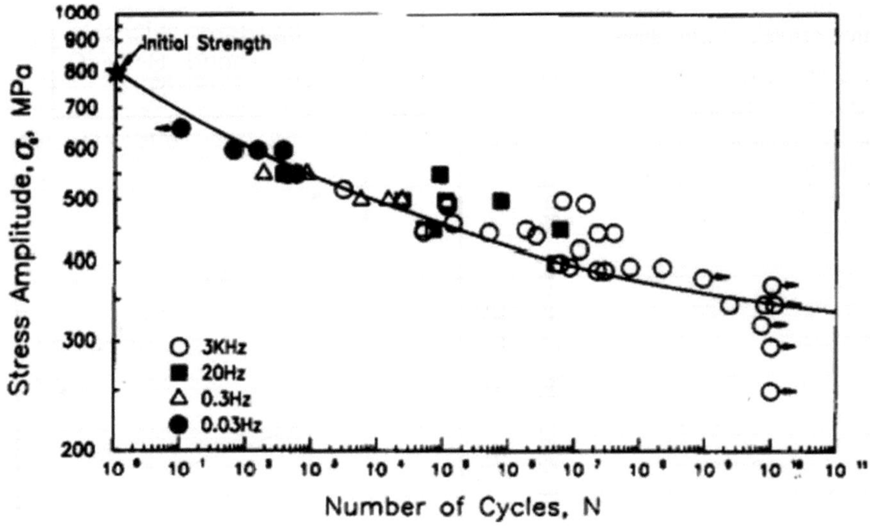

Fig. 10.33 S–N curve of cantilever beam specimens of SSN-A2 under alternating load, at frequencies from 0.03 Hz to 3 kHz. M. Masuda, T. Soma and M. Matsui, J. Eur. Ceram. Soc., 6, 253 (1990). With kind permission of Elsevier

Fig. 10.34 Mean stress versus stress amplitude for fatigue failure in SSN-A1 at 107 cycles measured by high cycle fatigue test at room temperature. The solid line represents tensile strength. M. Masuda, T. Soma and M. Matsui, J. Eur. Ceram. Soc., 6, 253 (1990). With kind permission of Elsevier

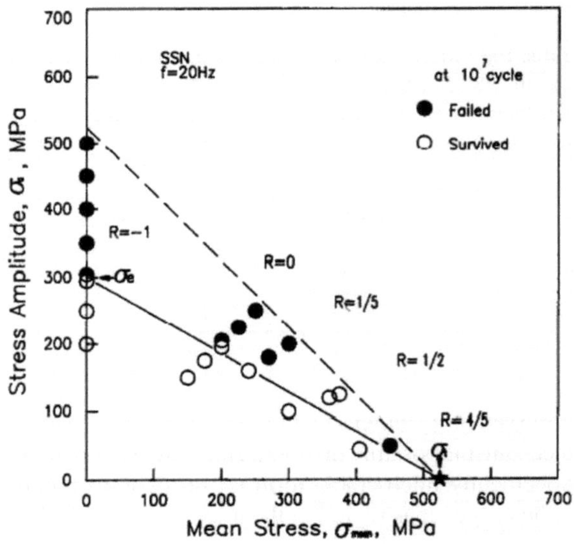

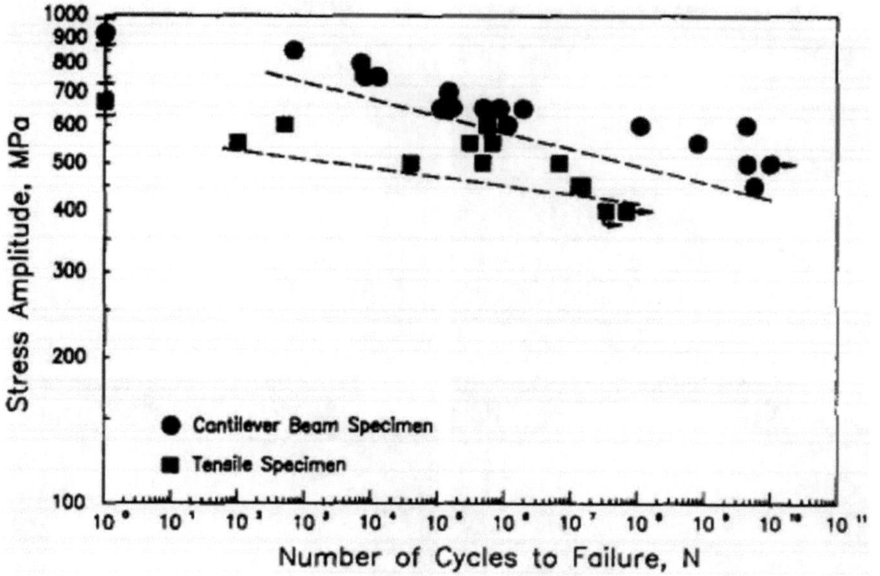

Fig. 10.35 The effect of volume on cyclic fatigue at room temperature. Cantilever beam specimens and the tensile specimens of SSN-B were used; the former have a small volume and the latter have a large volume. M. Masuda, T. Soma and M. Matsui, J. Eur. Ceram. Soc., 6, 253 (1990). With kind permission of Elsevier

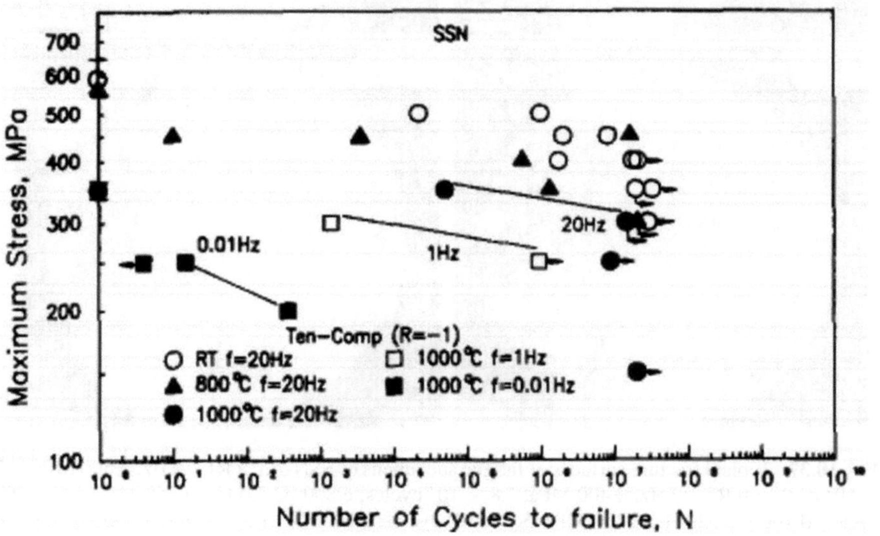

Fig. 10.36 S–N diagram of SSN-A3 using axial tension–compression load control at room temperature, 800 and 1000 °C. The symbols on the vertical axis represent the tensile strength of the materials used. Masuda, T. Soma and M. Matsui, J. Eur. Ceram. Soc., 6, 253 (1990). With kind permission of Elsevier

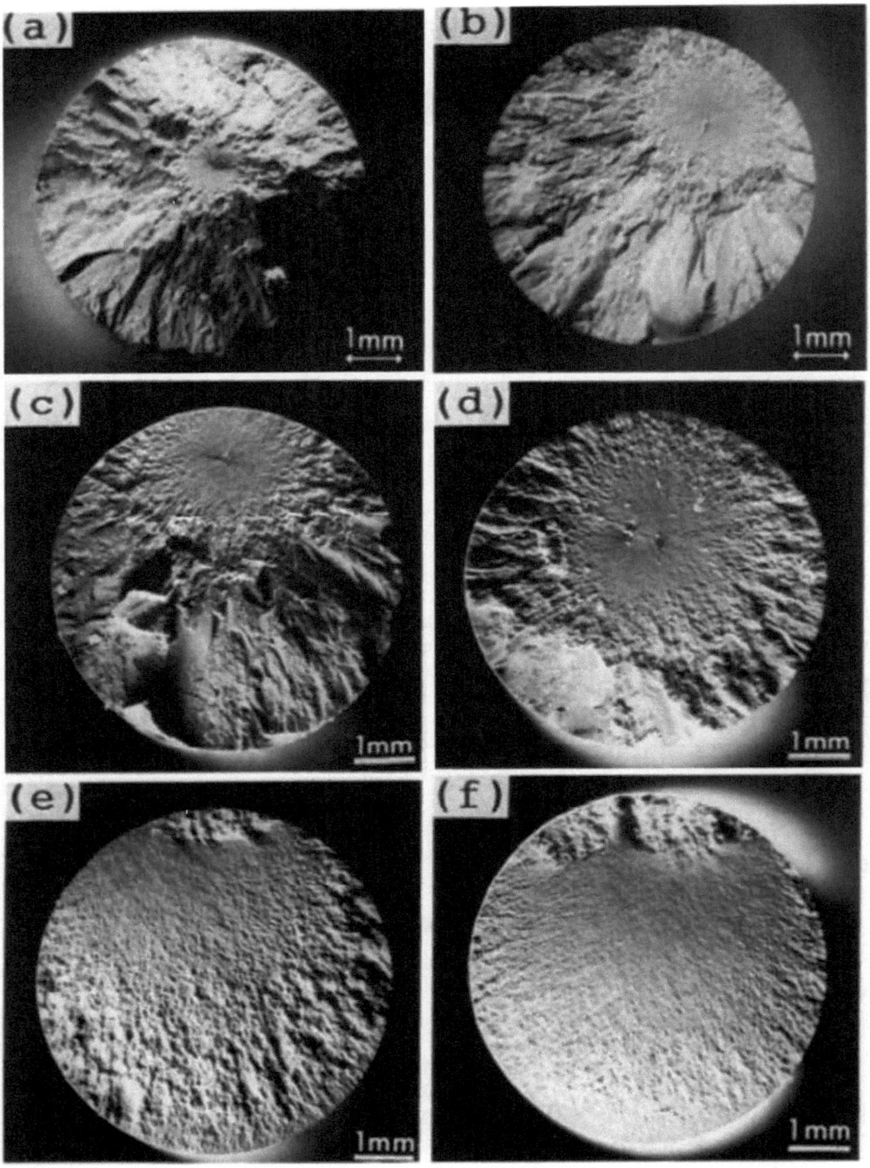

Fig. 10.37 Typical fracture surfaces of fatigue specimens of SSN-A3. **a** RT, 20 Hz, ±500 MPa, 2.2 × 10^4 cycles; **b** RT, 20 Hz, ±400 MPa, 1.8 × 10^7 cycles; **c** 800 °C, 20 Hz, ±350 MPa, 1.3 × 10^6 cycles; **d** 1000 °C, 20 Hz, ±350 MPa, 5.0 × 10^4 cycles; **e** 1000 °C, 1 × 10^2 Hz, ±200 MPa, 3.7 × 10^2 cycles; **f** 1000 °C, 150 MPa, 204 h. Masuda, T. Soma and M. Matsui, J. Eur. Ceram. Soc., 6, 253 (1990). With kind permission of Elsevier

dependent. The lower the frequency the less resistance to fatigue. Micrustructural observations suggest that the cyclic fatigue of sintered Si_3N progressed by microcrack formation at grain boundaries. The progress of cyclic fatigue at high temperatures is caused by slow crack growth.

10.2 Temperature Effect on Fatigue Fracture

10.2.1 Introduction

It is a common knowledge that the properties of material are temperature dependent. Clearly among the properties in this book the interest lies in the mechanical properties, and as expected and proven experimentally, tensile strength, yield strength, the elastic modulus decrease with increasing temperature. It is expected that fatigue is no exception and is also affected by temperature variation or change. The state of matter (solid, liqid and gas) is not a subject of this book although it is related to temperature variation, and when considering the properties, clearly it is tacitly understood that they refer to solid state.

Among the important experimental facts it is known, that high temperature reduces stiffness and strength while low temperature increases these properties. Further, one has to recall, that almost all materials creep over time under applied load if exposed to elevated temperatures. There is a fundamental relation of ceramic between its strucrure and properties which are directly affected by the temperature variation. Thus, it is revealed that in ceramics changes in density, grain size distribution, type of porosity, and second-phase content are correlated with temperature variations, which directly are connected with ceramic properties such as mechanical strength, hardness, toughness and also other physical properties such as the dielectric constant and optical properties exhibited by transparency. Ceramics are high temperature structural materials, but at conventional temperatures most of them are brittle and exhibit brittle fracture characterized by relativley little plastic deformation, crack is unstable and propagates rapidly without increase of applied stress. However, ductile fracture is preferred in most applications, because plastic deformation ahead of crack (which finally leads to facture by its propagation) is relatively stable, resist its further extension unless the applied stress is increased. At sufficiently high temperatures, however ceramics aquire some plasticity the magnitude of which is material (and temperature) dependent.

In this section considetion is given to oxide, carbide and nitride ceramics, limited only by the availibility in the open literature.

10.2.2 Temperature Effect on Fatigue Fracture in SiC

The objective of the author is not to use composites as far as possible. However, the pure monolithic SiC is seldom reported in the literature. Therefore, 2D SiC/SiC composite is used in this section.

Strength degradation of SiC fibers vs oxidation time in air and steam at high temperatures of 750, 1000, 1100, 1200 and 1300 °C is illustrated in Fig. 10.38 and predicted using Eq. (10.8). Where K_{IC} denotes the critical stress intensity factor; Y is a geometric parameter; and k is the parabolic rate constant. With the increasing cycle number, the interface shear

$$\sigma_0(t) = \sigma_0, t \leq \frac{1}{k}\left(\frac{K_{Ic}}{Y\sigma_0}\right)^4$$
$$\sigma_0(t) = \frac{K_{Ic}}{Y\sqrt[4]{kt}}, t > \frac{1}{k}\left(\frac{K_{Ic}}{Y\sigma_0}\right)^4 \tag{10.8}$$

stress and fiber strength decrease due to the interface wear and interface oxidation. When the fiber broken fraction approaches the critical value, the composite fatigue fractures. In the following the life prediction model is applied to evaluate fatigue life S–N curves at the mentioned temperatures in 0% and 60% in air. In Figs. 10.39, 10.40, 10.41 and 10.42 the fatigue life S–N curves are shown in the temperature range 750–1300 °C.

Basel), 10, 371 (2017). Open access

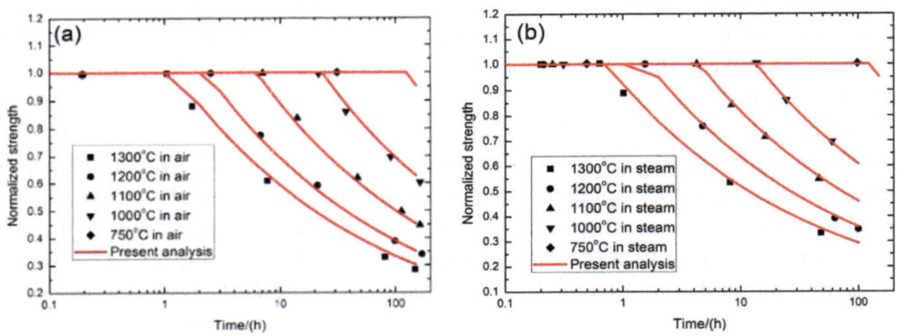

Fig. 10.38 The strength degradation of SiC fibers **a** in air and **b** in steam conditions at elevated temperatures. L. Li, Materials (Basel), 10, 371 (2017). Open access

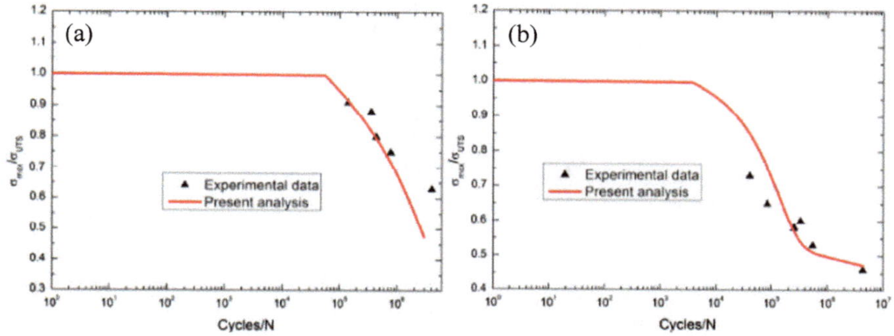

Fig. 10.39 **a** the fatigue life S–N curve with test environment of 0% moisture content condition; and **b** with test environment of 60% moisture content condition of 2D SiC/SiC composite at 750 °C. L. Li, Materials (Basel), 10, 371 (2017). Open access

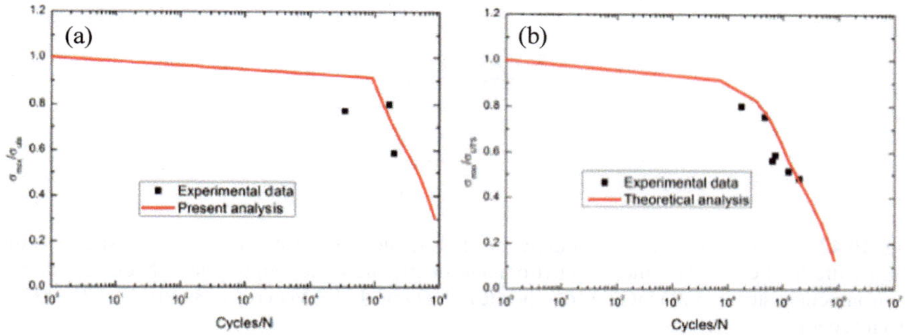

Fig. 10.40 **a** The fatigue life S–N curves of 2D SiC/SiC composite at 1000 °C in air; **b** the fatigue life S–N curves of 2D SiC/SiC composite at 1000 °C in steam. L. Li, Materials (Basel), 10, 371 (2017). Open access

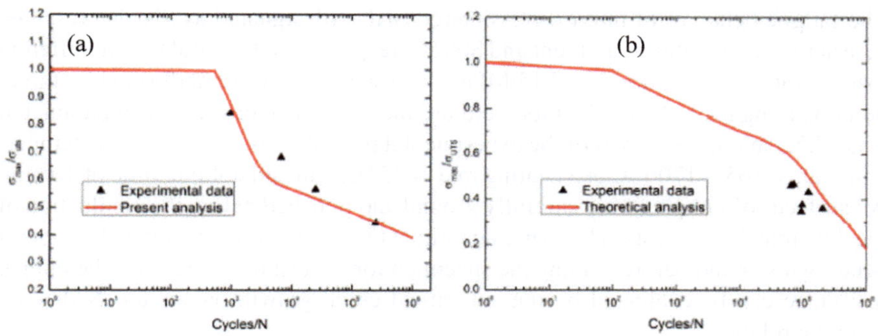

Fig. 10.41 **a** The fatigue life S–N curves of 2D SiC/SiC composite at 1100 °C in air; **b** the fatigue life S–N curves of 2D SiC/SiC composite at 1200 °C in air. L. Li, Materials (Basel),10, 371 (2017). Open access

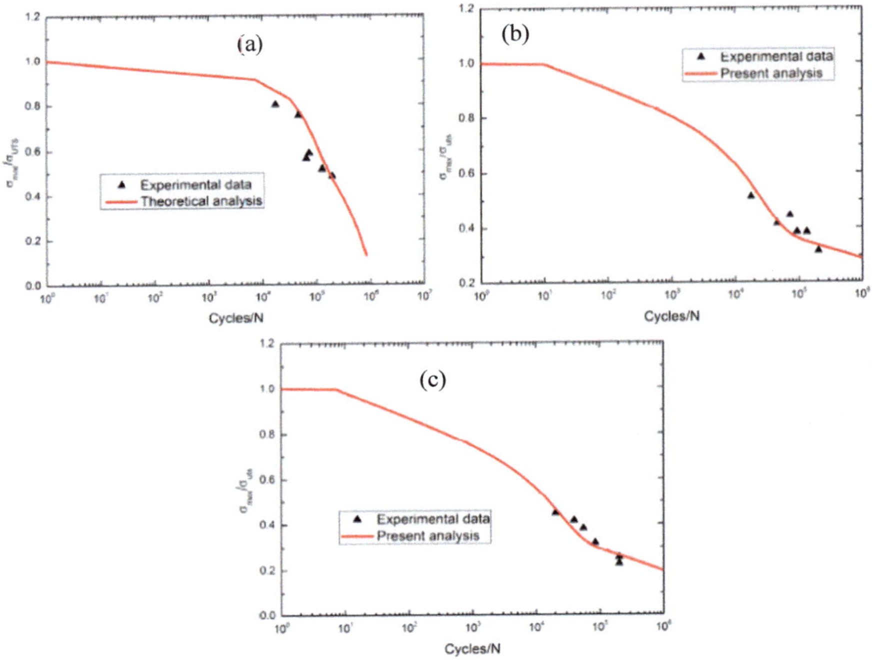

Fig. 10.42 **a** The fatigue life S–N curves of 2D SiC/SiC composite at 1200 °C in steam; **b** the fatigue life S–N curve with the test environment in air; and **c** the fatigue life S–N curve with the environment in steam of 2D SiC/SiC composite at 1300 °C. L. Li, Materials (Basel), 10, 371 (2017). Open access

10.2.3 Temperature Effect on Fatigue Fracture in Si_3N_4

The fatigue behavior of microwave-sintered Si_3N_4 (designated as SN) is presented here at two temperatures, ambient and 800 °C, respectively. Flexural tests at different loading rates of 0.05, 0.5, 5 and 15 MPa s^{-1}, respectively were performed by three-point bending tests. At ambient temperature the subcritical crack growth parameter, N is 57.5. An important part of the experimental procedure is sintering in the temperature range 1650–1700 °C at a heating rate of 25 °C/min and holding time of 10 min. All surfaces of samples were carefully ground and polished before the conduction of the dynamic fatigue tests. The dynamic fatigue tests were conducted in a three-point bending tester and before loading the specimen for the test it was heated. The fatigue resistance can be evaluated by the subcritical crack growth parameter, N derived from the relation

$$\ln \sigma_f = \frac{\ln \dot{\sigma}}{N+1} + \left[\frac{\ln B + \ln(N+1) + (N-2)\ln \sigma_i}{N+1} \right] \tag{10.9}$$

where σ_f is the flexural strength, σ_i is the inherent flexural strength, $\dot{\sigma}$ is the loading rate, B is a parameter associated with N, fracture toughness, crack geometry and loading configuration. From the linear relationship between the derivatives of flexural strength and loading rate, the subcritical crack growth parameter N can be calculated by measuring the value of slope.

The dynamic fatigue is illustrated in an Si_3N_4 (SN) σ_f plot versus $\dot{\sigma}$ in Fig. 10.43 and the detailed composition is shown in Table 10.6. SEM micrograph of tensile surface at a loading rate of 0.05 MPa s^{-1} is seen in Fig. 10.44 and the tensile surface was relatively smooth and oxidation was not obvious. Compare Fig. 10.43 with Fig. 10.45, the high temperature (800 °C) dynamic fatigue plot showing increased flexural strength with increasing loading rate. It had only a slight decrease at high temperature. This indicates that Si_3N_4 has a good thermal conductivity.

In summary, porosity is detrmintal to fatigue resistance and addition of of small amounts of a metal such as Ni for example, can improve the fatigue resistance of Si_3N_4 at room temperature, since it increases the density. The flexural strength showed

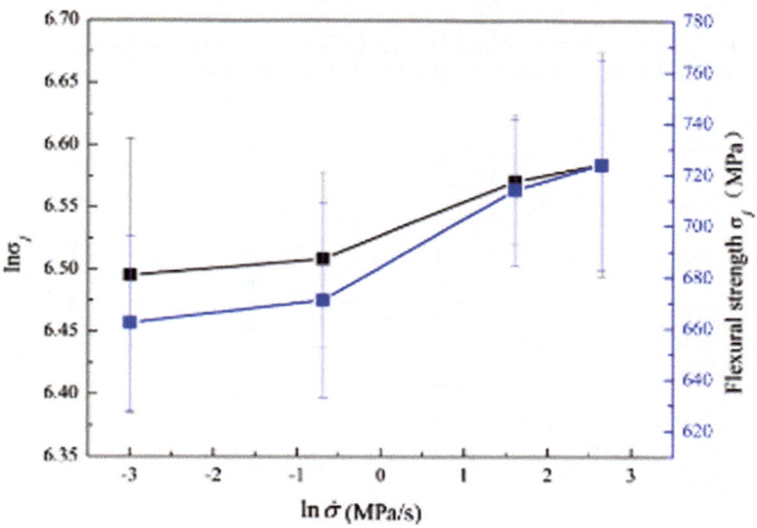

Fig. 10.43 Dynamic fatigue SN at ambient temperature. W. Xu, J. Yuan and Z. Yin, Ceram. Inter. 45, 21,572 (2019). With kind permission of Elsevier

Table 10.6 The composition ratios and properties of Si_3N_4 (wt.%). W. Xu, J. Yuan and Z. Yin, Ceram. Inter. 45, 21, 572 (2019). With kind permission of Elsevier

Symbol	α-Si_3N_4	(W, Ti)C	Ni	Sintering additives	Vickers hardness (GPa)	Fracture toughness (MPa $m^{1/2}$)	Relative density
SN	88%	/	/	12%	14.92	6.44	98.52%

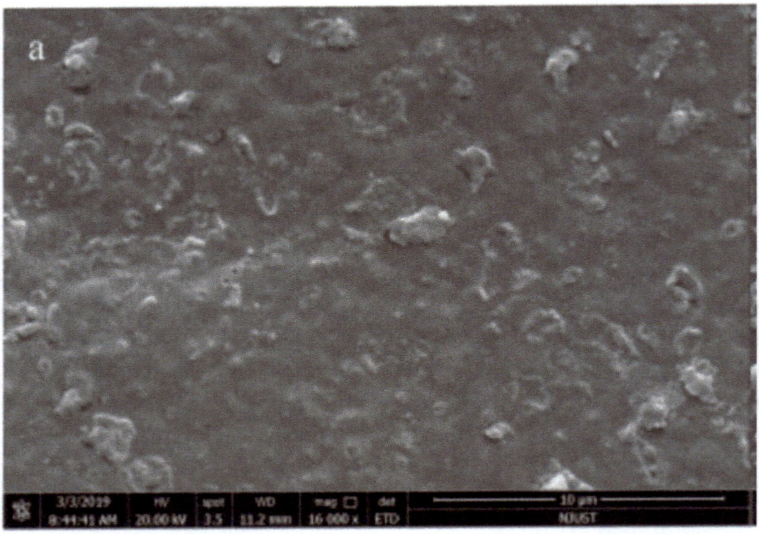

Fig. 10.44 The SEM micrograph of tensile surface of SN, at loading rate of 0.05 MPa/s. W. Xu, J. Yuan and Z. Yin, Ceram. Inter. 45, 21,572 (2019). With kind permission of Elsevier

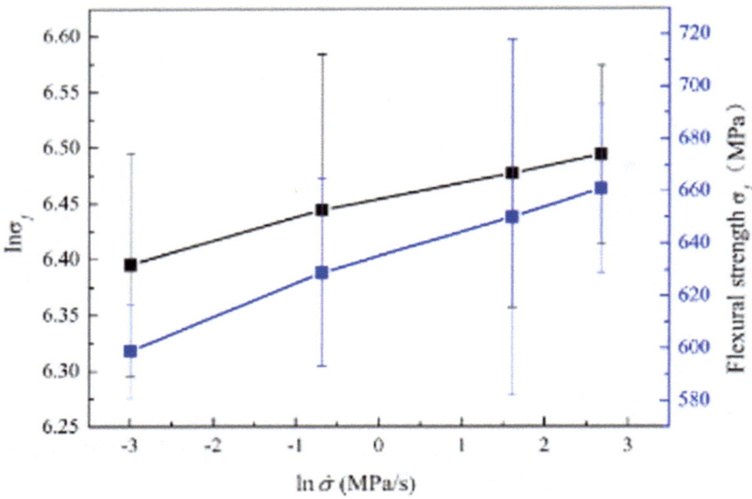

Fig. 10.45 Dynamic fatigue date of SN, at 800 °C. W. Xu, J. Yuan and Z. Yin, Ceram. Inter. 45, 21,572 (2019). With kind permission of Elsevier

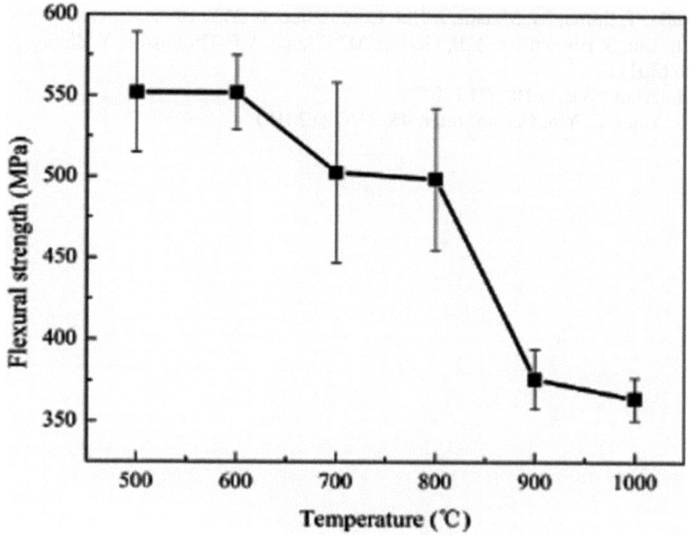

Fig. 10.46 The flexural strength of SNN at different temperatures. W. Xu, J. Yuan and Z. Yin, Ceram. Inter. 45, 21,572 (2019). With kind permission of Elsevier

only a slight decrease at the high temperature of 800 °C, which indicates that the flexural strength essentially unchanged compared with that of room temperature. The decrease of high temperature fatigue resistance is a consequence of grain boundary bonding strength which accelerates crack propagation which promotes fracture. The fracture surface exhibited typical fatigue characteristics of brittle materias and it was found that the fracture process can be divided into three stages (crack formation, propagation and rapid fracture stage).

It would be of interest to show the fatigue resistance of SNN (2 wt.% Ni and 15 wt.% (W,Ti)C additives) variation over a wide range of temperature with the expectation for a similar variation with temperature of Si_3N_4. This is shown in Fig. 10.46.

References

1. S. Maity, D. Vasu, B.K. Sarkar, Mater. Let. **21**, 63 (1994)
2. A. Hillerborg, M. Modéer, P.E. Peterson, Cem. Concer. Res. 6, 773 (1076)
3. H. Harmuth, R.C. Bradt, Refrac. Manual 6 (2019)
4. Y. Dai, D. Gruber, H. Harmuth, J. Eur. Cer. Soc. **37**, 5035 (2017)
5. A. Smirnov, J.I. Beltrán, T. Rodriguez-Suarez, C. Pecharromán, M.C. Muñoz, J.S. Moya, J.F. Bartolomé, Sci. Rep. **7** (2017). Article number: 44922
6. R.K. Govila, J. Mater. Sci. **26**, 1545 (1991)
7. S.-Y. Liu, I.-W. Chen, J. Am. Ceram. Soc. **74**, 1206 (1991)
8. S. Zhu, Y. Kagawa, J Stage Seisan Kenkyu **53**, 470 (2001)

9. M. Masuda, T. Soma, M. Matsui, J. Eur. Ceram. Soc. **6**, 253 (1990)
10. L. Ren, L. Liu, S. Bhowmick, Y.B. Gerbig, M.N. Janal, V.P. Thompson, Y. Zhang, J. Dent. Res. **90**, 1026 (2011)
11. L. Li, Materials (Basel) **10**, 371 (2017)
12. W. Xu, J. Yuan, Z. Yin, Ceram. Inter. **45**, 21572 (2019)

Chapter 11
The Notch Effect

Abstract It is possible to define a notch as a change in the cross-section of a component either in the form as a hole, grove or other cross-sectional variation. The stress distribution is uneven as a result of a notch, and one talks about a notch effect. Fracture toughness of a component is reduced and the rate of fatigue crack growth. The tendency of a material to behave as a brittle material in presence of a notch is termed as notch sensitivity. It is known that FCC metals are not notch sensitive. Notch sensitivity is influenced by many parameters such as notch size and notch geometry. Notches decrease the fatigue strength of structures. Alumina, SiC, and Si_3N_4 with notches were investigated but also yttrium and magnesium oxides stabilized zirconia were tested for fatigue fracture resistance in notched specimens. Microcracking and fracture (induced by cracking) along the grain boundaries are enhanced since grain boundary cracking in notched components is an important contributing mechanism for cyclic crack growth and eventual failure.

Keywords Notch effect · Notch sensitivity · Crack · Crack growth · Growth rate

11.1 Introduction

Several general comments on notch effect in materials will be considered here in order to understand the consequence of introducing a notch and the reason of testing specimens in the presence of a notch. It is possible to define a notch as a change in the cross-section of a component either in the form as a hole, grove or other cross-sectional variation. The stress distribution is uneven as a result of a notch, basically can lead to stress peaks which is known as the "notch effect". The notch, which is a deliberately introduced into the material is either V-shaped or U-shaped and it is intended to determine in fracture mechanics properties such as fracture toughness and the rate of fatigue crack growth. The tendency of a material to behave as a brittle material in presence of a notch is termed as notch sensitivity. It is an accepted understanding that FCC metals aren not notch sensitive. Notch sensitivity is influenced by many parameters such as notch size and notch geometry.

J. Pelleg, *Cyclic Deformation in Oxides, Carbides and Nitrides*, Structural Integrity 22,
https://doi.org/10.1007/978-3-030-86118-6_11

Notch is the common and the basically adopted method in impact test. The impact energy is a measure of the work done to fracture a test specimen. On striking the specimen by impact, the specimen will absorb energy until it yields and finaly breaking the specimen. The test specimen continues to absorb energy and work hardens at the plastic zone at the notch (in ductile material). Two kinds of test are common, the Izod and the Charpy methods. These impact tests indicate the ability of a material to withstand impact and are used by engineers to predict its behavior under actual conditions and are very important since material fail suddenly under impacr at flaws, cracks, or other imperfections and clearly at notches. The Izod impact test is commonly used to evaluate the relative toughness (called impact toughness) of materials and is an often used as a fast and economical test in quality control. The Charpy V-notch test is an indication of the ability to absorb energy at stress concentration or at a notch and is used to indicate the ductile-to-brittle transition in metals like steel. The presence of notch increases stress intensity, decreasing impact toughness. Factors affecting Charpy impact energy are yield strength, ductility, notch, temperature, strain rate and the fracture mechanism. It could be noted that in the Izod method, the test material is placed in a vertical position, while in the Charpy method, the test material is placed horizontally. In the Izod test a pivoting arm is raised to a specific height (constant potential energy) and then released. The arm swings down hitting a notched sample, breaking the specimen.

In the following notches in fatigue are considered. Notches and holes decrease the fatigue strength of structures. This phenomenon is called the fatigue notch effect and the fatigue strength depends on specimen size.

11.2 Notch Effect in Alumina

Alumina notched specimen with two different notch radii were fatigue tested by four-point bending tests. The cyclic life depends on the notch radii. It has been revealed by the tests that the experimental notch factor, K_f is about the same as the theoretical notch factor, K_t. Since ceramics are considered for applications in advanced engines, during operation they might be subject to cyclic loading and it is important to know the fatigue behavior of notched specimens. A notch may be considered as a stress concentration and as such be the root of fatigue crack initiation and and growth from the root of notches. Brittle material exposed to cyclic loads grow only short distances from the root of notches before failure sets in as fracture. The notch depth of the specimens was 3 mm and the notch radii were 0.5 and 1.5 mm, respectively. The frequencies of the test ~10–15 Hz with the load ratio $R = 0.15$. Al tests were performed at ambient temperature at three environments, among them in air. At a stress level $\sigma_{max} = 138$ MPa in air the test was interrupted when cycles reached 1.5×10^6 thout failure, survivor specimens were then tested for the residual strength. S–N curves shown in Fig. 11.1 where the maximum cyclic stress is plotted against failure cycles for the two notch radii. The 50% survivability curves are represented by

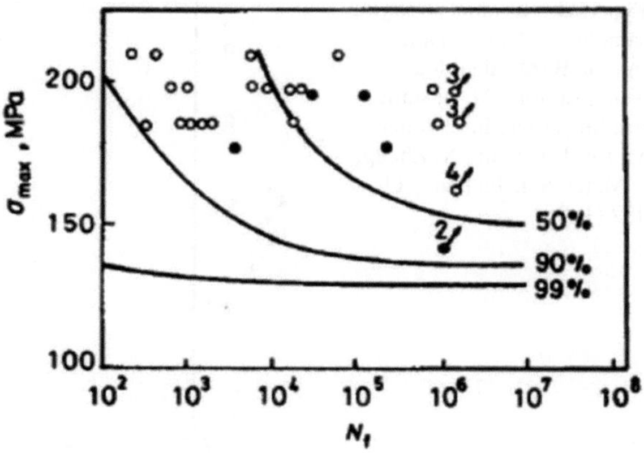

Fig. 11.1 Fatigue life with radius 1.5 mm (○), 0.5 mm (●). F. Wang, M. Lu, J. Wei and X. Zheng, J. Mater. Sci. Technol., 11, 452 (1995). The percents indicate specimen survivability

$$N_f = A(s - s_c)^{-2} \tag{11.1}$$

where A and σ_c are the fatigue resistance coefficient and the theoretical fatigue limit.

Two groups of fatigue results tested at $\sigma_{max} = 207$ MPa and $\sigma_{max} = 196$ MPa are plotted on log-normal probability paper in Fig. 11.2. Fatigue life follows a log-normal distributuion. σ_{mean}, the effective stress can be expressed in terms of the inner span, S_i, the outer span, S_o, and B, w, d the width, height and notch depth, repectively of the specimens.

$$\sigma_{mean} = \frac{3P(S_O - S_i)}{2B(w - d)^2} \tag{11.2}$$

The initial strength and residual strength are plotted in Fig. 11.3 on Weibull probability paper wher A represents residual strength of four specimens tested in room

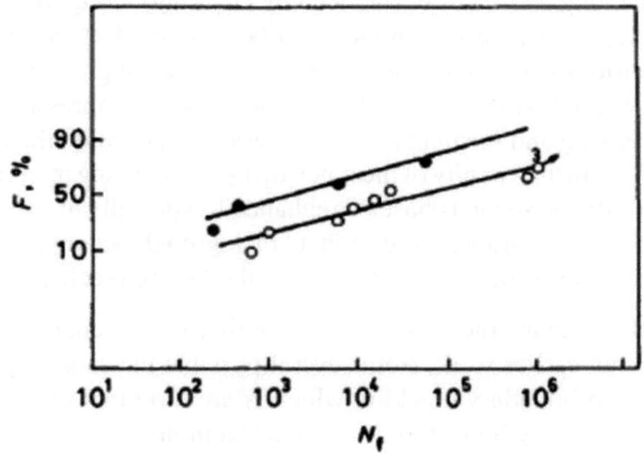

Fig. 11.2 Probability distributionof fatigue life (●) $\sigma_{max} = 207$ MPa, (○) $\sigma_{max} = 196$ MPa. F. Wang, M. Lu, J. Wei and X. Zheng, J. Mater. Sci. Technol., 11, 452 (1995)

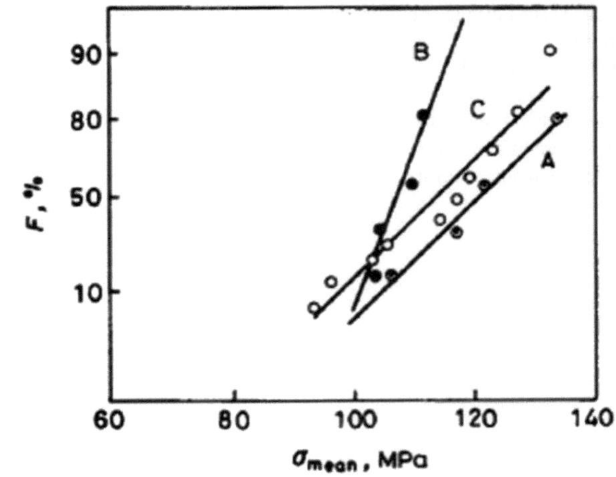

Fig. 11.3 Residual strength, sample A (⊙): room air, sample B (●): distilled water, sample C (○): static bending strength. F. Wang, M. Lu, J. Wei and X. Zheng, J. Mater. Sci. Technol., 11, 452 (1995)

air at $\sigma_{max} = 161$ MPa and B expresses the residual strength of specimen in distilled water at $\sigma_{max} = 138$ MPa. Sample C is the static bending strenght used as reference.

Further, in an earlier work on dynamic fatigue in polycrystalline aluminum, fatigue experimets under compression-compression loading conditions were carried out in single edge-notched specimens. The average grain size was about 18 μm. The material has the following mechanical properties (at 25 °C): tensile strength = 260 MPa; compressive strength = 2620 MPa: Young's modulus = 372GPa. The edge-notched specimens were machined to the following dimensions: height—37 mm, width = 15.9 mm, thickness = 9.4 mm, notch length = 6.3 mm and notch root radius = 0.75 mm. The application of compressive cyclic loads can lead to crack growth at room temperature. A series of optical micrographs are shown in Fig. 11.4, illustrating the progressive growth of a crack from the tip of the notch. The variation of the fatigue crack length, a, is shown as a function of compression cycles, N in Fig. 11.5.

Progressive crack growth is observed under the conditions of the test. Microcracking and fracture (induced by cracking) along the grain boundaries and progressive damage accumulation can be promoted by several mechanisms. Microcracks form because of residual stresses localized at grain boundaries as a result of thermal contraction anisotropy. The application of compressive loads induce further microcracking and the magnitude of stresses imposed on the material by the stress concentrain in the vicinity of the notch tip by pre-existing microstructural defects (inclusions, voids and second phases) is enhanced, especially by grain triple points. Scanning electron micrographs suggest that crack growth occur preferentiall along the boundaries of coarser grains which supports the failure mechanism resulting in fracture.

In summary, the cyclic life for alumina at different notch radii was presented. Further, in an earlier work, from several possible mechanisms, progressive failure is due to grain boundary cracking which is an important contributing mechanism for cyclic crack growth in ceramics such as alumina.

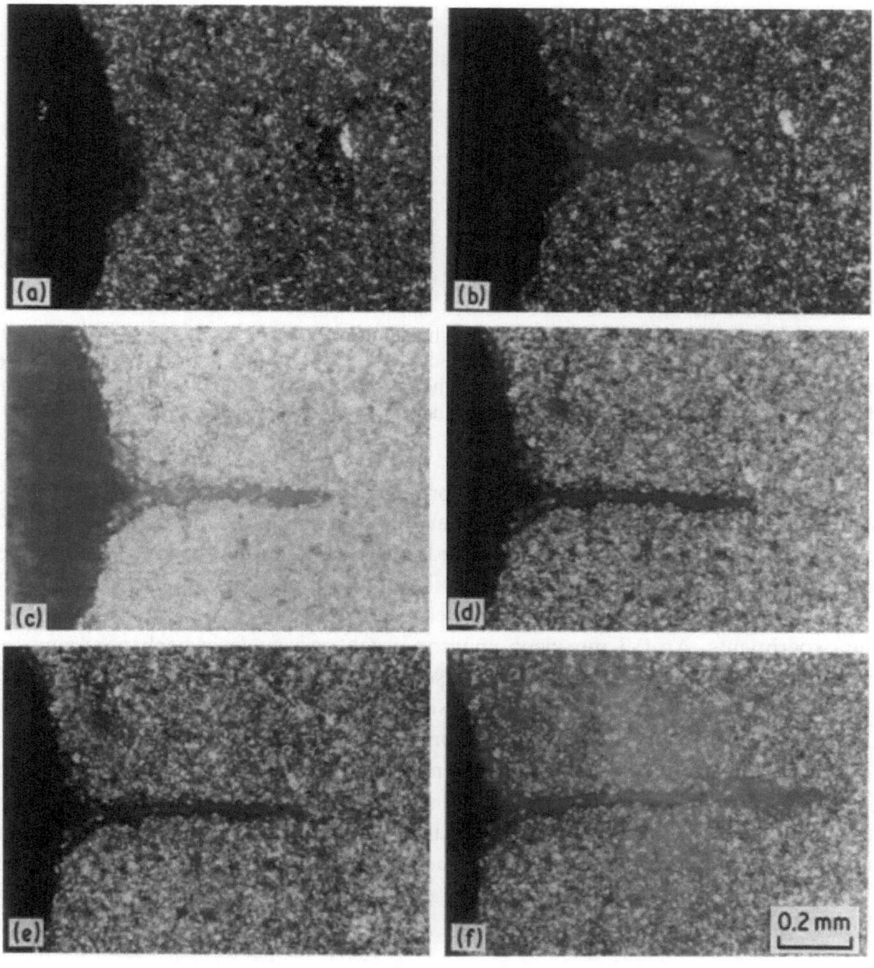

Fig. 11.4 Progressive increase in crack length with compression cycles. **a** 0 cycles and after, **b** 59,200 cycles, **c** 69,200 cycles, **d** 79,200 cycles, **e** 94,200 cycles and **f** 154,200 cycles. L. Ewart and S. Suresh, J. Mater. Sci. Lett., 5, 774 (1986)

11.3 Notch Effect in 3% Y-Zirconia

As stated several times in this book, zirconia based ceramics, such as 3% Y-Zirconia, have a wide range of different medical and technical applications such as listed as examples: endoprosthetics, thermal barrier coatings, stop valves, etc., due to their excellent physical–mechanical properties. Just to mention some of their properties are wear resistance, biocompatibility, high strength, fracture toughness, low thermal conductivity, etc. However, their application is limited due to considerable brittleness

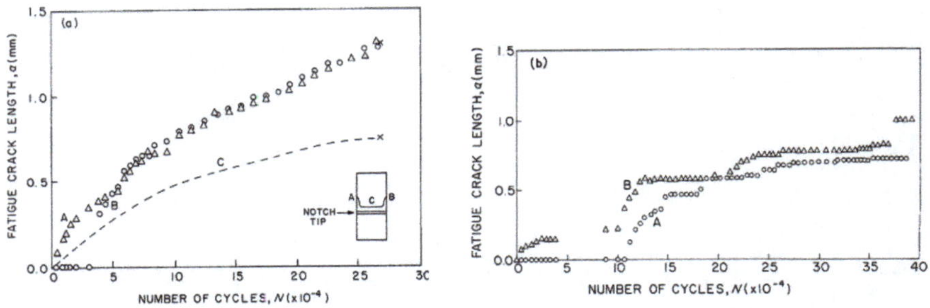

Fig. 11.5 A plot of fatigue crack lengths with the number of compression cycles N for **a** case a and **b** case b. L. Ewart and S. Suresh, J. Mater. Sci. Lett., 5, 774 (1986). In case a the particles are periodically remove by ultrasonically cleaning the specime after every 5000 cycles; in case b the debris were not removed and consequently the crack growth rate was smaller

which is intensified by unstable crack propagation. Double-cantilever specimens with chevron notches were used to evaluate the material toughening during a stable crack advancement. Finite element method (FEM) simulation is used to simulate the deformation and the fracture processes to interprete experimental observations.

Rectangular shaped specimens sintered at 1600 °C for 1 h were used used in the experiments. An 0.3 mm-deep chevron notch was made in each specimen by rotating a thin diamond-coated disc. A schematic of the specimen geometry and loading conditions is given in Fig. 11.6. Six specimens were tested using an Instron universal testing machine in a deformation-controlled mode with a velocity of $\nu_1 =$ 0.05 mm/min. In Fig. 11.7 the experimental loading results are plotted as applied force against cantilever deflection. It is seen in Fig. 11.7a a non-linear stage of deformation which is related to the phase tansformation, inelastic strain accumulation and crack propagation. The onset of nonlinearity and its duration vary in a wide range

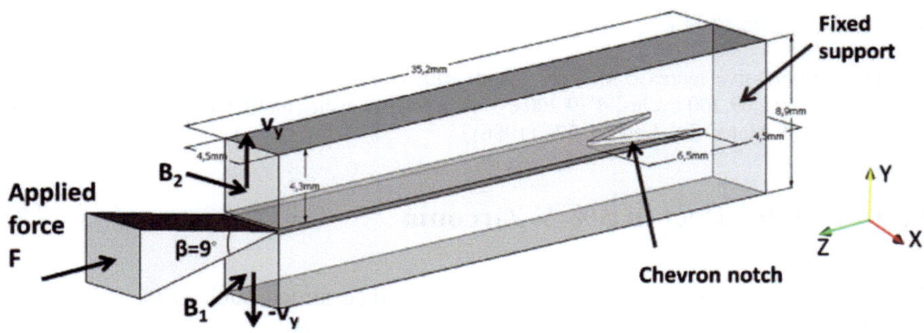

Fig. 11.6 Schematic of the double-cantilever specimen with a chevron notch and its loading conditions; the movement of the wedge tool is simulated by assigning the nodal velocities to the mesh nodes belonging to B1 and B2 surfaces, and the fixed support has all zero nodal velocity vector components. M. O. Eremin, E. E. Deryugin and S. Schmauder, Engineering Failure Analysis, 110,104,409 (2020). With kimd permission of Elsevier

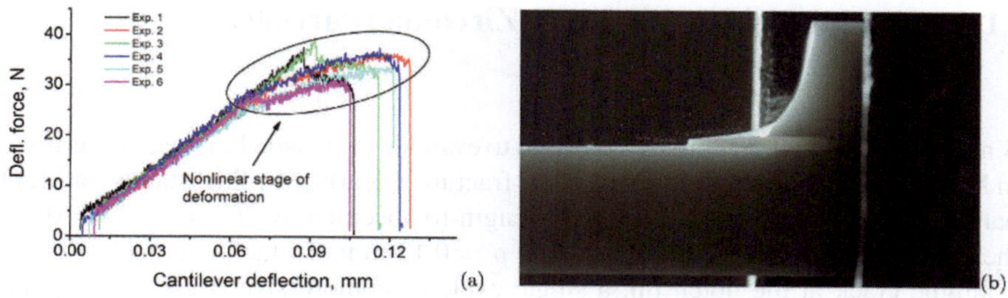

Fig. 11.7 Applied force versus cantilever deflection loading diagrams for 6 specimens **a**, a typical pattern of the loaded specimen failure demonstrating spalling of one of the cantilevers **b**. M. O. Eremin, E. E. Deryugin and S. Schmauder. Engineering Failure Analysis,110, 104,409 (2020). With kimd permission of Elsevier

due to imperfection of the notch production causing microcracks, which represent discontinuities decreasing the strength and elastic moduli of a material. Time to failure decreased linearly with increase in strain rate. FEM simulation was used to simulate the deformation process and to interprete the experimental observations. The results of numerical simulation are in a satisfactory agreement with the experimental data and also with the other available data. The numerical simulation data reveal that the higher the roughness, the lower the fracture toughness. The number of finite elements in the mesh was increased from ≈350,000 to ≈3,200,000 but increase in the number of element to more than ≈2,600,000 elements does not noticeably contribute to convergence. This can be seen in Fig. 11.8. The fracture surface roughness has to be taken into account in evaluating the fracture toughness, in order to avoid over estimating the fracture toughness, K_{Ic} resulting from the real crack surface area which is larger than its planar projection.

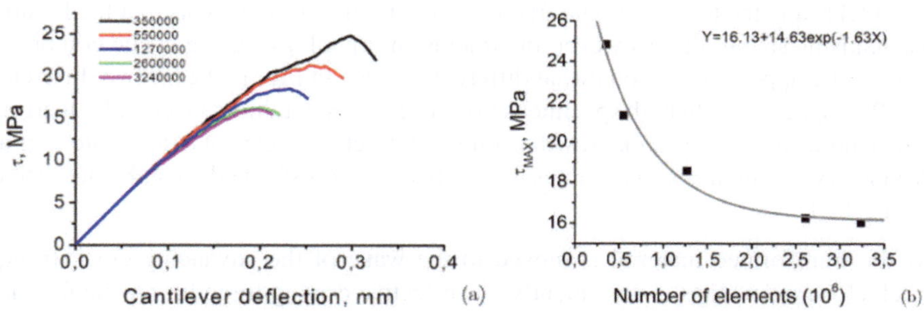

Fig. 11.8 Diagrams of convergence of numerical solutions: average equivalent stress versus cantilever deflection for the models with different numbers of elements **a** and exponential fit of the peak stress values **b**. M. O. Eremin, E. E. Deryugin and S. Schmauder. Engineering Failure Analysis, 110,104,409 (2020). With kimd permission of Elsevier

11.4 Notch Effect in 3% MgO-Zirconia (Partially Stabilized)

Single edge-notched specimens were used to evaluate the tensile fatigue crack growth in Mg-PSZ which might lead to failure by fracture. Clearly, Mg-PSZ stands for MgO partially stabilized zirconia. The notch length-to-specimen width ratio a/w = 0.36, the notch $\theta = 60°$ and notch-root radius $\rho = 0.12$ mm. To facilitate nucleation of a fatigue crack at the notch tip, a single cycle of compressive stress was applied to the specimen first, and subsequently cyclic tension in four-point bending in an electro-servohydraulic machine was applied. The length of the crack is up to 1 mm. With the increase of ΔK the length of the tensile fatigue crack increased from the threshold value to final fracture (for the illustrations see Fig. 1 of the work of Sylva and Suresh). Recall that the load ratio is

$$R = \frac{K_{\min}}{K_{\max}} \tag{11.3}$$

and

$$\Delta K = K_{\max} = K_{\min} \tag{11.4}$$

The fatigue crack growth rate is shown in Fig. 11.9. The average da/dN in Fig. 11.9 were calculated based on the crack advance over thousands of cycles, even though such fracture may have occurred during one cycle. Tensile fatigue surface at two magnifications is shown in Fig. 11.10. The failure mode is intergranular as seen in Fig. 11.10a, however at higher magnificataion some transgranular fracture can also be seen. Comparison of the failure mechanism of the quai-static tensile fracture of Fig. 11.11, and the tensile fatigue fracture seen in Fig. 11.10, reveals that the failure mechanisms are similar. However, the fracture morphology during cyclic compression failure appears to be somewhat different, as seen in Fig. 11.12 for two different Mg-PSZ specimens. In both specimens (a) and (b) considerable transgranular tearing can be noticed. Once a crack growth is initiated, fracture occurs at a rapid rate over a distance Δa^*, which is several times the width of the transformed zone, H illustrated in Fig. 11.13.

As the transformed material is moved to the wake of the advancing crack front, crack closure develops. Consequently, the effective driving force for tensile fatigue fracture, ΔK_{eff}, has a much smaller value than the nominal (far field) ΔK for constant amplitude of tensile loading at a fixed R. The threshold stress intensity range is ΔK_0 in Eq. (11.5); crack growth is reached when the maximum nominal stress intensity factor of the fatigue cycle K_{Imax} is equal to the

$$\Delta K \to \Delta K_0 \quad when\ K_{Imax} = \frac{\Delta K}{(1-R)} \to K_{Ic,m} \tag{11.5}$$

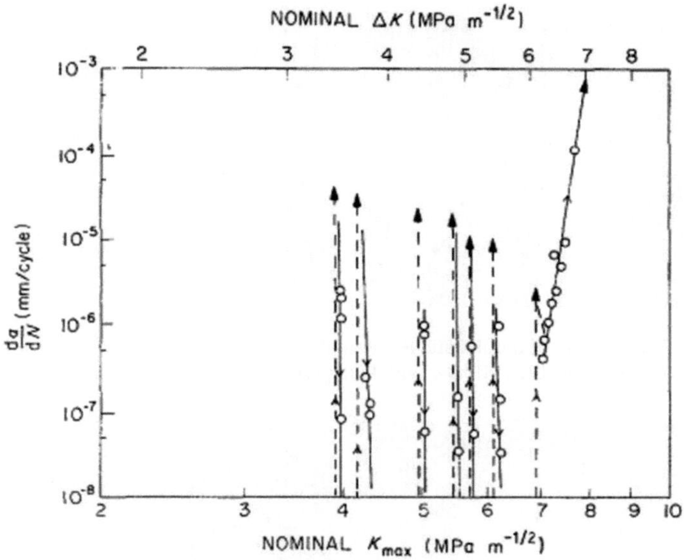

Fig. 11.9 Fatigue crack growth rate, da/dN, plotted as a function of the applied stress intensity factor range. L. A. Sylva and S. Suresh, J. Mater. Sci., 24, 1729 (1989). With kind permission of Springer Nature

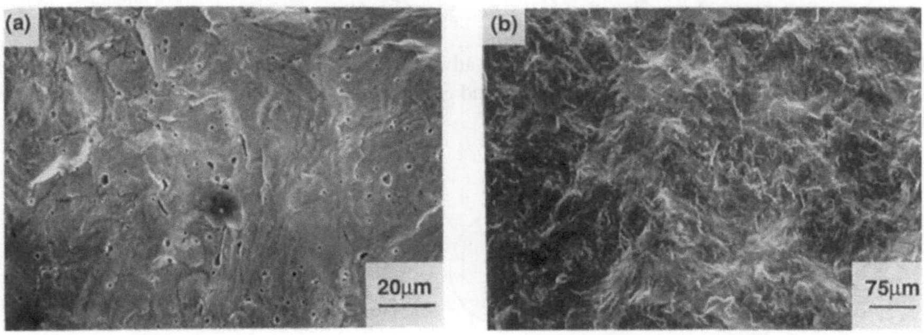

Fig. 11.10 a, b Scanning electron micrographs showing the fracture morphology during tension fatigue at different magnifications in, Mg-PSZ; $\Delta K \approx 5$ MPa $m^{1/2}$. L. A. Sylva and S. Suresh, J. Mater. Sci., 24, 1729 (1989). With kind permission of Springer Nature

fracture toughness of the fully stable cubic matix $K_{Ic,m.}$. The value of ΔK_{eff} mentioned earlier is given by

$$\Delta K_{eff} = \Delta K - \Delta K_{Ic} \tag{11.6}$$

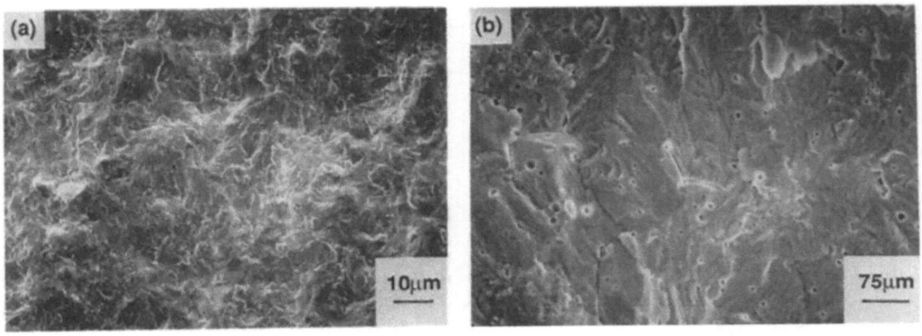

Fig. 11.11 Fractography of quasi-static tensile failure at $K_I \approx 6.5$ MPa $m^{1/2}$. L. A. Sylva and S. Suresh, J. Mater. Sci., 24, 1729 (1989). With kind permission of Springer Nature

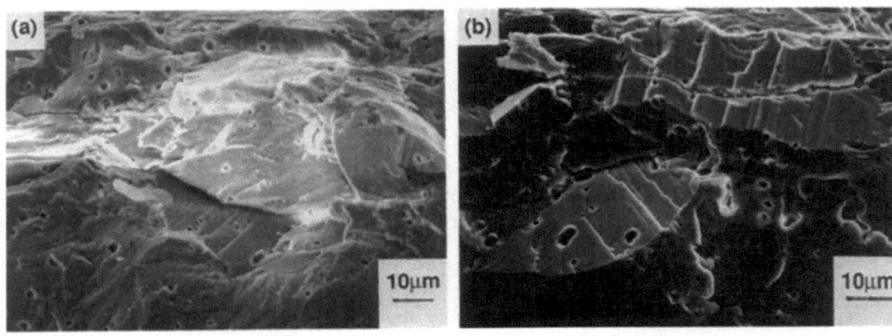

Fig. 11.12 Fracture morphology during crack advance under far-field cyclic compression in two different specimens of Mg-PSZ. L. A. Sylva and S. Suresh, J. Mater. Sci., 24, 1729 (1989). With kind permission of Springer Nature

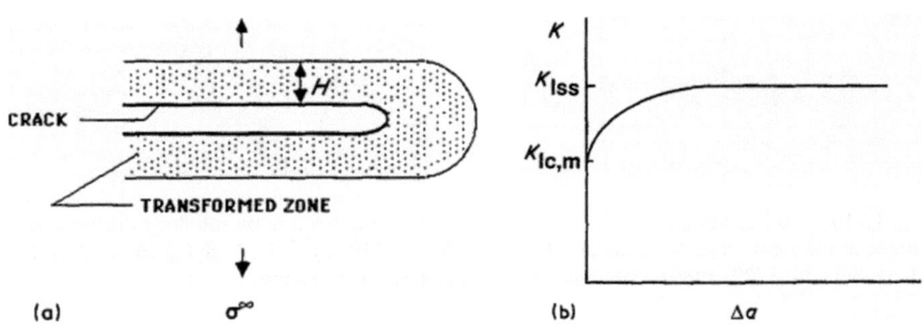

Fig. 11.13 A schematic diagram of **a** the transformed wake during steady state crack growth; and **b** the resistance curve resulting from transformation at the crack-tip. L. A. Sylva and S. Suresh, J. Mater. Sci., 24, 1729 (1989). With kind permission of Springer Nature

Fracture occurs catastrophically when $\Delta K = \Delta K_f$ when the maximum stress intensity factor of the fatigue cycle approaches the steady state toughness value, K_{Iss} as seen for the R curve in Fig. 11.13b, namely,

$$K_{max} = \frac{\Delta K_f}{(1 - R)} \rightarrow K_{Iss} \tag{11.7}$$

In summary, a possibility exists for stable crack growth under cyclic tension in tansformation toughened caramics. The basic features of short fatigue crack growth at room temperature under cyclic tensile loads in magnesia-partially stabilized zirconia is that single edge-notched specimens are fractured in the four-point bending configuration is similar to quai-static tensile loads. The threshold stress intensity factor, ΔK, for fracture initiation in cyclic tension at a load ratio of 0.1, is low as 3.4 MPa $m^{1/2}$ and catastrophic failure occurs at 6.6 MPa $m^{1/2}$. It is observed that crack growth rate is very discontinuous. The growth of cracks in transformation-toughened ceramics, including MgO is found to be strongly influenced by the crack size and shape, stress state and specimen geometry.

11.5 Notch Effect in SiC

Usually and as indicated along this book, SiC by itself is seldom used, since the interest lies in SiC/SiC composites for potential applications as a structural material (for example in fusion systems). Here the composite comprises SiC matrix and Nicalon SiC fibers. The two types of specimens showing the notch configuration tested are shown in Fig. 11.14. As seen in the illustration, one of them is a center notched panel specimen, which was used for tensile/tensile cyclic load tests at 25 °C, while the second type is a chevron notched beam specimen, which was used for three-point flexure static load and fracture toughness tests at 25–1400 °C. Starting with the tensile/tensile cyclic load tests at 25 °C, a matrix crack develops at an early statge test, and grows until enough fibers bridge the crack wake to balance the applied load. When this steady-state configuration is reached, the crack length becomes constant and the crack surface displacement is also constant except near the crack tip. Subsequently during further cycling, the bridging fibers in the wake progressively rupture and results in crack extension. The maximum strain was used to calculate half crack length for the center notched specimens. The maximum stress intensity factor, K_{max} was calculated using the expression (11.8) and the results are plotted in Fig. 11.15

$$K_{\max} = \sigma\sqrt{W \tan(\pi a/W)} \tag{11.8}$$

for two load ratios, R = 0.10 and R = 0.50. The crack growth depends on the load ratio, and the R = 0.10 is more damaging than the R = 0.50 indicating that cyclic

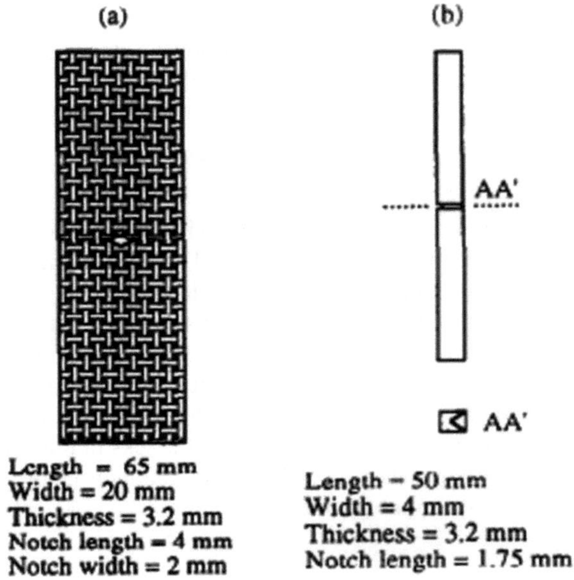

Fig. 11.14 Specimens used for crack growth testing; **a** center notched specimen, **b** chevron notched beam. S. Raghuraman, J. F. Stubbins, J. Nuclear Materials, 212, 840 (1994). With kind permission of Elsevier

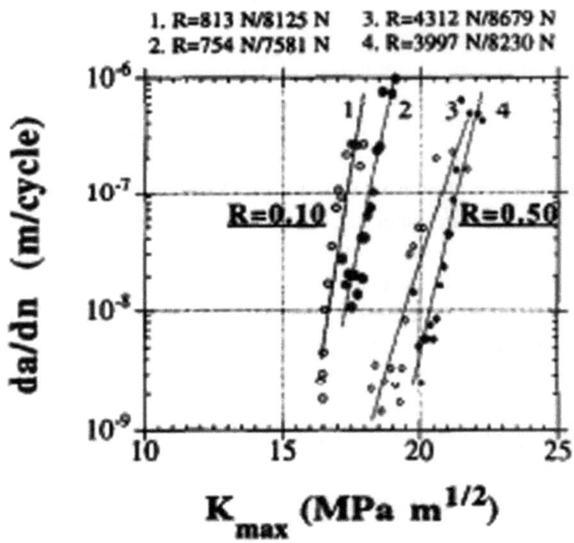

Fig. 11.15 Maximum stress intensity versus crack velocity for two load ratios, under tensile/tensile cyclic load. S. Raghuraman, J. F. Stubbins, J. Nuclear Materials, 212, 840 (1994). With kind permission of Elsevier

fatigue is occurring and accelerating the process (see Fig. 11.15). The results of the fracture toughness test performed by three-point flexure fracture toughness test is shown in Fig. 11.16. Increasing crack resistance with increasing crack length is due to the crack tip shielding effect of the bridging fibers in the crack wake. This increase is attributed to fiber processes, such as delamination, fiber fracture and fiber

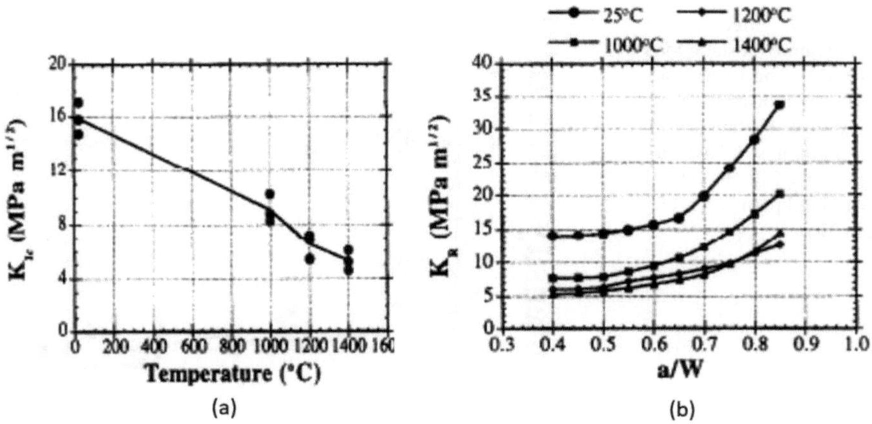

Fig. 11.16 Results of fracture toughness tests at 2% 1400 °C; **a** fracture toughness versus temperature, **b** rising crack resistance curves. S. Raghuraman, J. F. Stubbins, J. Nuclear Materials, 212, 840 (1994). With kind permission of Elsevier

pullout (against interface friction). The amount of crack shielding depends on the fibe pullout length. An extensive fiber pullout is at 25 °C, but a gradual reduction in the fiber pullout length from 1000 to 1400 °C (see Fig. 11.16). This trend is seen in Fig. 11.17. For comparison under static loads at 25 °C the material did not exhibit the applied stress intensity factor at elevated temperatures as shown in Fig. 11.18.

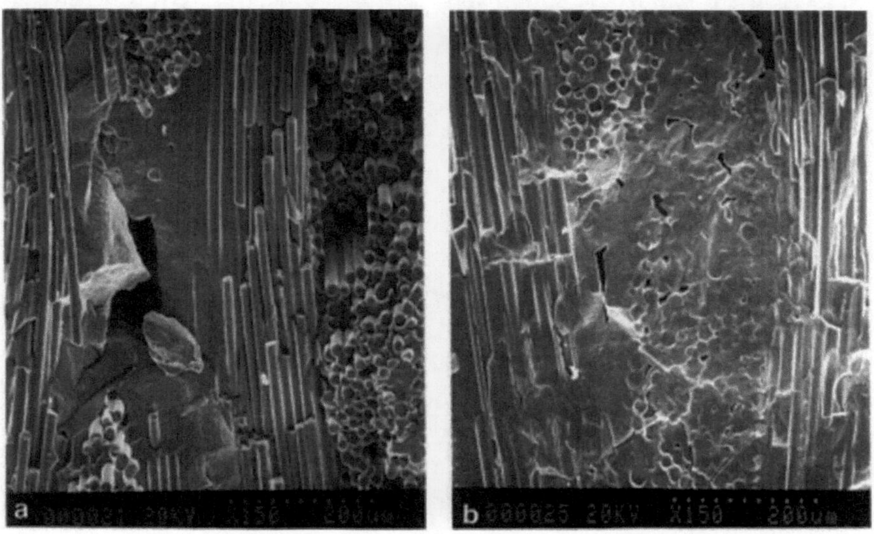

Fig. 11.17 Fracture surface of specimens tested at various temperatures; **a** 25 °C, (b) 1400 °C. S. Raghuraman, J. F. Stubbins, J. Nuclear Materials, 212, 840 (1994). With kind permission of Elsevier

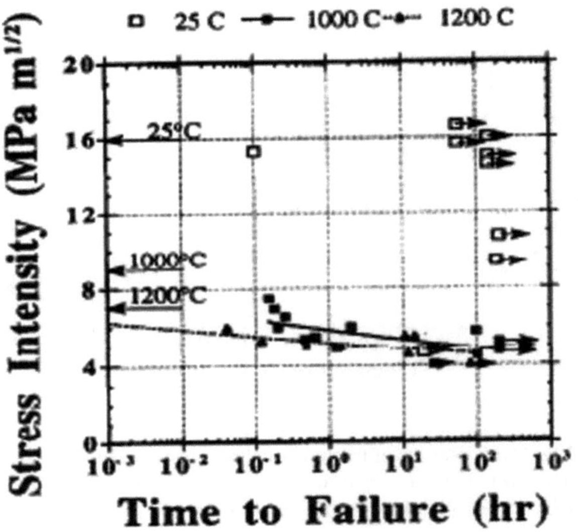

Fig. 11.18 Effect of temperature on static load behavior. The arrows on ordinate indicate fracture toughness. S. Raghuraman, J. F. Stubbins, J. Nuclear Materials, 212, 840 (1994). With kind permission of Elsevier

SEM at high temperature fractures show two distinct regions as seen in Fig. 11.19. No fiber pullout is seen at the tip of the notch where the crack initiated, but at the opposite end, the surface shows extensive fiber pullout. At the fiber-matrix interface oxidation causes embrittlement by development of a strong bond, which prevents

Fig. 11.19 Fracture surface of a static fatigue specimen; **a** near notch tip, **b** opposite end. S. Raghuraman, J. F. Stubbins, J. Nuclear Materials, 212, 840 (1994). With kind permission of Elsevier

fiber pullout and produces a flat appearance. A second region is relatively free from oxidation probably due to passivation protection, thus allowing fiber pullout.

In summary, cyclic fatigue in SiC_f/SiC composites is observed at RT tension tests. The load ratio affects the crack growth rate and the damage depend on R. No static fatigue was observed in RT flexure tests. High fracture toughness is achieved at RT due to fiber bridging of crack wake. At high temperatures oxidation of interface and chemical bonding of fibers to matrix occurs. With increasing temperature there is a gradual reduction in fiber pullout with the consequent decrease in the fracture toughness. Oxidation plays an important role in crack growth.

11.6 Notch Effect in Si_3N_4

The Si_3N_4 considered here has a β-crystalline structure and Table 11.1 indicates its Young's modulus, E and the Vickers hardness, H_v. Fatigue precracked specimens in a single-edge notched beam (SENB) configuration was obtained by cyclic fatiguing in four-point bend at an elevated temperature and were used to evaluate the fracture toughness. The specimen designated as HIP-Si_3N_4 is pressurelessly pre-sintered at 2073 K and sintered at 1973 K, resulting in almost a 100% density by hot isostatic press. The dimensions of the specimens are listed in Table 11.2. Here, B is the thickness, W is the width and L is the support span length and l are the loading span length. Precracking by fatigue is carried out in air at 1473 K using a servohydraulic testing machine. This was done for a four-point bend under a load control of the stress ratio R $(= P_{min}/P_{max}) = 0.1$ at 1 Hz. The results are seen in Fig. 11.20 where HIP-Si_3N_4 is plotted as the relationship between net section stress amplitude and the number of cycles to failure. The fatigue precrack was introduced at a loading stress amplitude of 65 MPa. The precracing was performed by constant K_{fmax} and decreasing K_{fmax} methods as shown in Fig. 11.21a; after the initial 800-cycle block test at constant stress amplitude of 65 MPa the test was interrupted to observe the

Table 11.1 Young's modulus E and Vickers hardnesses H_v of silicon nitrides. Y. Mutoh, K. Tanaka, T. Niwa and N. Miyahara, J. Mater. Sci., 23, 3939 (1988). With kind permission of Springer Nature

	E (GPa)	H_v (GPa)
HIP-Si_3N_4	320	15.67

Table 11.2 Specimen dimensions. Y. Mutoh, K. Tanaka, T. Niwa and N. Miyahara, J. Mater. Sci., 23, 3939 (1988). With kind permission of Springer Nature

	Kind of bend test	*B*	*W*	a_0	*L*	*l*
HIP-Si_3N_4	Bending strength : 3-point	4	3	–	27	–
	Fatigue precracking : 4-point	4.85	10	4	40	20
	Fracture toughness : 3-point	4.85	10	4	40	–

fatigue crack. The test was interrupted every 4000 cycles to measure the crack length. When the length of the crack attained 0.4–0.5 mm, the test was stopped. Fracture toughness tests were carried out on the precracked specimens at various temperatures from RT to 1473 K in the same way as the bending strength test. The bending strength test results is shown in Fig. 11.22. The temperature dependence of the fracture toughnes, K_c, is illustrated in Fig. 11.23. The figure also includes S-Si_3N_4 which is an alloyed Si_3M_4. Included also are notched specimens with root radius r = 0.15 and with values much higher than the precracked specimens. An optical micrograph of the fatigue crack at the notch root is shown on the specimen surface in Fig. 11.24. Detailed view of the fracture surface is seen in a SEM fractograph in Fig. 11.25, comparing fatigue fracture to a static fracture region. In Fig. 11.26 the relationship between the indentation crack length, a, and the applied load, P is indicated according to Eq. (11.9), and again the results of S-Si_3N_4 are included.

$$K_c = 0.0725\,P/a^{3/2} \tag{11.9}$$

K_c is the indentation fracture toughness.

Summarizing this part of this section, one observes an experimental technique for determining fracture toughness in ceramics (in our case Si_3N_4) by a method involving fatigue precrack in SENB specimens introduced by cyclic fatiguing in a four-point bend at elevated temperature. The values at room temperature were almost equal but higher than those obtained by indentation technique (see Fig. 11.26). The slightly

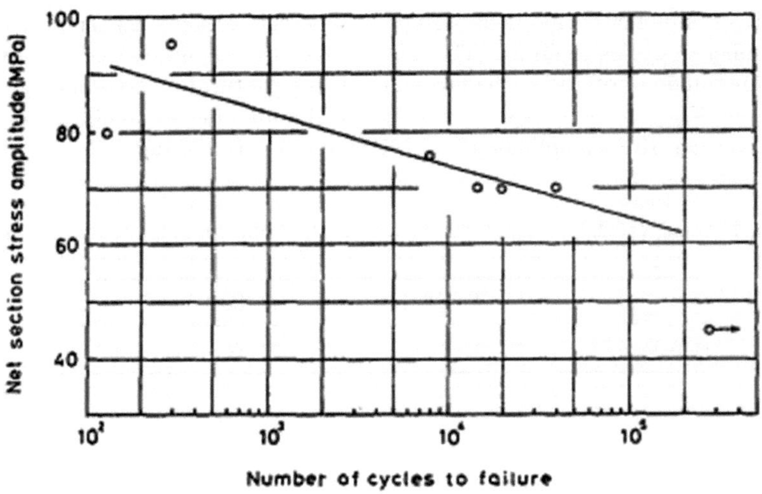

Fig. 11.20 S–N curve at 1473 K for HIP-Si_3N_4 SENB specimen. R = 0. Y. Mutoh, K. Tanaka, T. Niwa and N. Miyahara, J. Mater. Sci., 23, 3939 (1988). With kind permission of Springer Nature

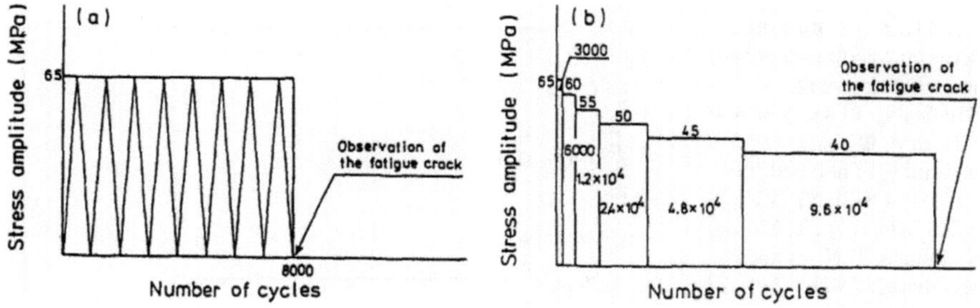

Fig. 11.21 The sequence of block loading for precracking. **a** The constant K_{fmax} method. **b** The decreasing K_{fmax} method. Y. Mutoh, K. Tanaka, T. Niwa and N. Miyahara, J. Mater. Sci., 23, 3939 (1988). With kind permission of Springer Nature

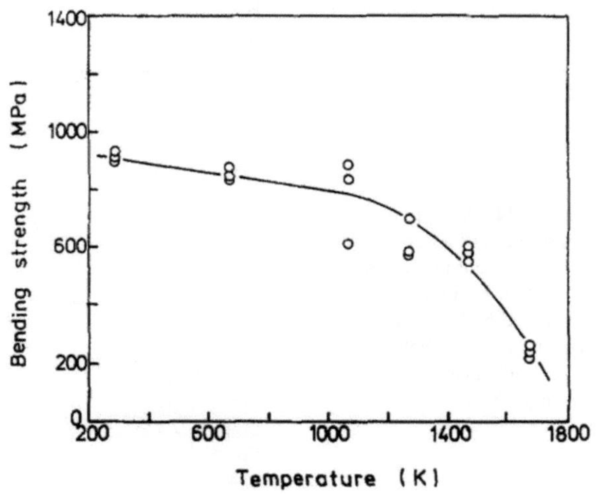

Fig. 11.22 Relationship between bending strength and temperature for HIP-Si_3N_4. Y. Mutoh, K. Tanaka, T. Niwa and N. Miyahara, J. Mater. Sci., 23, 3939 (1988). With kind permission of Springer Nature

higher values are attributed to crack closure resulting from the formation of oxide film on the fatigue fracture surface.

In a recent paper, in situ observation of crack propagation in notched Si_3N_4 single crystal specimens in four-point bending inside a SEM is recorded. Si_3N_4 is used for rolling elements in hybride bearing due to to the superior mechanical properties, as hardness, stiffness and low weight. Monotonic and cyclic loading experiments are focused on intergranular versus transgranular crack propagation, and on the analysis of the material resistance to cracking in terms of R-curves, respectively. Specimen loading for the four-point bending test is presented in Fig. 11.27a and the micrograph of the etched surface near the notch is seen in Fig. 11.27b. In situ visualization of a crack propagation is seen in Fig. 11.28. The crack is seen to propagate intergranularly and transgrunalarly as seen in (b) and in (a) and (c), respectively. Microstructure

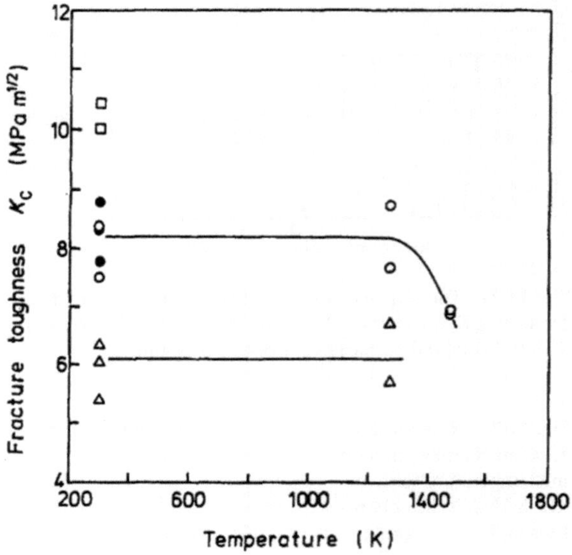

Fig. 11.23 Relationship between fracture toughness and temperature for HIP-Si_3N_4. ○ $K_{fmax} = 4.4$ MPa m^{12}, ● $K_{fmax} = 7.7$ MPa $m^{I/2}$, □ notched (r = 0.15). For S-Si_3N_4 Δ K_{fmax} = 2.8 MPa $m^{1/2}$. Y. Mutoh, K. Tanaka, T. Niwa and N. Miyahara, J. Mater. Sci., 23, 3939 (1988). With kind permission of Springer Nature

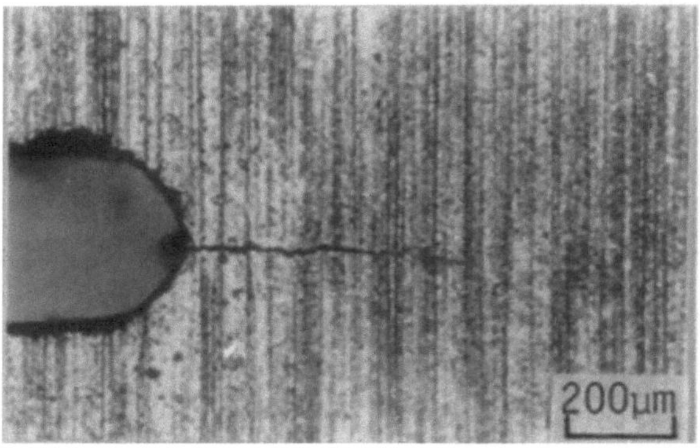

Fig. 11.24 An optical micrograph of a fatigue crack at a notch root in HIP-Si_3N_4. Y. Mutoh, K. Tanaka, T. Niwa and N. Miyahara, J. Mater. Sci., 23, 3939 (1988). With kind permission of Springer Nature

affects the R curve. R curves are shown in Fig. 11.29 and the corresponding fine and coarse grained microstructures are shown in Fig. 11.30.

Crack propagation under cyclic loading of a specimen with rough microstructure is illustrated in Fig. 11.31. The experiments were performed in two stages. First, the crack was initiated from a notch under cyclic load which was progressively increased by 1 N after each cycle (a). In the second stage the peak load was kept constant till the specimen fractured. The crack size versus the number of cycles is shown in Fig. 11.31b, which is the second stage. SEM micrographs of the second stage demonstrate the propagating crack after 5, 6, 7 and 8 cycles. This is seen in Fig. 11.32.

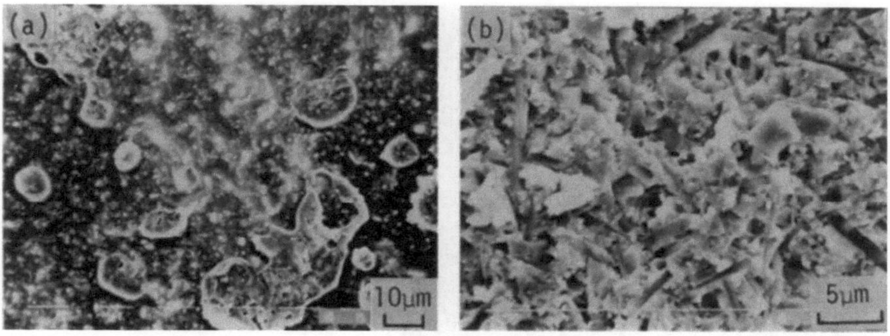

Fig. 11.25 SEM fractographs for a HIP-Si_3N_4 fracture toughness specimen. **a** Fatigue fracture region. **b** Static fracture region. Y. Mutoh, K. Tanaka, T. Niwa and N. Miyahara, J. Mater. Sci., 23, 3939 (1988). With kind permission of Springer Nature

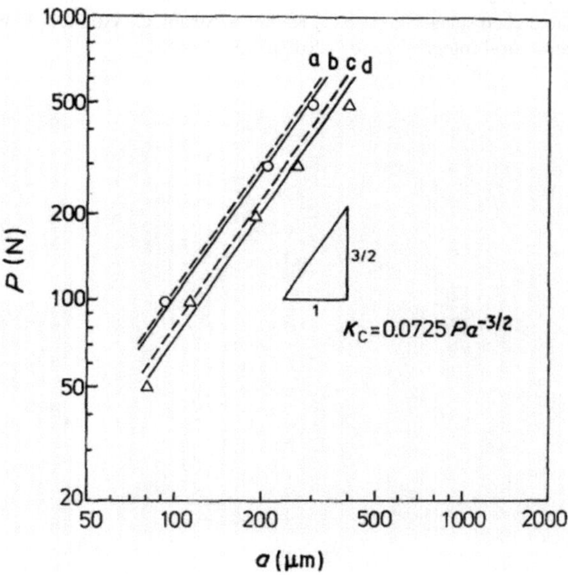

Fig. 11.26 Relationship between indentation crack length and applied load. ○ HIP-Si_3N_4, Δ S-Si_3N_4,—fatigue precracked SENB. **a** $K_c = 8$ MPa $m^{1/2}$, **b** $K_c = 7.3$ MPa $m^{1/2}$, **c** $K_c = 6$ MPa $m^{1/2}$, **d** $K_c = 5.2$ MPa $m^{1/2}$. Y. Mutoh, K. Tanaka, T. Niwa and N. Miyahara, J. Mater. Sci., 23, 3939 (1988). With kind permission of Springer Nature

In summary, fracture in Si_3N_4 under monotonic and cyclic loading is presented in this secion. Emphasis was given for crack propagation which occurs intergranulary and transgranularly. Further, the resistance to cracking, thus eventually to fracture, is analyzed in terms of R curve. The resistance of crack propagation as indicated by the R curve is microstructure dependent, indicating that fine grained microstructure has lower resistance than the coarse one. The cyclic loading revealed that Si_3N_4 has limited potential to resist further fatigue growth under high loading which can initiate a crack from the notch.

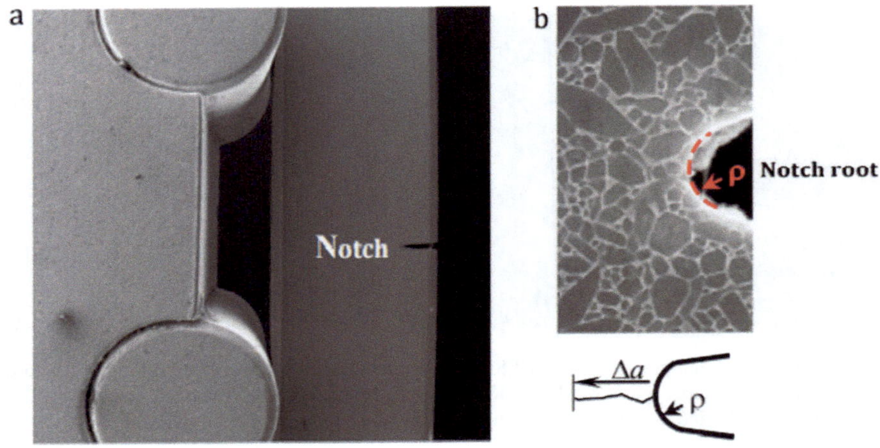

Fig. 11.27 Notch specimen supported in the four-point bending setup **a** and the micrograph of the etched surface zone **b**. Y. Kadin, S. Strobl, C. Vieillard, P. Wijnbergen and V. Ocelik, Procedia Structural Integrity 7, 307 (2017). Open access

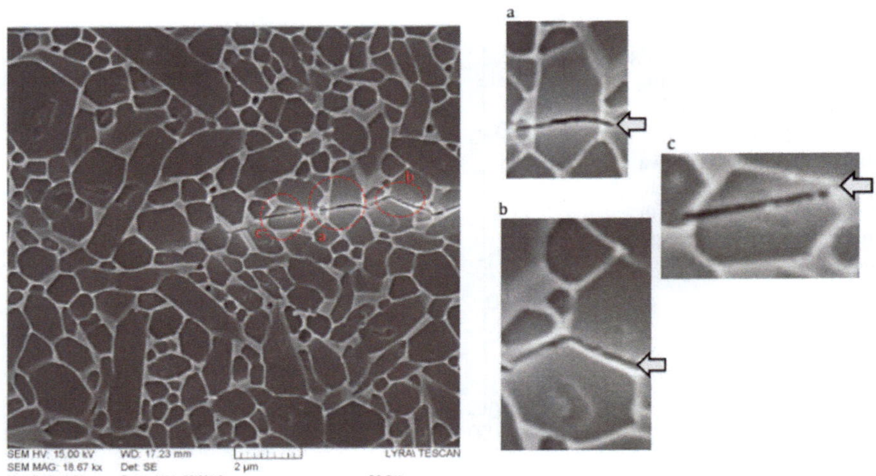

Fig. 11.28 A typical micrograph of a crack propagating in a Si_3N_4 specimen. The trans-granular and intergranular fragments of the crack are zoomed and presented in **a**, **c** and **b**, respectively. Arrows indicate the direction of crack propagation. Y. Kadin, S. Strobl, C. Vieillard, P. Wijnbergen and V. Ocelik, Procedia Structural Integrity 7, 307 (2017). Open access

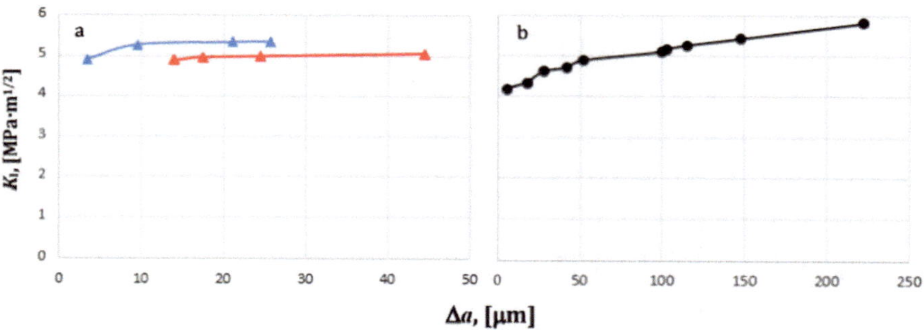

Fig. 11.29 R-curves for the ceramics with the fine **a** and the coarse **b** microstructures. Y. Kadin, S. Strobl, C. Vieillard, P. Wijnbergen and V. Ocelik, Procedia Structural Integrity 7, 307 (2017). Open access

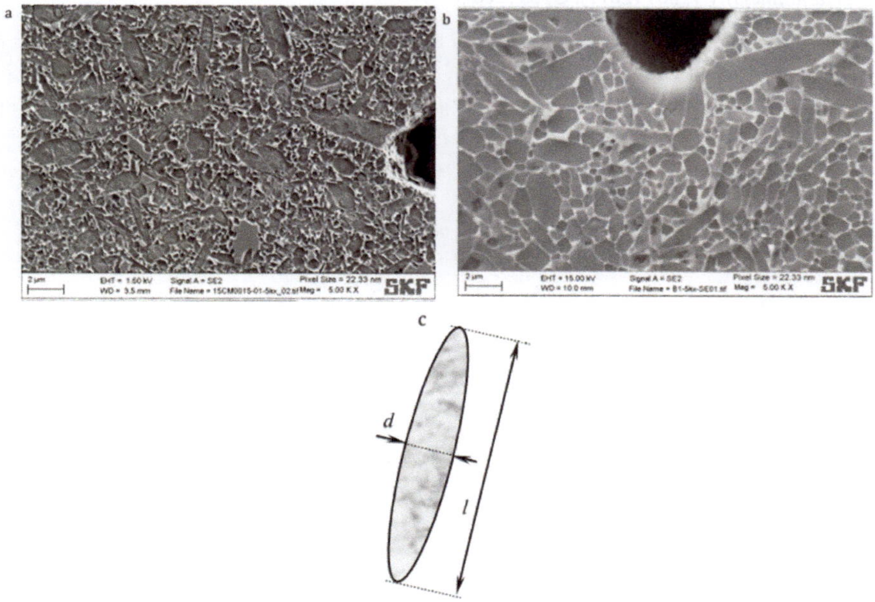

Fig. 11.30 Si_3N_4 ceramics with the fine **a** and the coarse **b** microstructure. Schematic of grain and its dimensions **c**. Y. Kadin, S. Strobl, C. Vieillard, P. Wijnbergen and V. Ocelik, Procedia Structural Integrity 7, 307 (2017). Open access

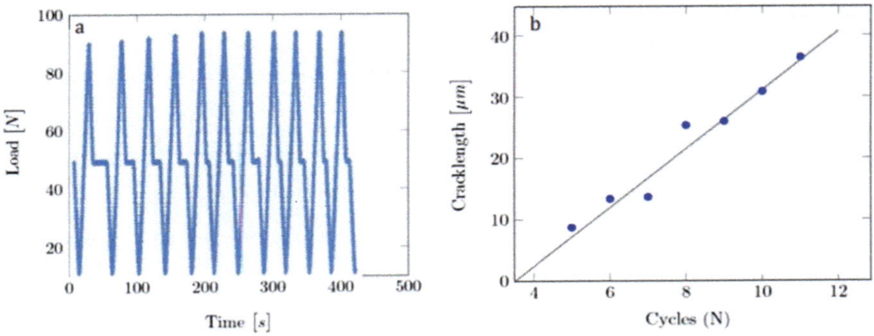

Fig. 11.31 Low cycle fatigue experiment with rough microstructure ceramics. Cyclic force with progressively increasing peak load **a** and crack extension versus number of cycles during the crack propagation stage **b**. Y. Kadin, S. Strobl, C. Vieillard, P. Wijnbergen and V. Ocelik, Procedia Structural Integrity 7, 307 (2017). Open access

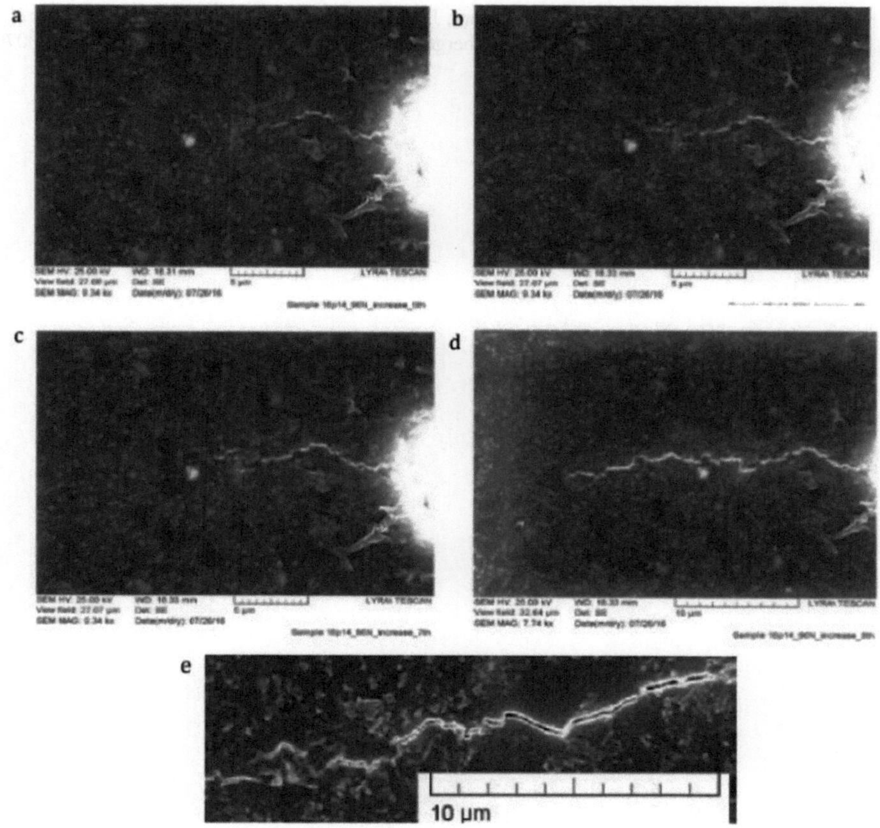

Fig. 11.32 Crack propagating under cyclic loading. Micrographs **a–d** correspond to the crack propagation after 5, 6, 7 and 8 cycles, respectively. Fatigue crack under higher magnification **e**. Y. Kadin, S. Strobl, C. Vieillard, P. Wijnbergen and V. Ocelik, Procedia Structural Integrity 7, 307 (2017). Open access

References

1. S. Maity, D. Vasu, B.K. Sarkar, Mater. Let. **21**, 63 (1994)
2. A. Smirnov, J.I. Beltrán, T. Rodriguez-Suarez, C. Pecharromán, M.C. Muñoz, J.S. Moya, J.F. Bartolomé, Sci. Rep. **7**, Article number: 44922 (2017)
3. A. Hillerborg, M. Modéer, P.E. Peterson, Cem. Concer. Res. **6**, 773 (1076)
4. R.K. Govila, J. Mater. Sci. **26**, 1545 (1991)
5. H. Harmuth, R.C. Bradt, Refrac. Manual **6** (2019)
6. S.-Y. Liu, I.-W. Chen, J. Am. Ceram. Soc. **74**, 1206 (1991)
7. F. Wang, M. Lu, J. Wei, X. Zheng, J. Mater. Sci. Technol. **11**, 452 (1995). L. Ewart, S. Suresh, J. Mater. Sci. Lett. **5**, 774 (1986)
8. M. O. Eremin, E.E. Deryugin, S. Schmauder, Eng Failure Anal. **110**,104409 (2020)
9. L.A. Sylva, S. Suresh, J. Mater. Sci. **24**, 1729 (1989)
10. S. Raghuraman, J.F. Stubbins, J. Nuclear Mater. **212**, 840 (1994)

11. Y. Mutoh, K. Tanaka, T. Niwa, N. Miyahara, J. Mater. Sci. **23**, 3939 (1988)
12. Y. Kadin, S. Strobl, C. Vieillard, P. Wijnbergen, V. Ocelik, Procedia Struct. Integrity **7**, 307 (2017)

Chapter 12
The Effect of Environment

Abstract Fatigue is initiated at the surface where local imperfections are present and a location where environment acts on the material, and its influence is thus more pronounced. Whatever the environment might be, liquid or gaseous, its effect generally is a reduction in fatigue life. The liquid environments are: water, seawater, liquid metals; those of gases are: hydrogen, water vapor (steam), helium, oxygen and air. Vacuum should also be considered as environment. Some of these environments were applied in the experimental investigations of alumina, yttria stabilized zirconia, SiC and Si_3N_4. In particular comparison of the mechanical properties in water and air should be emphasized, although tests were investigated in vacuum also.

Keywords Environment · Air · Water · Dry · Moist air · Steam · Hydrogen · Nitrogen · Vacuum · Alumina · Zirconia · SiC · Si_3N_4

12.1 Introduction

Recall that fatigue is initiated at the surface where local imperfections (flaws, inclusions, cracks etc.) are present. An important surface discontinuity such as crack where fatigue is initiated—and eventually may lead to failure—is a location where environment acts on the material, thus a place on the surface where the influence of environment is more pronounced. Whatever the environment might be, liquid or gaseous, its effect manifests itself generally in a reduction in fatigue life, associated with chemical and diffusion processes including adsorption. Of the many environmental variables—chemical and mechanical—one can mention pressure (or partial pressure) of gas, electrochemical potential, solution chemistry, temperature, frequency, stress, stress ratio etc.

The need for improved structural reliability to improve operation in service in a wide variety of systems to mention some as an example, like nuclear reactors (liquid metal or gas cooled), heat transfer systems, bridges etc.; it is essential to understand the effect of liquids and gases. The liquids of interest are: water, seawater, liquid metals; and gases: hydrogen, water vapor (steam), helium, oxygem and air. Vacuum should also be recalled as a medium under which systems are operating.

J. Pelleg, *Cyclic Deformation in Oxides, Carbides and Nitrides*, Structural Integrity 22,
https://doi.org/10.1007/978-3-030-86118-6_12

In this chapter environmental effects in fatigue deformation of oxides, carbides and nitrides are considered, within the framework of the availability of supporting data in the literature.

12.2 The Effect of Environment in Alumina

12.2.1 In Moist Air and Dry Environment-N_2

Below, a comparison is made between cyclic fatigue resistance and factors dominating fracture in moist air and dry environment (nitrogen) of polycrystalline alumina at ambient temperature. Experimentally measured intrinsic toughness (crack tip) from crack-opening profiles is less than for single crystal and about 30% lower in moist air than in dry N_2 (it means that the grain boundary strength is higher in dry N_2). Micronotches were introduced in the specimens to initiate crack extension at the lowest possible driving forces. R-curves (recall R-curve is defined as the plot of the resistance to fracture versus crack extension) are compared at room temperature in humid room air (moist air) and dry environment of N_2 in Fig. 12.1. Vacuum data are also included in the comparison, but it will be illustrated later. Although inially the line of the tested sample in dry N_2 environment rises much more steeply (see inset in figure) later when the crack reaches 2–3 mm, the lines of moist air and dry N_2 coincide becoming indistinguishable. Fatigue-crack growth for long cracks of $\Delta a_f > 2$ mm is shown in Fig. 12.2 as da/dN versus ΔK. Although in steady state no clear change was observed in the two environments, however in tests of individual

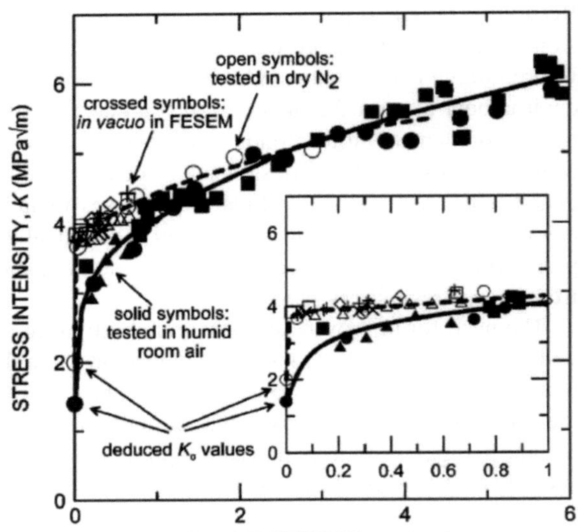

Fig. 12.1 R-curves for Al_2O_3 C(T) samples tested in dry N_2, vacuum in a field-emission scanning electron microscope (FESEM), and moist air. Inset magnifies behavior at initial crack extensions. Note the much more steeply rising, Kruzic et al. [1]. With kind permission of John Wiley and Sons. C(T) stands for compact tension for samples tested in moist air and dry N_2

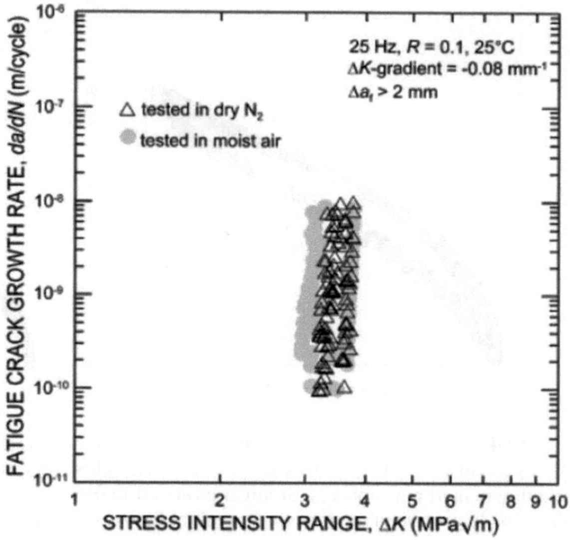

Fig. 12.2 Comparison of steady-state fatigue-crack growth rates for long cracks (>2 mm) in Al_2O_3 tested in moist air and dry N_2. Growth rates are essentially identical in both environments, Kruzic et al. [1]. With kind permission of John Wiley and Sons

specimens when the environment is suddenly changed during crack propagation at a constant ΔK a definite difference exists as seen ion Fig. 12.3. As seen in (a), indicating crack length versus number of cycles, where initially a long fatigue crack in dry N_2 was suddenly exposed to moist laboratoy air, an abrupt increase in crack growth rate has occurred (by about 3 orders) from 10^{-10} to 10^{-7} m/cycle. This sudden increase in fatigue-crack growth rate is attributed to a decrease of K_0 upon exposure to moist air. K_o is the intrinsic toughness of a material and it refers to the inherent

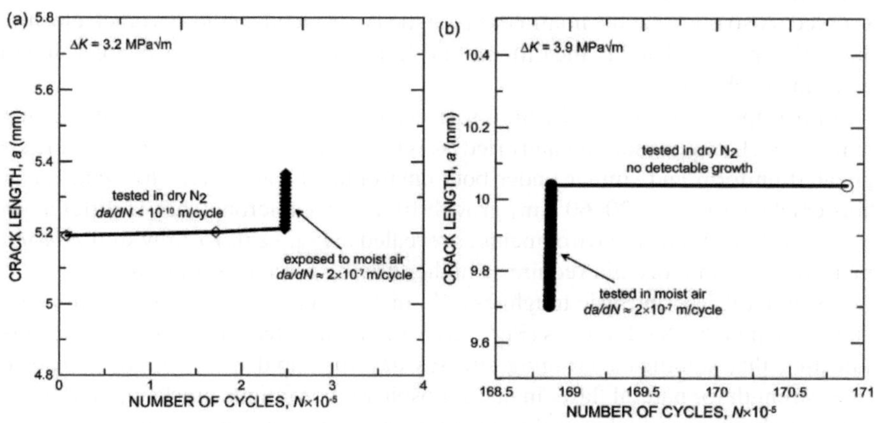

Fig. 12.3 Results show more than three orders of magnitude immediate change in fatigue-crack growth rates at a constant ΔK upon changing the environment from **a** dry N_2 to moist air and **b** moist air to dry N_2, Kruzic et al. [1]. With kind permission of John Wiley and Sons

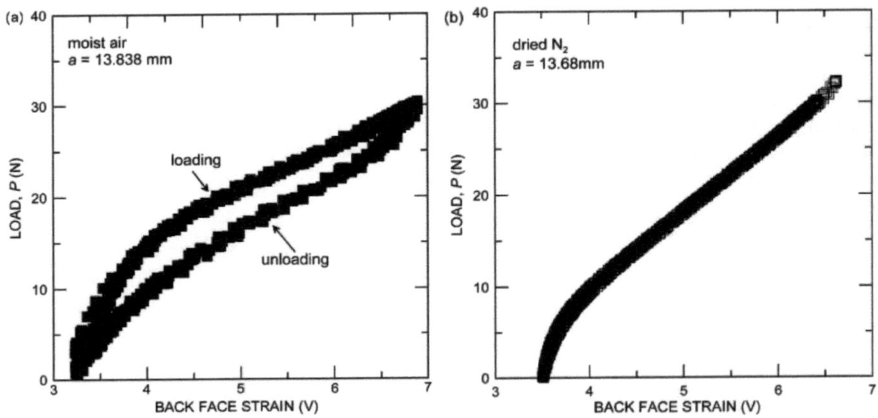

Fig. 12.4 Comparison of compliance curves for fatigue cracks grown in moist air and dry N_2 shows an absence of hysteresis for the dry case. This implies that there is much less frictional bridging in inert atmospheres. Strain units are arbitrary voltages. Crack lengths are computed from the unloading curve in the long linear region, Kruzic et al. [1]. With kind permission of John Wiley and Sons

resistance to fracture, while extrinsic toughening refers to mechanisms that act in the crack wake to lower the local driving force experienced at the crack tip, e.g., via grain bridging or phase transformations. In contrast, when a sample initially exposed to moist air is switched into dried N_2 a greater than three orders of magnitude decrease in transient crack growth rate has occurred from < 10^{-10} to 2 × 10^{-7}, and again it is attributed to K_o, but this time an increase occurred in the dry environment. Further change in fatigue behavior with change in the environment is the significant difference in the compliance curves shown in Fig. 12.4. Note that in dry air no hysteresis is observed. Also note that the nonlimearity at the lowest loads indicates crack closure. This difference is related to the different bridging mechanism in the samples due to environment change.

Similar appearance of the fracture surfaces of the cycled samples in alternating environment between moist air and dried N_2 is seen in Fig. 12.5. The fractured surface appeared uniform and similar under both environments. However, the behavior of short cracks ($\Delta a_f =$ ~ 20–60 μm) growth from razor micronotches is different in moist air than in dried N_2 environments as revealed in Fig. 12.6. To grow similar-sized cracks in N_2 as in moist air, requires a higher driving force (60–70% higher).

In summary, The intrinsic toughness. K_0 in alumina in the presence of moisture is lower than in dry N_2. R curves (Fig. 12.1) are initially steeper in dry environments indicating that significant bridging stresses develop rapidly, which could lead to larges strength for natural flaws in dry atmosphere. For fatigue cracks of 20–60 μm long growth rates are higher and fatigue thresholds lower in moist air (since crack bridging is diminished).

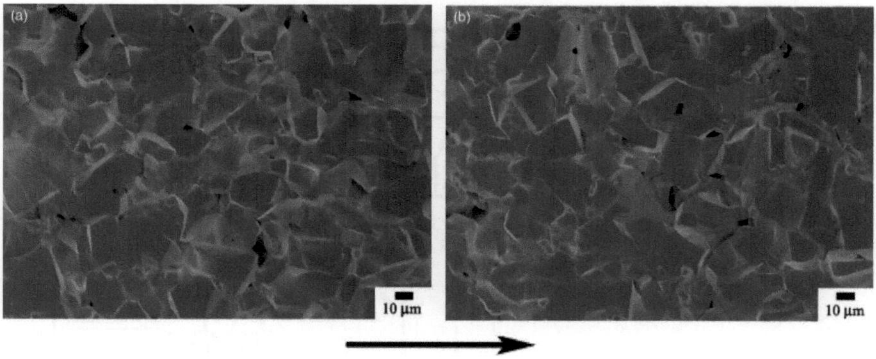

Fig. 12.5 Scanning electron micrographs of fatigue fracture surfaces for samples tested in **a** moist air and **b** dry N_2. Arrow indicates the general direction of crack growth, Kruzic et al. [1]. With kind permission of John Wiley and Sons

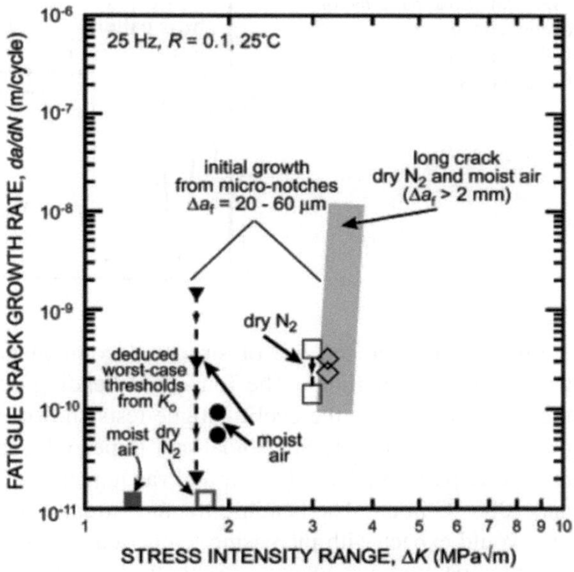

Fig. 12.6 Comparison of fatigue-crack growth rates from short (20–60 μm) cracks emanating from razor micronotches in moist air and dry N_2; each symbol represents an individual specimen. Although velocities for steady-state cracks (>2 mm) exhibit no apparent environmental effect, a significant effect exists for short cracks. Predicted worst-case fatigue thresholds are based on deduced K_0 values, Kruzic et al. [1]. With kind permission of John Wiley and Sons

12.2.2 In Vacuum and Water Vapor

The effect on the strength properties of water vapor in air and vacuum in alumina at high temperatures is the subject of this section. Fatigue fracture tests were carried out by the bending conditions. Also, static fatigue tests were performed. As known, strength of ceramics is influenced by many factors to mention some are the size of crystals, segregation of impurities if present, pores and clearly by environmental conditions. Environmental effects on the fatigue behavior must be evaluated because

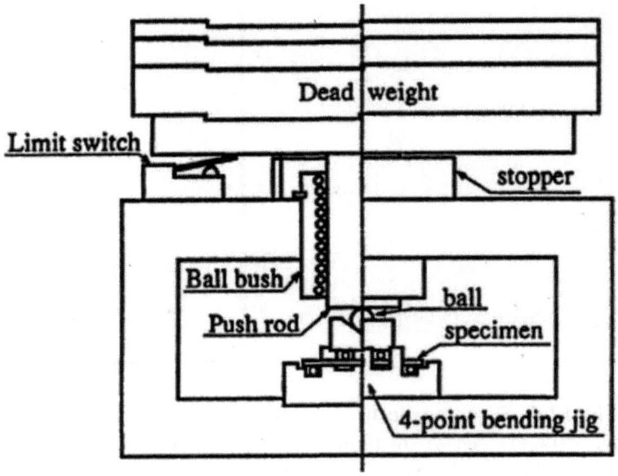

Fig. 12.7 Static four-point bending device, Nakayama and Ohira [2]. Open access

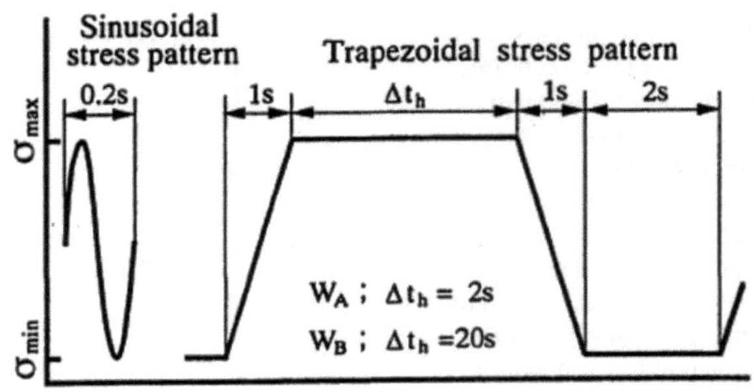

Fig. 12.8 Load patterns for cyclic fatigue tests, Nakayama and Ohira [2]. Open access

it might determine the life of structural components and its strength reliability under service conditions. The four-point bending device is seen in Fig. 12.7. In the load patterns for the cyclic fatigue tests shown in Fig. 12.8 the stress ratio R ($= \sigma_{min}/\sigma_{max}$) was 0.1. Sinusoidal and trapezoidal stress patterns were used. The tests were performed at RT in air and vacuum and at high temperatures of 600, 800 and 1000 °C in air. The bending fracture strength seen in Fig. 12.9 decreases—as one would expect-with increasing temperature. Also, it is seen that the strength at room temperature in vacuum is higher than in air by about 70 MPa. The distribution of the fracture strength at RT in air and in vacuum shown in Fig. 12.10 is plotted on normal probability paper, rather than on logarithmic normal distribution and/or Weibull distribution, because the interest was to determine the mean strength and scatter conveniently. It can be seen in this figure that the bending strength in vacuum is higher than in air. Tables 12.1 and 12.2 list the physical and mechanical properties and the statistical evaluation of the four-point bending fracture strength at room and high temperatures, respectively. S–N relationships under cyclic load at room temperature in air and in vacuum is shown in Fig. 12.11. As seen no fatigue degradation has occurred in tests performed in vacuum, unlike in air, where σ_{max} becomes lower

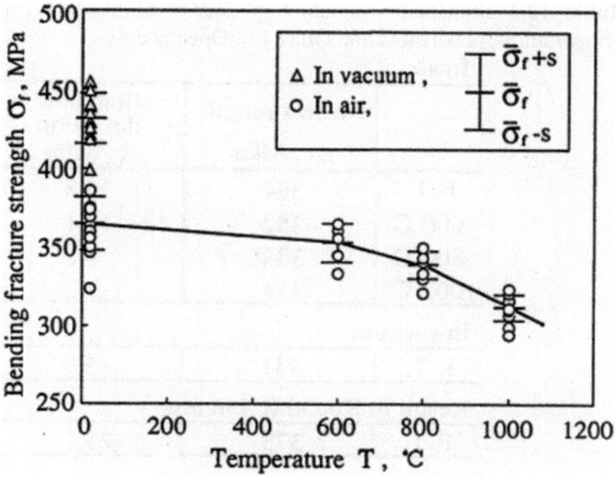

Fig. 12.9 Temperature dependence of bending fracture strength, Nakayama and Ohira [2]. Open access

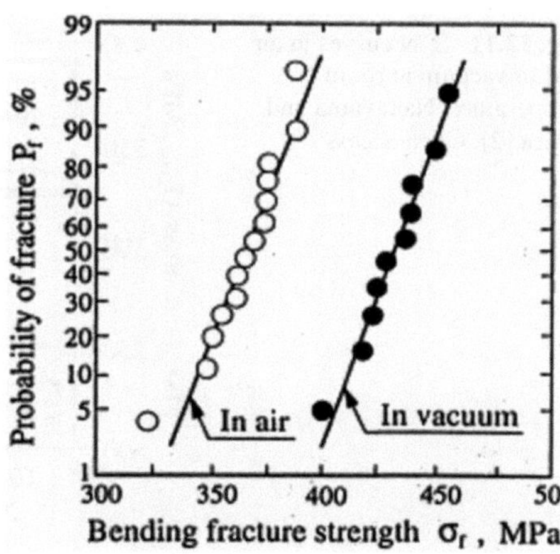

Fig. 12.10 Distribution of bending fracture strength of alumina in air and in vacuum plotted on normal probability paper, Nakayama and Ohira [2]. Open access

Table 12.1 Physical and mechanical properties of the material, Nakayama and Ohira [2]. Open access

Density (Mg/m^3) *1	3.92
Fracture toughness ($MPa \cdot m^{1/2}$) *2	4.4
Hardness (HV 10)	1670

*1 Archimedes method
*2 SEPB method (JIS R1607)

Table 12.2 Statistical evaluation of four-point bending fracture strength at room and high temperatures, Nakayama and Ohira [2]. Open access

	Mean strength σ_f, MPa	Standard deviation s, MPa	Coefficient of variation c.v.
In air			
R.T.	364	16.8	0.046
600°C	352	13.1	0.037
800°C	338	9.2	0.027
1000°C	311	9.2	0.029
In vacuum			
R.T.	431	15.9	0.037
Result in Round Robin test			
R.T.	373	23.1	0.062

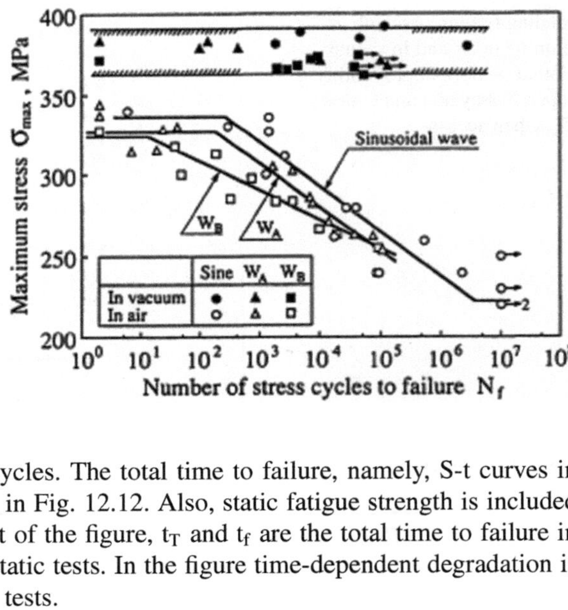

Fig. 12.11 S–N curves in air and in vacuum at room temperature, Nakayama and Ohira [2]. Open access

with increasing number of cycles. The total time to failure, namely, S-t curves in vacuum and air is illustrated in Fig. 12.12. Also, static fatigue strength is included in the figure. In the subscript of the figure, t_T and t_f are the total time to failure in cyclic tests and t_f is that in static tests. In the figure time-dependent degradation is observed in the static fatigue tests.

Experimental results at high temperatures of 600, 800 and °C are shown in Figs. 12.13, 12.14 and 12.15. In each figure, results of cyclic fatigue tests at room temperature in air and in vacuum are also included. Note that in Fig. 12.8 it is indicated that for the trapezoidal pattern two cases, namely, Wa (Δt_h 2 s) and W_B (20 s) are indicated which is expressed in the two S–N curves of 12.14 and 12.15.

In summary, fatigue four-point bending tests and fracture tests at high temperature in vacuum and air are presented in this section. The fracture strength at room temperature in vacuum is higher than in air at the same temperature. This is a consequence of the presence of moisture in air, which enlarges the virtual defect size. Cycle

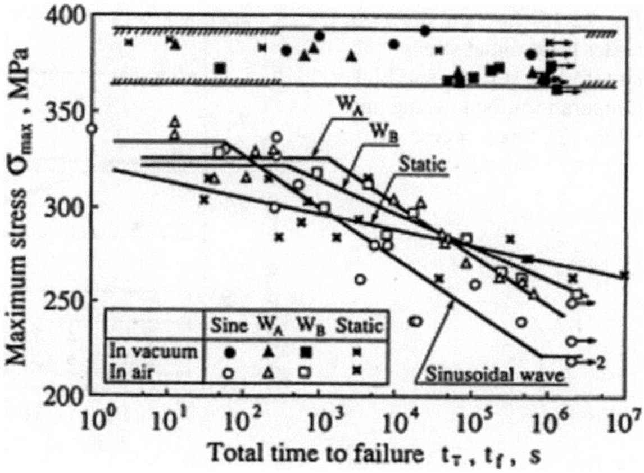

Fig. 12.12 S-t curves in air and in vacuum at room temperature, Nakayama and Ohira [2]. Open access

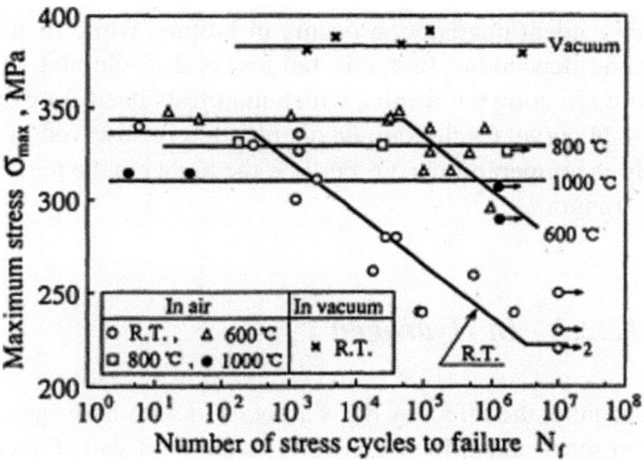

Fig. 12.13 S–N curves under sinusoidal stress pattern at room and high temperatures, Nakayama and Ohira [2]. Open access

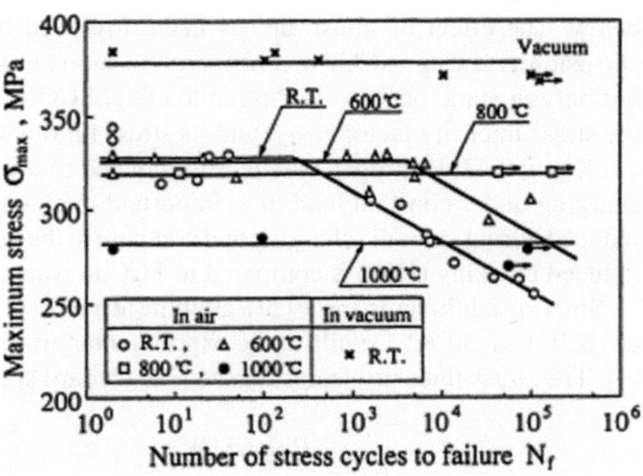

Fig. 12.14 S–N curves under trapezoidal stress pattern W_A at room and high temperatures, Nakayama and Ohira [2]. Open access

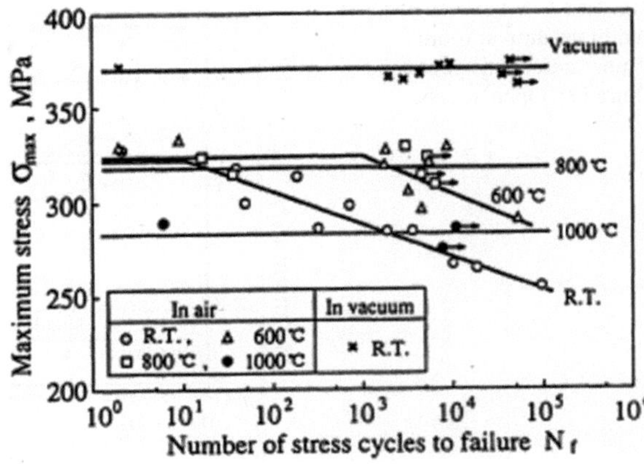

Fig. 12.15 S–N curves under trapezoidal stress pattern W_B at room and high temperatures, Nakayama and Ohira [2]. Open access

dependent degradation occurs in fatigue, while in static fatigue the degradation is time dependent. In cyclic fatigue, both cycle and superimposed time dependence are affecting the results, which manifests itself for example in a steeper slope of the S–N curve. Cyclic fatigue degradation is observed in the fatigue tests carried out at high temperature in air. Further, the higher is the temperature the lower is the fatigue strength.

12.2.3 In Hydrogen

Usually, the effect of H_2 is associated with hydrogen embrittlement in steel, but its presence and influence in other materials is also of interest. Thus, alumina the subject currently considered is also affected in someway by hydrogen which is discussed below. The effect of moist air has been discussed in Sect. 12.1 and 12.2. Stress corrosion cracking (SCC) was observed in moist atmosphere at room temperature not only in static but also in fatigue loading. SCC occurred also in water affecting the stress intensity factor K_{ISCC} and its stress ratio with the stress intensity factor in air, K_{IC} is 0.43. Hydrogen can be aslo induced to alumina ceramic during cathodic charging under constant load. It is important to characterize the effect of hydrogen induced during cathodic charging and compare to the one in water. Thus, the hydrogen induced cracking (HIC) is compared to SCC in water.

Sinering additives to the 99.5% alumina at a level of 4.5% were added for sintering at 1650 °C + 30 °C. Details of the experimental procedure can be found in Shi et al. [3]. The stress intensity factor for the nonstandard specimen is calculated by

$$K_I = \frac{PS\varphi(a/W)}{4BW^{3/2}} \sec\left(\frac{\pi a}{2W}\right)\sqrt{\tan\left(\frac{\pi a}{2W}\right)} \tag{12.1}$$

Table 12.3 Q φ(a/W) of the informal three-point bend specimen with S/W = 3.0. C, Shi et al. [3]. With kind permission of Springer

a/W	0.20	0.25	0.30	0.35	0.40	0.45
$\varphi(a/W)$	7.65	7.53	7.46	7.40	7.37	7.36

where P is the load, a = 1.8 mm is the crack length, S is the span (21 mm), B is the thickness (1.74 mm) and W is the width (7 mm) of the specimen. The function φ(a/W) for S/W = 3.0 is shown in Table 12.3. Substituting the fracture load into (12.1) for specimen loaded in air, the fracture toughness K_{IC} is obtained. K_I was measured in solution with Ni plated alumina as the cathode and Pt wire served as anode during dynamic charging. Current densities were i = 2 mA/cm^2 and 10 mA/cm^2 (as seen in Fig. 12.16 also). The time to fracture for HIC under different K was recorded. The streshold stress intensity factors for SCC and HIC were obtained according

$$K_{ISCC}\,or\,K_{IH} = \frac{1}{2}\left(K_{Iy} + K_{In}\right) \tag{12.2}$$

$$K_{Iy} - K_{In} \leq 0.1\left(K_{Iy} - K_{In}\right) \tag{12.3}$$

where K_{Iy} is the minimum K_I under which SCC or HIC occurred and K_{In} is the maximum K_I under which SCC of HIC would not occur within the fixed time (e.g. 100 h).

The average fracture toughness of the alumina is K_{IC} = 4.8 MPa m$^{1/2}$. The normalized stress intensity factor is plotted in Fig. 12.16.

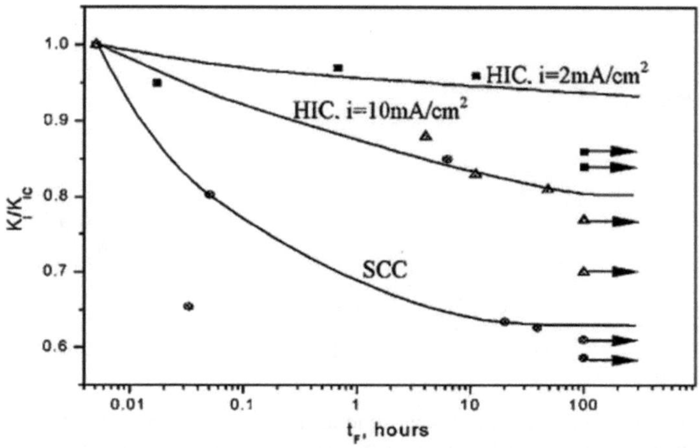

Fig. 12.16 Normalized stress intensity K_I/K_{IC} versus time to fracture t_F during SCC and HIC, Shi et al. [3]. With kind permission of Springer

In summary, hydrogen could enter alumina during cathodic charging and HIC occurred during charging under constanr load. The normalized threshold stress intensity of SCC in water is $K_{ISCC}/K_{IC} = 0.64$ and is smaller than that of HIC during dynamically charging with $i \leq 10$ mA/cm^2 which is K_{IH}/K_{IC} 0.79–0.92 (see Fig. 12.16).

12.3 The Effect of Environment in Yttria Stabilized Zirconia (Y-TZP)

12.3.1 In Water

Fatigue testing was performed on several zirconia materials in a water tank illustrated in Fig. 12.17. This material has become an attractive high-toughness core material for fixed dental restorations. A representive microstructure of the Y-TZP specimens (four) is illustrated in Fig. 12.18. S–N fatigue data of the as rteceived soecimens is shown in Fig. 12.19. The fatigue limit was set at 10^6 cycles for the three-point bending tests at 10 Hz in water. Least square linear regression fit was used to evaluate the trend of the lines in Fig. 12.18. Failed specimens were analyzed fractographically for determination of the failure origin using a stereo microscope and SEM. Figure 12.20 displays the two fractured halves of a zirconia specimen that failed under fatigue loading. The smooth surfaces are the radiating hackle lines and clearly visible on the tensile aspects of the samples. The mirror-like smooth surface surrounds the origin of failure. Processing flaws associated with fabrication and sintering and griding defects are seen by SEM im Fig. 12.21. Note that sandblasted specimen (A) is also included. Sand blasted specimens improve fatigue. The flaws act as stress concentrations (stress

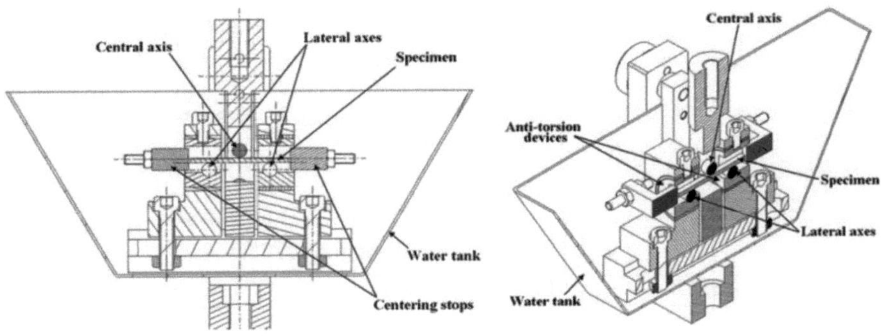

Fig. 12.17 Technical drawing of the 3 point-bending assembly used for fatigue testing. The specimen in the center is supported by lateral axes (rollers) incorporated in anti-torsion devices allowing rotation, thus compensating for minor unevenesses in the surfaces, Scherrer et al. [4]. With kind permission of Elsevier

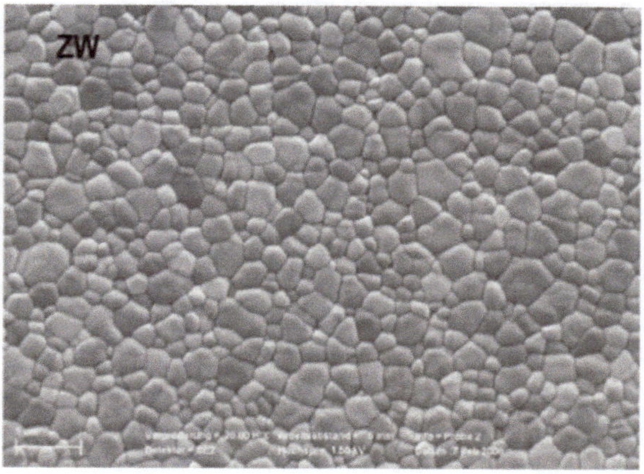

Fig. 12.18 Microstructure of Y-TZP (one of four similar structures) ceramics, × 20,000. The average grain size according to the linear interceptive count method (383 ± 47 nm), Scherrer et al. [4]. With kind permission of Elsevier

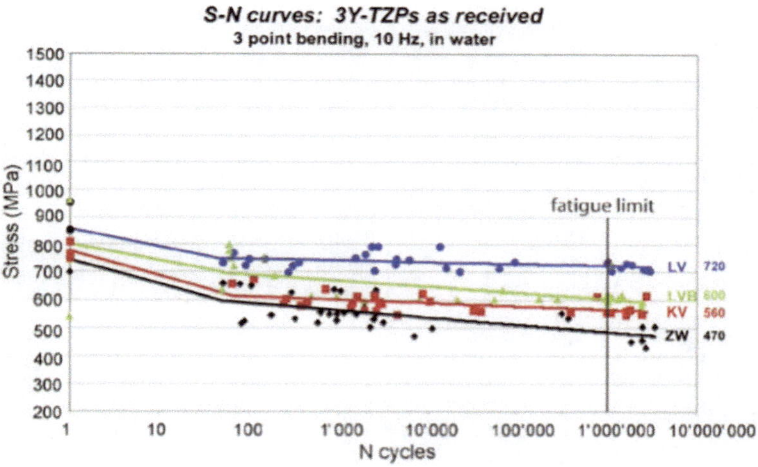

Fig. 12.19 S–N fatigue data for as received specimens. Four specimens and their designation on the far-right of the figure, Scherrer et al. [4]. With kind permission of Elsevier

raisers) during crack nucleation. The stresses and the number of cycles leading to failure at the respective flaw sizes are listed in the legend of Fig. 12.21. Despite the concern of using zirconia ceramic in the oral environment due to exposure to water (saliva is 98% water) and mechanical stress over prolonged periods, still Y-TZP is currently the strongest cerams in clinical use.

The types of flaws observed in all Y-TZP tested are illustrated in Fig. 12.22. As seen in the figure, below 2000 cycles the blank pressing (processing) flaws are dominant, but beyond it grinding flaws are the signicant. Grinding flaws resulting from machining act as sress raisers. The flaws are shallow (between 7 and 20 μm deep) and elongated (>100 μm) along the surface. Stable crack extension prior to

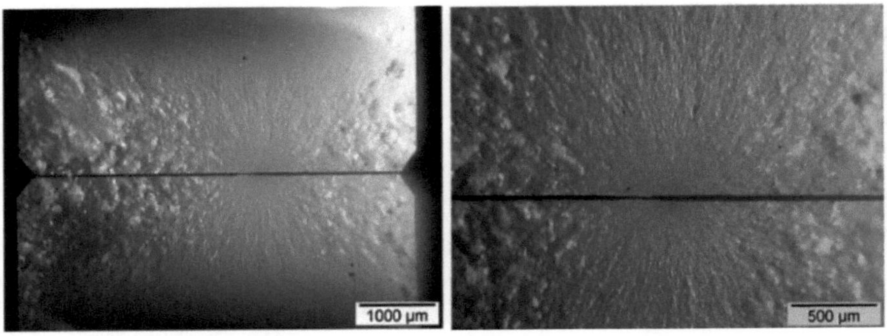

Fig. 12.20 Stereomicroscope views (32 × and 57 × magnification) of two broken halves of a fatigue-failed zirconia specimen. The mirrors (smooth surfaces) and the radiating hackle lines surround the failure origin located on the tensile surface of this specimen, Scherrer et al. [4]. With kind permission of Elsevier

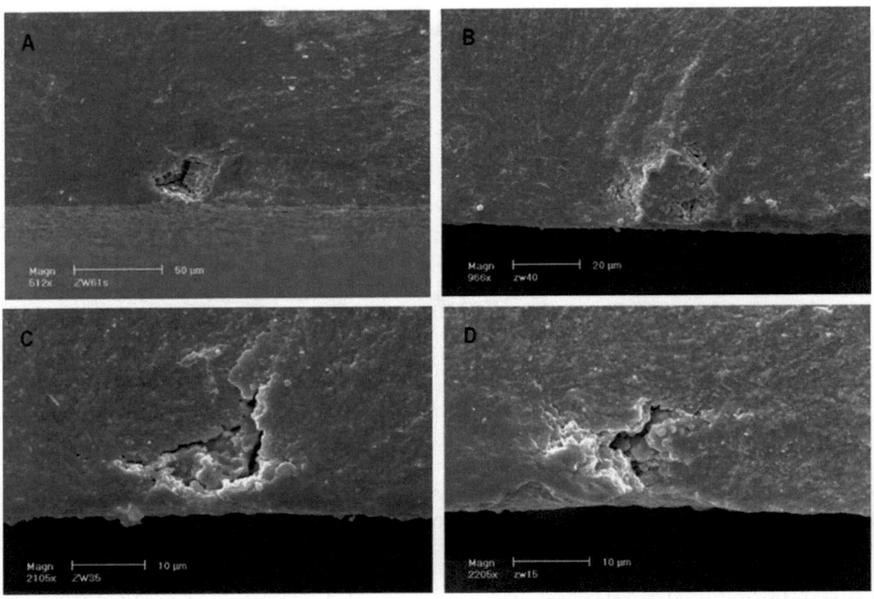

Fig. 12.21 Processing flaws located at-or near the tensile surface, which acted as stress concentrators during fatigue testing. These flaws were incorporated into the ceramic during cold pressing of the blanks. The number of cycles and stress level for each image were as follows: **a** "Sandblasted", 501,550 cycles at 581 MPa; **b** "as received", 900 cycles at 635 MPa; **c** "as received", 51 cycles at 658 MPa; **d** "as received", 1577 cycles at 548 MPa, Scherrer et al. [4]. With kind permission of Elsevier

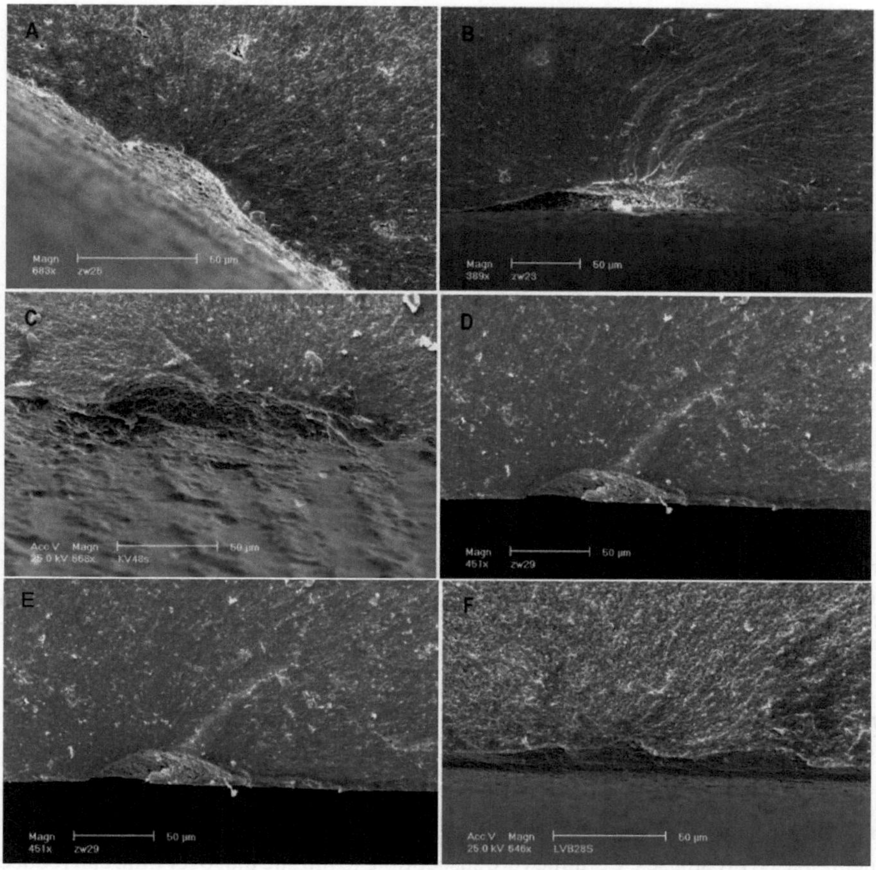

Fig. 12.22 Examples of critical grinding flaws located at the tensile surface and acting as stress concentrators during fatigue testing. **a** Zeno ZW (Wieland), failure after 2473 cycles at 532 MPa; **b** Zeno ZW (Wieland), failure after 89 cycles at 524 MPa; **c** Everest sandblasted KV (KaVo), failure after 233 cycles at 723 MPa; **d** Zeno ZW (Wieland), failure after 591 cycles at519MPa; **e** Lava colored LVB (3 M Espe), failure after 21,458 cycles at 602 MPa; **f** Lava colored sandblasted LVB (3 M Espe), failure after 2576 cycles at 850 MPa, Scherrer et al. [4]. With kind permission of Elsevier

fracture is seen in Fig. 10.22 for all types of zirconia tested (their designations are indicated). Caution must be excersised when determining toughness based on failure stress and crack size in a wet environment.

In summary, cyclic fatigue test (sinusoidal loading/unloading) was performed in 3-point bending in a water tank. Stress levels were lowered from the initial static value until surviving 10^6 cycles and from the trend lines the fatigue limits were determined for several Y-TZP zirconias (Fig. 12.23).

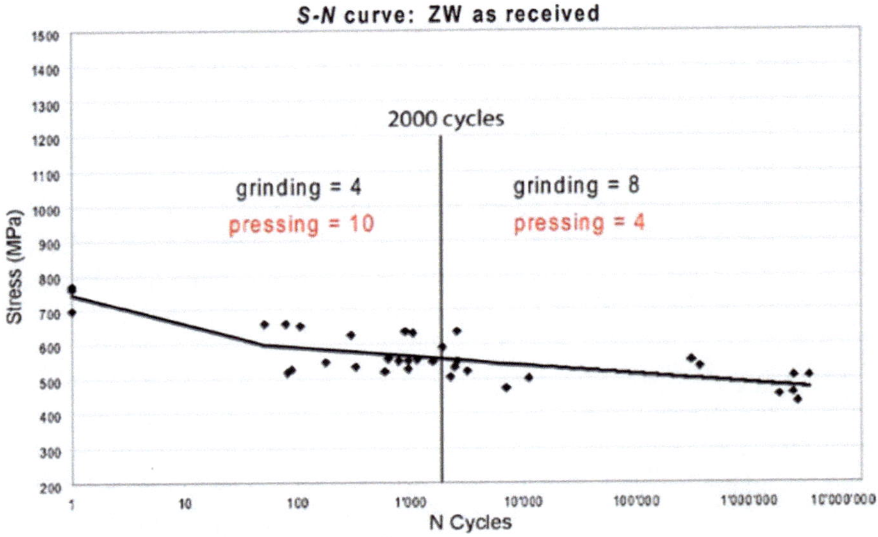

Fig. 12.23 Grinding and blank pressing failure flaws on the fatigue curve for ZW "as received" specimens. For specimens failing below 2000 cycles, the pressing flaws outnumbered the grinding flaws (4 versus 10). After 2000 cycles, the grinding flaws prevailed (8 versus 4), Scherrer et al. [4]. With kind permission of Elsevier

12.3.2 In Air

One of the exctensive uses of 3Y-TZP is for dental restorations because their improved biocompatibility and esthetical appearance compared with other material in particular metal- based prostheses. Ceramics are susceptible to fatigue that over time can reduce their strength and load bearing capacity. The reduction of mechanical strength due to fatigue is caused by the propagation of natural cracks initially present in the component's microstructure. Ceramics such as yttria-stabilized zirconia (3Y-TZP) has widened the application of ceramics in dentistry, enabling the use of all-ceramic restorations also in posterior regions where high-strength structures are required. Crack propagation under subcritical conditions ($\sigma < \sigma_c$; $K_I < K_{IC}$) results from the stress-assisted reaction of water molecules with the metal-oxide bonds at the crack tip. Clearly, crack propagation can be further promoted when cyclic stresses are applied. Cyclic fatigue behavior of 3Y-TZP is considered in this section in air compared to that in water to evaluate lifetime. Schematic illustration of the bending testing is shown in Fig. 12.24. The lifetime experiments were performed by applying a maximum stress level, σ_{max}. Fatigue samples fail only when their natural (subcritical) cracks grow to the critical size given by σ_{max} applied, which is lower (by ~50%) than the inert strength of the specimen illustrated in Fig. 12.24. The tensile stress, σ developed on the sample surface upon the application of the force F can be calculated based on the bending equation.

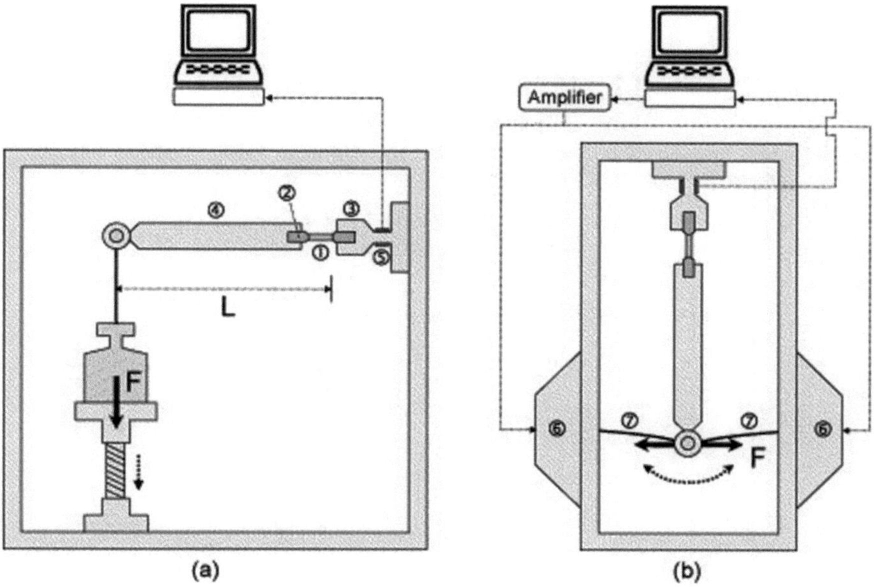

Fig. 12.24 Schematic drawing of the mechanical testing device used in this work for measuring **a** the initial mechanical strength (simple bending mode) and **b** the lifetime of the zirconia specimens (oscillatory bending mode). The specimen (1), specimen adapter (2), specimen holder (3), cantilever (4), strain gage system (5), coil–magnet system from loudspeakers (6) and flexible wires (7) are indicated in the drawing, Studart et al. [5]. With kind permission of Elsevier

$$\sigma = \frac{Mz}{I} = \frac{FLz}{I} \tag{12.4}$$

where z is the distance between the surface and the neutral axis of the specimen (in this case h/2) and I is the moment of inertia (in this case $wh^3/12$). The height of the specimen is h and its widh is w. The moment is given by M = FL where F is the force applied. L is illustrated in Fig. 12.23. Weibull distribution describes the lifetime of the TZP specimen and can be expressed by

$$\ln \ln\left(\frac{1}{1-F}\right) = m^* \ln N_f - m^* \ln N_{f,0} \tag{12.5}$$

where N_f is the number of cycles until failure, $N_{f,0}$ the characteristic number of cycles at 63% failure probability and m^* is the lifetime Weibull modulus. The failure probability is calculated assuming the relation: $F_i = (i - 0.5/p)$, where F_i is the failure probability of the ith specimen (in a strength-increasing rank) and p is the total number of specimens tested. The Weibull parameters $\sigma_{c,0}$ and m indicate the magnitude and scattering, respectively, of the measured strength data. In Fig. 12.25 the strength results in a double logarithmic Weibull plot are presented, where the slope and the

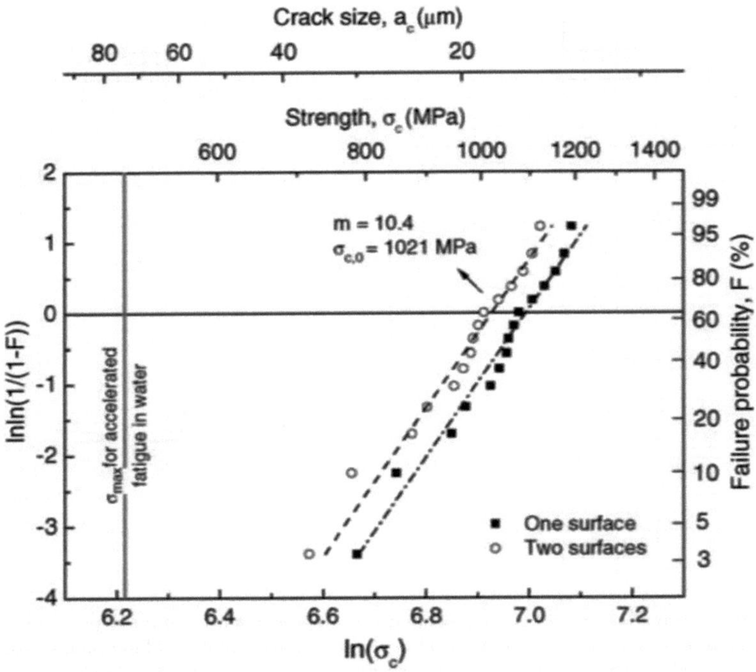

Fig. 12.25 Weibull distribution for the inert strength of the TZP specimens taking into account only one surface (measured data) or two surfaces of the sample (calculated data). The corresponding critical crack sizes for different strength levels, as well as the maximum stress chosen for the fatigue experiments in water are also indicated in the graph, Studart et al. [5]. With kind permission of Elsevier

x-axis intercept of the fitting curve corresponds to the Weibull modulus, m, and the characteristic inert strength, $\sigma_{c,0}$, respectively. Lifetime Weibull plots obtained for specimens are compared in (a) air and (b) water are illustrated in Fig. 12.26. Here also as in Fig. 12.25, the slope and the *x*-axis intercept of the fitting curves shown in Fig. 12.26 correspond to the characteristic number of cycles $N_{f,0}$ and the lifetime Weibull modulus m^*. Preliminary fatigue tests indicated that the lifetime of the TZP samples in water is significantly lower than that obtained in air. The lifetime tests in air were therefore performed at a higher maximum stress ($\sigma_{max} = 550$ MPa) than that applied in water ($\sigma_{max} = 500$ MPa), so that all the samples could be evaluated within a reasonable amount of time.

The empirical law of Paris is used to describe the subcritical growth of zirconia, which correlates the propagating crack, ν with the stress intensity factor at maximum and minimum stress intensity factors $K_{I,\ max}$ and $K_{I,\ min.}$

$$\upsilon = A\Delta K_I^n = A(1-R)^n K_{I,\max}^n = A^* K_{I,\max}^n \quad (12.6)$$

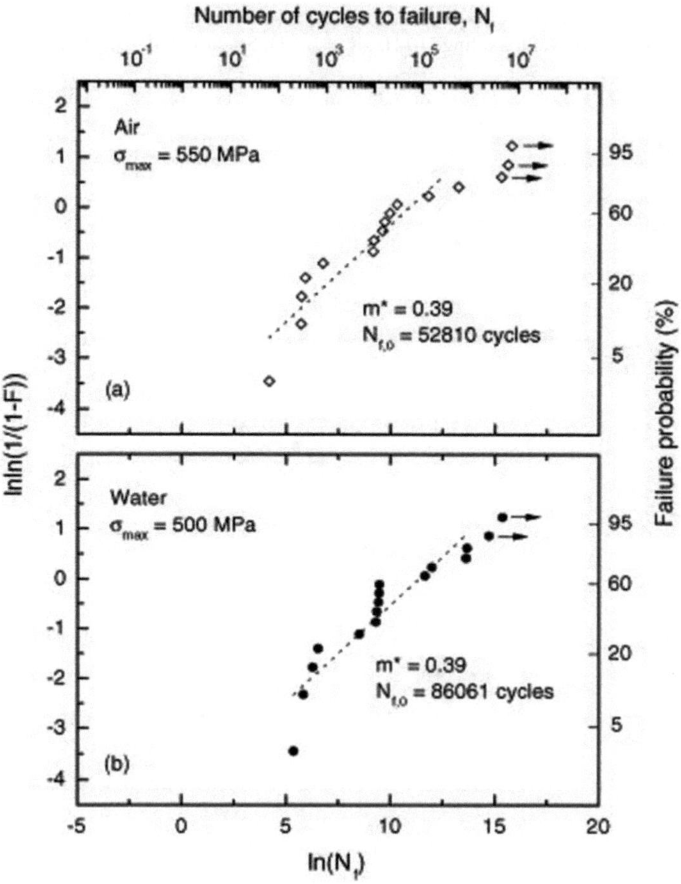

Fig. 12.26 Weibull distribution for the lifetime of TZP specimens measured in **a** air ($\sigma_{\max}$ = 550 MPa) and **b** water ($\sigma_{\max} = 500$ MPa), Studart et al. [5]. With kind permission of Elsevier

ΔK_I is the amplitude of the stress intensity factor applied ($K_{I,\max} - K_{I,\min}$), R is the stress ratio $K_{I,\min}/K_{I,\max}$, and A and n are the subcritical crack growth parameters for cyclic fatigue. The critical growth parameters A, B and n of the cyclic loading are listed in Table 12.4. In the case of cyclic loading the crack velocity, ν is given by the increament of crack size, da per number of cycles applied, dN as da/dN. $K_{I,\max}$ and ΔK are directly related to the maximum stress $\sigma_{\max}$ and to the stress amplitude Δσ (Griffith law) as Y is a geometrical constant that depends on crack location and shape (for surface cracks = 1.3). The faster crack growth observed in the presence of water can be attributed solely to the higher concentration of corrosive water molecules surrounding the propagating crack. The higher the concentration in the environment is the higher crack growth rate is. The subcritical crack growth behavior of TZP in air and in water is illustrated in Fig. 12.27 as velocity versus K_I, max.

Table 12.4 Subcritical crack growth parameters A, B and n obtained in this work for the TZP specimens submitted to cyclic loading (10 Hz, R = − 1), Studart et al. [5]. With kind permission of Elsevier

3Y-TZP	A	n	A	B
Air	5.73×10^{-34}	29.00	3.07×10^{-25}	4.83×10^{11}
Water	6.07×10^{-33}	28.95	3.15×10^{-24}	4.97×10^{10}

The values of A, A* and B were obtained using stress intensity factor and velocity data in MPa $m^{1/2}$ and m/cycle, respectively.

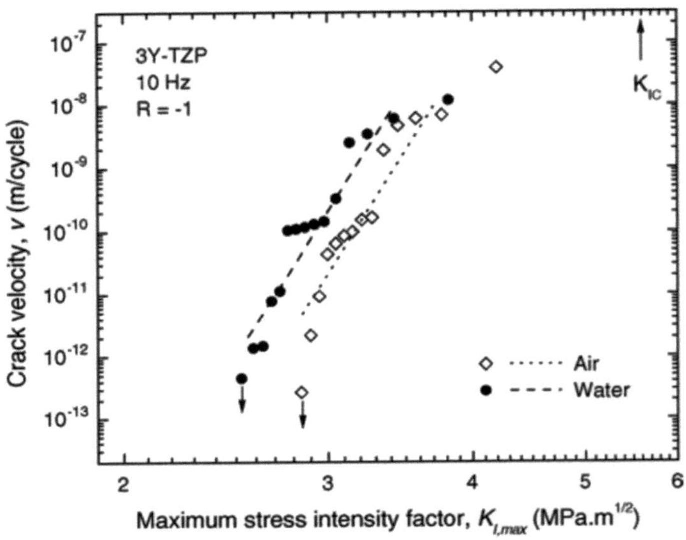

Fig. 12.27 Crack velocity as a function of the maximum stress intensity factor applied to TZP specimens tested under cyclic loading (10 Hz, R = − 1) in air and in water, Studart et al. [5]. With kind permission of Elsevier

$$K_{I,\max} = \sigma_{\max} Y \sqrt{a} \tag{12.7}$$

$$\Delta K_I = \Delta \sigma Y \sqrt{a} \tag{12.8}$$

In summary, 3Y-TZP growth parameters for dental application was presented. Subcritical crack growth parameters under cyclic loading in air and water were evaluated. It was observed -and as expected- cyclic fatigue crack growth in water is faster than in air and adversely affects lifetime of 3Y-TZP.

12.4 Effect of Environment in SiC

12.4.1 In Air

As indicated along this book, SiC is used as a composite rather than as a pure constituent. Of the composites, SiC/SiC is the more relevant system to be considered as also done in this section.

High fracture toughness and thermal shock resistance are among the important features of SiC/SiC ceramic matix composite (CMC). Fatigue resistance of continuous fiber (Nicalon) reinforced SiC matrix is therefore of interest for various applications at elevated temperatures. Environmental factors may cause a degradation, a topic to be considered when the expected improvement of the system by the fibers at high temperaturtes is compared to the standard SiC/SiC.

The cyclic fatigue life of three SiC/SiC systems are compared and shown in Fig. 12.28 as the maximum stress vs, cycles to failure. Hi-NicalonTM/SiC is like that of Enhanced SiC/SiC, but much longer than that of the Standard SiC/SiC at 1300 °C in air. The evolution of the hysteresis loop during fatigue at 1300 °C in air is shown in Fig. 12.29. As seen the slope decreases and the width of the loops increases with cycles. The former indicates the decrease of the modulus and the latter means the decrease of the interfacial sliding resistance. Note that the hysteresis loops move to the right along the strain axis, which is known as ratchetting due to time dependent deformation (e.g. creep). The modulus normalized by the value from the linear part during the first loading versus cycles is shown in Fig. 12.30. Fatigue cracks are always found at the large pores between fiber bundles as seen in Fig. 12.31a, b. The fiber pullout under fatigue is in Fig. 12.32. Ther longer fiber pullout of fatigue fracture (Fig. 12.32) demonstrates that the cyclic unloading promotes the debonding of the interfaces by decreasing the interfacial sliding resistance, more so than in creep for example under te same conditions.

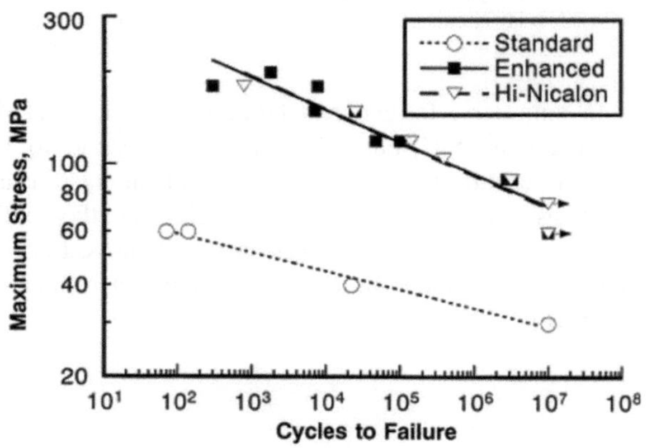

Fig. 12.28 The maximum stress versus cycles to failure for fatigue in Hi-NicalonTM/SiC, Enhanced SiC/SiC, and Standard SiC/SiC in air at 1300 °C, Mizuno et al. [6]. With kind permission of Elsevier

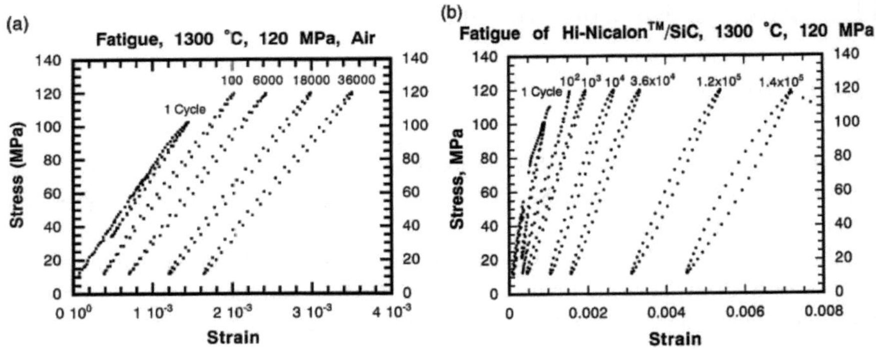

Fig. 12.29 Evolution of the hysteresis loops during fatigue at 1300 °C under the maximum stress of 120 MPa in air: **a** Enhanced SiC/SiC; **b** Hi-NicalonTM/SiC, Mizuno et al. [6]. With kind permission of Elsevier

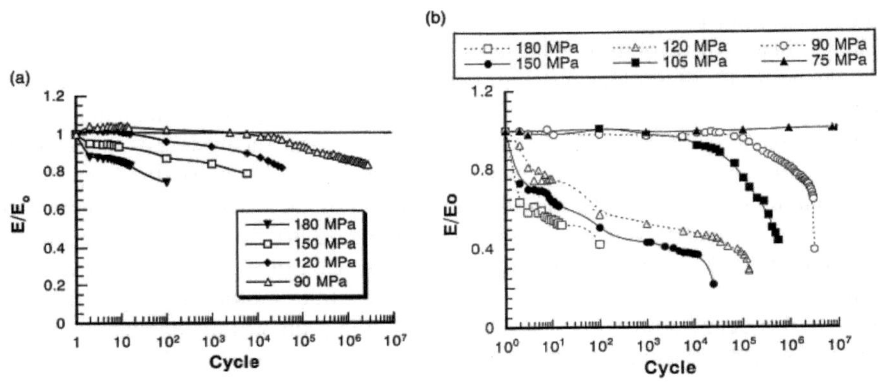

Fig. 12.30 The elastic modulus normalized by the value of the modulus under the first loading (E/Eo) versus cycles for fatigue in air at 1300 °C under the diferent maximum stresses. **a** Enhanced SiC/SiC; **b** Hi-Nicalon™/SiC, Mizuno et al. [6]. With kind permission of Elsevier

In summary, fatigue resistance of Hi-NicalonTM/SiC was similar to that of enhanced SiC/SiC, but much better than Standard SiC/SiC. The slope decreases with increasing the number of cycles of the hysteresis loop evolved during fatigue indicates a decrease in Young's modulus, and the width increase of the loops means that the interfacial sliding resistance decreases. Further, the effect of air on fatigue behavior is presented in the next section, where its influence is compared to that of steam (Fig. 12.31).

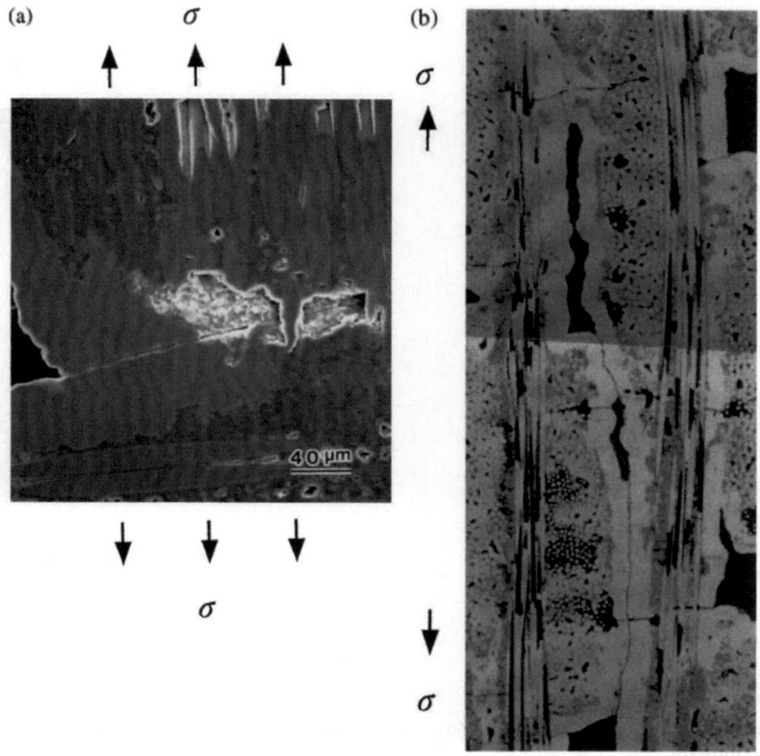

Fig. 12.31 **a** A crack initiated at a large pore in the specimen of the Enhanced SiC/SiC fatigued in air at 1300 °C and 75 MPa for 10^7 cycles; **b** cracks in the specimen of Hi-Nicalon™/SiC fatigued in air at 1300 °C and 150 MPa for 2.5×10^4 cycles, Mizuno et al. [6]. With kind permission of Elsevier

12.4.2 In Steam and Laboratory Air

The fatigue behavior of SiC/SiC ceramic matrix composite (CMC) at 1200 °C is explored in laboratory air and in steam rnvironment. The SiC matrix is reinforced with laminated woven Hi-Nicalon fibers. Tension-tension fatigue tests were performed at frequencies of 0.1, 1.0 and 10 Hz for fatigue stresses in the range 80–120 MPa in air and in the range 60–110 MPa in steam. The fatigue performance significantly degraded in the presence of steam. In air and steam test environments the fatigue limit and fatigue lifetime decreased with increasing frequency. The material retained 100% of its tensile strength, yet modulus loss up to 22% was observed. The maximum stress vs, time plots to failure are shown in Fig. 12.33. The fatigue performance increases with decreasing loading frequency. It can be seen in the figure that at 1200 °C in air for a given maximum stress, time to failure decreases with increasing frequency. The presence of steam causes noticeable degradation in fatigue performance of the

Fig. 12.32 Fiber pullout in Hi-NicalonTM/SiC at the maximum stress of 120 MPa and 1300C in air: **a** fatigue, Mizuno et al. [6]. With kind permission of Elsevier

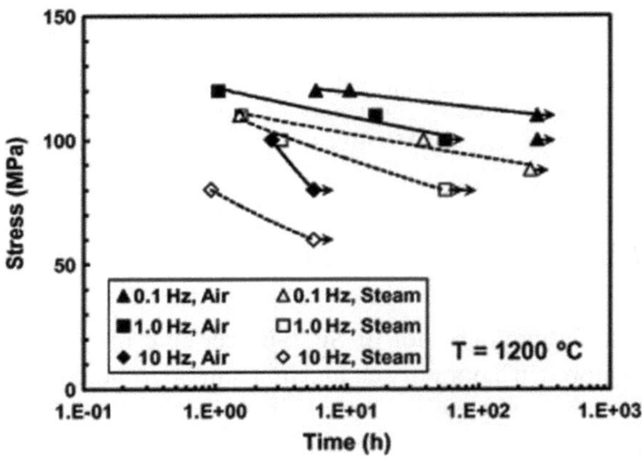

Fig. 12.33 Stress versus time to failure for Hi-Nicalon™/SiC ceramic composite at 1200 °C in air and in steam. Arrow indicates that failure of specimen did not occur when the test was terminated, Ruggles-Wrenn et al. (2011). With kind permission of Elsevier

composite and the fatigue run-out decrease was frequency dependent. Time decreases to failure with increasing frequency and for a given time the stress to failure also decreases. Cyclic strains as functions of fatigue cycles for fatigue tests conducted at 1200 °C in air and in steam are presented in Fig. 12.34 a, b, respectively. The damage development during fatigue cycling manifests itsel also in the reduction in modulus (hysteresis modulus determined from the maximum and minimum stress–strain data points during a load cycle) at 1200 °C in air and in steam, respectively. This drop

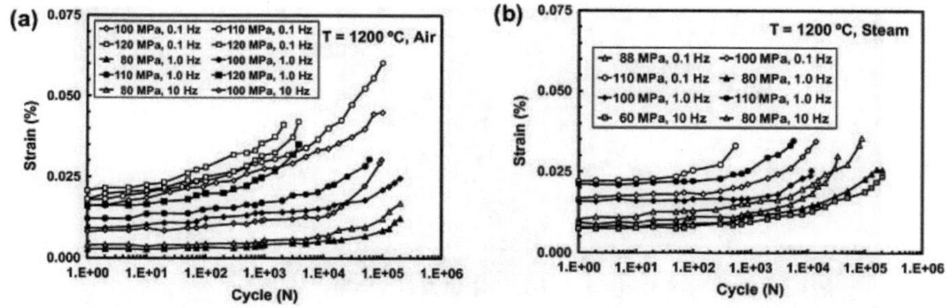

Fig. 12.34 Maximum strain versus fatigue cycles for Hi-Nicalon™/SiC ceramic composite at 1200 °C: **a** in air and **b** in steam, Ruggles-Wrenn et al. (2011). With kind permission of Elsevier

in modulus is thought to be due to a low level microcracking. Generally, loss in the modulus increases with increasing fatigue stress. The continuous drop in modulus observed at higher fatigue stress levels suggests progressive damage accumulation with continued cycling. Change in the normalized modulus (i.e., modulus normalized by the modulu obtained in the first cycle) with fatigue cycles at 1200 °C in air and steam, respectively, is shown in Fig. 12.35. This drop in modulus is believed to be the result of microcracking. The continued drop in modulus at higher fatigue stress levels suggests progressive damage accumulation with continued cycling. Further, full retention of tensile strength after fatigue testing suggests that no damage occurred to the fibers. The reduction in stiffness is most likely due to additional matrix cracking. SEM micrographs of the fracture surfaces of specimens tested in fatigue in air and in steam are presented in Figs. 12.36, 12.37 and 12.38.

In summary, while the same damage and failure mechanisms appear to be operating in steam and in air, damage development and failure are accelerated in the presence of steam. For a given loading frequency damage development in steam proceeds at lower applied stress levels than in air. In air the fatigue performance improves with

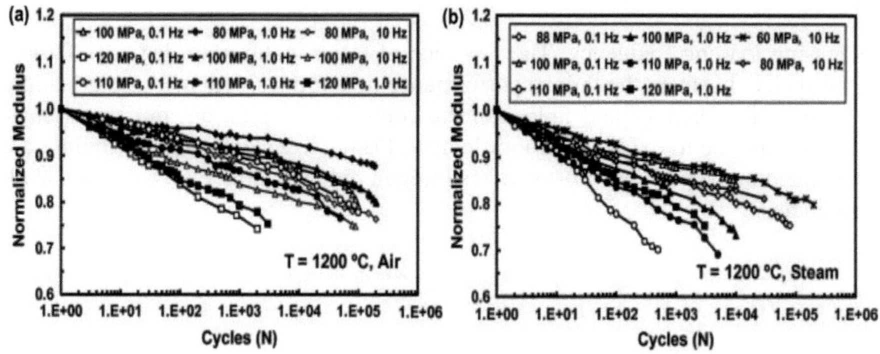

Fig. 12.35 Normalized modulus versus fatigue cycles for Hi-Nicalon™/SiC ceramic composite at 1200 °C: **a** in air and **b** in steam, Ruggles-Wrenn et al. (2011). With kind permission of Elsevier

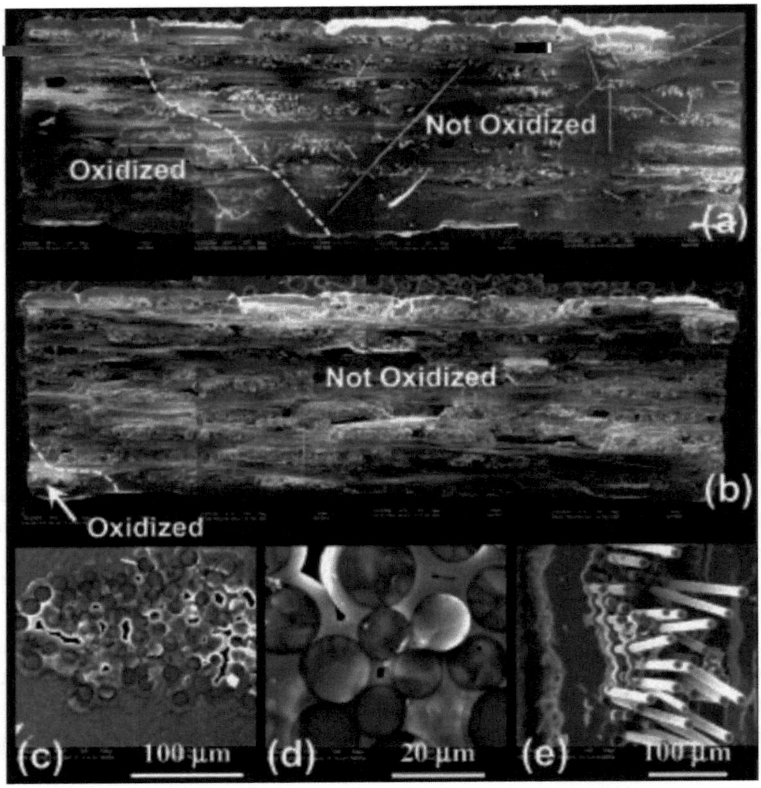

Fig. 12.36 SEM micrographs of fracture surfaces produced in fatigue tests conducted at 1.0 Hz at 1200 °C in air: **a** $\sigma_{max} = 120$ MPa, $N_f = 3766$ cycles, **b** $\sigma_{max} = 80$ MPa, $N_f > 200{,}000$ cycles, **c** higher magnification image showing oxidation of fibers and matrix, **d** higher magnification image showing glassy phase in the oxidized region, **e** higher magnification image showing fiber pull-out typical in the not oxidized region, Ruggles-Wrenn et al. (2011). With kind permission of Elsevier

decreasing loading frequency. The presence of steam noticeably degrades fatigue performance. In steam the fatigue performance deteriorates with increasing loading frequency. The damage and failure of the composite in both air and steam environments are due to oxidation embrittlement. Damage development and failure are accelerated in the presence of steam, resulting in reduced fatigue performance at all frequencies.

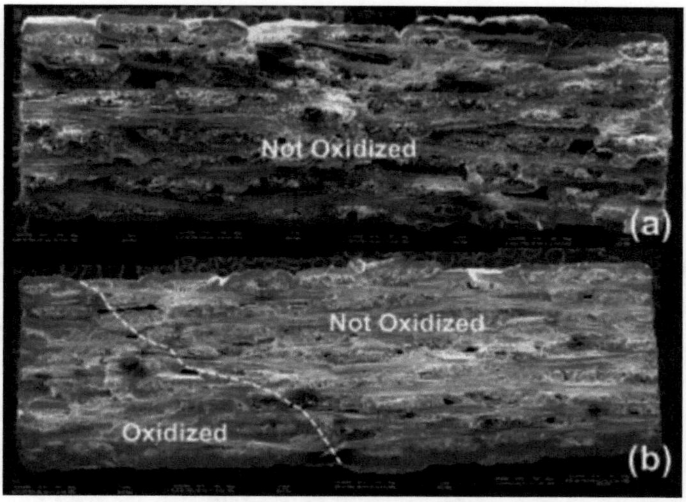

Fig. 12.37 SEM micrographs of fracture surfaces produced in fatigue tests conducted at 1200 °C in air: **a** at 0.1 Hz, $\sigma_{max} = 100$ MPa, $N_f > 100{,}000$ cycles, **b** at 10 Hz, $\sigma_{max} = 100$ MPa, $N_f = 96{,}721$ cycles, Ruggles-Wrenn et al. (2011). With kind permission of Elsevier

12.5 Effect of Environment in Si_3N_4

12.5.1 In Water

Fatigue tests in various environments are presented in the following sections. All tests except those in moist air were performed inside an environment chamber. As the title indicates we shall start our presentation with fatigue tests performed in water. In Fig. 12.39 the normalized cyclic fatigue is indicated and deionized, and double distilled water (DI H_2O) is included. The plot is according to Paris' law given as

$$\frac{da}{dN} = A(\Delta K)^n \tag{12.9}$$

In the equation and in the figures, da/dN is the crack growth rate (m/cycle) and ΔK is the difference between the maximum and minimum applied stress intensity (K_{max} and K_{min}), n is an exponent and A is a constant which depends on the R-ratio (K_{min}/K_{ax}). Unnormalized and normalized figures for the four Si_3N_4 materials indicated in the curves are shown in Fig. 12.40 a, b respectively. The normaliztion is performed to allow comparison of various test results for various toughness materials. Crack growth rates in DI H_2O were at most twice as fast as in moist air as shown in Fig. 12.41a. Cyclic fatigue proceeded fastest in these four environments compared to the "dry" environments shown in Fig. 12.41b. The microstructure of one of the nitrides (SN1) is shown in Fig. 12.42a. The specimen is fine grained having diameters

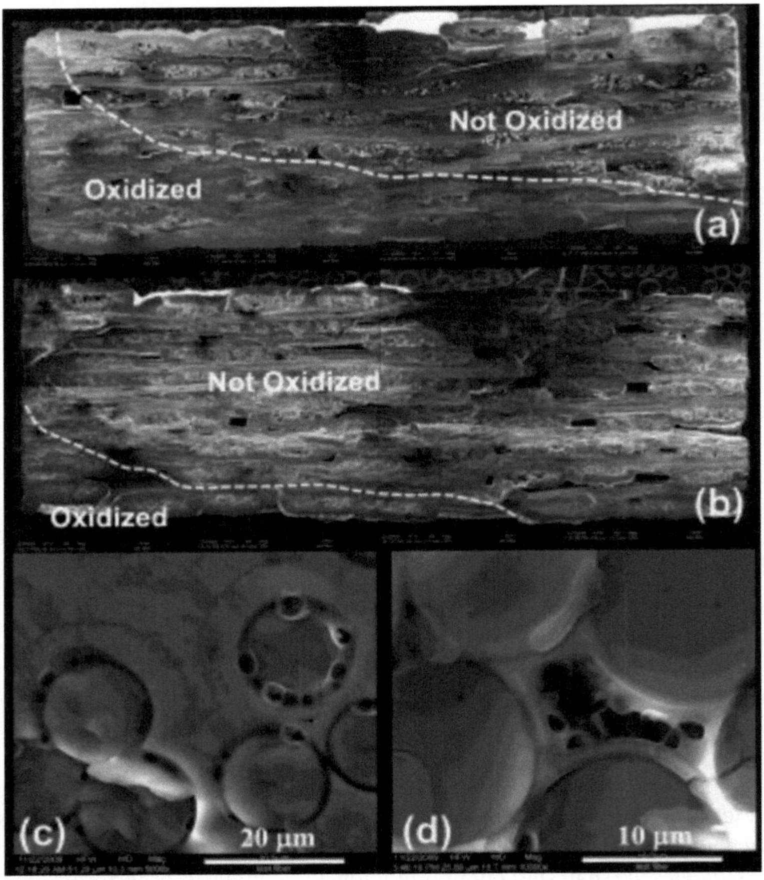

Fig. 12.38 SEM micrographs of fracture surfaces produced in fatigue tests conducted at 1200 °C in steam: **a** at 0.1 Hz, σ_{max} = 88 MPa, N_f > 88,575 cycles, **b** at 1.0 Hz, σ_{max} = 80 MPa, N_f > 200,000 cycles, and (c and d) higher magnification images showing increased glass formation during oxidation in steam, Ruggles-Wrenn et al. (2011). With kind permission of Elsevier

in the range 0.1–0.5 μm and length up to 1.5 μm. In (b) the fracture surface represent cyclic strain (not stictly in water, but in moist air) and the fracture seems to be intergranular (but also with large amounts of a glass coating the grains). Fracture was largely through the glass where also the environmental interaction occurred. The microstructures of another nitride specimen (SN2-4) are very similar to those of Fig. 12.42a, b, but coarser (grain length ~4–5 μm as seen in Fig. 12.43a) and the fracture is intergranular as shown in Fig. 12.43b.

In summary, Si_3N_4 is very sensitive to water environmnet. Intergranular glass appears to be the site of environmental (i.e., water) interaction. The mechanism of cyclic fatigue involves damage to the crack wake shielding zone.

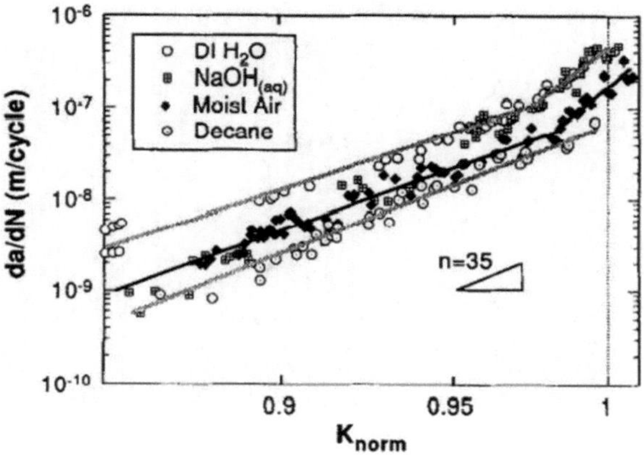

Fig. 12.39 **a** Normalized cyclic fatigue in moist environments. Jacobs and Wei Chen, Jacobs and Wei [7]. Open access

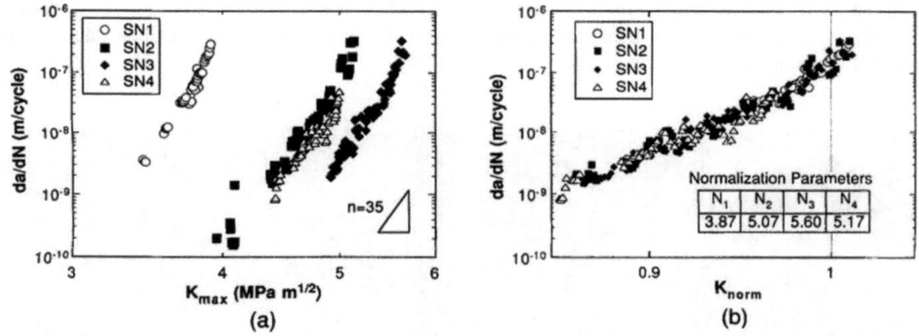

Fig. 12.40 **a** Variation of cyclic fatigue behavior in moist air environment for all four Si_3N_4 materials. **b** Data normalized so curves coincide. Jacobs and Wei Chen, Jacobs and Wei [7]. Open access

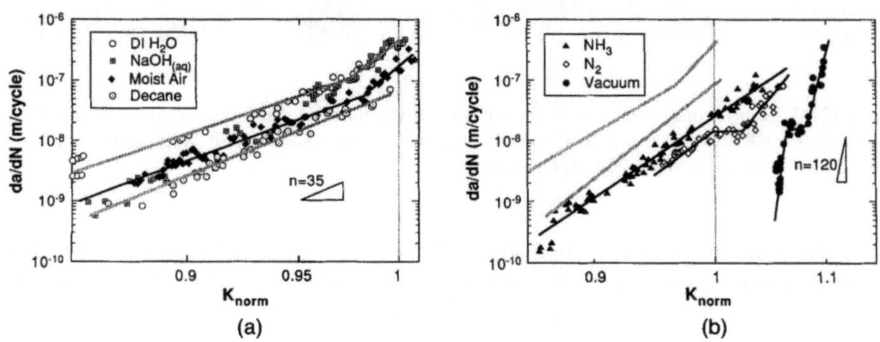

Fig. 12.41 **a** Normalized cyclic fatigue in moist environments. **b** Normalized cyclic fatigue in "dry" environments (shaded lines indicate bounds of data in moist environments). Jacobs and Wei Chen, Jacobs and Wei [7]. Open access

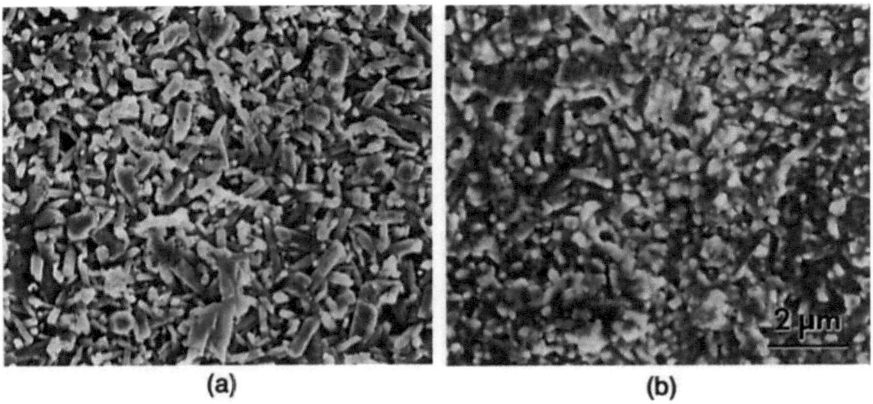

Fig. 12.42 Fracture surfaces of SN1: **a** etched in molten NaOH for 15 s at 350 °C to reveal underlying grain structure; **b** surface produced by cyclic fatigue. Jacobs and Wei Chen, Jacobs and Wei [7]. Open access

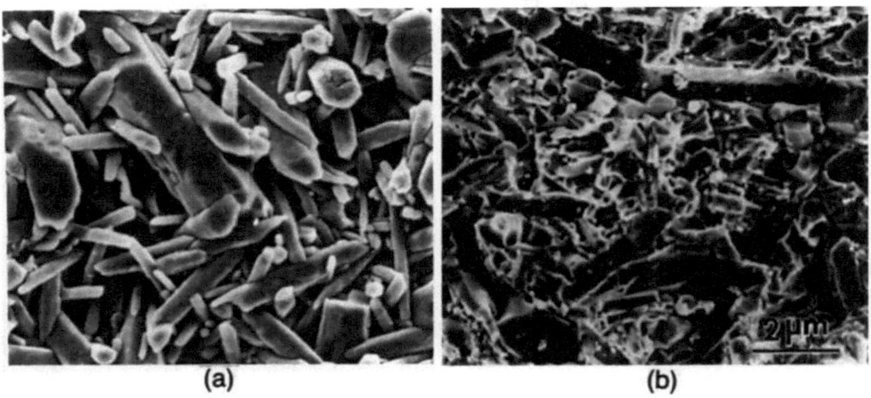

Fig. 12.43 Representative microstructure of SN2, SN3, and SN4. Surfaces produced by **a** etching in molten NaOH for 356 s at 350 °C to reveal underlying grain structure, **b** cyclic fatigue in moist air (SN3), Jacobs and Wei Chen [7] Open access

Further, fatigue tests were performed in Si_3N_4 in water (and other environments) to understand its behavior and the meaning of n value in bending experimemnts. The applied stress intensity factor at the tip of a surface crack in a specimen subjected to bending is given by

$$K = \phi\sigma\, d^{1/2} \tag{12.10}$$

Rewriting Eq. (12.9) as Eq. (12.11)

Table 12.5 Propertiea of materials used in these experiments. Qiao et al,. Open access

Materials		Density (g cm^{-3})	Bending strength[a] (MPa)	Fracture toughness[b] (MPa $\sqrt{m}$)
Si_3N_4	SN2	3.12	750	6.75
	SA5	3.25	624	5.78
	ZSN4	3.33	657	7.50

[a] 3-point bend.
[b] 3-point SENB with notch width of ~0.3 mm.

$$V = da/dn = AK^n \tag{12.11}$$

where σ is the applied bending stress, a is the half crack length and ϕ is the crack shape factor. Under such a driving force, the slow crack growth rate may be assumed to follow a power law (Eq. (12.11)). K is clearly the stress intensity factor at the crack tip and A and n are constants. Assuming this, n is large, Eqs. (12.10) and (12.11) can be integrated to obtain the lifetime, t_f for static and cyclic fatigue.

$$\text{Static fatigue : } \ln t_f = \ln C - n\ln \sigma_a \tag{12.12}$$

$$\text{Cyclic fatigue: } \ln t_f = \ln C' - n\ln \sigma_{max} \tag{12.13}$$

Cyclic fatigue for nitrides tested (designation listed in Table 12.5) are listed in Table 12.5 and the figures for cyclic fatigue and static fatigue are compared in Figs. 12.44 and 12.45.

In summary, crack growth in ceramic materials occurs under both static and cyclic loading conditions. Static fatigue proceeds by subcritical crack growth under constant loads and has been recognized to be an environmentally actived process. Cyclic fatigue, on the other hand, is classically thought of as a mechanical phenomenon in which load oscillation leads to crack propagation. Both types of fatigue have been observed in Si_3N_4. Lifetime prediction depends on accurate determination of fatigue parameters such as n. A typical *n* value may range from 8 to 120 for ceramics, compared with typical values of 2–5 for metals. In most cases, the values of n determined by static fatigue are higher than those by cyclic fatigue as seen in Tabe 12.6. The cyclic fatigue n values are about the same for the three nitrides as seen in Table 12.6.

12.5.2 In Air

Two Si_3N_4s commercially available and designared as AS800 and SN282 are presented below. These toughened materials by the microstructure were processed into elongated grain structures. Prior to testing, all AS800 test specimens were annealed at 1200 °C in air for 2 h to eliminate or minimize damage and/or residual

Fig. 12.44 Cyclic fatigue curves of Si_3N_4 materials: **a** SN2, **b** SA5 and **c** ZSN4, Qiao et al. Open access

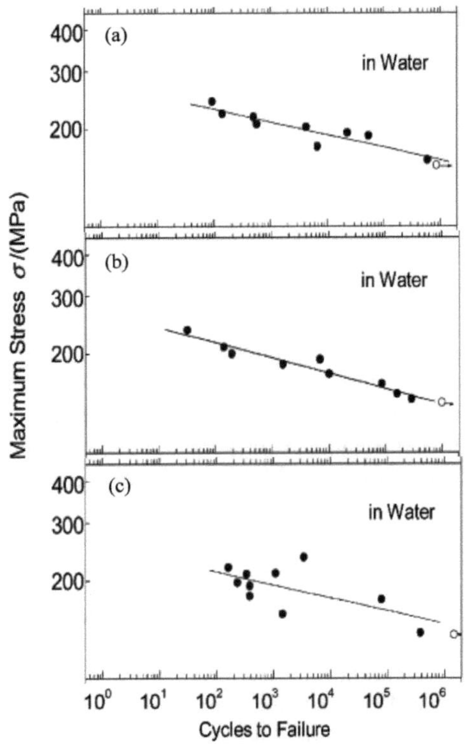

stresses presumably associated with machining materials. Both materials are toughened silicon nitrides with microstructures tailored into elongated grain structures. The degree of elongation and the size of grains were greater in AS800 than SN282, as shown in Fig. 12.45. The Weibull Distribution of the flexural strength determined at 1050 °C in air for AS800 and SN282 silicon nitrides is shown in Fig. 12.47. Figures 12.45 and 12.46 are illustrated to indicate the basic microstructure and strength (flexural) of the nitrides. However below dynamic fatigue data are illustrated, and the flexural results can be compared with those of Fig. 12.47. Note that the dynamic fatigue test is one of the methods to evaluate the slow crack growth behavior by measuring relationship between stress rate and fracture stress. Result of dynamic fatigue testing for AS800 and SN282 silicon nitrides at 1050 °C in air are shown in Fig. 12.47, where flexure strength is plotted as a function of applied stress rate. For AS800, between applied stresses of 50 and 0.05 MPa/s, there was little strength degradation. However, at applied stress rates $\leq$0.05 MPa/s, strength degradation took place with further decreasing applied stress rate, showing a susceptibility to slow crack growth. By contrast, no appreciable strength degradation occurred for SN282, indicative of significant resistance to slow crack growth.

Weibull distributions of strength determined at different stress rates in dynamic fatigue testing at 1050 °C in air are shown in Fig. 12.48. The Weibull modulus

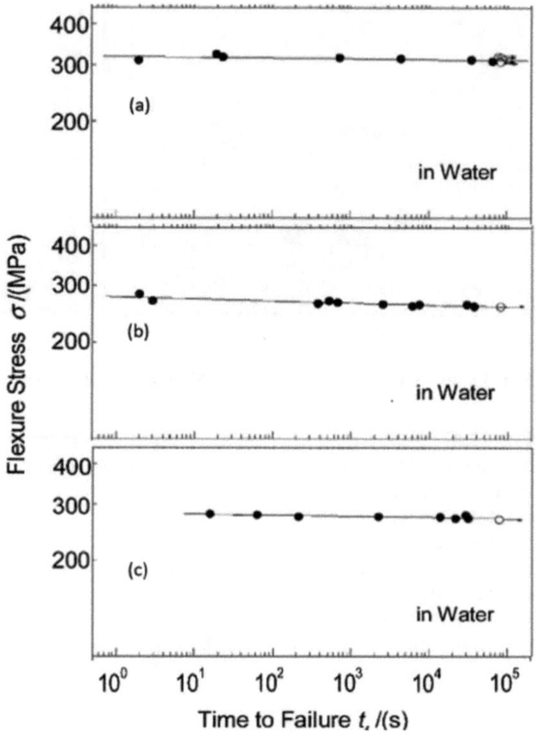

Fig. 12.45 Static fatigue curves of Si_3N_4 materials **a** SN2, **b** SA5 and **c** ZSN4, Qiao et al,. Open access. Open access

Table 12.6 Values of *n* determined by fatigue tests, after, Qiao et al,. Open access

Si_3N_4	Fatigue	Environment	n
SN2	Cyclic	Water	23.9
SN5	Cyclic	Water	20.8
ZSN4	Cyclic	Water	31.9
SN2	Static	Water	325.3
SN5	Static	Water	271.8
ZSN4	Static	Water	525.8

ranged from m = 14 to 26 for AS800 with a somewhat decrease at lower applied stress rates $\sigma \leq 0.005$ MPa/s. Weibull modulus for SN282 ranged from m = 14 to 18, almost independent of test rate. Figure 12.49 shows typical fracture surfaces of AS800 specimens tested at 0.05 and 0.0005 MPa/s. At 0.05 MPa/s, a region of fracture mirror including fracture origin was well defined, whereas, at 0.0005 MPa/s, the fracture mirror region became significant, indicating the occurrence of extensive slow crack growth. No noticeable evidence of slow crack growth was observed from fracture surfaces of SN282 specimens tested at the lowest test rate of 0.0005 MPa/s.

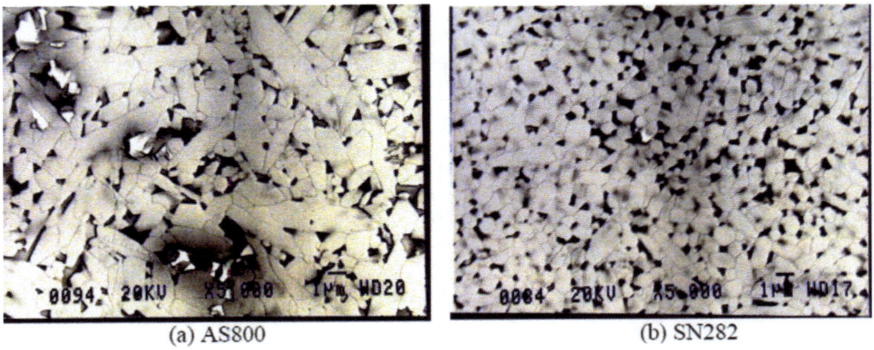

Fig. 12.46 Microstructures of AS800 and SN282 silicon nitrides, Choi and Krause [8]

Fig. 12.47 Weibull distributions of flexure strength determined at1050 °C in air for AS800 and SN282 silicon nitrides. m = Weibull modulus, σθ = characteristic strength and *F* = failure probability, Choi and Krause [8]

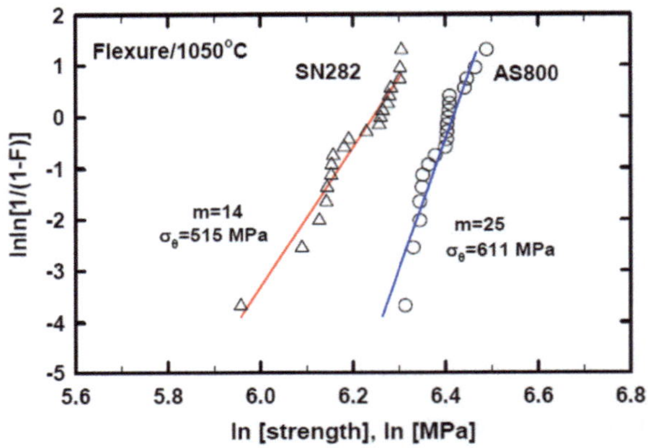

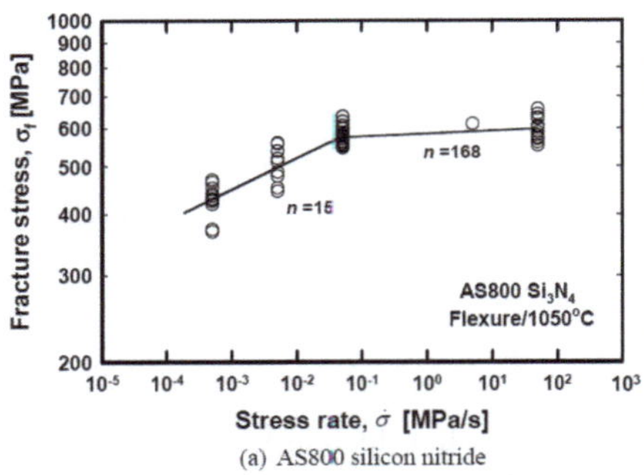

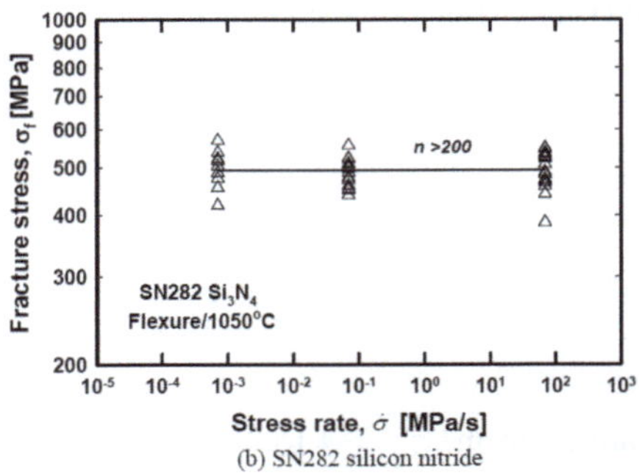

Fig. 12.48 Summary of flexure strength as a function of applied stress rate, determined for AS800 and SN282 silicon nitrides in dynamic fatigue testing in flexure at 1050 °C in air. Slow crack growth parameter n is shown for each material, Choi and Krause [8]

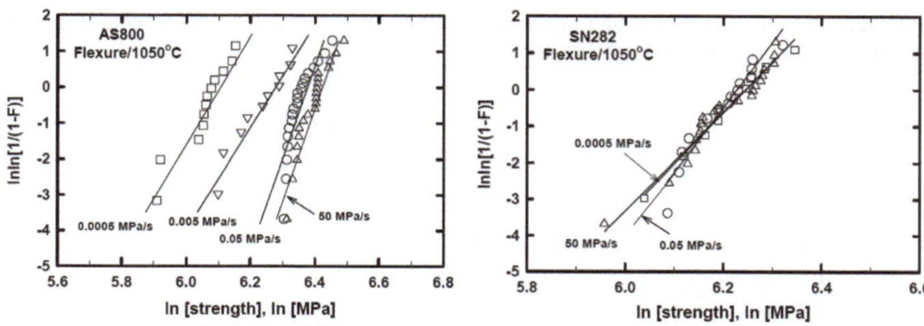

(a) AS800 silicon nitride (b) SN282 silicon nitride

Fig. 12.49 Weibull distributions of flexure strength determined for AS800 and SN282 silicon nitrides in dynamic fatigue testing at 1050 °C in air, Choi and Krause [8]

Life prediction is seen in Fig. 12.50 using the same geometrical and dimensional configurations of test specimens that were employed in dynamic fatigue testing.

It might be of interest to show the basic ambient-temperature mechanical and physical properties of both AS800 and SN282 silicon nitrides (such as elastic modulus, Poisson's ratio, density, Vickers hardness, flexurestrength, both uniaxial and biaxial, fracture toughness, and R curve parameters. These are seen in Table 12.7. The R-curve was estimated by the indentation technique and can be expressed as

$$K_R = ka^q \tag{12.14}$$

where K_R is fracture resistance (MPa $m^{1/2}$), a is the crack size (μm) and q are R-curve parameters which are listed in Table 12.7. The basic formulation of slow crack growth (SCG) is (Fig. 12.51).

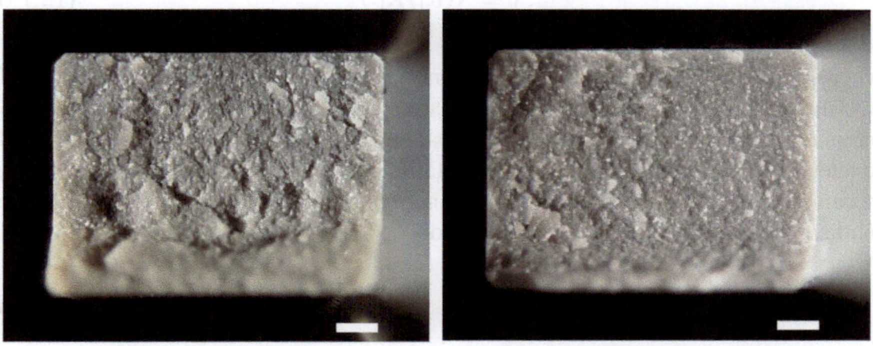

(a) Tested at 0.05 MPa/s (b) Tested at 0.0005 MPa/s

Fig. 12.50 Typical examples of fracture surfaces of specimens tested at 0.005 **a** and 0.0005 MPa/s **b** in dynamic fatigue testing for AS800 silicon nitride at 1050 °C in air. Bars in 500 μm, Choi and Krause [8]

Table 12.7 BASIC mechanical and physical properties of As800 and Sn282 silicon nitrides at ambient temperature, Choi and Krause [8]

Material	Elastic modulus[1] E (GPa)	Poisson's ratio[1] ν	Density[2] ρ (g/cm³)	Hardness[3] H (GPa)	Flexure Strength[4]				Fracture toughness[5] K_{Ic} (MPa√m)	R-curve Parameters[6]	
					Type of loading	Mean strength (MPa)	Weibull modulus	Characteristic strength (MPa)		*k*	*q*
AS800 Si_3N_4	309	0.27	3.27	13.6(1.4)	Uniaxial	775(45)	21	795	8.1(0.3)	3.02	0.13
					Biaxial	678(45)	18	698			
SN282 Si_3N_4	304	0.28	3.32	15.3 (0.2)	Uniaxial	595(64)	11	623	5.5(0.2)	2.56	0.10
					Biaxial	426(60)	8	451			

Notes:

1. By the impulse excitation technique, ASTM C 1259 (ref. 20)
2. By mass/volume method
3. By Vickers microhardness indentation, ASTM C 1327 (ref. 23)
4. Uniaxial testing: by four-point flexure with 20/40 mm spans with 3 by 4 by 45 mm bars, ASTM C 1161 (ref. 24) (Number of test specimens used: 20 for each material). Biaxial testing: by ring-on-ring configuration with 20/40 mm rings with 2-mm-thick and 45-mm-diameter disks (Number of test specimens used: 10 for AS800 and 21 for SN282).
5. By single-edge-precracked-beam (SEPB) method, ASTM C 1421 (ref. 25).
6. See the text for descriptions.

The numbers in the parentheses indicate ±1.0 standard deviations.

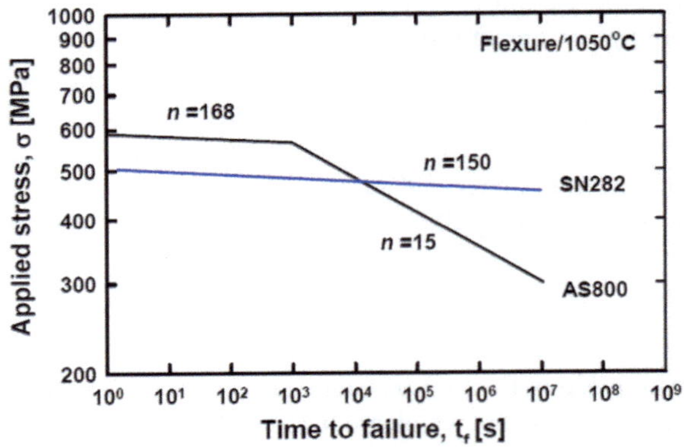

Fig. 12.51 Simplified life prediction diagram, converted from the dynamic fatigue data for AS800 and SN282 silicon nitrides at 1050 °C in air, Choi and Krause [8]

$$\nu = A(K_I/K_{Ic}) \tag{12.15}$$

where v, K_I, and K_{Ic} are crack velocity, mode I stress intensity factor, and mode I fracture toughness, respectively. A and n are material/environment dependent SCG parameters. In case of dynamic fatigue loading, a constant stress rate, $\dot{\sigma}$ is applied to a test specimen until the test specimen fails. The corresponding fracture strength, σ_f can be derived from Eq. (12.15) as a function of applied stress rate, $\dot{\sigma}$ with some mathematical manipulations to give relation

$$\log \sigma_f = \frac{1}{n+1} \log \dot{\sigma} + \log D \tag{12.16}$$

where D is another SCG parameter associated with A, n, and K_{Ic}, inert strength, and crack geometry factor. Figure 12.47 is plotted according to Eq. (12.16) and the SCG parameters n and D can be determined from the slope and intercept by a linear regression analysis. Equation (12.16) is the basis commonly used in dynamic fatigue testing, which has been adopted to determine SCG or life prediction parameters. For AS800, SCG parameters were found to be n = 15 and D = 699 between $\dot{\sigma} = 0.05$ and 0.0005 MPa/s, whereas n = 168 and D = 585 and between $\dot{\sigma} = 50$ and 0.05 MPa/s. For SN282, SCG parameters were n = 16,750 and D = 493. It is seen that significant SCG occurred for n < 30, imtermediate SCG for 30 > n < 70, and insignificant for n > 70. In the dynamic fatigue tests, in the experimental procedure, preloads were applied at the lowest test rate of 0.0005 MPa/s. The fracture stress as a function of preload is given as

$$\sigma^* = \left(1 + \alpha_p^{n+1}\right)^{\frac{1}{n+1}} \tag{12.17}$$

σ^* is the normalized fracture stress, in which the fracture stress with preload is normalized with respect to the fracture stress without preload, α_p is preloading factor ($0 \le \alpha_p \le 1$), in which the applied preload stress is normalized with respect to the fracture stress without preload. The fracture stress as a function of preload, applied at 0.0005 MPa/s for AS800 and SN282, is shown in Fig. 12.52. Also included is the theoretical prediction based on Eq. (12.17) together with values of n = 15 and 150, respectively for the nitrides AS800 and SN282.

In summary, strength and life assessment for two nitrides (AS800 and SN282) conducted in air at 1250 °C are presented in this section. Slow crack growth that controls life with a low life prediction parameter n = 15 was apparent in the AS800 nitride, while SN282 the other nitride exhibited an increased resistance to life limiting phenomena. Flexure strength and Weibull modulus of AS800 are greater than those of SN282.

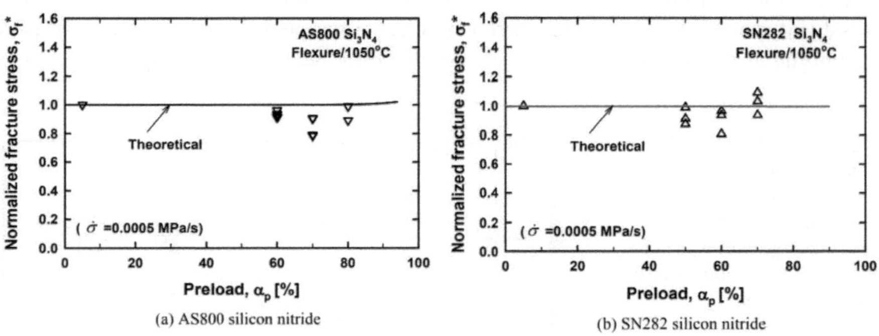

(a) AS800 silicon nitride (b) SN282 silicon nitride

Fig. 12.52 Normalized fracture stress as a function of preload for AS800 and SN282 silicon nitrides in dynamic fatigue testing at 1050 °C in air at the lowest stress rate of 0.0005 MPa/s. The theoretical prediction based on Eq. (12.17) is also presented, Choi and Krause [8]

12.5.3 In Vacuum

It is a general concept that fatigue fracture of ceramics results from a slow ctrack growth of preexisting flaw. It is thus obvious that understanding of the slow crack growth behavior in ceramics is essential for their lifetime predictions. In order to understand the degradation process in nitrides due to the slow crack growth process in cyclic fatigue, clartifying its mechanism is essential. In the following dynamic fatigue behavior in Si_3N_4 in vacuum is considered and compared with that in air.

The dynamic fatigue test is one of the methods to evaluate the slow crack growth behavior by measuring relationship between stress rate, $\dot{\sigma}$ and fracture stress, σ_f. The relation (see Eq. (12.16)) is approximated by

$$\sigma_f = C\dot{\sigma}^{\frac{1}{n+1}} \tag{12.18}$$

where C is a constant. Fracture of ceramics at room temperature occurs in elastic deformation range and the stress rate, $\dot{\sigma}$ is proportional to the crosshead speed υ and Eq. (12.18) can be rewritten as

$$\sigma_f = C'\upsilon^{\frac{1}{n+1}} \tag{12.19}$$

where C' is a constant. The dynamic fatigue tests in vacuum (10^{-5} Torr) at room temperature were carried out under the same testing conditions as in air. The results of the dynamic fatigue tests are shown below in a set of Figs. 12.53, 12.54, 12.55 and 12.56. The three-point bending test method with 30 mm span was used as a failure test, where five different crosshead speeds such as 5, 0.5, 0.05, 0.005 and 0.0005 mm/min were chosen. Three specimens for each material were tested at each crosshead speed. (Author's note: currently at least five specimens are commonly used for fatigue tests). The relationship between the crosshead speed and the mean value

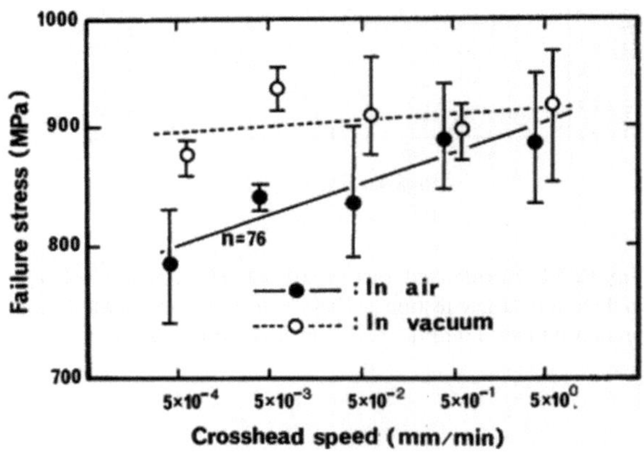

Fig. 12.53 Dynamic fatigue behavior of normal sintered Si_3N_4 (N_2) at room temperature. In vacuum, slow crack growth was not observed., Choi and Krause [8]. Free access

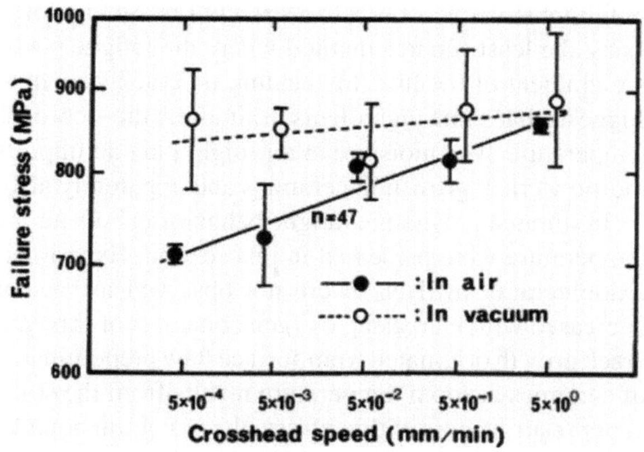

Fig. 12.54 Dynamic fatigue behavior of normal sintered Si_3N_4 (N_3) at room temperature. In vacuum, slow crack growth was not observed, Yamauchi et al. [9]. Free access

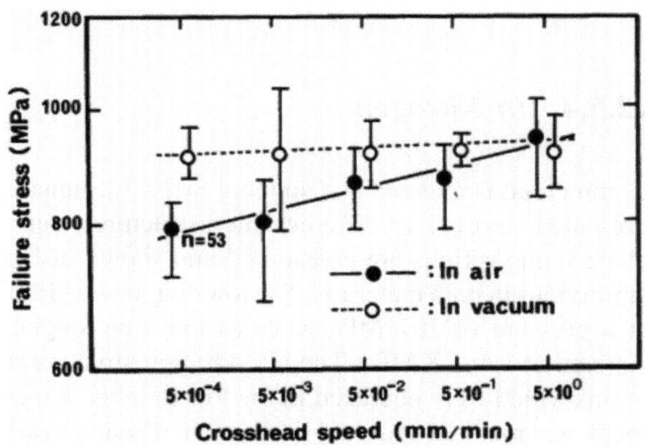

Fig. 12.55 Dynamic fatigue behavior of normal sintered Si_3N_4 (N_4) at room temperature. In vacuum, slow crack growth was not observed, Yamauchi et al. [9]. Free access

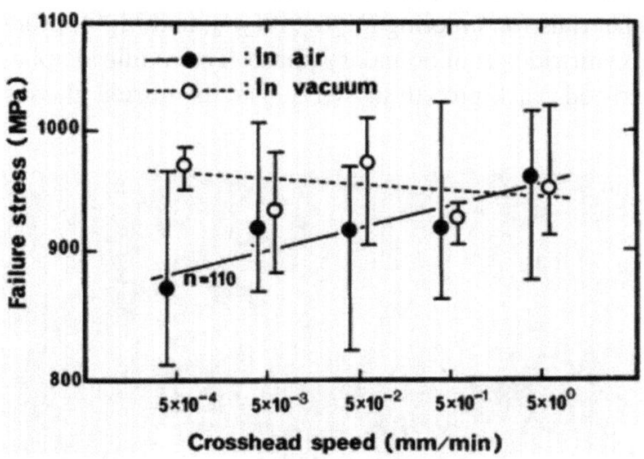

Fig. 12.56 Dynamic fatigue behavior of hot-pressed Si_3N_4 (N5) at room temperature. In vacuum, slow crack growth was not observed, Yamauchi et al. [9]. Free access

of failure stress was fitted to a straight line on log–log representation (see Eq. 12.16) using the least squares method where the fatigue parameter "n" was estimated from the gradient of the line. In vacuum, no crack growth was observed in all tests. This suggests that some ingredients in air affect the slow crack growth of ceramics at room temperature. The moisture in air might play an important role in the acceleration of the slow crack growth in ceramics and apparently also in silicon nitride.

In summary, dynamic fatigue behavior in both air and vacuum (10^{-5} Torr) at room temperature were presented in this section. Experimental slow crack growth in non oxide ceramics (Si_3N_4) were not observed in vacuum. In air, slow crack growth decreased with increasing oxygen content (not shown in this presentation). The slow crack growth originated from surface flaw predominated in air. The slow crack growth of ceramics at room temperature results from the stress corrosion induced by water vapor in air (occurs within glassy phase at grain boundaries). The presence of oxygen affects the fatigue parameter of non-oxide (Si_3N_4) ceramics.

12.5.4 In Nitrogen

Commercial isopressed and sintered Si_3N_4 (designated by the producer as GS44) is presented below. Two different microstructures with similar glassy grain boundary phase compositions are considered and shown in Fig. 12.57. The second phase is amorphous in both materials. The low-toughness GS-44 was hot isostatically pressed at a pressure of 207 MPa whereas the high-toughness version was gas pressure sintered at 1.4–2.8 MPa. The GS-44 microstructure consisted of primarily β-Si_3N_4 grains which were equiaxed in the low toughness material and acicular in the high-toughness material as a result of controlled grain growth. The high-toughness material had an average grain diameter of ~1.5 μm and an aspect ratio of 4.0. Additive materials for sintering (Y_2O_3, Al_2O_3, and MgO) produce about 10 vol% of vitreous oxynirtride grain boundary phase. The oxinitride phase affects the environmentally existed crack growth in Si_3N_4. The oxynitride glasses reside as thin films (1–5 nm)

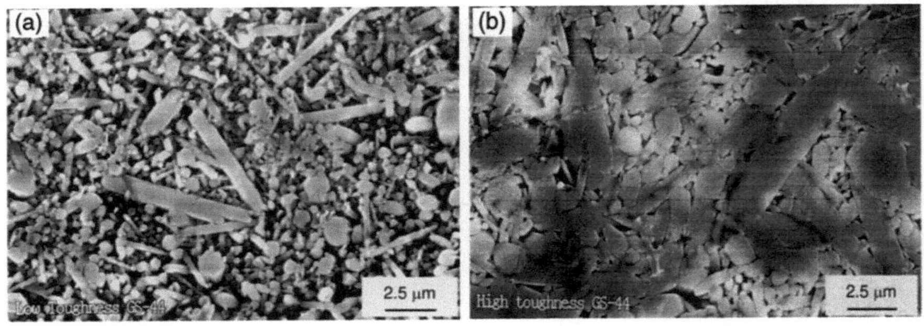

Fig. 12.57 Microstructures of the **a** low-toughness and **b** high-toughness GS-44 silicon nitride microstructures, Bhatnagar et al. [10]. Full access

between grains and as pockets at triple grain junctions. Fractographic examination of mating fracture surfaces under both, subcritical and critical loading conditions reveal that the crack path is primarily intergranular through the intergranular oxinitride glassy phase as seen in Fig. 12.58. The presence of these phases leads to significant toughening by promoting debonding of favorably oriented elongated Si_3N_4 grains to produce grain bridging. Such grain bridging results in considerable resistance curve (known as R-curve) behavior as it forms in the crack wake. Long-crack-growth data for oxynitride glasses measured in dry nitrogen (and moist air) environments is shown in Fig. 12.59a, b the normalized stress intensity is indicated. As mentioned,

Fig. 12.58 Representative SEM micrographs showing the intergranular nature of the Si_3N_4 fracture surface under both **a** monotonic and **b** cyclic loading conditions. Note the oxynitride glassy wear debris produced under cyclic loading. Bhatnagar et al. [10]. Full access

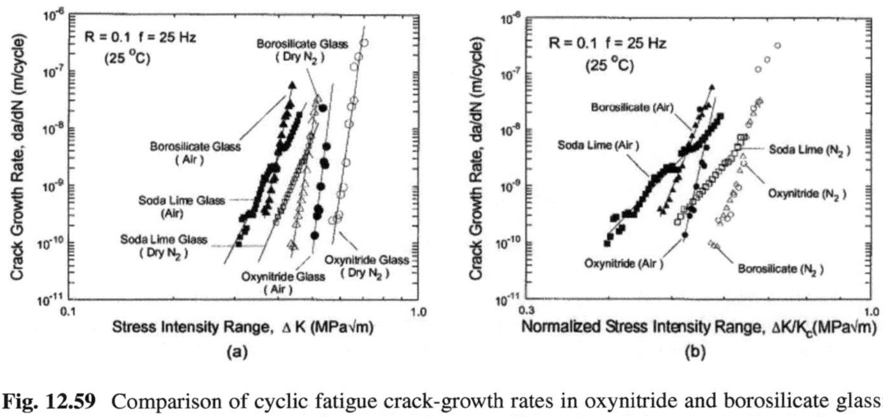

Fig. 12.59 Comparison of cyclic fatigue crack-growth rates in oxynitride and borosilicate glass: **a** in inert and moist environments, **b** obtained by normalizing glass fatigue curves by the fracture toughness of the respective glasses, Bhatnagar et al. [10]. Full access

the crack-growth rates are sensitive to the concentration of H_2O as evidenced by the shift of the entire crack-growth curve to higher stress intensities compared to the dry N_2 environment. The presence of oxynitride phase strengthens the Si_3N_4. Note that additional phases are included in Fig. 12.59a, b. The effect of environment on the subcritical crack-growth behavior of the low-toughness GS-44 ceramic obtained under fatigue loading is shown in Fig. 12.60a and compared with static loading (Fig. 12.60c).

In summary, even at room temperature, fracture in Si_3N_4 is predominantly intergranular through the oxynitride glassy phase. Oxynitride is inherently present at the grain boundaries of advanced Si_3N_4 ceramics. Crack growth in bulk Si_3N_4 ceramics occurs through the oxynitride grain boundary phase. The presence of these phases leads to significant toughening by promoting debonding of favorably oriented elongated Si_3N_4 grains to produce grain bridging. Crack-growth rates are accelerated in the presence of moist air environments compared to dry nitrogen environments.

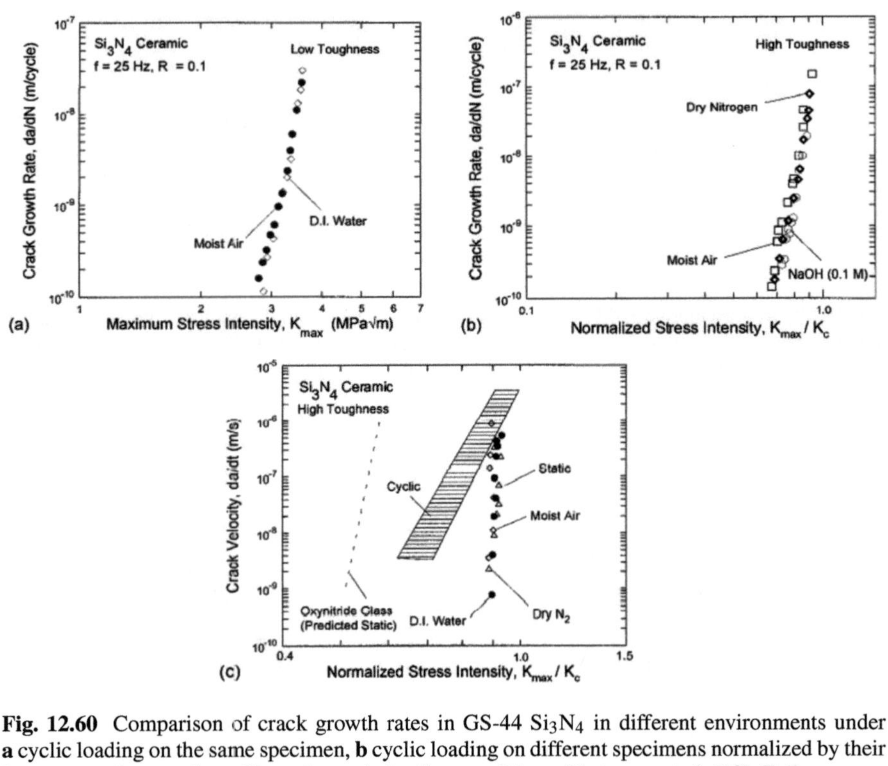

Fig. 12.60 Comparison of crack growth rates in GS-44 Si_3N_4 in different environments under **a** cyclic loading on the same specimen, **b** cyclic loading on different specimens normalized by their fracture toughness values, K_c, and **c** static loading conditions, Bhatnagar et al. [10]. Full access

References

1. J. Kruzic, R.M. Cannon, R.O. Ritchie, J. Am. Ceram. Soc. **88**, 2236 (2005)
2. H. Nakayama, K. Ohira, Mater. Sci. Res. Inter. **4**, 212 (1998)
3. B. Shi, Y.J. Su, L.J. Qiao, W.Y. Chu, J. Mater. Sci. Lett. **20**, 2047 (2001)
4. S.S. Scherrer, M. Cattani-Lorente, E. Vittecoq, F. de Mestral, J.A. Griggs, H.W. A. Wiskott, Dental Mater. **27**, e28 (2011)
5. R. Studart, F. Filser, P. Kocher, L.J. Gauckler, 23, 106 (2007). Dent. Mater. **23**, 106 (2007)
6. M. Mizuno, S. Zhu, Y. Kagawa, H. Kaya, J. Eur. Cer. Soc. **18**, 1869 (1998)
7. D.S. Jacobs, I.-W. Chen, J. Amer. Ceram. Soc., **77**(1), 155 (1994)
8. S.R. Choi, D.L. Krause, Available from NASA Center, NASA/TM—2006–214089, January 2006
9. Y.Yamauchi, S. Sakai, M. Ito, T. Ohji, W. Kanematsu, S. Ito, Nippon-Seramikkusu-Kyokai-Gakujutsu-Ronbunshi **96**, 885 (1988)
10. A. Bhatnagar, M.J. Hoffman, R.H. Dauskardt, J. Am. Cer. Soc. **83**, 585 (2004)

References

Chapter 13
Surface Treatment

Abstract Fatigue lifetime of surface treated specimens can be improved, which can be performed by either by shot peening, sandblasting or laser treatment. These treatments were selectively performed on some of the oxides, carbides or nitrides. Some of the technics applied in surface protection of ceramics are: plasma spraying, chemical vapor deposition, i.e., CVD, physical vapor deposition-PVD, ion implantation and laser treatment. Further, in laser treatment local melting occurs which flow and fill pores and cracks. This process effectively seals porous material and enables their use in vacuum. Shot peening was applied to PSZ, Si_3N_4/SiC composite and sandblasting was applied to alumina and yttria stabilized zirconia, while laser treatments was used for alumina, YSZ, SiC, Si_3N_4 and Si_3N_4/SiC composite. Compressive residual stress is generated by these treatments with the consequent improvement of the properties.

Keywords Surface treatment · Shot peening · PSZ · Si_3N_4 · SiC · Sandblasting in alumina and YSZ · Laser treatment

13.1 Introduction

Fatige lifetime in metals is known to be improved by various mechanical treatment. Such treatments are shot peening, sand blasting and laser treatments to mention the most common. Regarding laser treatment, shock peeming or deep rolling processes can induce several favorable surface effects, such as work hardening, low surface roughnesas etc. The common of all these trearments is the induction of compressive residual stresses. The limitations of using ceramics relate to the surface flaws which are the fracture initiation sites in fatigue. Further, to the mentioned and most popular treatments indicated, there are a large number of surface engineering processes which are used in the treatment of metals compared to the relatively few techniques which are applicable to ceramics. Some of the technics applied in surface protection of ceramics are: Plasma spraying (for high temperature protection and reducing possible interactions), chemical vapor deposition, i.e. CVD (sealing surface pores and cracks), physical vapor deposition-PVD, ion implantation and laser treatment (sealing cracks and pores, improved properties of friction, wear, toughness and fatigue). The improvement of toghness by laser occurs via local melting to flow and

J. Pelleg, *Cyclic Deformation in Oxides, Carbides and Nitrides*, Structural Integrity 22,
https://doi.org/10.1007/978-3-030-86118-6_13

fill pores and cracks. This process effectively seals porous material and enables their use in vacuum. In the following shot peening, sand blasting and laser treatment are considered as the preferred surface treatments in this chapter.

13.2 Shot Peening Alumina

Laser shock processing (LSP) is a surface modification process that utilizes high-energy laser pulses to irradiate material surfaces. In this process a plasma is formed, and the explosive expansion of the plasma generates shock waves which penetrate into the bulk material and induce significant compressive residual stresses in a range from 100 MPa to several GPa.

LSP and mechanical shot peening are generally comparable in generating compressive residual stress in the surface of materials. Mechanical shot peening is a traditional process in which a stream of steel, ceramic, or glass balls with velocities up to 100 m/s is blasted at the workpiece material to mechanically prestress its surface. However, LSP is superior to mechanical shot peening because it offers many advantages such as deeper penetration and higher magnitude of compressive residual stresses, fast speed (nanoseconds for one laser pulse), easy control, and no contamination of shot peening materials. Pfeiffer and Frey conducted early investigations on the mechanical shot peening of structural ceramics such as Si_3N_4 and Al_2O_3.

The benefits of surface compression on the mechanical properties of ceramics include increased fracture resistance, Weibull modulus, thermal shock resistance, wear resistance, and contact damage resistance. Therefore, LSP is used here insteafd the mechanical shot peening. Compressive surface strengthening depends on the magnitude of the compressive stress and the ratio of the compressive layer depth to flaw size. The increased fracture resistance may be derived from a reduction of the crack-driving force by the presence of compressive stresses. Experimental investigations have revealed that the benefit of the LSP is related to the microstructural evolution near the surface of materials during the LSP process.

Commercially available polycrystalline α-Al_2O_3 ceramics were used. The ceramic samples are 99.6% Al_2O_3 with a little impurity of SiO_2 and Fe_2O_3, a porosity of $<$ 0.5 vol%, a tensile strength of 275 MPa, and a compressive strength of 2600 MPa. Thermal etching was performed at 1550 °C to reveal the grain morphologies. Figure 13.1 shows the x-ray characterization of the LSP indication compressive residual stresses in α-Al_2O_3. The negative value of residual stress is used to indicate a compressive residual stress, while the positive value indicates a tensile stress. The surface of as-received α-Al_2O_3 ceramics has a small magnitude of compressive residual stress (−76 MPa), which is likely a result of manufacturing and machining processes. The compressive residual stress induced by LSP is in a range from −172 to −671 MPa on the surface of α-Al_2O_3 ceramics. The changes in the microstructure of α-Al_2O_3 by LSP are characterized by SEM, FIB, and TEM and is illustrated

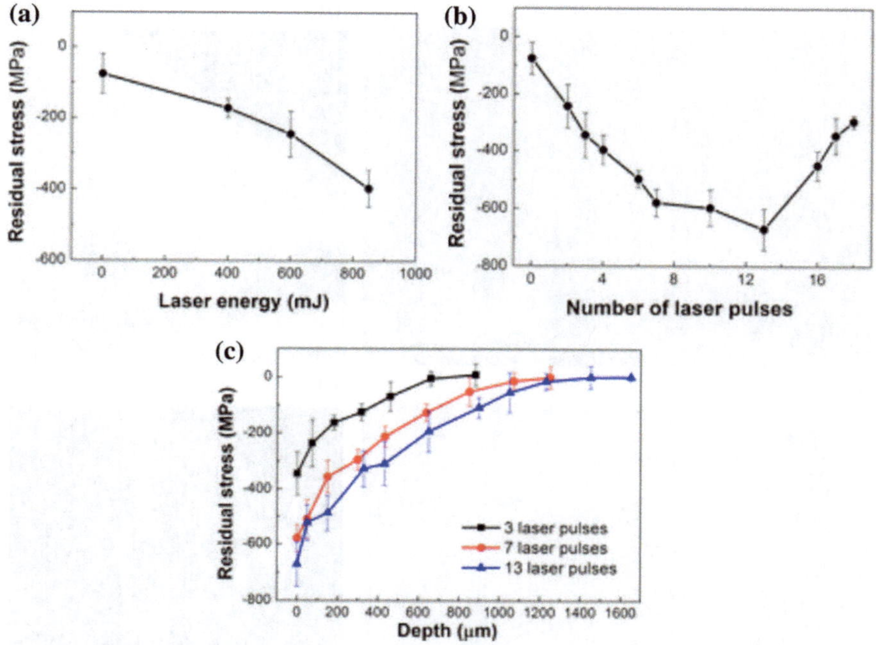

Fig. 13.1 Compressive residual stress in laser shock processing (LSP) treated α-Al_2O_3 ceramics: **a** residual stress on the surface vs laser pulse energy at four laser pulses; **b** residual stress on the surface vs laser pulses at a laser energy of 850 mJ; **c** stress distribution along the depth vs laser pulses at a laser energy of 850 mJ, Wang et al. [1]. Full access

in Fig. 13.2. Vickers indentations were made on the surface of untreated and LSP-treated α-Al_2O_3 ceramics to evaluate their resistance to indentation cracking. It is observed that at a given load (P), the radius of radial cracks (c) was around the Vickers indentations indentations significantly reduced in LSP-treated samples (Fig. 13.3b) than untreated samples (Fig. 13.2a), which suggests that LSP-treated samples have an improved resistance to the mechanical impact.

In summary, LSP generates compressive residual stress on the surface of α-Al_2O_3, which can extend to a depth of more than 1.2 mm from the surface. Microstructural observations indicate limited plastic deformation in the α-Al_2O_3 grains, while it might occur along the grain boundaries. The compressive residual stresses on the surface induced by LSP can improve the resistance of α-Al_2O_3 ceramics to indentation cracking.

Further, shot peening in alumina was reported in an earlier work. Cemented carbide beads with a diameter of (650 ± 40) μm was shot on the surface of polished alumina using dfferent pressures between 2 and 4 bar and treatment times between 280 and 840 s. The residual atress states of the near surface layer of the samples was measured by X-ray (Fig. 13.3) diffraction. The mean stress values within the penetration depth in the range of 10 μm were evaluated using $\sin^2 \psi$ method. The

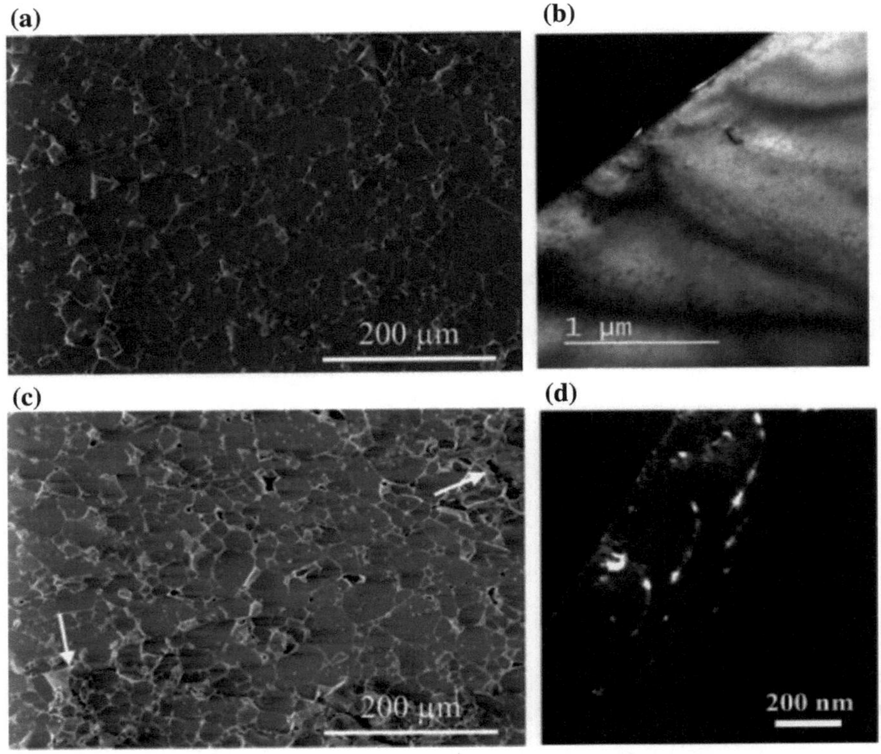

Fig. 13.2 Microstructures of α-Al_2O_3 ceramics treated by laser shock processing (LSP). (**a**, **b**) Scanning electron microscopy image of the surface morphology of LSP-treated α-Al_2O_3 ceramics by **a** a laser pulse energy of 400 mJ and 4 laser pulses; **b** a laser pulse energy of 850 mJ and 16 laser pulses. White arrows indicate the damage region where grains were pulled-out. **c** Bright-field transmission electron microscopy (TEM) micrograph of the α-Al_2O_3 grain underneath the intact region. No crystal defects such as dislocations are present in the microstructure. **d** Weak-beam dark-field TEM micrograph of the α-Al_2O_3 grain underneath the damage region. Dislocations arrays on the prism-plane slip system [1 − 100] (11 − 20) are revealed, Wang et al. [1]. Full access

fracture load versus the residual stress is illustrated in Fig. 13.4 for alumina, and Si_3N_4 is also included. Here, the near surface stress states and the load caoaxities of differently shot peened samples are correlated. A compressive residual stress up to 1.2 GPa was introduced by the shot peening. For this to occur the load needed to fracture the surface layer shifted from 3 kN to more than 9 kN. The increase in the near-surface strength was so high that the load limit of the testing device (ball-on-plate) was reached and in some cases the samples passed the test without observed crack formation. Figure 13.5 shows the mean values of the diameter and depth of single dip as a function of peening pressure. The mean values were calculated from the measurement of typically 25 dips per peening pressure. Higher pressures leads to

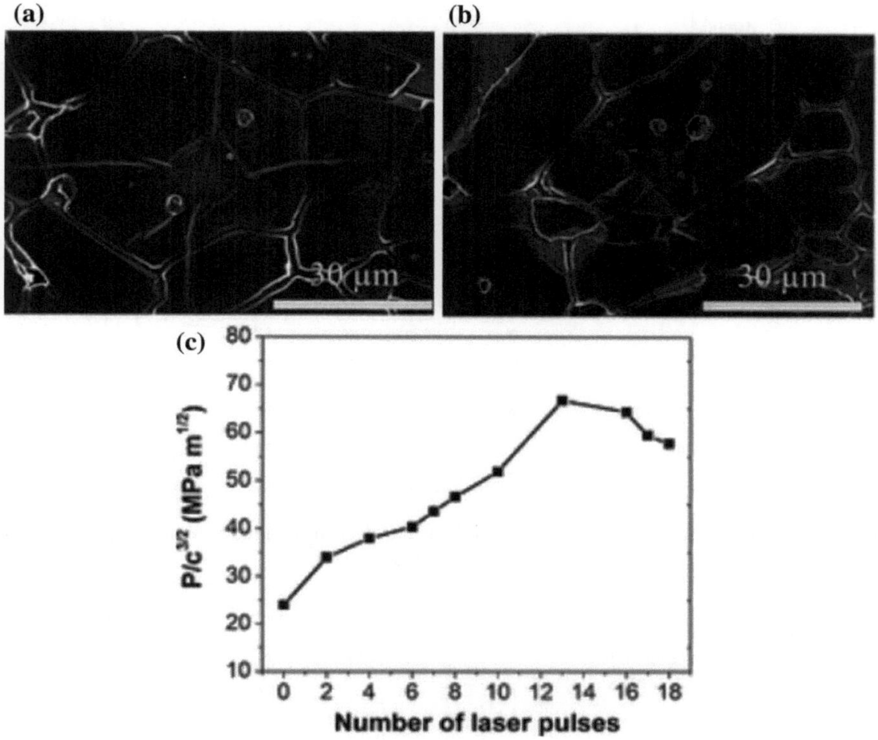

Fig. 13.3 Comparison of the Vickers indentation in **a** untreated α-Al_2O_3 ceramics; and **b** laser shock processing (LSP)-treated α-Al_2O_3 ceramics by a laser pulse energy of 850 mJ and 13 laser pulses. **c** The values of $P/c^{3/2}$ vs the number of laser pulse in LSP-treated α-Al_2O_3 ceramics by a laser pulse energy of 850 mJ, Wang et al. [1]. Full access

a higher velocity and to a larger deformation and to somewhat higher plastic deformation. Silicon nitride is included. In case of peening pressure of 3 and 4 bar, very few cone cracks could be obtained near some single hits. Such cone cracks -rarely detectable- are seen in Fig. 13.7 of an alumina. Sometime ago, almost up to quite recently, it has not been applied to ceramics (Fig. 13.6).

In summary, shot peening is a common technique to improve the strength of metal components. Almost up to quite recently, the technique has not been applied to ceramics which generally are brittle, amd has been assumed not to show significant plastic deformation on mechanical loading. Hence would not develop any residfual stresses for improving the strength. However, under specific shot peening conditions also in brittle materials like silicon nitride and alumina high compressive stresses up to more than I GPa can be introduced near the surface. For this to happen, the loading situations should be characterized by a steep near-surface stress gradient induced by shot peening.

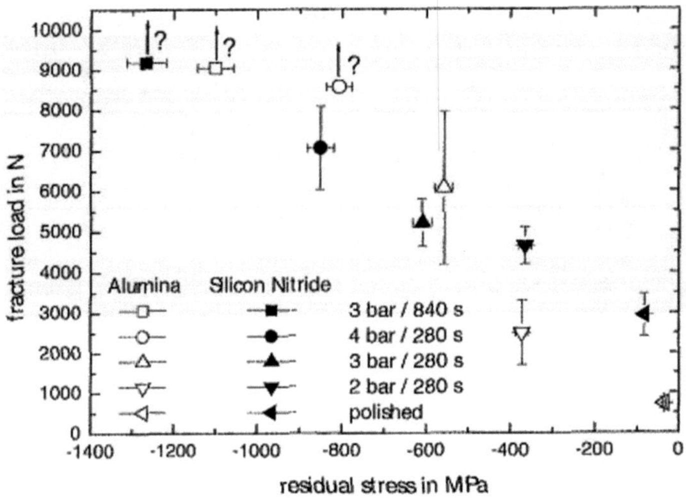

Fig. 13.4 Fracture load vbersus stress of silicon nitride and alumina samples in polished and different shor peened conditions, Frey et al. [2]. Full access

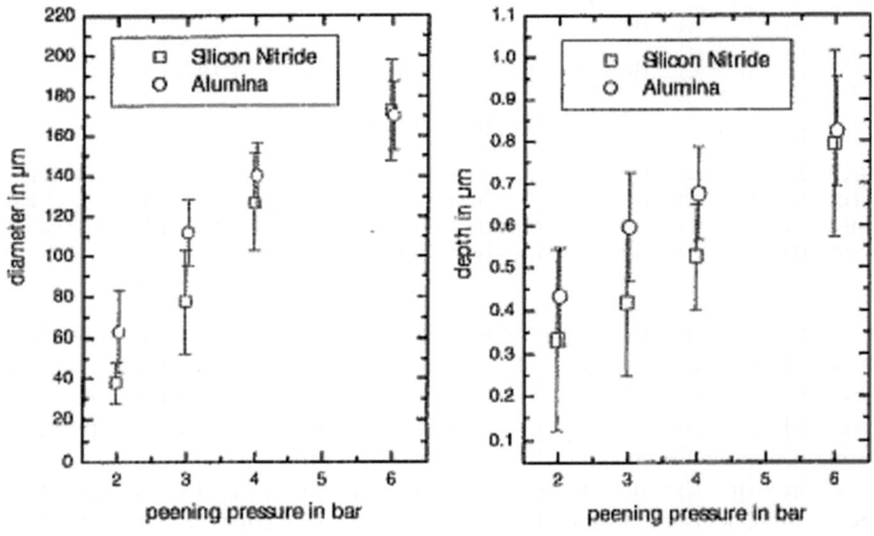

Fig. 13.5 Diameter and depth of single hits by different peening pressures for silicon nitride and alumina, Frey et al. [2]. Full access

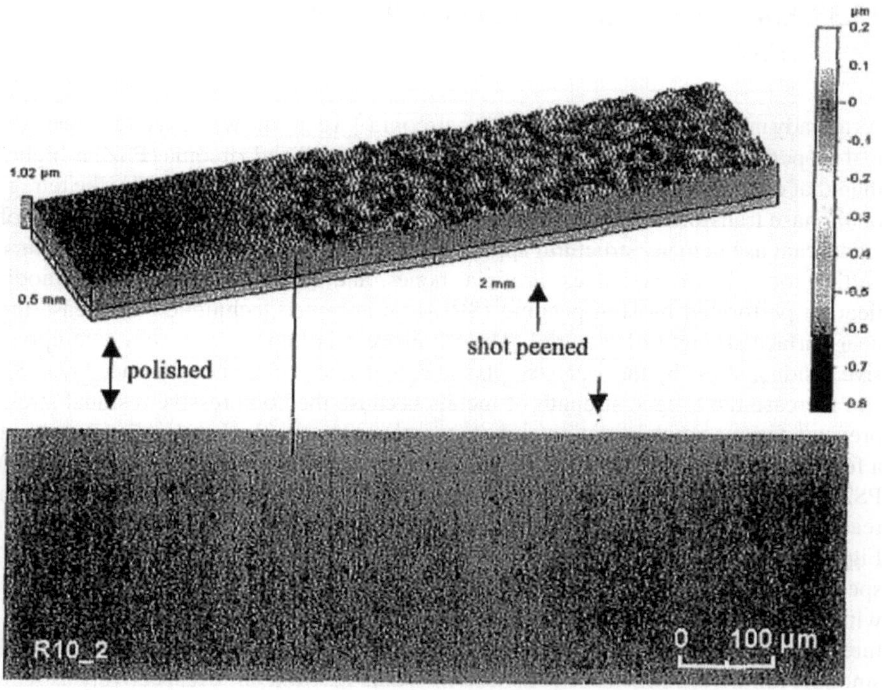

Fig. 13.6 Topographical map (top) and optical micrograph near the boarder between a polished and a shot peened (2 bar) area of a silicon nitride sample, Frey et al. [2]. Full access

Fig. 13.7 Paillally developed cone clack in alumina due to a simple hit, Frey et al. [2]. Full access

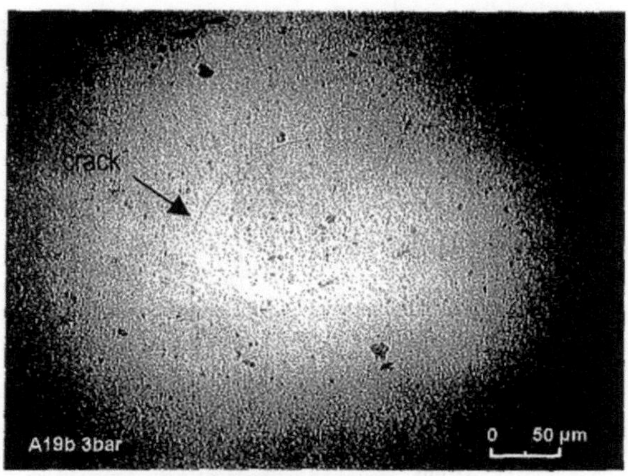

13.3 Shot Peening in Yttria Partially Stabilized Zirconia (PSZ)

As already indicated in previous chapters zirconia is a ceramic with excellent mechanical properties. The fracture toughness of partially stabilized zirconia (PSZ) is higher than that of other ceramics, owing to transformation toughening whereby the tetragonal phase transforms to a stable monoclinic phase under stress. Therefore, it is of significant use in many structural applications among them for medical applications, cutting tools, bearings, valves, artificial bones, and artificial joints. Surface modification performed by shot peening (SP) is an accepted technique to increase the near-surface strength of material. The SP process is known to generate compressive residual stress by the localized plastic deformation of the near-surface layer. SP can increase the fatigue strength of metals because the compressive residual stress prevents fatigue crack initiation and propagation. The residual stress distribution of a few ceramics is illustrated in Fig. 13.8, among them SPned and Non-SP (smooth) PSZ specimens. A compressive residual stress of 1400 was on the surface which reached a maximum of 1800 MPa at a depth of 20 μm in the SPned PSZ as seen in Fig. 13.8. Only a small compressive residual stress is observed on the Non-SP PSZ specimens. Also is seen in the figure that the compressive residual stress decreases with further increasing depth of the PSZ. The effect of SP on the apparent fracture toughness, K_C is seen in Fig. 13.9. Compare the values of K_C of the SP and smooth Non-SP specimens of 16.0 MPa $m^{1/2}$ and 4.5 MPa $m^{1/2}$, respectively as seen in Fig. 13.9. SP suppresses the initiation and propagation of surface cracks which

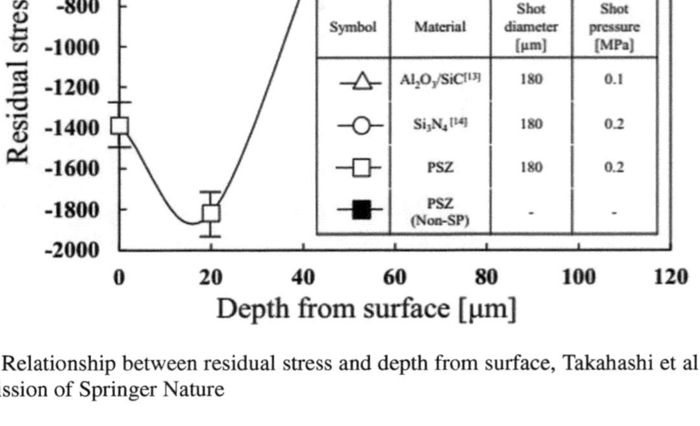

Fig. 13.8 Relationship between residual stress and depth from surface, Takahashi et al. [3]. With kind permission of Springer Nature

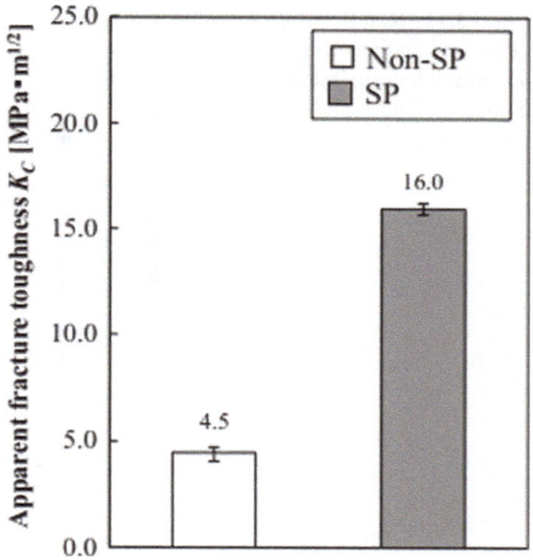

Fig. 13.9 K_C values of the different surface-treated specimens. The error bar indicates the standard deviation, Takahashi et al. [3]. With kind permission of Springer Nature

is illustrated in Fig. 13.10 illustrating Vickers indentations. Bending tests for each specimen is shown in Fig. 13.11. The bending strength, σ_B of precracked specimens decreased significantly with increased crack length. The bending strength improved significantly after SP as seen in Fig. 13.11. Specifically, the Precrack + SP specimens with crack length of ~ 140 μm or less were comparable to the Smooth + SP specimens. Optical micrographs of the surface of the Pre-ctack + SP specimen is seen in Fig. 13.12. SEM images of the fracture surface of the Pre cracked and Precrack + SP specimens, respectively are shown in Fig. 13.13. The Pre-cracked specimen fractured in the pre-cracked area, as shown in Fig. 13.13a. However, Pre-crack + SP specimens with a crack length of les than 140 μm fractured outside the pre- cracked area, as shown by the arrow in Fig. 13.13b, c.

In summary, a large compressive residual stress can be introduced into the specimen by SP to a maximum level of 1800 MPa at a depth of 20 mm from the surface.

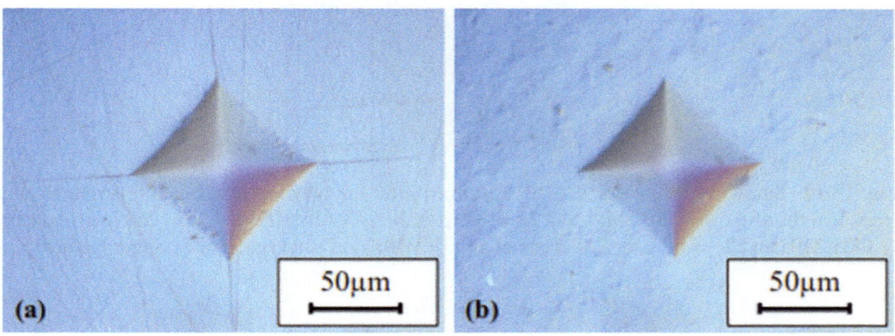

Fig. 13.10 Photographs of Vickers indentations and cracks in the, **a** Non-SP and. **b** SP specimens, Takahashi et al. [3]. With kind permission of Springer Nature

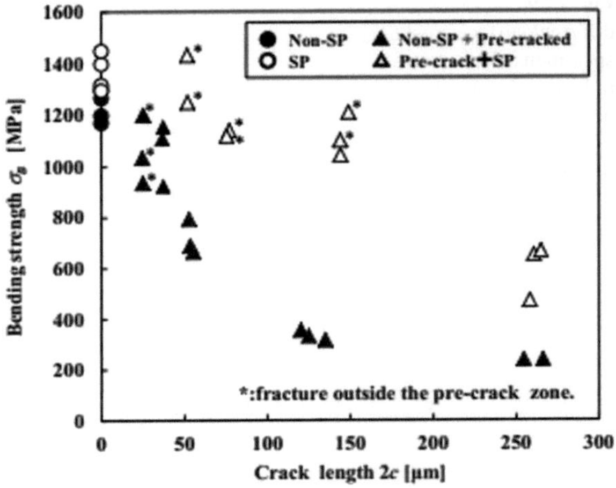

Fig. 13.11 Relationship between bending strength and crack length, Takahashi et al. [3]. With kind permission of Springer Nature

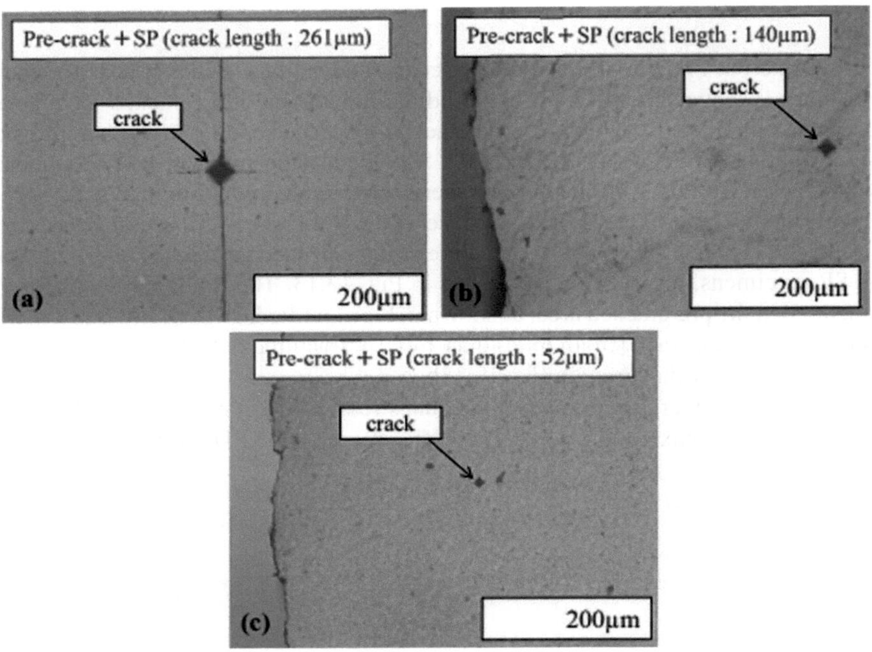

Fig. 13.12 Surface of a Pre-crack + SP specimen after the bending test; bending strengths and crack lengths of **a** $\sigma_B = 1100$ MPa and $2c = 140$ μm, **b** $\sigma_B = 651$ MPa and $2c = 261$ μm, and **c** $\sigma_B = 1431$ MPa and $2c = 52$ μm, Takahashi et al. [3]. With kind permission of Springer Nature

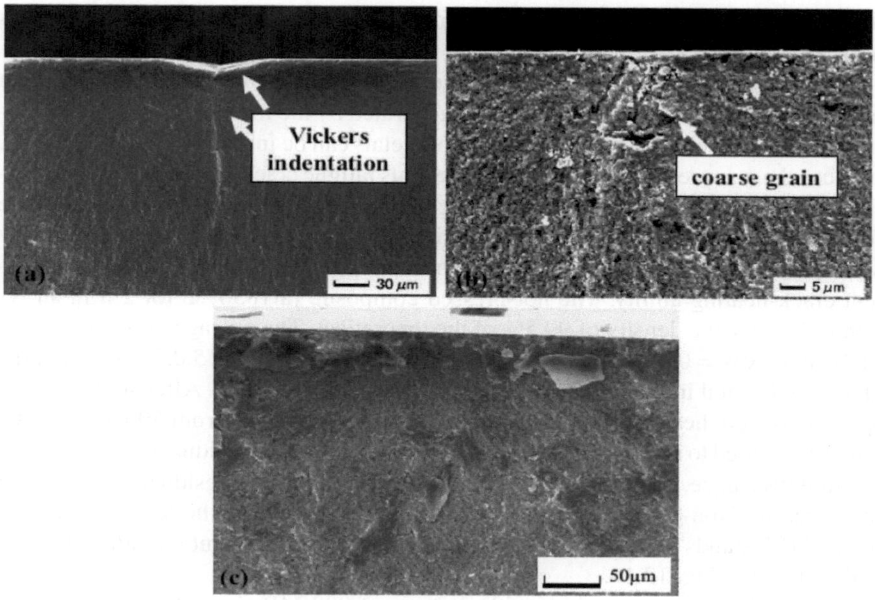

Fig. 13.13 SEM images showing the origins of fractures: **a** Pre-cracked specimen, 2c = 120 μm, σ_B = 354 MPa, **b** Pre-crack + SP specimens, 2c = 140 μm, σ_B = 1100 MPa, and **c** Pre-crack + SP specimens, 2c = 51 μm, σ_B = 1250 MPa, Takahashi et al. [3]. With kind permission of Springer Nature

The toughness, K_C has been improved by SP, which is atttributed to the compressive residual stress induced by SP. The σ_B value of the precracked specimens are significantly improved by SP. The fatigue limits of the pre-cracked specimens after SP were significantly increased, by 200–540%, compared to those of the pre-cracked specimens without SP. The compressive residual stress induced by the local plastic deformation and the transformation of zirconia from the tetragonal to the monoclinic phase contributed to the improvement in the fatigue limits. SP is a useful surface treatment method for improving the fatigue properties of PSZ and rendering detrimental surface cracks harmless.

13.4 Shot Peening in Si_3N_4/SiC Composite

Information on surface treatment by shot peening, neither in pure or monolithic SiC or in Si_3N_4 exist, but it should not be unexpected. Both materials are technologically important structural components, and as such, the main interest is in their strength properties. Having said this, the tendency is in improving their strength ptroperties, which is generally achieved by appropriate additives, namely alloying. Therefore, in this section a composite is presented where the major constituent is Si_3N_4, which is alloyed with 20% SiC.

Shot peening is a common procedure used to increase the near-surface strength of metals. Compressive residual stress is generated by the localized plastic deformation of the surface layer. Fatigue strength of metals can be increased by shot peening, because the compressive residual stress prevents fatigue crack initiation and propagation. Not only metals, but the shot peening of ceramics is also a promising technique for increasing strength in the surface region. Silicon nitride (Si_3N_4) reinforced by silicon carbide (Si_3N_4/SiC) is not only reinforced, but the composite has also excellent crack-healing ability. The hot-pressed composite (at1850 °C for 2 h in an N_2 atmosohere) had a density of 99.5% of the theoretical. The average grain size of the Si_3N_4 matrix is = 0.44 μm, and the average aspect ratio is about 5.0. Most SiC particles were located in grain boundaries and distributed uniformly. After shot peening, specimens were heat-treated in air for 1 or 5 h at temperatures from 500 to 1400 °C, which permitted to investigate the effects of heat treatment on residual stress, apparent fracture resistance, contact strength and bending strength. The residual stresses were measured for Non-SP, SP and SP + heat-treated specimens by the X-ray diffraction method (SP stands for shot peening). The residual stress distribution induced by SP is illustrated in Fig. 13.14.

The maximum compressive residual stress of 880 MPa is at the surface which is decreasing with depth. The in-depth residual stress distribution, is obtained by removal of the surface layers by polishing (see Fig. 13.14). The compressive residual stress was induced approximately up to 40μm. The residual stress at the surface as a function of heat treatment in the range of 800–1100 °C (heat-treatments are

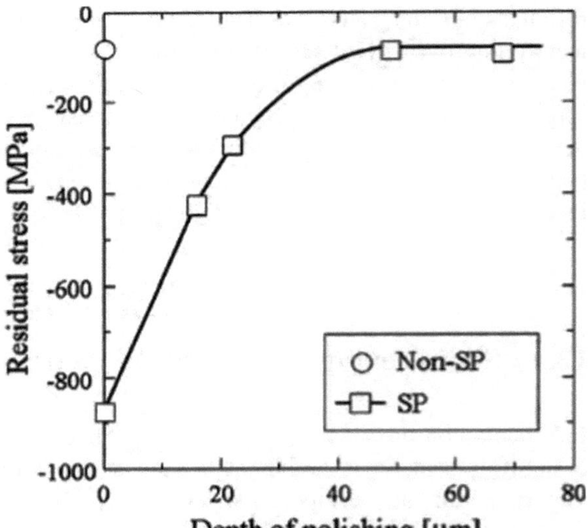

Fig. 13.14 Relationship between residual stress and depth from surface, Takahashi et al. [5]. With kind permission of Elsevier

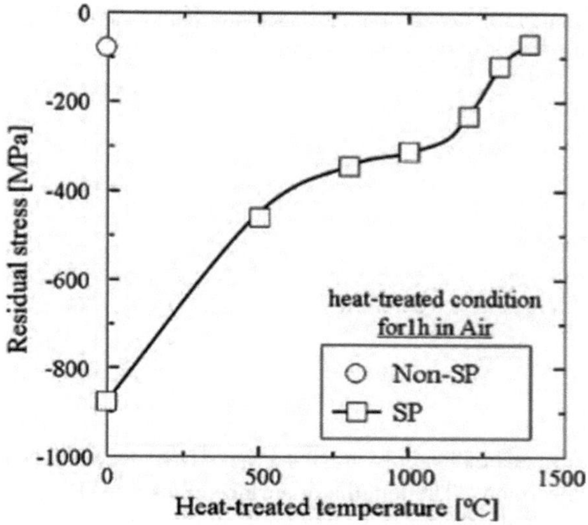

Fig. 13.15 Residual stress at the surface as a function of heat-treatment temperature, Takahashi et al. [5]. With kind permission of Elsevier

performed in air for 1 h at 900–1300 °C.) showed a compressive residual stress of 300–400 MPa as illustrated in Fig. 13.15. Above 800 °C crack healing occurs.

Thus, a dual effect acting simultaneously, namly, generation of a residual compressive stress at the surface and crack healing. Increasing heat treatment temperature relieves lattice strain due to thermal expansion, while crack healing occurs due to oxidation of SiC above 800 °C (accompanied with ~ 80% volume expansion). Volume expansion at the surface occurs due to filling microcracks by the oxidation products. Also during the process lattice strain was eliminated due to softening the grain boundary glassy phase above 1200 °C. Optical micrograph of Vickers indentation and cracks for Non-SP and SP specimens are seen in Fig. 13.16. Cracks are clearly observed in non-SP specimens, as seen in Fig. 13.16a, whereas the surface cracks are very short in SP specimens, as shown in Fig. 13.16b. Table 13.1 shows the length of

Fig. 13.16 Optical photograph of Vickers indentation and cracks (indented load: 98 N): **a** Non-SP specimen and **b** SP specimen, Takahashi et al. [5]. With kind permission of Elsevier

Table 13.1 Length of surface cracks for Non-SP and SP specimens, Takahashi et al. [5]. With kind permission of Elsevier

Vickers indentation load [N]	Length of the surface cracks *2c* [μm]		
	Non-SP + cracked specimens	SP + cracked specimens	SP + heat-treated + cracked specimens
19.6	107.4	63.5	87.0
49.0	188.4	106.8	117.1
98.0	300.1	156.7	179.6

surface cracks for Non-SP + cracked, SP + cracked and SP + heat-treated + cracked specimens. The lengths of surface cracks in SP + cracked andSP + heat-treated + cracked specimens were much smaller than that of SP + cracked specimens. It is clear that the compressive residual stress increases the resistance to crack initiation at sub-surface. The apparent fracture resistance, K_C at subsurface was measured by the indentation-fracture (IF) method and is illustrated in Fig. 13.17. Note the fracture resistance of the SP specimen which is higher than the Non-SP specimen (~ 2–3 times). The apparent fracture resistance, K_C is evaluated by the indentation-fracture (IF) method, and can be estimated by the following equation

$$K_C = 0.026E^{1/2}\frac{a}{c^{3/2}} \tag{13.1}$$

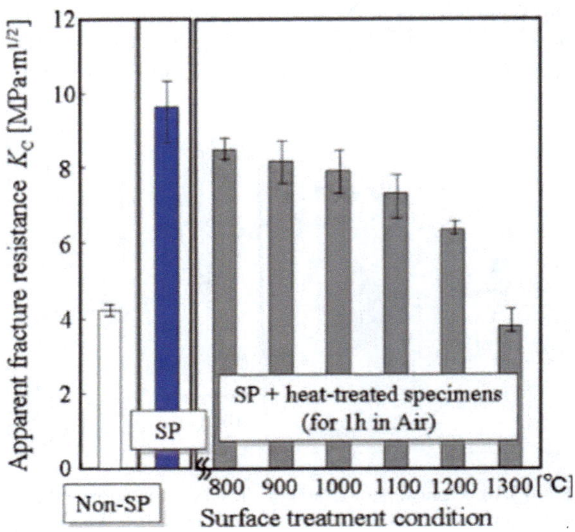

Fig. 13.17 Relationship between surface treatment conditions and apparent fracturer esistance, Takahashi et al. [5]. With kind permission of Elsevier

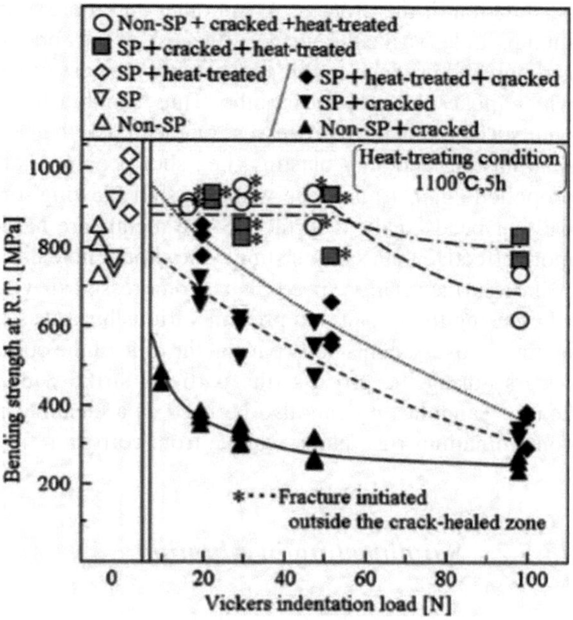

Fig. 13.18 Relationship between indentation load and bending strength, Takahashi et al. [5]. With kind permission of Elsevier

where P is the indent load, a is half of the diagonal length, c is half the surface crack lenght and clearly E is Young's modulus. The bending strength as a function of Vickers indentation load is shown in Fig. 13.18.

In summary, compressive residual stress is generated by the localized plastic deformation of the surface layer. Fatigue strength of materials can be increased by shot peening, because the compressive residual stress prevents fatigue crack initiation and propagation. A compressive residual stress is inducesd at the surface. The apparent fracture resistance is improved by shot peening. Simultaneously with strength improvement by SP, crack healing also occurs. The increase of bending strength was due to healing of small surface cracks produced by the shot-peening process. Shot peening in combination with crack healingis a useful technique to improve the strength and reliabilityof ceramics.

13.5 Sandblasting

13.5.1 Introduction

Sandblasting is also known as abrasive blasting and it is a surface finishing process. In this method abrasive particles are sprayed under high pressure agains a surface. Since the abrasive paricles usually is sand, the term sandblasting is an appropriate designation of the process. A smoother and more even texture is obtained after the sand particles strike the surface. Because of the abrasive properties of sand (the sand is rough and gritty) it is able to wear away excess or unwanted material on a surface, which makes the surface smoother.Thus, sandblasting is just one of several blasting-type surface finishing processes. One can ask then what is the difference between sandblasting and shot blasting, i.e., shot peening exactly? In sandblasting, sand is propelled against a surface while in shot blasting, on the other hand, small metal balls or beads strike a surface. Since metals are harder than sand, shot blasting is more effective than sandblasting. One should note that in sandblasting sand particles strike a surface at high speeds using compressed air. Great precaution should be taken when using this process to prevent various hazards such as silicosis and usually the operator covers all his body parts at the time of the operation. Further, heat generation occurs during the process (the particles strike with high velocity) that can harm health. Sandblasting can also be used as a cleaning method to effectively remove contaminations rust, paint, residue from corrosion from the surface of materials.

13.5.2 Sandblasting in Alumina

In addition to aesthetics in dental clinical practice, the essential criterion for selection of crown materials is the resistance to fracture and deformation under long-term cyclic conditions. Alumina (and zirconia) is an attractive candidate in dentistry because its high modulus, hardness and strength. The effect of sandblasting on the strength of alumina layers joined to polymeric subbstrates is considered in this section for the purpose of simulating occlusal contact in ceramic crowns on tooth dentin. The sandblast treatment is applied to the ceramic bottom surface before bonding to the substrate, as in common dental practice. Specimens with polished surfaces are used as a control. Tests are conducted with monotonically increasing (dynamic) and sinusoidal (cyclic) loading on the spherical indenter, up to the point of initiation of a radial fracture at the ceramic bottom surface immediately below the contact.

A dense fine-grained alumina (and zirconia) was chosen as the ceramic test material because it is representative of that used in dental and biomechanical applications. Micrograph of the microstructure is shown in Fig. 13.19. Pertinent properties of alumina, measured in earlier studies, is given in Table 13.2. The sandblasting was performed with 50 μm Al_2O_3 particles for 5 s at a standoff distance 10 mm and a comppressed air pressure 276 kPa. Small-scale nanoindentations were placed in selected as-polished and sandblasted surfaces, and effective Young's moduli E evaluated from instrumented load–displacement data. A relatively high nanoindentation load of 1 N was chosen, corresponding to 15–20 μm contact diameter, that is, considerably greater than the grain size. Cyclic tests were carried out on a hydraulic testing machine at a frequency of 10 Hz. Plots of maximum stress, S versus effective time t_R to radial fracture from dynamic fatigue and cyclic fatigue tests for polished Y-TZP and alumina bilayers is shown in Fig. 13.20.

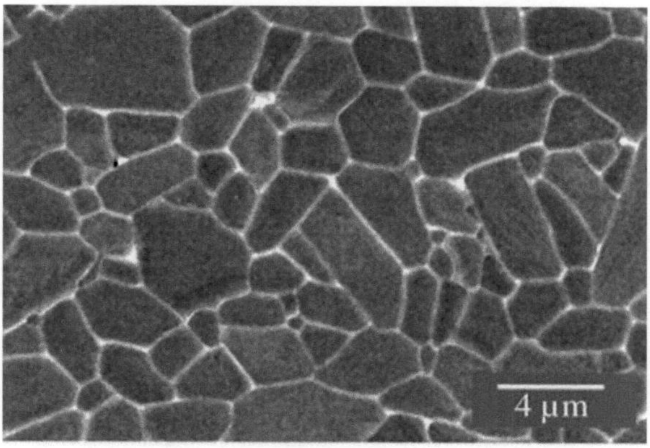

Fig. 13.19 Microstructures of alumina (Al_2O_3). SEM image. Surface is thermally etched, Zhang et al. [6]. With kind permission of John Wiley and Sons

Table 13.2 Properties of Dental Materials, Zhang et al. [6]. With kind permission of John Wiley and Sons

Material	Name	Supplier	Modulus *E* (GPa)	Hardness *H* (GPa)	Strength (10 year) *S* (MPa)	Velocity Exponent *N*
Core ceramic						
Alumina (dense)	ADS-95 -R	CoorsTek	372	19.6	722	26
Zirconia (Y-TZP)	Prozyr	Norton	205	14.0	2325	25
Substrate/Adhesive						
Polycarbonate	Hyzod	AIN plastic	2.3			
Epoxy	RT Cure	Master bond	3.5			
Dentin			16			

Data points are individual test results, arrows indicate runouts after 10^7 cycles. Solid lines are best fits to the data using the radial fracture relations. Test durations over the range 0.1–106 s were accomplished by varying the stressing rates in the dynamic tests and the maximum stresses in the cyclic tests. Corresponding strengths S were calculated from the critical loads P_R using the relation

$$S = \left(P_R/Bd^2\right)log(E_c/E_s) \tag{13.2}$$

where E_c and E_s are moduli of the ceramic plate and substrate, d is plate thickness, and B = 1.35 is a dimensionless constant. In Fig. 13.20, the plot is of maximum stress vs effective time t_R to radial fracture from dynamic fatigue and cyclic fatigue for polished Y-TZP and alumina bilayers and is given as

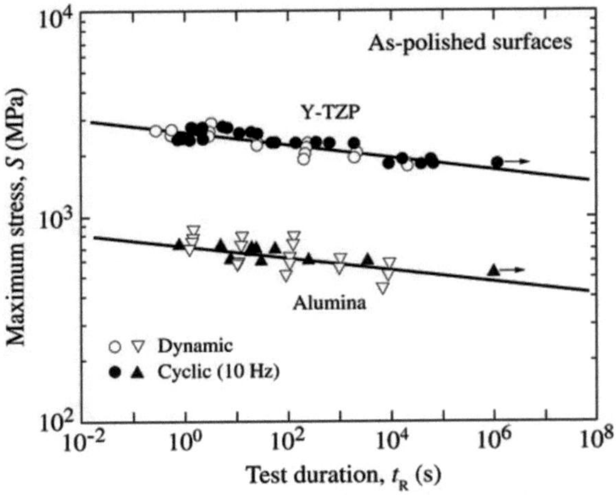

Fig. 13.20 Maximum tensile stress S in ceramic layer versus effective time to radial fracture t_R for as-polished Y-TZP and alumina plates bonded to polycarbonate substrates. Data represent individual tests at constant monotonic stressing rates (unfilled symbols) and in cyclic loading at 10 Hz (filled symbols). Solid lines are data fits in accordance with slow crack growth relations. Arrows indicate runouts, Zhang et al. [6]. With kind permission of John Wiley and Sons

$$S^N t_R^c = 2AN^{0.47} \quad \text{(cyclic)} \tag{13.3}$$

$$S^N t_R^d = A(N+1) \quad \text{(dynamic)} \tag{13.4}$$

Figure 13.21 is a similar plot for sandblasted specimens. The solid line is from Fig. 13.20 as zero damage base line for comparison. Solid lines are best fits to the data using the radial fracture relations based on slow crack growth model. Here, t_R^c and t_R^d

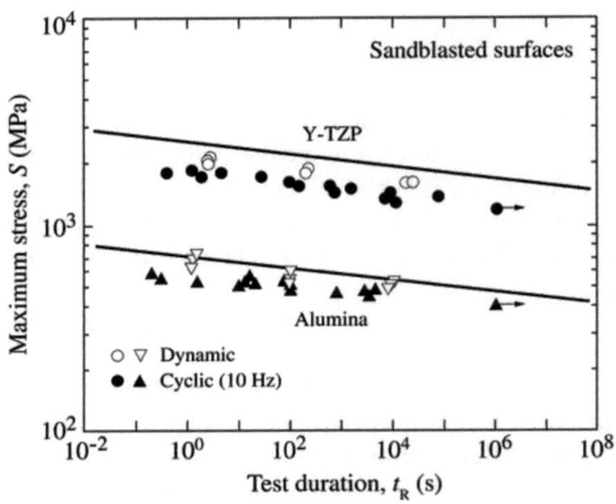

Fig. 13.21 Same as Fig. 13.20, but for sandblasted Y-TZP and alumina plates. Solid lines are fits from Fig. 13.20 for as-polished surfaces, Zhang et al. [6]. With kind permission of John Wiley and Sons

are actual fracture times, N is a crack velocity exponent, A is a load independent, time independent and thickness-independent constants. Data from cyclic and dynamic test can be reduced to a common curve by defining effective fracture times as zero damage base line for comparison. Solid lines are best fits to the data using the radial fracture relations based on slow crack growth model. Here, t_R^c and t_R^d are actual fracture times, N is a crack velocity exponent,

$$t_R = t_R^c \quad \text{(cyclic)} \tag{13.5}$$

$$t_R = \left[\frac{2N^{0.47}}{(N+1)}\right] r_R^d \quad \text{(dynamic)} \tag{13.6}$$

Equation (13.6) is obtained from Eq. (13.3) by (13.4).

The reduction in strength of both alumina and zirconmia is a consequence of introduction of larger microcrack flaws in the sandblast specimens. The sand blast damage is seen in Fig. 13.22. From the lifetime point of view, it is more appropriate to consider in dental crowns in the sandblast state sustainable load P rather than stress, S, because of clinical tendency to relate oral conditions to biting forces. Such loads can be obtained from Eq. (13.2) and allowing to take into account the dental cement of modulus E_i and thickness, h between crown and dentin, and replacing substrate modulud E_s in Eq. (13.2) with an effective substrate modulus E_* given as

$$E_* = E_i(E_s/E_i)^L \tag{13.7}$$

Fig. 13.22 Micrograph showing partial top and cross-section view of sandblast damage (50 μm Al_2O_3 particles) in Y-TZP, Zhang et al. [6]. With kind permission of John Wiley and Sons

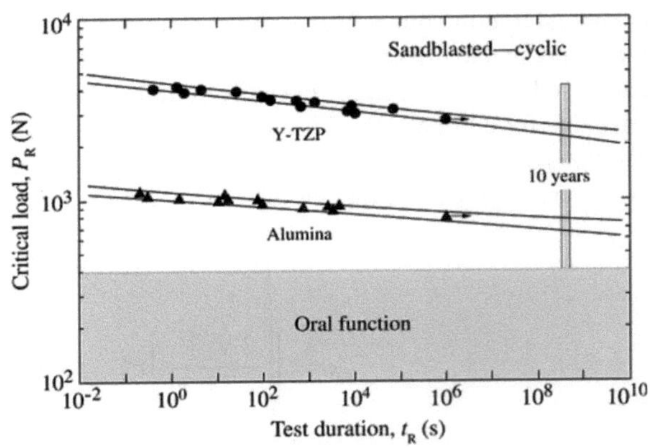

Fig. 13.23 plots corresponding to cyclic fatigue data in Fig. 13.22 for sandblasted alumina and Y-TZP of thickness 1.5 mm, but in terms of critical loads instead of stress and for dentin-like substrate with intervening dental cement of thickness 100 μm, using Eqs. 13.2, 13.7, and 13.9 to convert the data, Zhang et al. [6]. With kind permission of John Wiley and Sons

Where

$$L = L(h/d) \tag{13.8}$$

is an empirical Weibull function.

$$L = \exp\{-[\alpha + \beta log\left(\frac{h}{d}\right)]^{\gamma}\} \tag{13.9}$$

with $\alpha = 1.18$, $\beta = 0.33$, and $\gamma = 3.13$. By converting the sandblasted cyclic fatigue data of Fig. 13.21 to equivalent critical load is seen in Fig. 13.23 for ceramic plate of thickness d = 1.5 mm on thick dentin substrates with E_s = 16 GPa and cement interlayer of thickness h = 1100 μm and modulus E_i = 5 GPa. The plot includes extrapolation to lifetime at t_R = 10 years. The force P_R = 0–400 N corresponds to normal oral function (shaded area in Fig. 13.23). For guaranteed long lifetime the operational load required should be above be the shaded region.

13.5.3 Sandblasting in Yttria

The initial strength and survival rate of a dental yttria stabilize polycrystal (Y-TZP) to fatigue is found to be highly dependent upon surface preparation and to various hydrothermal exposure. Metal-based fixed partial dentures in restorative dentistry are replaced with all-ceramic crowns and bridges, which became a popular practice. The ceramic of choice for multi-unit posterior bridges and implant supported crowns is Y-TZP because its superior strength, fracture toughness and damage tolerance. Following grinding, sandblasting is commonly used to improve the adhesion between the luting cement and Y-TZP framework. The surfaces of Y-TZP after grinding and sandblasting, are partially transformed to ther monoclinic phase and are also damaged. The heavily damaged surfaces are the consequence of the partial plastically deformed surfaces due to grinding and airborne particles abrasion. Grinding-induced surface flaws are the strength-determining factor because their length commonly exceeds the depth of grinding-induced surface compressive layer, while substantial strength improvement occurs with sandblasting because the impact-induced flaws are comparable in size to pre-exixting flaws. It is expected that dental resorations under clinical conditions will be in service for at least 7–10 years, despite being exposed to cyclic mechanical and thermal loadings in chemically aggressive emvironment of the oral cavity. The strength tends to diminish steadily with time, from stress corrosion and fatigue. Recent reports have shown that sandblasting with 30 μm alumina particles at 2.5 bar, not only enhances the monotonic strength but also improves the fatigue limits and survival probabilities of the dental Y-TZP ceramics.

Tests were performed both in air and in artificial saliva. For biaxial fatigue testing, the specimens were subjected to a sinusoidal cyclic loading ranging from 50 to 850 N at a frequency of 15 Hz. The lower and upper loads corresponded to a stress of 36 and 620 MPa, respectively. For each specimen, the number of cycles to failure was registered. A fatigue cycle limit of 10^6 was set and the specimens that survived the mechanical fatigue testing were subsequently monotonically loaded to fracture. This procedure allows to explore the effects of all clinically relevant surface treatments. XRD analysis of surface treated specimens shown in Table 13.3 reveal monoclinic peaks with preference for the M ($\bar{1}11$) orientations and a substantial T (111) peak broadening accompanied by a reversed intensity of the tetragonal (200) (002) peaks. The results of the bi-axial flexural strength measurements in air and in artificial saliva obtained by monotonic loading are presented in Fig. 13.24. The mean strength and standard deviation (SD) of the as sintered samples at RT in air and in artificial saliva stored at 37 °C was 1012 ± 92, and 886 ± 108 MPa, respectively. When the samples were ground, the strength values obtained by testing in air (804 ± 106 MPa) and when tested in artificial saliva (either before or after accelerated ageing in an autoclave) were significantly lower ($P < 0.05$) compared to all other groups tested as indicated in Fig. 13.24. In contrast to grinding, sandblasting increased the strength in air (1289 ± 49 MPa) and, more importantly, there was no strength reduction when tested in artificial saliva either before or after short- or long-term accelerated ageing at 134 °C.The results of fatigue testing are summarized in Table 13.4. Whereas none of the sandblasted samples failed during the fatigue testing, neither nor in

Table 13.3 Relative amounts of X_m (monoclinic phase) on the surface of as-sintered and surface treated biomedical-grade Y-TZP ceramics before and after ageing in artificial saliva, Oblak et al. [7]. With kind permission of Springer Nature

Material	X_m before ageing (%)	X_m before ageing (%)		
		37 °C/24 h	134 °C/2 h	134 °C/24 h
As sindered	<1	15 ± 0.2	5.1 ± 0.4	25.6 ± 4.1
Sandblasted	14.5 ± 2.1	14.0 ± 1.6	14.5 ± 1.4	23.7 ± 2.4
Sandblasted + annealed 920 °C	1.5 ± 0.5	2.5 ± 0.5	2.9 ± 1.2	9.4 ± 2.6
Ground	4.3 ± 0.8	6.3 ± 1.5	7.2 ± 2.6	24.8 ± 2.1

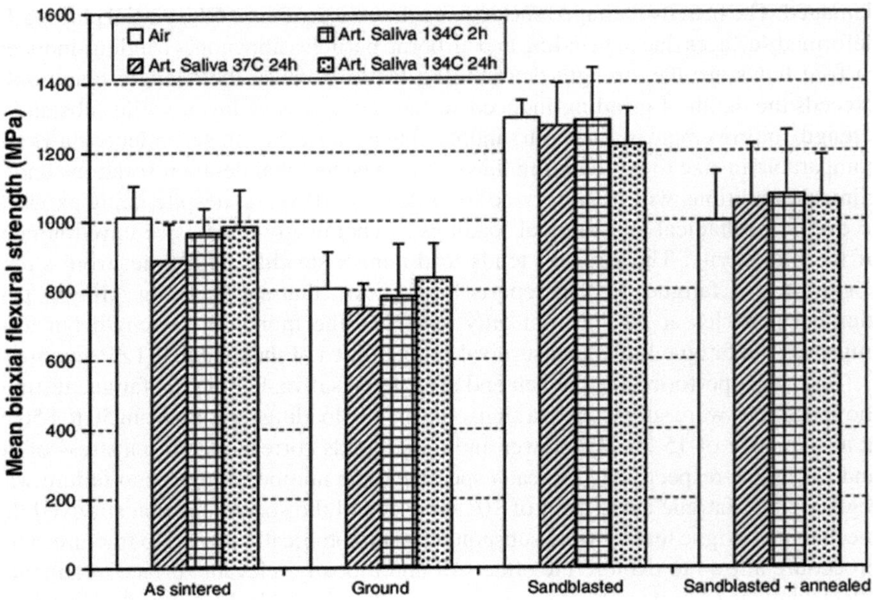

Fig. 13.24 Mean biaxial flexural strength values and SD after monotonic loading for as-sintered and surface-treated Y-TZP ceramics: in air, in artificial saliva, after accelerated ageing in artificial saliva for 2 h at 134 °C and after accelerated ageing in artificial saliva for 24 h at 134 °C, Oblak et al. [7]. With kind permission of Springer Nature

Table 13.4 Fatigue behavior of as-sintered and surface treatedY-TZPceramics in air and in artificial saliva, Oblak et al. [7]. With kind permission of Springer Nature

	As sintered (control group)	Dry ground (150 mm)	Sandblasted	Sandblasted + annealed
Air				
Survival rate[a] (%)	6 (60 %)	2 (20 %)	10 (100 %)	9 (90 %)
Survival strength + SD (MPa)	1,016 (105)	926 (2)	1,266 (107)	1,037 (50)
Art.Saliva, 37 °C 24h				
Survival rate[a] (%)	5 (50 %)	1 (10 %)	10 (100 %)	6 (60 %)
Survival strength + SD (MPa)	886 (98)	940	976 (339)	981 (167)
Art.Saliva, 134 °C 2h				
Survival rate[a] (%)	6 (60 %)	1 (10 %)	8 (80 %)	9 (90 %)
Survival strength + SD (MPa)	1,038 (68)	816	1,125 (193)	979 (165)
Art.Saliva, 134 °C 24h				
Survival rate[a] (%)	5 (50 %)	1 (10 %)	8 (80 %)	6 (60 %)
Survival strength + SD (MPa)	877 (188)	722	1,159 (177)	1,004 (91)

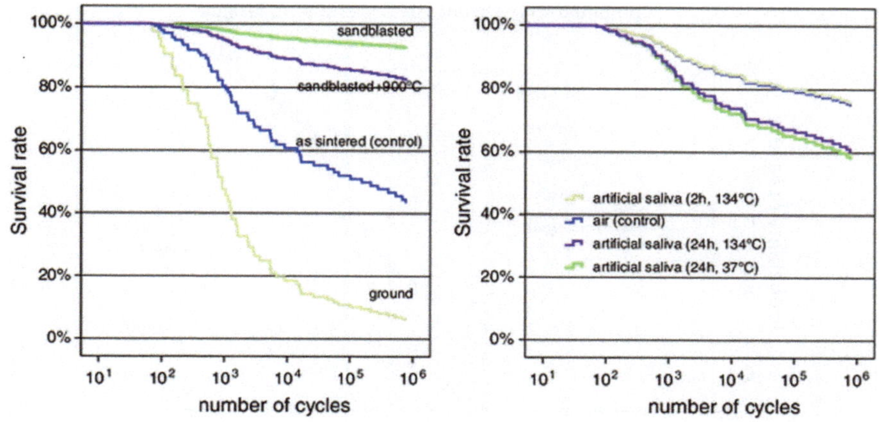

Fig. 13.25 The Cox survival curves after mechanical fatigue testing: left various surface treatments (each line includes together specimens treated in air and aged in artificial saliva), right various (pre)testing conditions (each group includes together as sintered and surface treated specimens), Oblak et al. [7]. With kind permission of Springer Nature

artificial saliva, after grinding, the survival rate dropped down to only 20% in air and to 10% in the artificial saliva solution. The survival rate of the sandblasted and annealed samples is 90% in air, but this dropped down to 60% in artificial saliva. The survival rate analysis after mechanical fatigue testing for various surface treatments and (pre)testing are presented in Fig. 13.25. The probability of failure vs strength is plotted in Fig. 13.26. Also, the strength of the specimens that survive cyclic fatigue testing for 10^6 cycles to 620 MPa are shown.

The estimated dc/da versus K at the max cyclic stress of fractured samples during loading, which showed monotonic strength reduction is shown in Fig. 13.27 for all samples. Very similar trends are seen for specimens following various treatments for crack growth of ~ 2–2.5 MPa $m^{1/2}$ where K_{max} is the threshold stress intensity factor.

13.6 Laser Treatment

13.6.1 Introduction

There has been a growing interest in recent years in the improvement of fatigue resistance through laser surface treatment (LST). The mechanical surface treatments can significantly be improved by processes such as deep rolling, shot peening and laser shock peening. Laser surface treatment carried out on a specimen imparts higher fatigue strength as a result of inducing compressive residual stress. Laser surface treatment offer a wide range of possibilities to achieve desired surface mechanical properties. The method can be used to reduce wear and improve fatigue resistance in particular when hard surface layers are required on tough and cheap base materials without undesirable thermal distortion. An important advantage of the mentioned

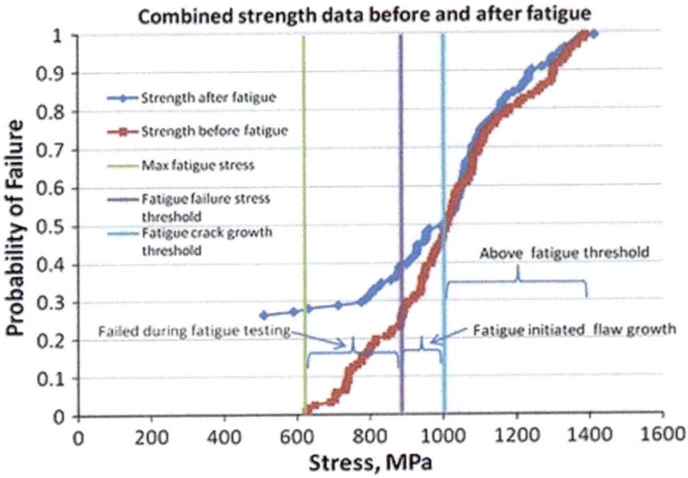

Fig. 13.26 A plot of the fracture strength of the monotonic strength results for all Y-TZP specimens versus the probability of failure. Also included are the results of the monotonic strengths for those specimens that survived cyclic loading. Three regions are identified: (i) samples with initial strengths between 620 and 890 MPa that failed during 1 million cycles at 620 MPa, (ii) samples between 890 and 1000 MPa initial strength that displayed strength reduction, and (iii) samples with initial strengths above 1000 MPa that did not show strength degradation, Oblak et al. [7]. With kind permission of Springer Nature

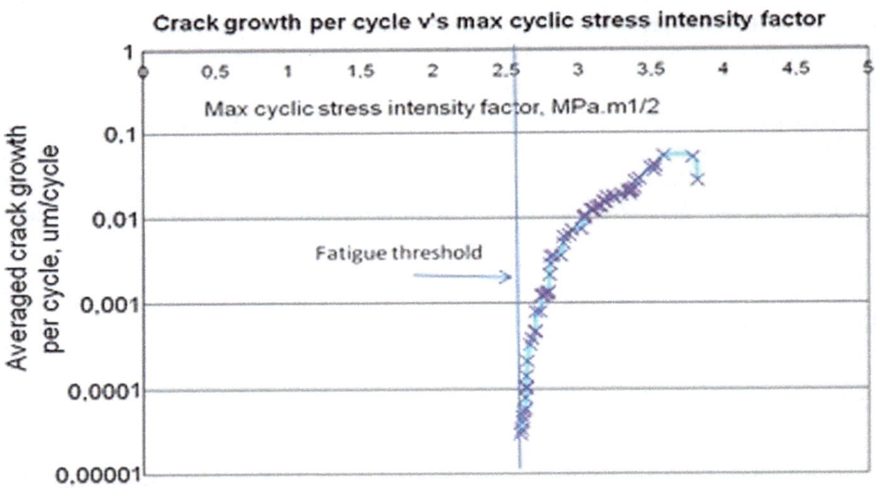

Fig. 13.27 Plot of the averaged fatigue crack growth rate (μm/cycle) during stressing at 620 MPa for the specimens that failed and for those that exhibited monotonic strength degradation after surviving 1 million cycles. Previous observations by Chevalier et al. and others indicate that the threshold value for cyclic fatigue is ~ 0.45 K_{Ic} suggesting that the Y-TZP material had a fracture toughness of 5.5 MPa $m^{1/2}$, Oblak et al. [7]. With kind permission of Springer Nature

hardened surface achieved by laser treatment is the self quenching of local overheated surface layer through heat conduction to the cold bulk of the material. Since laser treatment is done without quenching media, the procedure is clean without any further clean of the finished workpiece after laser treatment (such as heat treatment). It is possible to heat treat extremely small or large parts with complex shapes and such which include small bores or treating inaccessible areas. The energy input can be adapted as required by changing the laser source power and different focusing lenses (having different focuses) with different position of the lense focus with respect to the workpiece.The optical system can be adapted to the shape and complexity of the product by means of different shapes of lenses and mirrors. The laser treated surface layer depth or width are accurately reproducible. All the various laser surface treatment (LST) processes are based on heat effects controlled by process parameters causing microstructural changes. The LST technological parameters are: applied power, size and shape of beam spot, power density distribution on the treated surface and scanning velocity of beam. One of the advantages of LST to retard crack propagation since in the process local heating is used and because it is a surface treatment. Further advantage of the LST is flexibnility and by enhancing tribological properties improvement of metals and ceramics. Under laser irradiation the surface temperature is increased.

13.6.2 Laser Surface Treatment of Alumina

Alumina (Al_2O_3) is a widely used ceramic with various applications in industry. This is because of the superior mechanical and electrical properties that the material offers. However, its crack sensitivity and brittleness prevent ceramics such as Al_2O_3 to use its full potential (like metals offer) and therefore, strengthening ceramics in general is desirable and is a requirement for applications for example in medical, automotive, and nuclear industries. Various laser surface can modify the properies, thus laser shock peening introduces residual treatments compressive stress in laser treated coatings which is a potential strengthening mechanism which in metals generate local plastic deformation, which probably is likely to occur in ceramics also. Further, compressive residual stresses induced by laser peening prevents failures from fatigue and other surface deterioration effects by inhibiting crack initiation. Laser peening possesses precise control and the ability to access hard-to-reach areas of a component. In addition, the presence of compressive residual stress in laser treated coating also enhances fatigue life for the coatings. Alumina, especially α-Al_2O_3, is a ceramic oxide which has been used as a biomedical implant material due to its excellent biocompatibility, corrosion resistance and phase stability. In addition, superior mechanical properties of alumina including high hardness, wear resistance, scratch resistance, low coefficient of friction and sufficient mechanical strength to resist fatigue make it one of the strongest candidates for implant applications among other ceramics.

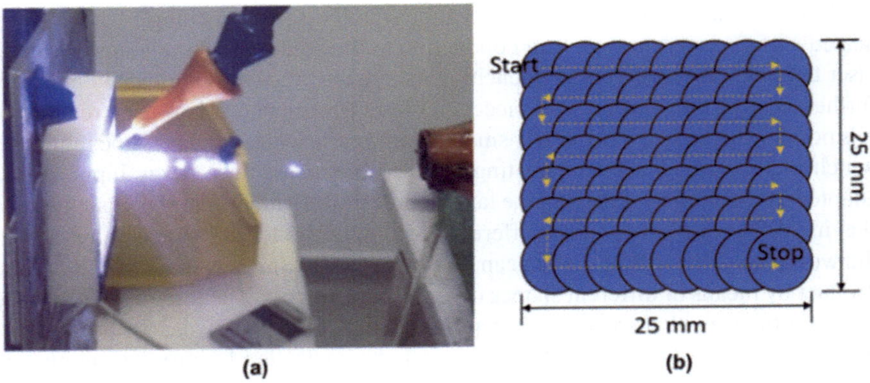

Fig. 13.28 illustrates with an optical image of the laser shock peening set-up in **a**, and **b** laser beam path with peening strategy for the Al_2O_3 ceramic, Shukla et al. [8]. With kind permission of Elsevier

Laser shock peening, using a Q-switch Nd-YAG laser with varying pulse energy of 1 J, 1.5 J to 1.7 J was applied. A 50 mm by 50 mm laser peened patch was programmed for the three respective energies applied with a 50% overlap on the Al_2O_3 armour grade ceramics. A pulse duration of 10 ns and a pulse repetition rate of 10 Hz was applied. Traverse speed was kept constant at 4.8 mm/s. The 99% fine grained armor plate as-sintered Al_2O_3 consisted of 0.56 μm (Ra) original surface roughness on average (Ra is surface roughness parameter). For full description of the laser shotpinning of alumina, the reader should consult the original publication, but some details, mainly related to the laser applied for alumina is presented below. The laser shock peening set-up is illustrated in Fig. 13.28. The laser parameters for the treatment of alumina are listed in Table 13.5.

Line-scans were performed normal to the laser shock peening direction across 2 mm with a step size of 40.8 μm. A confocal setup allowed autofocus, ensuring that the spot remained focused over long line scans is illustrated in Fig. 13.29. Confocal fluorescence microscopy is a technique used to acquire high-resolution optical sections within a thick sample or to reduce background fluorescence. The surface characteristics of Al_2O_3 before and after laser shock peening is indicated in Table 13.6. A byproduct of laser shot peening is material removal with consequent increase in roughness. This is shown in Fig. 13.30, indicatinfg that the surface roughness, Ra prior to laser shock peening of Al_2O_3 is 0.989 (Fig. 13.30a) μm while the roughness of laser shock peened surfaces increased to 1.08 μm at 1 J (Fig. 13.30b), 1.64 μm at 1.5 J (Fig. 13.30c), and 1.92 μm at 1.7 J (Fig. 13.30d). The root mean square roughness Rq was the lowest for the as-received surface, and as the energies increased, the Rq values began to increase as shown in Table 13.6. In Table 13.6 R_v, R_p and R_t are the profile valley depth, sharp Vickers indenter and the maximum profile height, respectively. The maximum compressive stress on the surface after laser shock peening is -276 MPa and the maximum tensile stress is 39.53 MPa, which provided an average stress over a length of 2 mm as illustrate in

Table 13.5 showing the experimental parameters for laser shock peening of Al_2O_3 ceramic, Shukla et al. [8]. With kind permission of Elsevier

Parameters	Values
Pulse energy (J)	1J, 1.5J, 1.7J
Laser power intensity (Gw/cm^2)	19.9 @ 1J 29.85 @ 1.5J 33.83 @ 1.7J
Shock pulse pressure (GPa)	7.58 @ 1J 9.29 @ 1.5J 9.89 @ 1.7J
Radiance density (W/Cm^2/Sr/μm)	4.42 @ 1J 6.63 @1.5J 7.52 @ 1.7J
Spot diameter (mm)	0.8
Pulse repetition rate (Hz)	10
Over lapping rate (%)	50
Pulse duration (ns)	10
Wavelength (nm)	1064
Traverse speed (mm/s)	4.8
Hugoniot elastic limit (GPa)	7.98

Fig. 13.31. The maximum compressive stress of −494 MPa is the highest from the three energies applied.

In summary, laser shock peening can be used to strengthen ceramics such as Al_2O_3. The use of laser pulses is to strengthen difficult to process advanced ceramics. One of the major advantages of the laser as a tool for material processing is the ability to precisely control where in the material and at what rate energy is deposited. This control is exercised through the proper selection of laser processing parameters to achieve the desired material modification. As the laser energy increased the shock pulse pressure, the compressive stress also increased. Consequently, the dislocation density after laser shock peening considerably increased locally in bands (at 1 J the density of dislocations was found to be 6.11×10^{11} at 1.5 J was 2.67×10^{14} and at 1.7 J 2.10×10^{13}, respectively).Further, the results have shown that laser shock

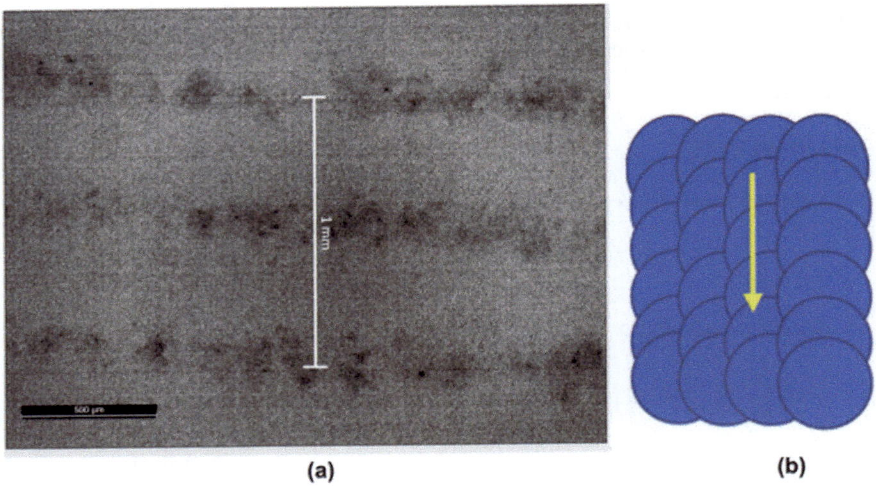

Fig. 13.29 **a** An optical image from the confocal microscope showing the peened area and scanned region; **b** schematic of the path of the fluoresence line-scan over the laser peened areas, Shukla et al. [8]. With kind permission of Elsevier

Table 13.6 Presents 2-D surface roughness characteristics of the Al_2O_3 ceramic prior-to and post laser shock peening, Shukla et al. [8]. With kind permission of Elsevier

Surface condition	Surface roughness (Ra)				
	Ra (μm)	Rp (μm)	Rq (μm)	Rt (μm)	Rv (μm)
As-received	0.989	6.59	1.33	27.24	-20.64
1J	1.082	5.71	1.65	28.46	-22.74
1.5J	1.64	6.079	2.73	29.45	-23.37
1.7J	1.92	5.19	3.10	30.54	-25.35

peening (LSP) is a way to generate local plastic deformation. One might note that LSP is a process that was developed as a substitution for conventional shot peening. it is a surface treatment that is used to improve fatigue.

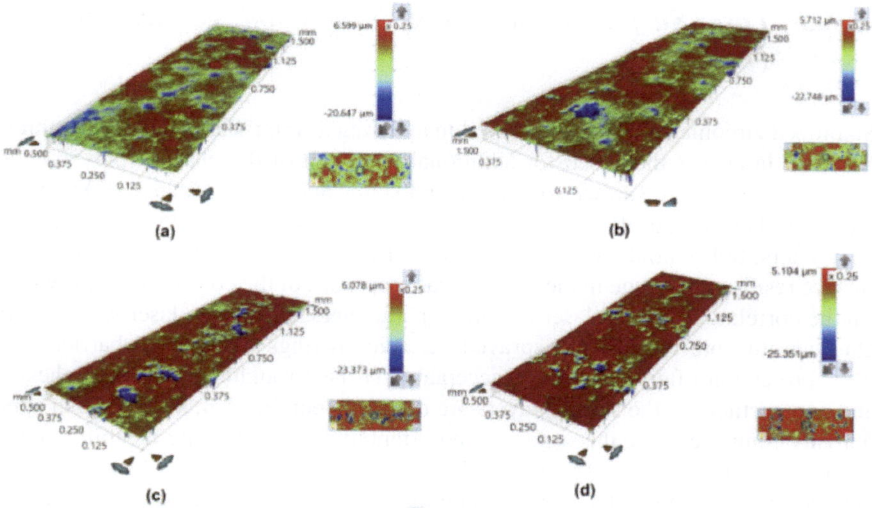

Fig. 13.30 Showing an illustration of a 3-D profile of the untreated Al_2O_3 ceramic in **a**; the laser shock peened surface with 1 J in **b**; laser shock peened surface with 1.5 J in **c** and the laser shock peened surface with 1.7 J in **d**

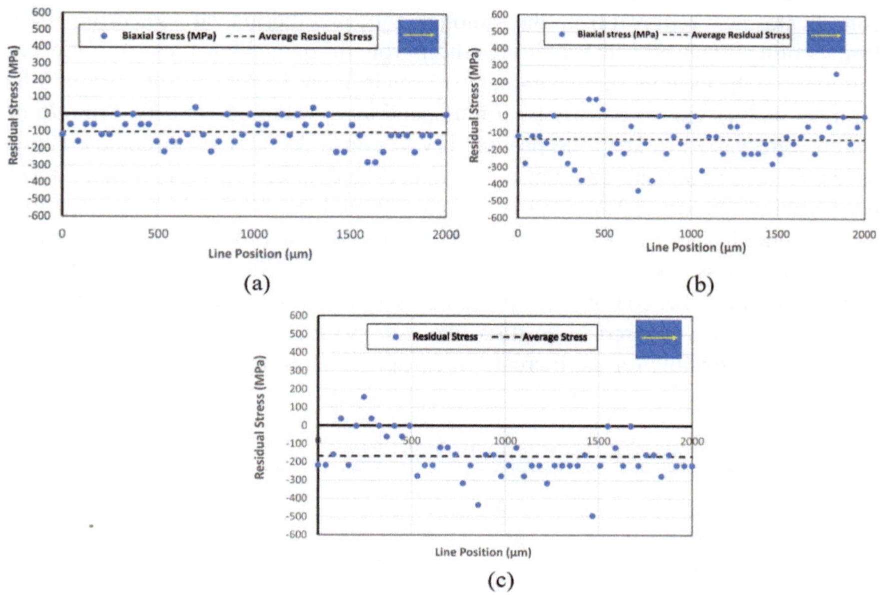

Fig. 13.31 Illustrating the residual stress after laser shock peening Al_2O_3 ceramics at 1 J in **a**; at 1.5 J in **b**; and 1.7 J in **c**, Shukla et al. [8]. With kind permission of Elsevier

13.6.3 Laser Surface Treatment of Yttria Stabilized Zirconia (YSZ)

Stabilized zirconia was plasma sprayed in air using a water-stabilized plasma torch. Nd-YAG laser was then used for additional treatment of the plasma sprayed coatings. Microhardness, surface roughness and slurry abrasion resistance (SAR) were measured before and after the laser treatment and the results vary with the laser parameters. Substantial changes of the structure up to a complete re-melting of the surface resuls in a change in the mechanical properties of the coated materials which can be correlated with the laser treatment parameters such as the laser wavelength. One has tto note that plasma sprayed ceramic coatings for thermal barriers and wear protection often exhibit non-acceptable surface roughness and the cohesion and compactness of the surface layers are not sufficient. Further, a decrease in open porosity connected with the surface is important to minimize the permeability of the coating for various oxidizing agents. Thus, the laser post-treatment of sprayed coatings ensures melting of the surface ceramic layer to a certain depth while, further below the surface, the layer remains unmelted and maintains the original coating structure. The rapid crystallization after the laser treatment modifies the structure from lamellar to a dense equiaxed crystal structure.

The ZrO_2 is stabilized by 8 wt.% of Y_2O, but its very low thermal conductivity complicates heat removal from the remolten zone and thermal stresses originationg from cooling induces verical cracks resulting from the quenching stresses. Table 13.7 summarizes the laser working parameters. Seen in Table 13.8 the labeling of the YSZ samples such as energy density E (J/mm^2), and laser scanning velocity v [mm/min] which were varied to keep the product E·v = const. As-sprayed zirconia coatings have uneven surface and contains a lot of pores and defects as seen in Fig. 13.32a. Average microhardness HVm 0.1 is 727 as seen in Table 13.9, which is lower than the microhardness of bulk YSZ. As laser treatment leads to a significant increase of microhardness for all used laser parameters. The roughness values of the laser treated zirconia YSZ coatings are listed in Table 13.10. Generally, the laser treated YSZ coatings are covered by a glass-like (glassy) layer containing many defects which are responsible for the increase of R_y and to a lower extent in R_a. Recall that

Table 13.7 Laser parameters, Ctibor et al. [9]. Open access

Working wavelength (μm)	1.06
Power (kW)	2
Focus distance (mm)	55
Working distance (mm)	35
Beam width (mm)	10
Protective gas	Argon

Table 13.8 Labelling of the samples, Ctibor et al. [9]. Open access

Sample label	Energy density E (J/mm²)	Laser scanning velocity v (mm/min)	Plasma spray distance SD (mm)
YSZ	-	-	300
YSZ 5	5	2400	300
YSZ 15	15	800	300
YSZ 20	20	600	300
YSZ 25	25	475	300
YSZ 30	30	400	300

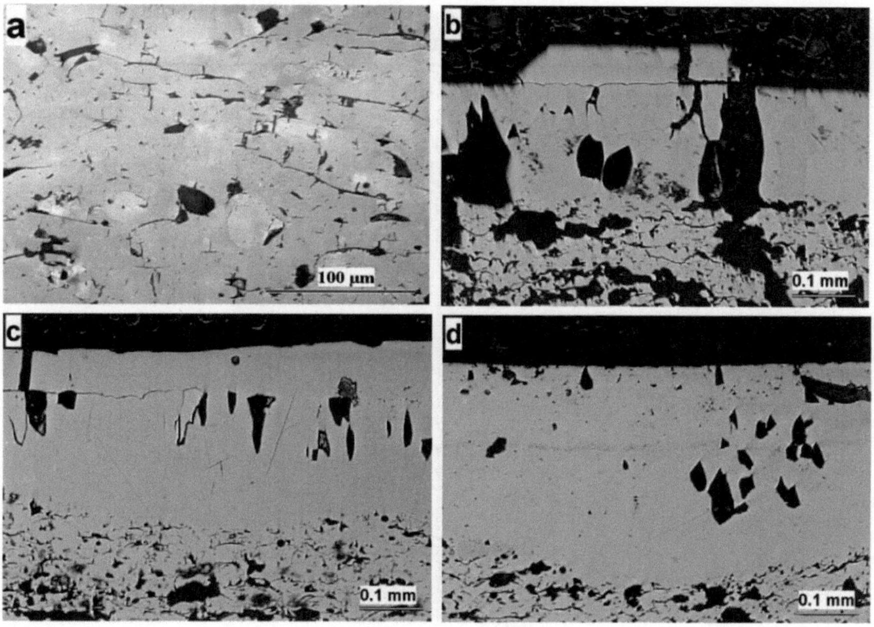

Fig. 13.32 Microstructure of zirconia samples, light microscopy**a** YSZ, **b** YSZ 15, **c** YSZ 25, **d** YSZ 30, Ctibor et al. [9]. Open access

R_a in the surface roughness is the mean arithmetic deviation and R_y is the maximum height of the profile. The wear resistance of YSZ is illustrated in Fig. 13.33.

Specimens YSZ 5 and YSZ 15 indicated that after the first two cycles their wear loss is higher than for the as-sprayed coatings, but then their wear loss becomes lower. This can be seen in Fig. 13.33. This can be probably explained by wearing-off the surface glass-like layer.

Table 13.9 Microhardness values, Ctibor et al. [9]. Open access

Sample label	$HVm_{0.1}$	Standard deviation	Increase of $HVm_{0.1}$ (%)
YSZ	727	161	-
YSZ 5	n.a.	n.a.	n.a.
YSZ 15	1729	55	58
YSZ 20	1464	101	50
YSZ 25	1717	59	58
YSZ 30	1793	77	60

Table 13.10 Roughness values, Ctibor et al. [9]. Open access

Sample label	R_a (µm)	R_y (µm)	Relative change of	
			R_a (%)	R_y (%)
YSZ	14.7	61.1	-	-
YSZ 5	15.6	103.1	5.77	40.74
YSZ 15	17.1	109.6	14.04	44.25
YSZ 20	12.6	111.9	-16.67	45.40
YSZ 25	15.0	100.6	2.00	39.26
YSZ 30	8.2	55.3	-79.27	-10.49

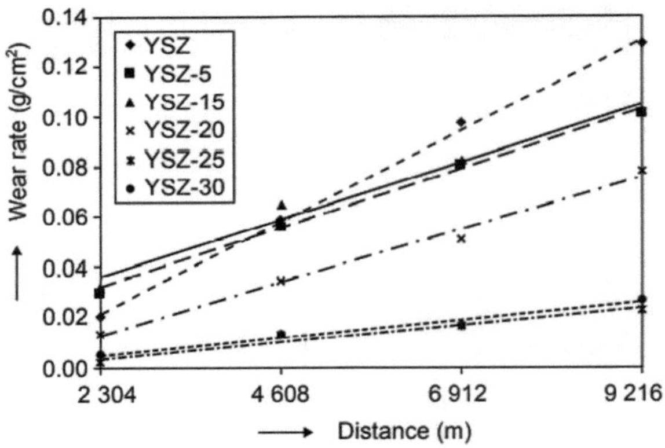

Fig. 13.33 Wear rate of zirconia samples, Ctibor et al. [9]. Open access

In summary, additional laser treatment was used of plasma-sprayed coatings of stabilized zirconia. Microhardness, surface roughness and slurry abrasion resistance were measured before and after the laser treatment. All these properties are enhanced after the laser treatment. Increase of the laser energy density leads to an improvement of the wear resistance and microhardness. Nevertheless, the coating application is somewhat limited by the formation of large horizontal cracks. Thermal cycling up to 1270 °C is reported to show that there is no zirconia coatings improvement of the treated vs, untreated YSZ. Finally, the laser shot peening (LSP) process offers many benefits such as, improvement in fatigue and hardness; reduced wear rates; increase in compressive stresses just to mention a few.

13.6.4 Laser Surface Treatment of SiC

The silicon carbide (SiC) produced by cold isostatic pressing (CIP) was caracterized mechanically and microstructuraly before the surface reatment by laser shock peening. The CIPed sample sintered at 1200 °C for 5 h had an average grain size of 17.1 μm. The laser surface treatment is expected to improve fatigue and hardness and reduce wear rate. More so, longer lifetime is expected in the various operations of SiC.

The schematic set-up of the LSP of SiC is illustrated in Fig. 13.34. The pressute of the plasma was calculated by using the equation

$$P_{\max}(GPa) = 0.01\sqrt{\frac{\alpha}{2\alpha + 3}}\sqrt{Z}\sqrt{I_0} \tag{13.10}$$

Here, P_{max} is the maximum peak pressure induced by the plasma in GPa, α is the ratio of the thermal to internal energy, which was 0.1, Z is the reduced shock impedance between the SiC and the confinement medium (water) in this case (926,000), I_0 is the constant absorbed laser power density in the confined ablation mode (Kim and Lee). The laser parameters are presented in Table 13.11. The model of the plastic deformation due to the shock wave produced by the LSP surface treatment can be determined by Eq. (13.11) as given below:

$$HEL = \frac{1 - \upsilon}{1 - 2\upsilon}\sigma_y^d \tag{13.11}$$

This relation is used to determine the plastic deformation produced by the LSP surface treatment of SiC. The Poisson ratio, ν for SiC is 0.14 and σ_y^d is the dynamic yield strength at high strain rates. Furthermore, as the shock wave propagates into

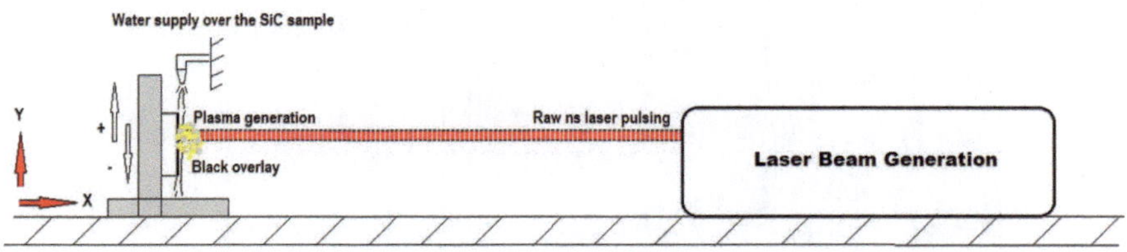

Fig. 13.34 A schematic of the laser beam delivery and the experimental set-up of LSP SiC ceramic, Shukla et al. [10]. With knd permission of Elsevier

Table 13.11 Presents values of both pressure and HEL for the parameters applied for this work, Shukla et al. [10]. With knd permission of Elsevier. HEL stands for Hugoniot Elastic Limit

Laser Energy (mJ)	Laser Intensity - I_0 (Gw/cm²)	Plasma Pressure (GPa)
215	0.063	0.427
600	0.1763	0.7136
800	0.235	0.8246
966	0.2838	0.904

the material (metals), plastic deformation occurs up to a depth at which the peak stress induced on the material would equal to its HEL. The HEL is related to the dynamic yield strength σ_y^d at high strain rates. SEM micrograph of the as received SiC is shown in Fig. 13.35. Striations-like machine marks are indicated in this figure in some areas and also voids and micro-pores.

The average surface roughness, R_a of the as received and the LSPed surfaces are listed in Table 13.12. The effect of increasing the laser energy density on SiC is shown in Figs. 13.36a–d. The machining marks and striations on the SiC surface which was observed in Fig. 13.35 disappeared with increasing laser energy density

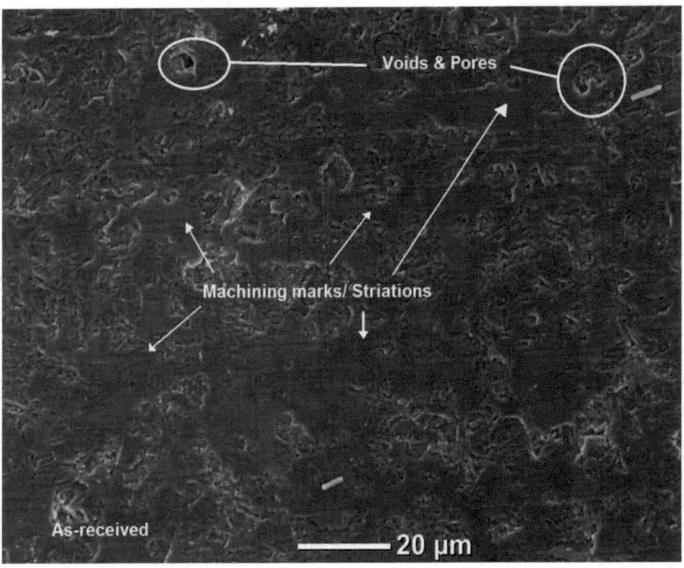

Fig. 13.35 As-received SEM micrograph of the SiC ceramic, Shukla et al. [10]. With knd permission of Elsevier

Table 13.12 The average surface roughness of the SiC as received and the laser shock peened surfaces of the SiC ceramic, Shukla et al. [10]. With kind permission of Elsevier

Surfaces	As-received 0 J/cm²	Laser shock processing @ 0.379 J/cm²	Laser shock processing @ 1.057 J/cm²	Laser shock processing @ 1.702 J/cm²
Average Surface Roughness (Ra)	1.55 μm	3.50 μm	4.35 μm	5.30 μm

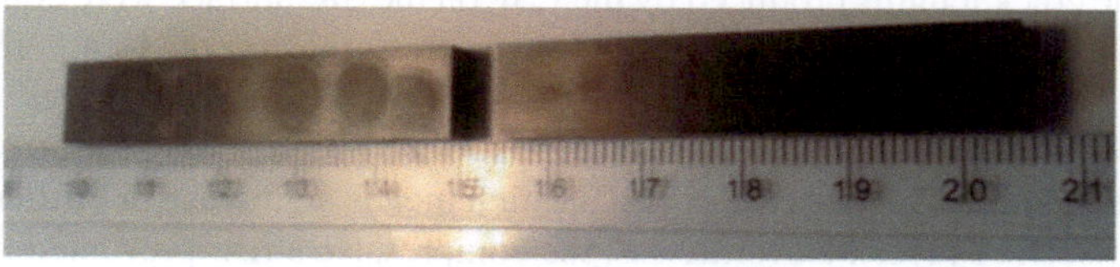

(a)

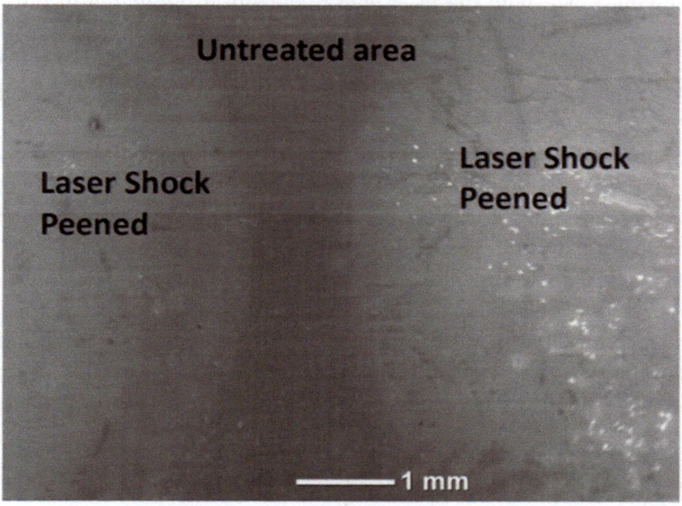

(b)

Fig. 13.36 Optical and SEM images of the SiC ceramic with distributed pulses in (a); and (b) LSP conducted at 1.702 J/cm^2, 10 Hz, for 6 ns pulse duration with a single pulse, Shukla et al. [10]. With kind permission of Elsevier

as seen in Fig. 13.36. SEM micrograph cross sections of untreated and laser treated SiC surfaces are shown in Fig. 13.38. Table 13.13 lists the crystallite size and micro strain measured on the untreated and the LSPed SiC with change in the power density. Further, the residual stress developed on the surface of the as received and laser shock peened SiC ceramic are listed in Table 13.14. It is indicated in the table, that LSP alteres the residual stress from tensile (as received) to compressive, which depends on the amount of energy delivered onto the surface as well as its ability to generate plastic deformation. An overall reduction of ~ 198 Pa is noted (from tensile to compressive). Introducing a compressive residual stress enhances the mechanical properties of SiC, such as hardness, K_{Ic} and reduces the crack length. But also, other properties such as wear resistance, bending strength would improve from this type of surface treatment, but of special importance is the fatigue performance enhancement. The LSP surface treatment increases the surface hardness of SiC compared to the as received material, which is attribured to the residual compressive stress induced by LSP. This can be seen in Fig. 13.39 where the Vickers indentation size is related to the laser energy intensity. Further, the crack lengths produced by the LSP on the SiC surface were considerably smaller than those in the as-received SiC surface as seen in Fig. 13.40. The lowest flaw size found was 152 μm, compared to 217 μm for the as-received SiC. The lowest flaw size of 152 μm, indicates that the surface became less prone to cracking, despite the increased hardness after LSP (which means an incrase in brittlenes in the top or near surface layer). Nevertheless, the compressive stress prevented the cracks from expanding whilst being indented by Vickers indentation. The variation of the fracture toughness, K_{Ic} with the laser energy density is compared in Fig. 13.41 for the as-received and LSP treated SiC. The comparisons show an increase in K_{Ic} from 2.32 MPa $m^{1/2}$ for the as-received surface (0 J/cm^2) to 2.60 MPa $m^{1/2}$, at 0.379 J/cm^2 and 3.29 MPa $m^{1/2}$ at 1.702 J/cm^2 respectively for the laser shock peened surfaces (Fig. 13.37).

In summary, it was observed that in the brittle SiC treated by LSP the applied laser energy induces residual stresses and the consequent changes modify the surface roughness, microstructure, hardness, crack geometry and the fracture toughness, K_{Ic}. The LSP process can effectively enhance the performance and functional lifetime of ceramics such as SiC for industrial applications. Considerable microstructural modification occurs on applying the process, could prove to be effective for enhancing the performance and functional life of many industrial ceramic parts. In particular, the results of this work showed an increase in surface roughness after LSP compared to the as-received surface. This indicated a change in surface integrity at sub-micron

Table 13.13 Geometric parameters for the as received and laser shock peened SiC ceramic with changes in the power density, Shukla et al. [10]. With knd permission of Elsevier.

Sample Type J/Cm^2	Crystallite Size (μm)	Strain (×10^{-4})
0	1.65 ± 0.03	1.85
0.379	1.55 ± 0.09	1.23
1.057	1.61 ± 0.04	1.4
1.702	1.80 ± 0.07	1.5

Table 13.14 Residual stress on both the as-received laser shock peened SiC ceramic, Shukla et al. [10]. With knd permission of Elsevier

Sample Type J/cm^2	Residual stress (MPa)
0	101 ± 16
0.379	51 ± 11
1.057	-92 ± 9
1.702	-97 ± 7.5

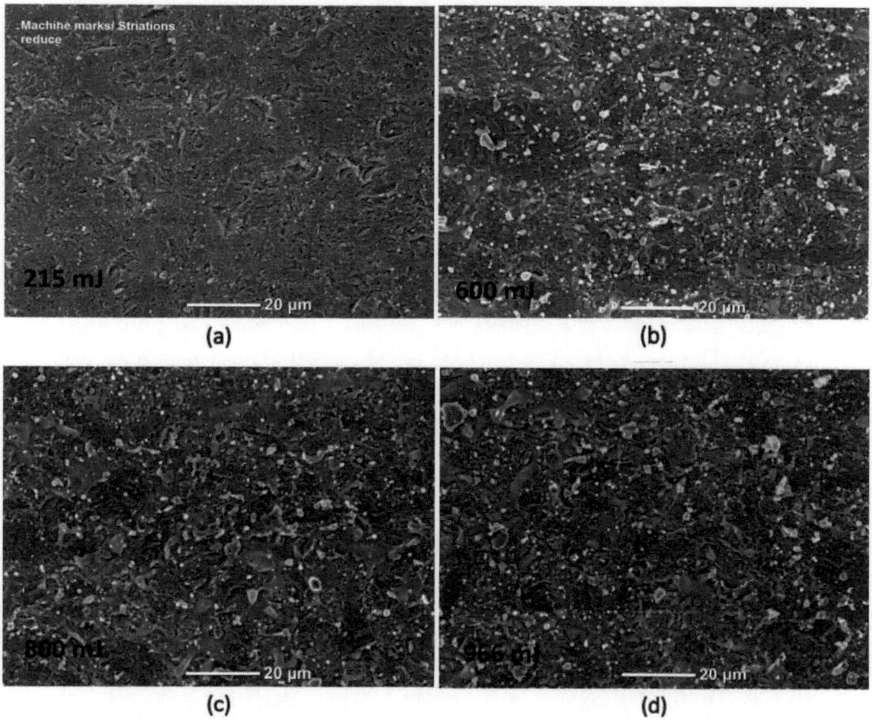

Fig. 13.37 SEM micrograph of the laser shock peened SiC ceramic at 0.379 J/cm^2 in **a**; 1.057 J/cm^2 in **b**; 1.410 J/cm^2 in **c**; and 1.702 J/cm^2 in (d), around the centre of the spot, Shukla et al. [10]. With kind permission of Elsevier

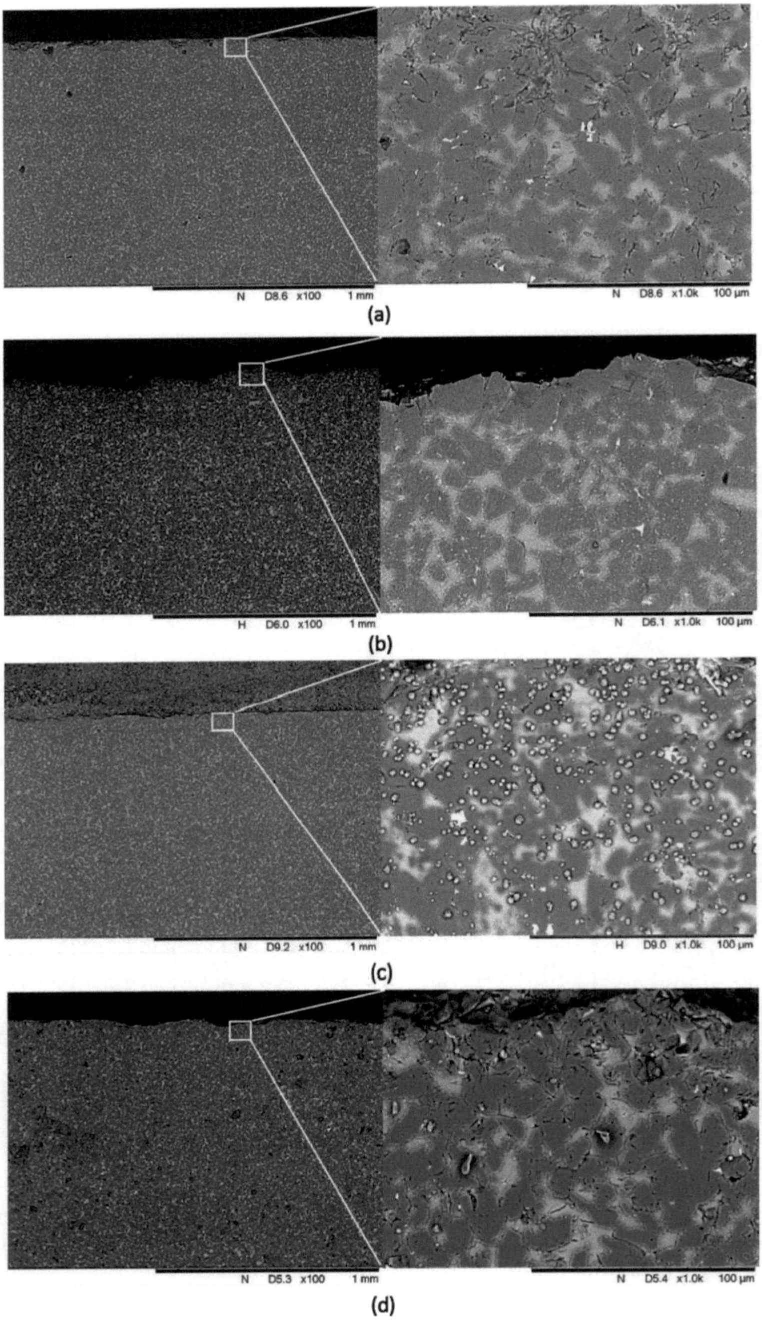

Fig. 13.38 SEM micrographs of cross section of untreated SiC in (**a**), LSP SiC with laser energy density of 0.379 J/cm^2 in (**b**), LSP SiC with laser energy density of 1.057 J/cm^2 in (**c**), and LSP SiC with laser energy density of 1.702 J/cm^2 in (**d**), Shukla et al. [10]. With knd permission of Elsevier

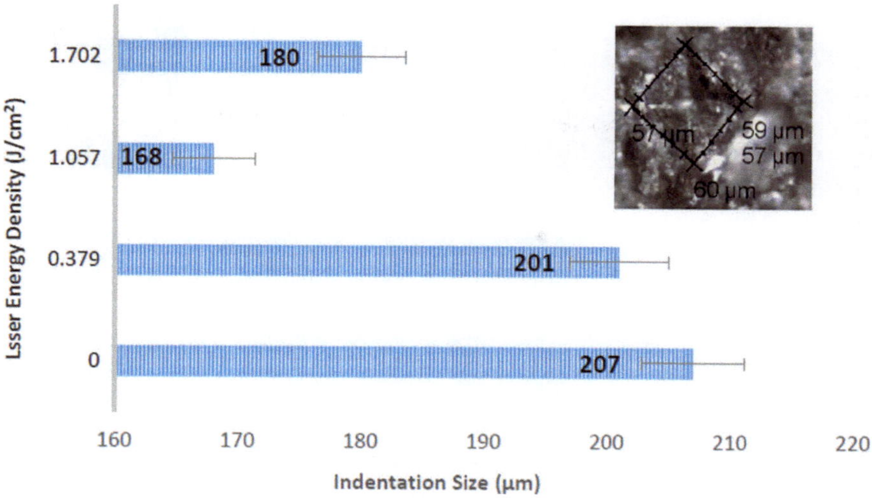

Fig. 13.39 A graphical illustration showing Vickers indentation size of both the as-received (0 J/cm^2) and the laser shock peened surfaces of SiC ceramic. The diamond indentation footprint accompanied by the graph demonstrate the size of the diamond footprint, Shukla et al. [10]. With kind permission of Elsevier

level at the highest applied laser energy. Considerable microstructural modification also occurred on the laser shock peened surfaces as pre-existing LSP to SiC. Surface defects were eliminated with increasing laser power, but the extent depends on the laser power applied. Both the increase in hardness, a reduction in the flaw size and the increase in the surface K_{Ic} indicated that the laser shock peened surface may have undergone a level of plastic + elastic deformation. A key factor in the surface modification technique of SiC (and other ceramics) is the compressive stress induced by LSP.

13.6.5 Laser Surface Treatment of Si_3N_4

Si_3N_4 is a relatively dense and light weight ceramics with good corrosive properties. However, it has a relatively low fracture toughness, K_{Ic}. In some of the industrial application of Si_3N_4 (such as valves; pistons; exhaust manifold; seals; turbo chargers; bearings; turbine blades, rocket nozzles and rotor), K_{Ic} is an essential property since low fracture toughness in comparison to metals and alloys is one of the disadvantages of the ceramics in general and Si_3N_4 in particular. Crack sensitivity and low K_{Ic} can limit the use of Si_3N_4, particularly for the applications mentioned above. With this in mnd, and due to its desirable physical, mechanical, and thermal properties, and longer lifetime in practical use, it would therfore be of interest to increase its K_{Ic}.

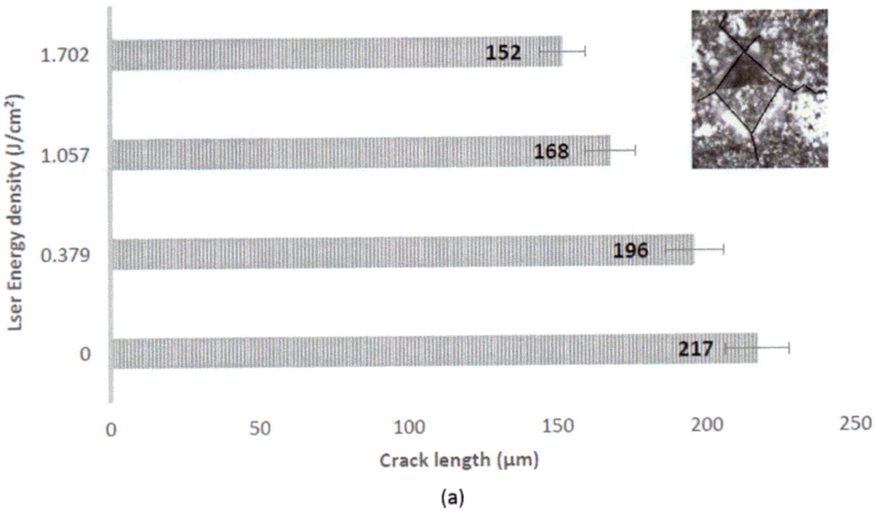

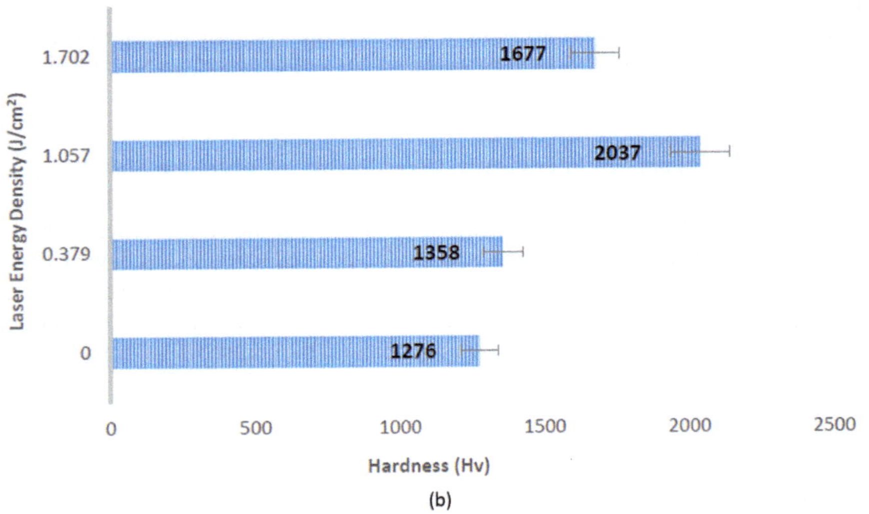

Fig. 13.40 The change in flaw size in **a**; and **b** the change in hardness for both the as received (0 J/cm^2) and the laser shock peened surfaces of SiC ceramic applied at various energy densities, Shukla et al. [10]. With kind permission of Elsevier

Surface properties of ceramics can be modified by laser treatment, in particular LSP affects the surface hardness and K_{Ic}.

High power Nd-YAG laser was applied on cold isostatically pressed (CIPed) Si_3N_4. The Si_3N_4 ceramic comprised of 90.5wt% Si_3N_4, 6 wt% yttria (sintering additive), and 4 wt% unspecified content. The Si_3N_4 was CIPed at 455 bar and followe by sintering at 1200 °C for 5 h. The Si_3N_4 was mechanically and microstructurally

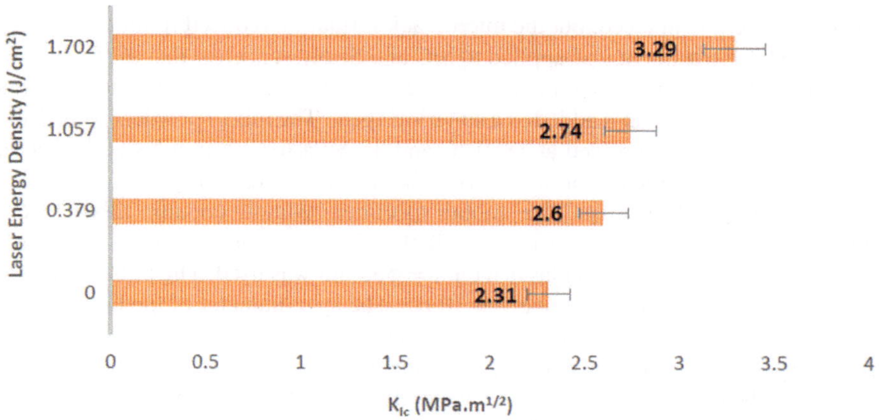

Fig. 13.41 The fracture toughness parameter (K_{Ic}) for both the as-received (0 J/cm^2) and the laser shock peened surfaces at different laser energy densities applied to the SiC ceramics, Shukla et al. [10]. With knd permission of Elsevier

characterized before to all experiments. The average as-received surface finish was Ra 1.50 μm. The surface hardness was measured to be 1467 H_V using 10 kg indentation load, and a plane strain fracture toughness, K_{Ic} was measured to be 2.91 MPa.m$^{1/2}$. In Fig. 13.41a a comparison of the hardness of the as-received Si_3N_4 with that of the LSPned Si_3N_4 showed a slight change in the surface hardness. The LSPned surface comprised 4.5% lower hardness. The fracture toughness was improved because the reduction in hardness and brittleness. This was an indication that the Si_3N_4 surface exhibits some ductility.

In summary, reduction in surface Vickers hardness ater LSP decreases the brittleness and impart some ductility to Si_3N_4. Further, an improvement in K_{Ic} by LSP makes Si_3N_4 applicable for practicle use.

13.6.6 *Laser Surface Treatment of Si_3N_4/SiC*

It has been mentioned in the earlier section that ceramic materials are brittle. Thus, their fracture toughness is much lower than that of metals. It was also indicated that laser peening, LP is a technique that utilizes the impact force of plasma generated by the irradiation of a pulsed laser in order to prevent many forms of damage, such as fatigue, wear, and stress corrosion cracking (SCC) in metals and applied to ceramics (such as partially stabilized zirconia, silicon nitride, alumina, etc.) also. The improvement is a consequence of the compressive residual stress induced by SP. However, the application of LP to ceramics has been limited because of the concerns regarding the formation of micro-cracks under the strong LP conditions required to introduce large compressive residual stresses. Therefore, the LP has to be such that a residual stress is induced, but the microcrack formation reduced if not completely eliminated. LP treatment was applied to SiC and Si_3N_4, it is clear that the Si_3N_4 can be improved by incorporating in the Si_3N_4 reinforcing SiC. The composite contains 20 wt % SiC and yttria is is added as a sintering agent. The mixed powders were dried and subsequently hot-pressed at 1850 °C and 35 MPa for 2 h in a N_2 atmosphere. Following surface preparation to a mirror-like finish, the resulting average surface roughness was Ra 0.025 μm. The smooth specimens without coating (LPwC) were esxposed to laser peening in water and Table 13.15 lists the conditions of the LP process. Q-switched Nd-YAG laser with λ = 532 nm was used. The pulse duration, t was 6.2 ns and the repetition rate was 10 Hz. The spot diameter of the laser, D was fixed as 0.5 mm. The pulse energy, Ep indicates the energy contained per pulse of the laser. The irradiation density, N_p indicates the number of pulses irradiated per unit area. The peak power density, G, which is the light intensity per unit area, can be calculated by the following equation using the pulse energy, Ep, and spot area, ($A_p = \pi d^2/4$):

$$G = \frac{E_p}{A_p t} \tag{13.12}$$

The power dewnsities selected were $G = 2$, 16, and 33 GW/cm². The lase peening area, the peening pattern and track to the peening area and the experimental setup are shown in Fig. 13.42. The pitch in Fig. 13.42 is defined as the distance between

Table 13.15 Conditions of laser peening, Saigusa et al. [13]. Open access

Specimens	Pulse energy, E_p [mJ]	Spot diameter, D [mm]	Power density, G [GW/cm²]	Irradiation density, N_p [Pulse/mm²]	Overlapping pitch [mm]	Overlap ratio [%]	Coverage, C_v [%]
Specimen 1	24	0.5	2	16	0.25	50	314
				51	0.14	72	1000
				153	0.081	84	3000
				255	0.063	87	5000
Specimen 2	194		16	16	0.25	50	314
				51	0.14	72	1000
				153	0.081	84	3000
				255	0.063	87	5000
Specimen 3	401		33	16	0.25	50	314
				51	0.14	72	1000
				153	0.081	84	3000
				255	0.063	87	5000

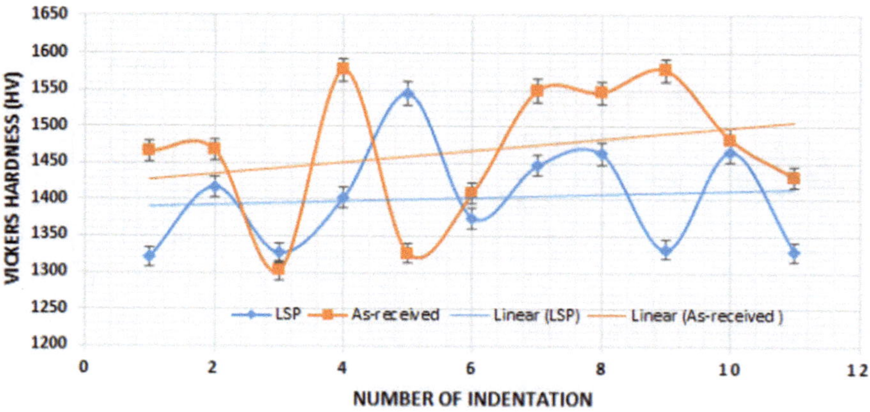

(a)

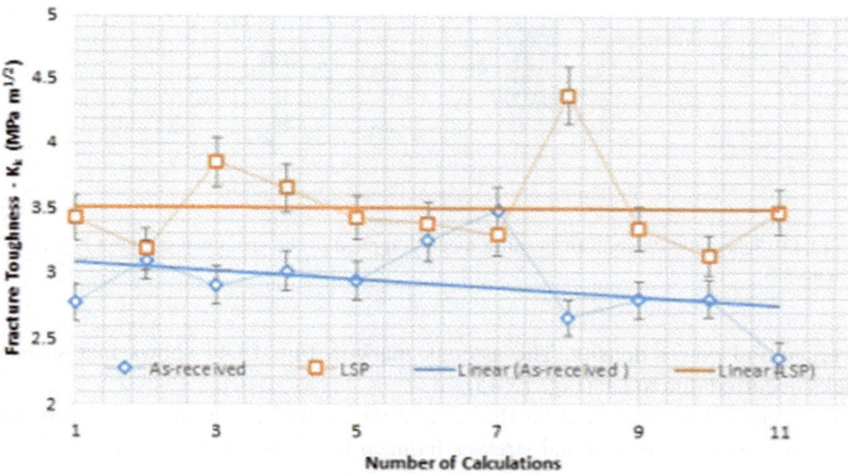

(b)

Fig. 13.42 Surface hardness in **a**; and the K_{Ic} in **b** of the LSPned Si_3N_4 advanced ceramic, Shukla et al. [12]. Free access

two laser-irradiated spots. The coverage, C_v, which is the amount of overlap per unit area, is calculated by the following relation

$$C_v = N_p A_p \tag{13.13}$$

N_p is the irradiation density and A_p is the spot area as indicated above. The surface of each specimen is shown in Fig. 13.43. The surface roughness, R_a is shown for each specimen in Fig. 13.44. R_a of the LPned specimens increases with increasing

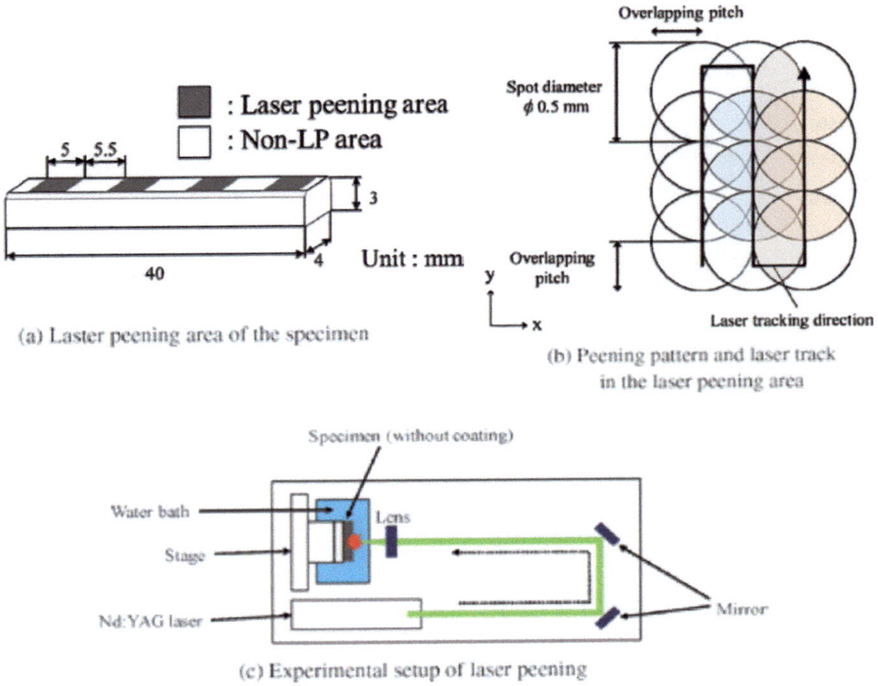

Fig. 13.43 Schematic diagram of peening method applied to the specimen, Saigusa et al. [13]. Open access

coverage and power density. The asterisk mark in Fig. 13.44 shows that surface cracks occurred due to LP (Fig. 13.45).

The cracks were observed on the sample when the LP and coverage were too high. The compressive residual stresses induced by LP is illustrated in Fig. 13.46. The residual stress distribution vs. coveragge is show in Fig. 13.46 for each power appllid. Basically, the compressive residual stress and the depth, namely the crossing point where the compressive stress changes to tensile tend to increase as the value of G increases. In Fig. 13.47 the residual stress is plotted vs. the depth from the surface and for different values of the power densities, GW/cm^2. The amount of heat increases as G increases, and therefore, as expected, the residual stress would become small. When SP was applied to Si_3N_4/SiC, the compressive residual stress on the surface was about 1500 MPa and the introduction depth was about 40 μm, as seen in Fig. 13.47. The residual stress of the SPned specimen greatly decreases with dpth, and approaches 0 MPa when te depth exceeds 20 μm.

In summary, increasing LP increases surface roughnes probably because of ablation. Also, when the power density increases surface cracks appear in the sample. Compressive residual stresses up to 230 MPa are introduced in the specimen by LP without peeling and surface chipping. Comparison of LP and SP indicate (Fig. 13.47)

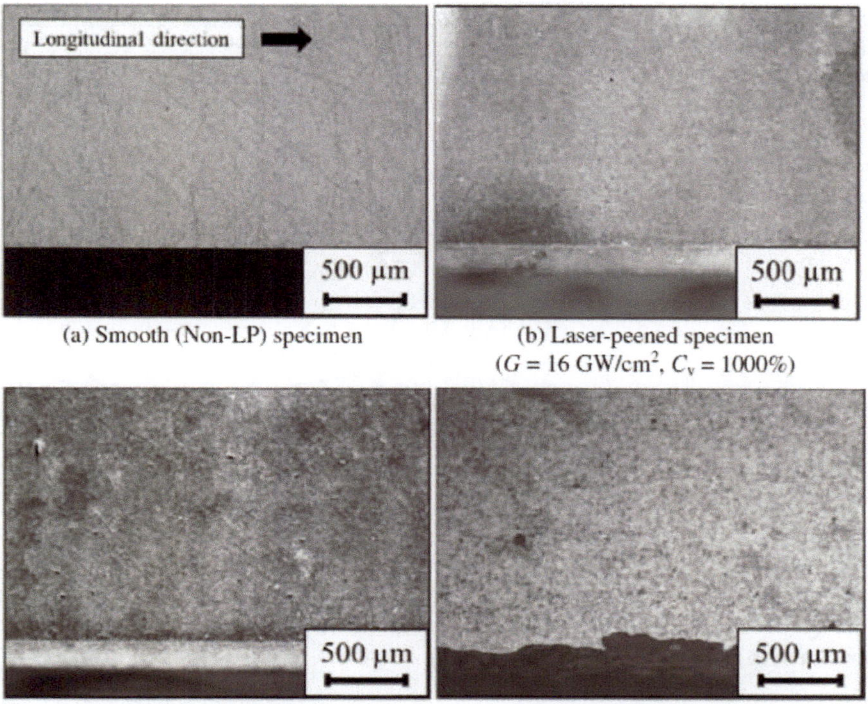

(a) Smooth (Non-LP) specimen

(b) Laser-peened specimen ($G = 16$ GW/cm^2, $C_v = 1000\%$)

(c) Laser-peened specimen

(d) Shot-peened specimen ($G = 16$ GW/cm^2, $C_v = 3000\%$)

Fig. 13.44 Optical micrographs of the surfaces of the specimens, Saigusa et al. [13]. Open access

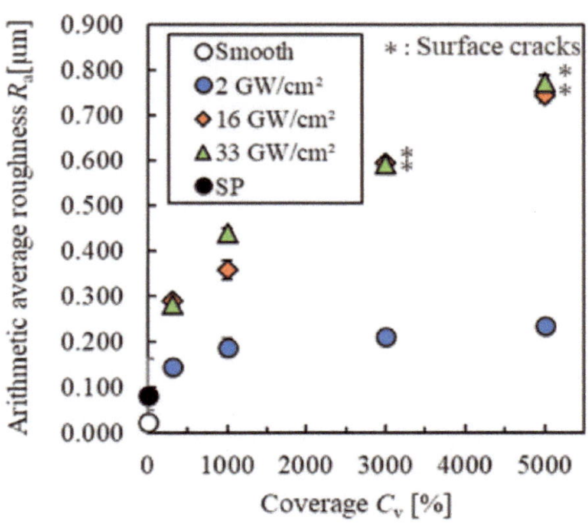

Fig. 13.45 Surface roughness (Ra) of each surface-treated specimen, Saigusa et al. [13]. Open access

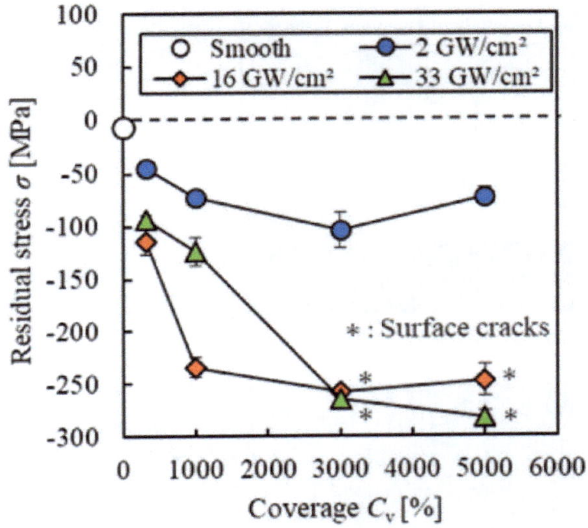

Fig. 13.46 Surface residual stress at each power density and coverage, Saigusa et al. [13]. Open access

that the depth of compressive residual stress of LPned specimens is ~ 50μm, while that of SP is smaller, but the surface compressive residual stress of the SPned specimen is larger than of the LP one (Fig. 13.48).

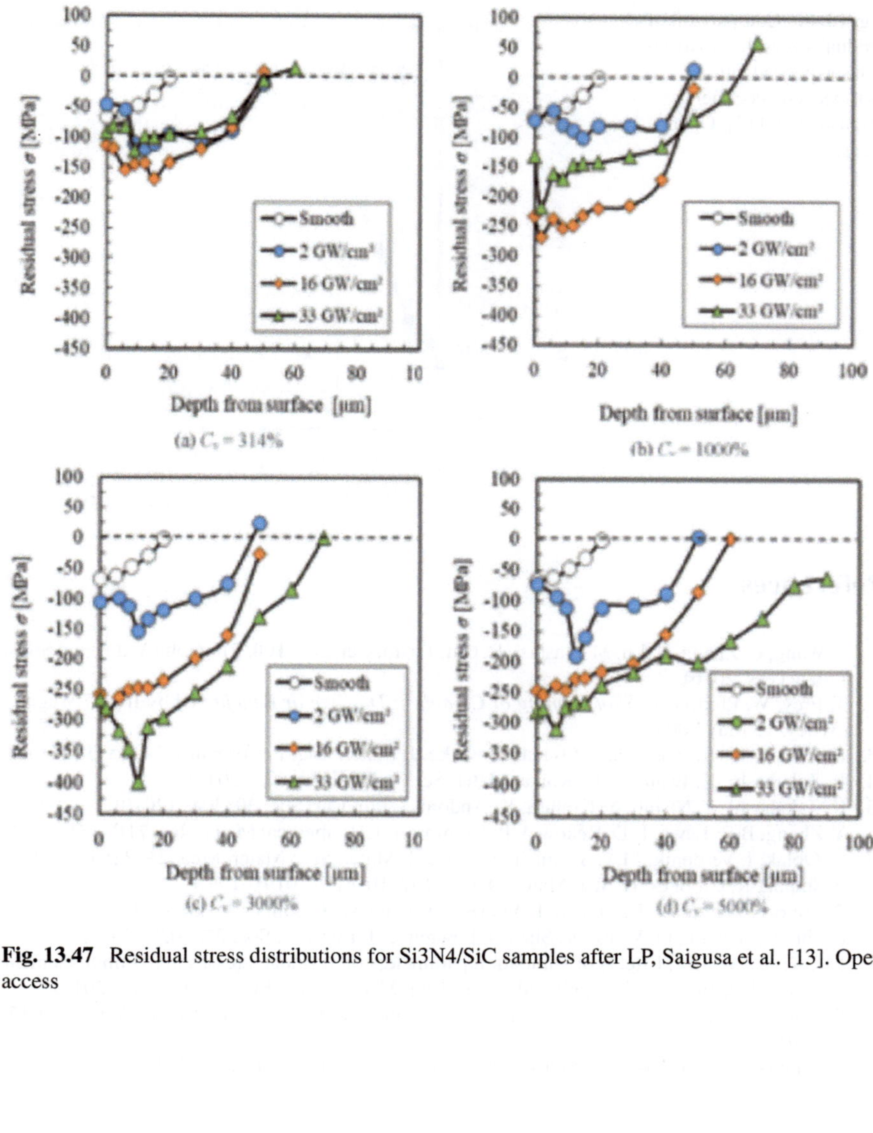

Fig. 13.47 Residual stress distributions for Si3N4/SiC samples after LP, Saigusa et al. [13]. Open access

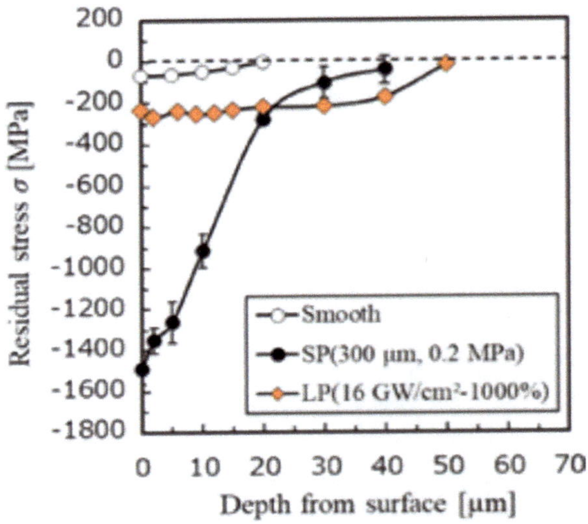

Fig. 13.48 Comparison of residual stress distributions of laser-peened and shot-peened specimens, Saigusa et al. [13]. Open access

References

1. F. Wang, C. Zhang, Y. Lu, M. Nastasi, B. Cui, J. Am. Cer. Soc. **100**, 911 (John Wiley and Sons, Full access, 2016)
2. T. Frey, W. Pfeiffer, in *Shot Peening of Ceramics: Damage or Benefit*? ed.by Ing L. Wagner, (Wiley, 12 May 2003)
3. K. Takahashi, K. Iwanaka, T. Osada, H. Koike, J. Mater. Eng. Performance **24**, 3573 (2015)
4. K. Takahashi, K. Iwanaka, H. Koike, Mater. Sci. Technol. **33**, 623 (2017)
5. K. Takahashi, Y. Nishio, Y. Kimura, K. Andom, J. Eur. Cer. Soc. **30**, 3047 (2010)
6. Y. Zhang, B.R. Lawn, E.D. Rekow, V.P. Thompson, J. Biomedical Mater. Res. **71B**, 381 (2004)
7. C. Oblak, I. Verdenik, M.V. Swain, T. Kosmac, J. Mater. Sci.: Mater. Med. **25**, 2255 (2014)
8. P. Shukla, R. Crookes, H. Wu, Mater. Design **167**, 107626 (2019)
9. P. Ctibor, L. Kraus, J. Tuominen, P. Vuoristo, P. Chráska, Ceramics—Silikáty **51**, 181(2007)
10. P. Shukla, S. Nath, G. Wang, X. Shen, J. Lawrence, J. Eur. Cer. Soc. **37**, 3027 (2017)
11. J.H. Kim, J.W. Lee, Effects of simulation parameters on residual stresses in 3D finite element laser shock peening analysis. Global J. Res. Eng. Mech. Mech. Eng. **13**(9), 1–10 (2013)
12. P. Shukla, J. Lawrence, Sci. Techn. Facilities Council (STFC) Annual Reports/article, 2015, 41 (2014)
13. K. Saigusa, K. Takahashi, N. Sibuya, Int. J. Peening Sci. Techn. **1**, 221 (2019)

Chapter 14
Cyclic Deformation in Nanostructures

Abstract Significant increase in strength of ultrafine (UF) and nanocrystalline (NC) material is observed. Deviation in their mechanical behavior from the known conventional concept of dislocation theories is by now a proven fact. Sever plastic deformation (SPD) by equal channel angular pressing is an accepted method to induce severe plastic deformation in a material. This technique enables the production of fully dense bulk specimens of macroscopic dimensions and is also of great interest in cyclic deformation of UF and NC grained structures. In the cyclic deformation by SPD cyclic hardening and cyclic softening has been observed and their behavior is associated with the observation of grain coarsening during cycling. Three major aspects are responsible for SPD effect, namely, (1) the microstructure stability; (2) the limitation of the cyclic lifespan; and (3) the imposed plastic strain amplitude. In the following cyclic deformation of oxides, carbides and nitrides is considered.

Keywords Nanostructures · SPD · Cyclic deformation · Fracture · Thermal cycling · Microstructure

14.1 Introduction

By now it is well-established that the small size grains in the ultra-fine (UF) and nanocrystalline (NC) grained metals and alloys provide a significant increase in strength and exhibit remarkable mechanical properties. Further, at the same time the small grain size in nanocrystalline materials lead to deviation in their mechanical behavior from the known conventional concept of dislocation theories. The deviation from the conventional dislocation theories is attributed mainly to the interaction of the stress fields surrounding dislocations with that surrounding the grain boundaries as a result of the small grain size. Interest in understanding the mechanical behavior of UF and NC grained materials has emerged in the past decades in order to safely put nanostructred materials into practical and technological applications. As such, it was essential to produce this structure on a commercial scale. A number of different technniques are used to produce UF and NC grain sizes which require severe plastic deformation (SPD), among them equal channel angular pressing (ECAP) is of particular importance. ECAP allows the production of fully dense bulk specimens of

J. Pelleg, *Cyclic Deformation in Oxides, Carbides and Nitrides*, Structural Integrity 22,
https://doi.org/10.1007/978-3-030-86118-6_14

macroscopic dimensions and is also of great interest in cyclic deformation of UF and NC grained structures. Another technique of interest is accumulative roll-bonding (ARB). As mentioned, there are many different processing techniques for producing UF and NC grained metals, each of them is with its own characteristics. ECAP is considered to be the most developed, and likely most popular, of the SPD techniques.

In the cyclic deformation by SPD cyclic hardening and cyclic softening has been observed depending on various factors. Specifically, for SPDed UF grained metals, their behavior has often been associated with the observation of grain coarsening during cycling. Thus, of the many factors that affect the cyclic deformation by SPD of materials (in particular metals) one can indicate three major aspects: (1) the microstructure stability; (2) the limitation of the cyclic lifespan; and (3) the imposed plastic strain amplitude. In the following oxides, carbides and nitrides are considered limited by the availibility of appropriate data.

14.2 Cyclic Deformation in Nanostructure (NS) Alumina

Oxide ceramics are promising materials for high temperature structural applications. Nevertheless, oxides like Al_2O_3, MgO, ZrO_2 suffer from relatively low fracture toughness and strength, significant strength degradation at high temperatures and poor creep, fatigue and thermal shock fracture resistances. Many attempts have been made to improve these problems as well as the brittle fracture by incorporating second phases. But also, fabrication of nanostructures has been observed to improve the mechanical properties. A strengthening second phase is SiC which is a promissing reinforcing agent. Our main interest in this book is to consider monolithic or pure unalloyed materials. However, the information of interest appears in pulications of alloys where comparison is made between the alloyed and unalloyed substances in order to evaluate the improvement by the second phase or phases. Such case is the Al_2O_3/SiC composite where pure alumina is compared with the properties of the alloy. It is expected that improvemnent of toughness, strength, high temperature fracture and fatigue behavior will occur at the same time being strength properties of the material in our case of alumina. The aim of addition of nanosize SiC dispersions is to improve not only the mechanical properties such as hardness, fracture strength and toughness and reliability at room temperature but also high-temperature mechanical properties such as hardness, strength, and creep and fatigue fracture resistances. The nano-size SiC dispersions (particles) restrain the grain growth of matrix and decrease the abnormal grain growth by pinnng and pile up the dislocations to make subgrain boundaries within the matrix grains.

Monolithic Al_2O_3 is included in Fig. 14.1 illustrating a dispersion of 5vol% nano-size SiC particles into the Al_2O_3 grains. The temperature dependence of Vickers hardness for Al_2O_3 is seen in Fig. 14.2. In this figure the hardness of alumina is compared to that of the composite Al_2O_3/16 vo% Si_3N_4. The sudden slope changes correspond to the brittle-to-ductile transition temperature (BDTT) in both materials. The variation of fracture strength with temperature of alumina indicates a significant

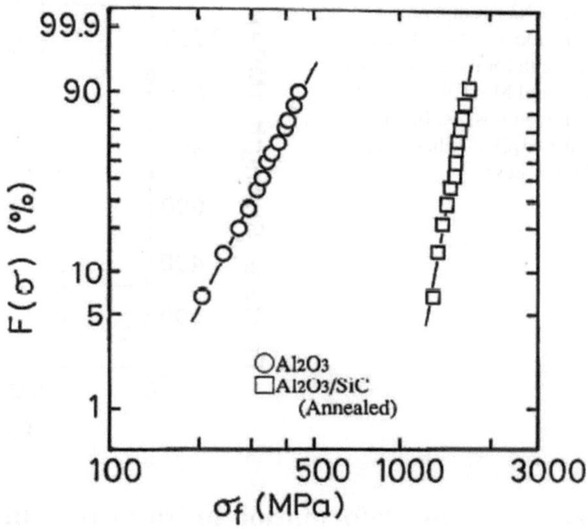

Fig. 14.1 The Weibull plot for the Al_2O_3/5vol% SiC nanocomposite and the monolithic Al_2O_3 ceramics, Niihara, [1]. Free access

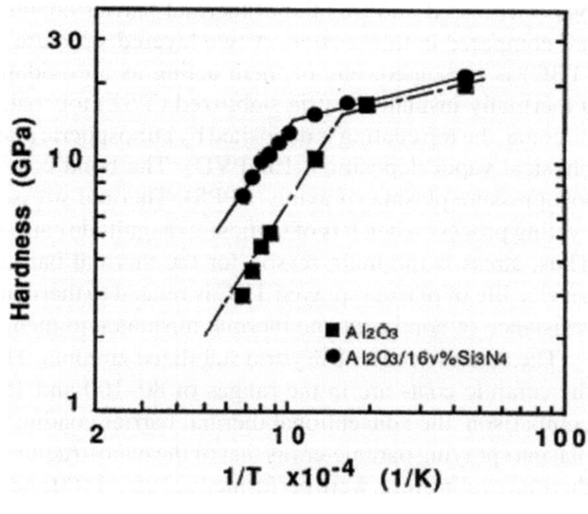

Fig. 14.2 The variation of Vickers hardness with temperature for the Al_2O_3/16 vol% Si_3N_4 nanocomposite and that of pure Al_2O_3, Niihara, [1]. Free access

degradation in strength of the monolithic Al_2O_3 at high temperatures as seen in Fig. 14.3. Fatigue properties are usually related to the microstructure. Although fatigue strength is not explicitly shown in the figures of this section, but the static properties such as hardnes and fracture strength indicate what is expected regarding the fatigue behavior also.

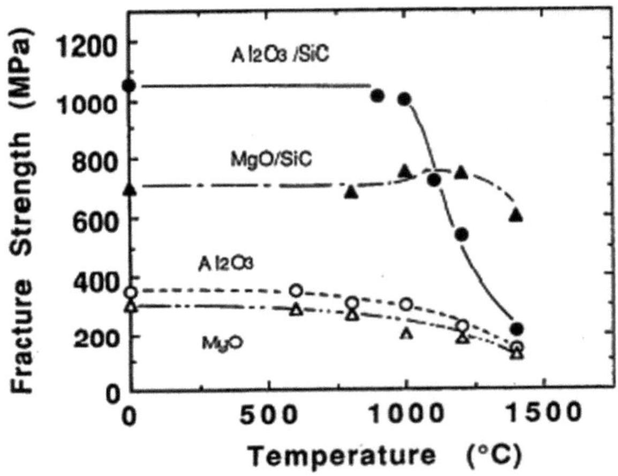

Fig. 14.3 Temperature dependence of the fracture strength for the Al_2O_3/5vol% SiC and MgO/30vol% SiC nanocomposites. Included is pure Al_2O_3, Niihara, [1]. Free access

14.3 Cyclic Deformation in Nanostructure Yttria

Nanostructured and traditional thermal barrier coating of yttria stabilized zirconia are compared in this section. A two layered structure in the thermal barrier coating (TBC) is presented, one of them acting as an oxidation protective bond coat and a thermally insulation yttria stabilized (YSZ) topcoat. The partially yttria stabilizd zirconia, the top coating is deposited by atmospheric plasma spraying or electron bean physical vapor deposition (EB-PVD). The bond coating, MCrAlY is produced by low pressure plasma spraying (LPPS). Thermal stresses induced during the thermal cycling process when it is of sufficient magnitude causes spallation of the top coating. Thus, stress is the main reason for the thermal barrier coating failure. Hence, the service life of plasma sprayed TBC is related to thermal cycling resistance, oxidation resistance of bond coat and thermal mismatch to mention some.

The ceramic coat is 8% yttria stabilized zirconia. The ranges of the bond coat and the ceramic coats are in the ranges of 80–100 and 150–200 μm, respectively. For comparison, the conventional thermal barrier coating was produced under the same plasma spraying parameters as that of the nanostructured thermal barrier coating. The thermal cycling tests were performed at 1050, 1100 and 1150 °C in static air. Fractured cross-section of as-sprayed nanostructured zirconia coating microstructure is seen in in Fig. 14.4. It is revealed in microstructure that there are two kinds of structures in the coating: the continuous molten phase and the loose microstructure. To investigate the cyclic oxidation (oxidation occurs during the exposure of the thermal barrier coating to high temperature in air), two tests were performed on two kinds of thermal barrier coatings. The weight change as a function of of the numcer of cycles occurring during the tests at 1100 °C in the static air is shown in Fig. 14.5 for nanostructured YSZ and traditional YSZ. In both specimens the thermal barrier coatings initially gained weight and showed only little variation up to 200 cycles. A large part of the top coating was spalled off from the substrate after thermal cycling test at 1100 °C for 782 cycles,

Fig. 14.4 Fractured cross-section of as-sprayed nanostructured zirconia coating, Zhou et al. [2]. With kind permission of Elsevier

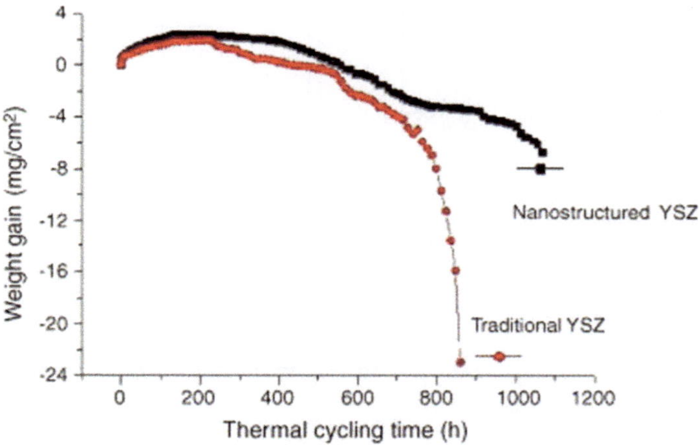

Fig. 14.5 Kinetics of nanostructured and common TBC during 1 h cycles at 1100 °C, Zhou et al. [2]. With kind permission of Elsevier

however, severe spallation occurred at 1068 cycles for the nanostructured coating. This indicates that the lifetimes of the two kinds of coatings were 782 and 1068 cycles, for the traditional and nanostructured, respectively. The nanostructured coating has longer thermal cycling lifetime. The cross-sectional micrograph of the nanostructured and traditional thermal barrier coatings after cyclic oxidation at 1100 °C for 1068 and 782 cycles, respectively is shown in Fig. 14.6. Thermally grown oxide (TGO) has grown between the top and bond coat, and there are some lateral cracks which are initiated near the interface of YSZ top layer and thermal growth oxide. The failure of both coatings occurs within the top coat and close to the YSZ/thermal growth oxide interface. Thermal cyclic as a function of temperature for both kind of specimens is seen in Fig. 14.7. The nanostructured zirconia coating has relatively longer life than

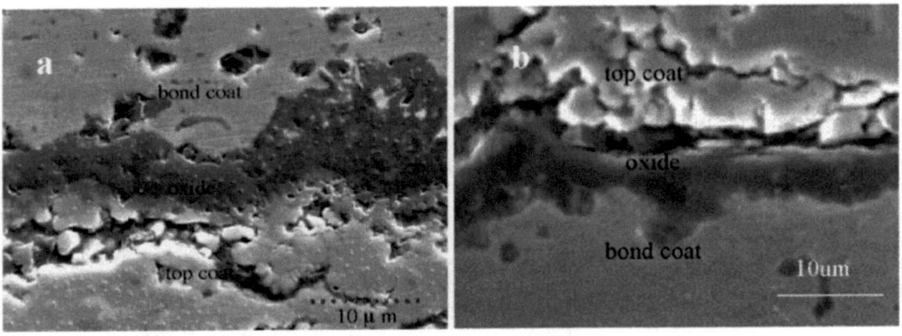

Fig. 14.6 Cross-sections of nanostructured **a** and traditional TBC **b** after cyclic oxidation at 1100 °C for 1068 and 788 cycles, respectively, Zhou et al. [2]. With kind permission of Elsevier

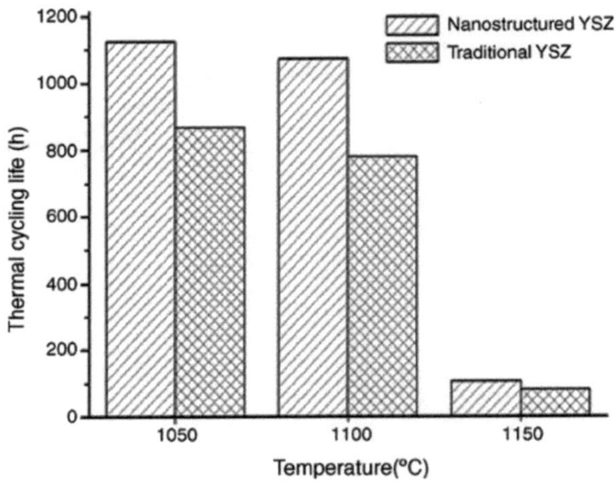

Fig. 14.7 Thermal cycling life of the nanostructured YSZ and traditional YSZ at different temperatures, Zhou et al. [2]. With kind permission of Elsevier

that of the traditional zirconia coating at different temperatures. The stress change of TGO with time during cooldown to 20 °C from a stress state at 1050 °C is illustrated in Fig. 14.8. The stress distribution in the top coat is important to the failure of the coating. The relation of the stress in the axial direction S_x with distance from the center axis to the horizontal edge for the two kinds of ceramic coatings is shown in Fig. 14.9. In both coatings the stress S_x is compressive. The relation of S_x with time during the course of coling to 20 °C within the ceramic coat is shown in Fig. 14.10. The change of stress S_y in the radial direction (i.e., y direction) at the vicinity of the interface between TGO and YSZ topcoat at the edge of the sample with time during the cooling down to 20 °C from a stress-free state at 1050 °C is shown in Fig. 14.11.

In summary, it could be concluded that the longer thermal cycling lifetime of the nanostructured thermal barrier coating can be attributed to the lower stress in the nanostructured ceramic layer. The stress S_x (axial direction) in the nanostructured YSZ is about 67% that of the traditional YSZ, while S_y (radial direction) in the

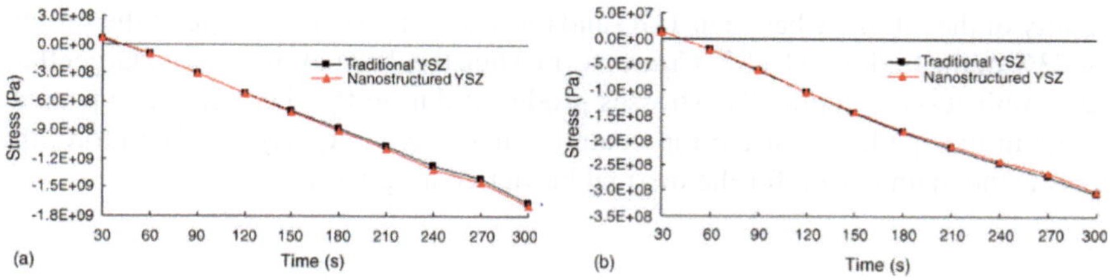

Fig. 14.8 Stress changing of TGO with time during the cooling to 20 °C from a stress-free state at 1050 °C **a** S_x and **b** S_y, Zhou et al. [2]. With kind permission of Elsevier. S_x and S_y refer to the axial and radial directions, respectively

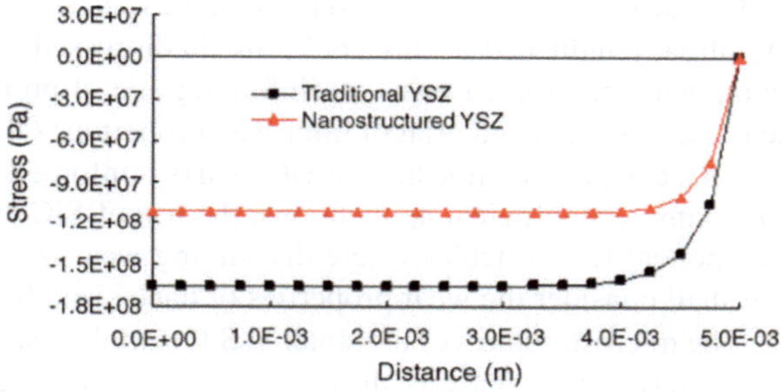

Fig. 14.9 Relation of S_x with distance from the center axis to the horizontal edge within the two kinds of ceramic coatings after the temperature cooling to 20 °C, Zhou et al. [2]. With kind permission of Elsevier

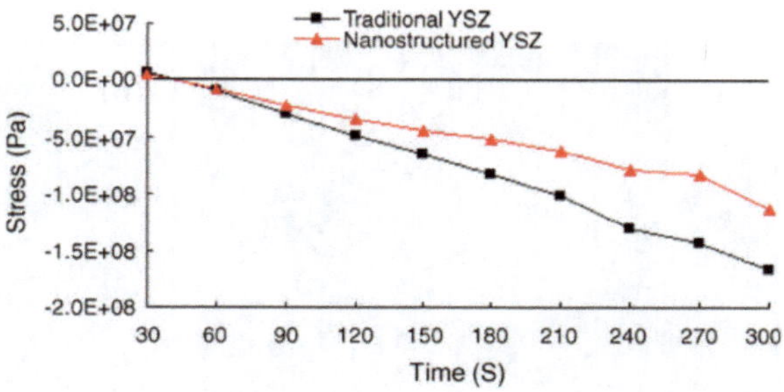

Fig. 14.10 Relation of S_x within the ceramic coat with time during the cooling to 20 °C from a stress-free state at 1050 °C, Zhou et al. [2]. With kind permission of Elsevier

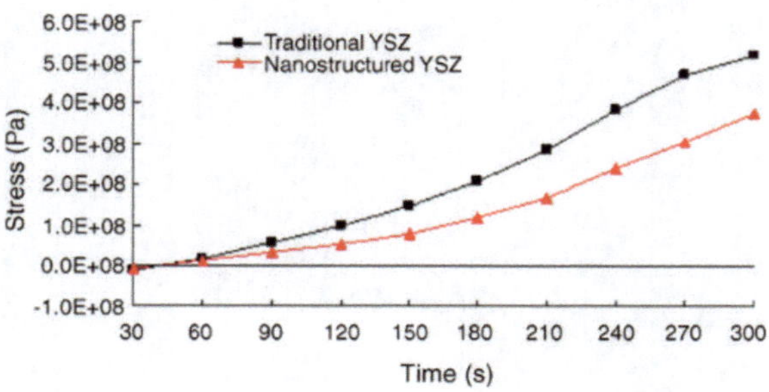

Fig. 14.11 S_y at the interface between the TGO and the YSZ topcoat at the edge of the sample with time during the cooling to 20 °C from a stress-free state at1050 °C, Zhou et al. [2]. With kind permission of Elsevier

vicinity of the interface between TGO and the YSZ topcoat at the edge of the sample is ~ 73% of the traditional YSZ. Cracks form when the thermal stresses are larger than the combination strength. The stresses produced during the thermal cyclic process results in the spallation of the top coating when it is large enough. Therefore, the stress is the main reason for the thermal barrier coating failure.

14.4 Cyclic Deformation in Nanostructure SiC

Nano-sized materials are considered as potential substitutes of the traditional materials in structural applications due to their expected enhanced mechanical behavior. As already indicated in this book, SiC is discussed in the literature generally as a composite (reinforced or as a reinforcing agent) probably because the technological interest is in a material of improved properties. Of the many composites such as Al/SiC, a popular composite is a SiC matrix reinforced with SiC particulates. Among the many technological applications, the use of SiC or SiC/SiC as a resisting wear component (in particular where dry sliding wear is critical) should be noted. Here we shall consider the wear properties of nano sized SiC pins against SiC disks.

The microstructures of the sintered SiC sample is shown in Fig. 14.12 after plasma etching. By image analysis, the mean grain size was calculated as 540 nm and used for

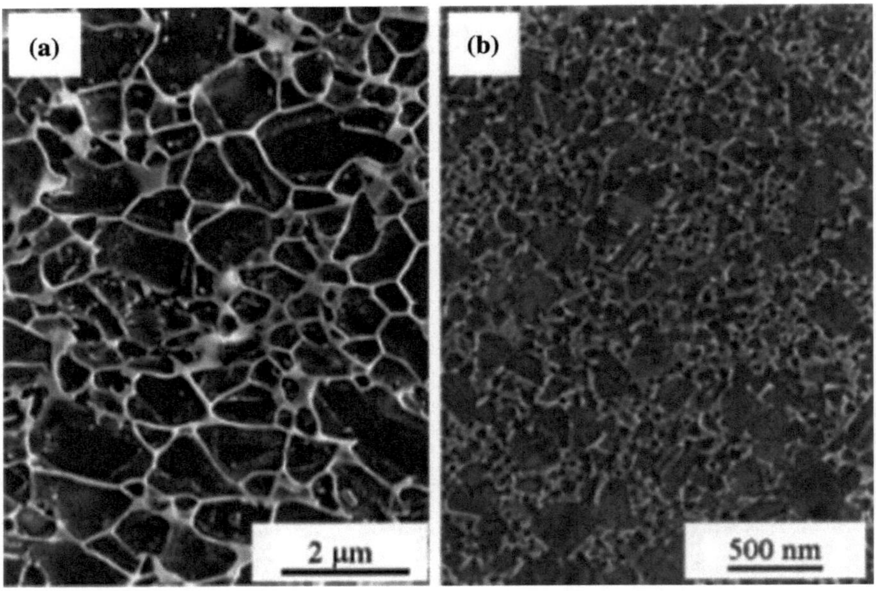

Fig. 14.12 Scanning electron micrograph of the SiC microstructures: **a** micro-sized silicon carbide (μ-SiC) used as pins and mating discs, **b** nano-sized SiC (n-SiC), used as pins, Guicciardi et al. [3]. With kind permission of Elsevier

the pin. The micron size SiC used as disk is also shown in this figure for comparison. The mechanical properties are listed in Table 14.1 and Si_3N_4 is also included. The coupling indicates a stationary period after the initial running in as shwn in Fig. 14.13 for the friction coefficient. The friction coefficient is almost constant during this stationary period. In the figure the coupling is between the μ-SiC pin and the disc material. The steady state friction coefficient is shown in Fig. 14.14, indicating that the coefficient decreased when the ceramics were not self-mated. For friction properties, the nano-sized microstructure does not seem to represent any improvement with respect to the micro-sized microstructure. The wear track of the disk is shown by SEM in Fig. 14.15a indicating disc/pin coupling of μ-SiC/μ-SiC. The EDS analysis is also shown. The pin worn surfaces were mainly clean from debris as shown in Fig. 14.16. The smooth worn surfaces indicate that fine abrasion is the main wear mechanism. Depth sensing indentation tests were applied to evaluate the hardness and Young's modulus as illustrated in Fig. 14.17. The hardness of the n-SiC shows no

Table 14.1 Mechanical properties of the tested ceramics, Zhou et al. [2]; Guicciardi et al. [3]. With kind permission of Elsevier

Material	Microhardness[a] (GPa)	Young's modulus[b] (GPa)	Fracture toughness[c] (MPa $m^{0.5}$)	Flexural strength (MPa)
μ-SiC	22.0 ± 0.8	386	3.0 ± 0.2	746 ± 46
n-SiC	21.9 ± 0.5	376	2.9 ± 0.1	660 ± 65
μ-Si_3N_4	18.9 ± 0.3	325	5.7 ± 0.1	941 ± 67

Mean values ± 1 S.D. when appropriate.
a Vickers hardness 9.81 N.
b Resonant frequency
c Indentation method at 98.1 N.

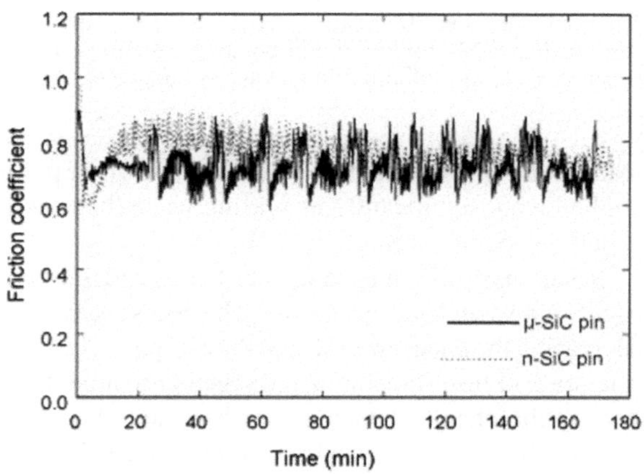

Fig. 14.13 Friction coefficient as a function of time and disc material: **a** μ-SiC, Guicciardi et al. [3]. With kind permission of Elsevier

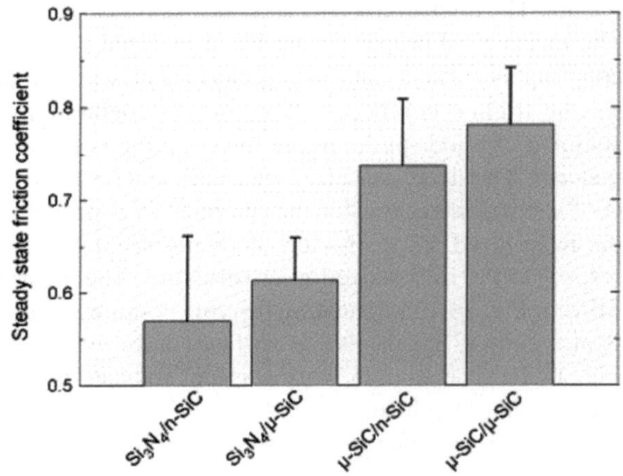

Fig. 14.14 Bar chart showing the mean values of friction coefficient as a function of disc/pin coupling. Error bars represent 1 S. D. Guicciardi et al. [3]. With kind permission of Elsevier

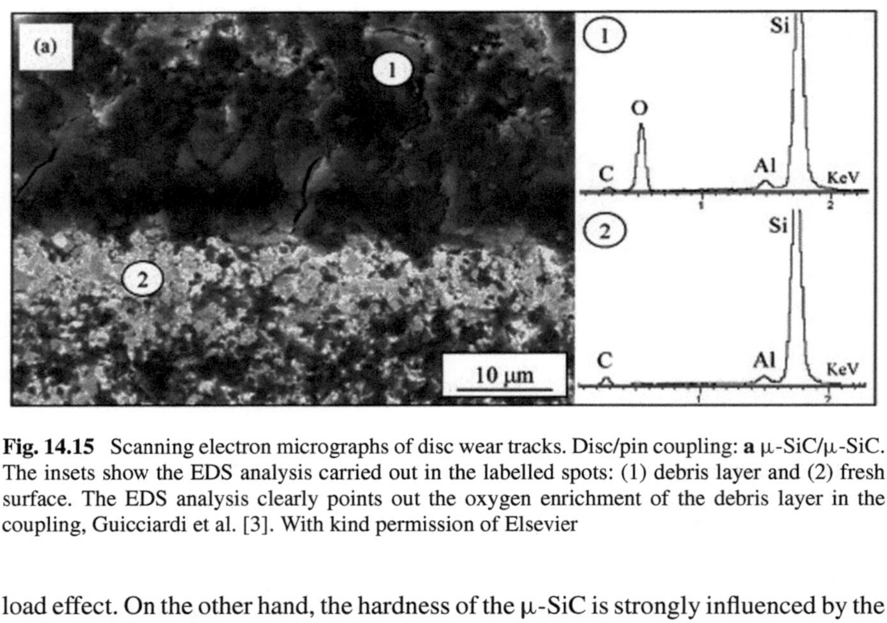

Fig. 14.15 Scanning electron micrographs of disc wear tracks. Disc/pin coupling: **a** μ-SiC/μ-SiC. The insets show the EDS analysis carried out in the labelled spots: (1) debris layer and (2) fresh surface. The EDS analysis clearly points out the oxygen enrichment of the debris layer in the coupling, Guicciardi et al. [3]. With kind permission of Elsevier

load effect. On the other hand, the hardness of the μ-SiC is strongly influenced by the applied load, significantly increasing when the peak load decreases, a phenomenon called indentation size effect (ISE).

In summary, the wear behavior of nano and micro sized SiC was evaluated by tests of SiC pin vs SiC disc wear tests. The disc wear rates were significantly improved by decreasing the mean grain size of the SiC pins.The basic wear mechanisms were grain fracture and fine abrasion. An oxygen enrichment was observed in the debris layer due to tribo-chemical reactions with air humidity. The very different wear behaviour induced by the nano-sized and micro-sized SiC pins could be explained by the

Fig. 14.16 Scanning electron micrographs showing details of the pin wear tracks. Coupling disc/pin: **a** μ-SiC/μ-SiC. Arrows indicate SiC debris. Whitish features (circled) are residual secondary phases of the starting SiC material, Guicciardi et al. [3]. With kind permission of Elsevier

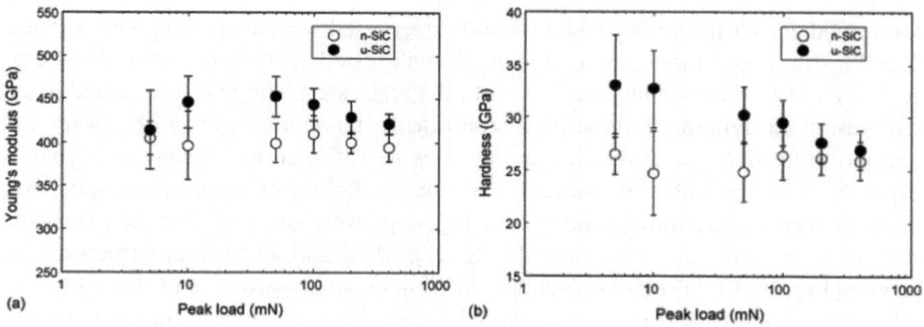

Fig. 14.17 Depth-sensing indentation Young's modulus **a** and hardness **b** as a function of peak load and material. Point and bars represent mean values ± S.D. Guicciardi et al. [3]. With kind permission of Elsevier

different mechanical properties that the two ceramics show when tested at microstructural level by depth-sensing indentation. Micro- and nano-sized hot-pressed silicon carbide pins have been characterized by room-temperature unlubricated disk-on-pin tribological tests on hot-pressed silicon carbide discs. The mean grain size was shown not to influence the steady state friction coefficient.

14.5 Cyclic Deformation in Nanostructure Si_3N_4

The dynamic fatigue strength of a hot-pressed silicon nitride/silicon carbide nanocomposite and reference monolithic Si_3N_4 have been tested by four-point bending at 1350 °C in air using different loading rates from 0.01 to 1 mm min^{-1}. The strength of the Si_3N_4-SiC nanocomposite was lower than that of the monolithic Si_3N_4, and the flexure strength significantly decreased with decreasing stress rate in both ceramics. Strength degradation is observed, which is associated with the formation and the growth of a stress-oxidation damage zone is responsible for the strength degradation of the monolithic ceramic, while in the composite intrinsic flaws are the main fracture origins, which extend by oxidation assisted slow crack growth and decrease the strength of the specimens significantly. Silicon nitride-based materials are one of the most promising candidates for structural application around the temperature of 1350 °C (for gas turbines for example). In such applications the main goal is to optimize their fracture toughness, strength, creep, oxidation resistance and probable candidates are monolithic or fiber reinforced materials.

For gas turbine engines applications sintering aids such as Y_2O_3, Mgo, etc. are most refractory at high temperatures and found to be useful with Si_3N_4 based ceramics. Damage regions usually starts at the surface of specimens during high temperature tests and seems to be associated with oxidation which affect stress corrosion cracking (SCG). This aspect has been tested in air and inert environment at 1370 °C on Si_3N_4 + 6 wt.% Y_2O_3 samples. The Y_2O_3 is a sintering agent. As mentiomed earlier the strength significantly decreased with decreasing stress rate in air and a stress-oxidation zone had formed in all tested specimens, with SCG being the dominant failure mechanism in air at all stress rates. The tests are aimed to get information on dynamic fatigue of silicon nitride nanocomposites with a reference of monolithic Si_3N_4 at different loading rates at 1350° in air. (From the dynamic fatigue tests, it is possible to predict the lifetime to failure by determining the distribution of flaws). Preparation and processing steps were the same for the two materials, the composite and the monolithic Si_3N_4. SEM and TEM microstructures are shown in Fig. 14.18. Porosity is found by SEM in the microstructure of the composite with pore diameters of 0.1–0.5 μm and with occasional clusters of these pores. The relatively homogeneously distributed SiC particles in the microstructure are located either intergranularly (approximately 30%) orintragranularly. The size of the intergranular SiC particles is approximately 150 nm, while the size of the intragranular SiC is approximately 30 nm (Fig. 14.18b). The monolithic material exhibits a similar microstructure as the composite, however it is more homogenous with less porosity and without pore clusters (Fig. 14.18c). The strength of the composite is lower than the monolithic S_3N_4 as illustrated in Fig. 14.19. A single loading rate of $N = 7.1$ characterizes the composite, but two values, namely, a lower and a higher loading rate $N = 4.7$ and $N = 19.8$ has to be applied for the monolithic Si_3N_4. This means that one and two SCG mechanisms, respectively are operating in the composite and monoloithic Si_3N_4. Three distinguishable areas can be observed in the fracture surface which failed at lower loading rates.

(i) An area with an oxidized amorphous character near the tensile surface of the specimen, seen in Fig. 14.20 (the original microstructure is hardly identifiable). (ii) An area in the SCG zone with intergranular fracture (without visible oxidation damage) and very rough fracture surface in the monolithic Si_3N_4 as seen in Fig. 14.21a and smooth surface covered predominantly by small particles in composite as seen in Fig. 14.21b. (iii) a fast fracture area, with mainly intergranular fracture and with a rougher fracture surface than the room temperature fracture surface in monolithic (Fig. 14.22a) and mixed inter/transgranular fracture in composite (Fig. 14.22b).

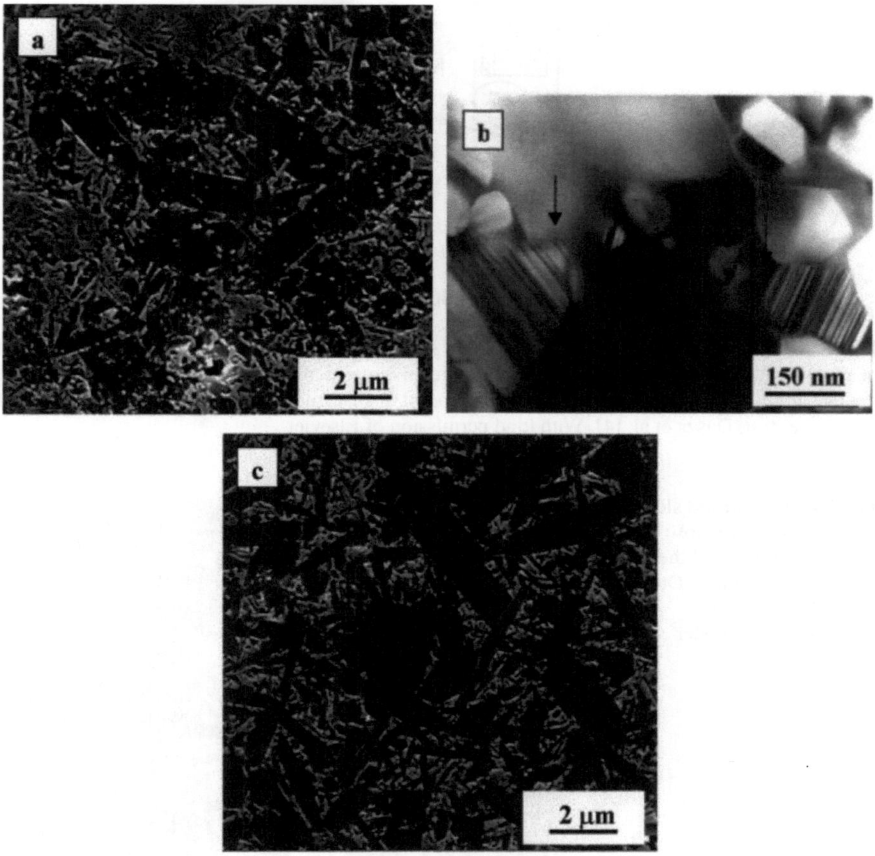

Fig. 14.18 Microstructure of the investigated materials: **a** characteristic microstructure of the composite material, SEM; **b** microstructure of the composite, illustrates blocking of the large silicon nitride grain (arrows) by intergranularly located SiC nanoparticles, TEM; (c) microstructure of the monolithic silicon nitride, SEM, Dusza et al. [4]. With kind permission of Elsevier

SCG area was clearly identified and, in all cases, originated from the tensile surface/corner of the specimen (Fig. 14.23). The size of this area increases with decreasing loading rate.

Failure initiated from an inherent flaw, probably a machining induced flaw (Fig. 14.23b). No additional cracks were identified in the specimens beside the crack causing fracture.

In summary, it has been confirmed that the mechanism of failure by fracrure is assisted at high temperatures by oxidation and the presence of SCG. Fracture in the composite at relativly low stresses results from the presence of defects on the tensile surface of the specimens (as a consequence of machining) and micropores present in the microstructure. Reconnection of the pores during the SCG contribute to the

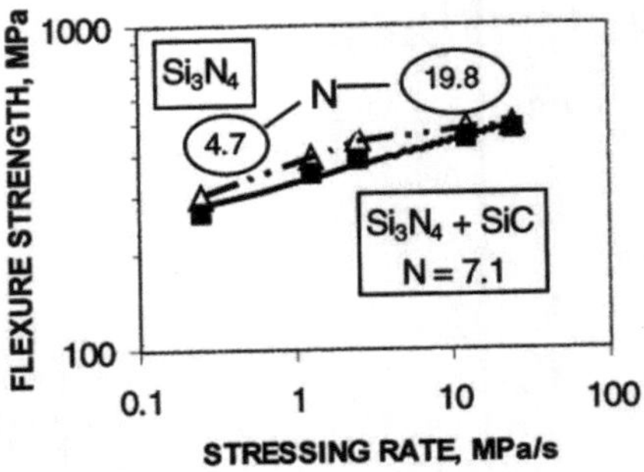

Fig. 14.19 Influence of the loading rate on the strength of the investigated materials (four-point bend with lower/upper span 40/20 mm and specimen dimension 3 × 4 × 45 mm). Please note that the test at low loading rates was 'constant load-point displacement rate' test, rather than 'constant stress rate' test, Dusza et al. [4]. With kind permission of Elsevier

Fig. 14.20 Oxidized slow crack growth area close to the tensile surface of the composite specimen, Dusza et al. [4]. With kind permission of Elsevier

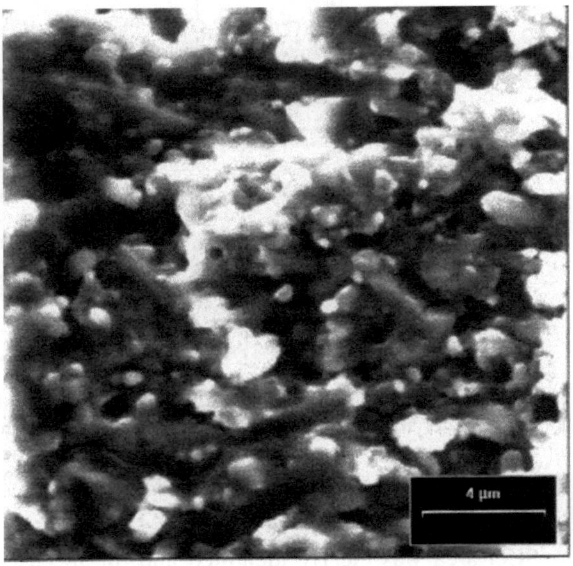

degradation of the slow crack growth resistance of the material. The strength of the composite is lower than that of the monolithic Si_3N_4 at all stress rates, but the flexural strength significanly decreases at 1350 °C in air. At low loading rates the mechanism responsible for specimen failure is stress corrosion cracking. Oxidation assisted slow crack growth has been identified as the dominant strength decreasing mechanism in the composite.

Fig. 14.21 Detail of the slow crack growth area in the monolithic silicon nitride **a** and composite **b** far from the tensile surface of the specimen, Dusza et al. [4]. With kind permission of Elsevier

Fig. 14.22 Detail of the fast fracture area in the monolithic **a** and composite **b** ceramics, Dusza et al. [4]. With kind permission of Elsevier

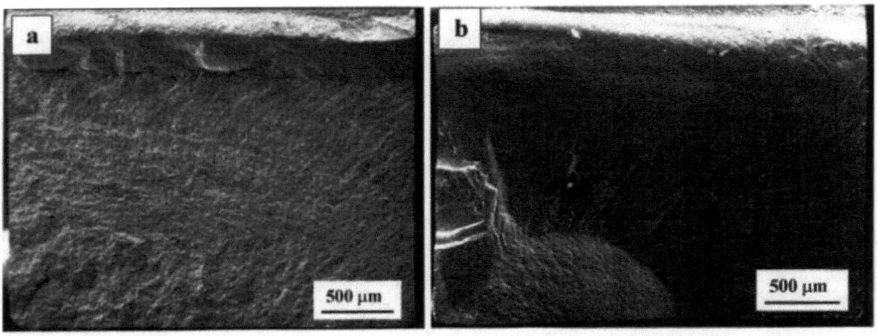

Fig. 14.23 Slow crack growth area arising from the tensile surface of the specimens, **a** in monolithic **b** in composite material tested at loading rate 0.01 mm min^{-1} at 1350 °C, Dusza et al. [4]. With kind permission of Elsevier

References

1. K. Niihara, The centennial memorial issue of the cer. Soci. Japan **99**, 974 (1991)
2. C. Zhou, N. Wang, H. Xu, Mater. Sci. Eng. A
3. S. Guicciardi, D. Sciti, C. Melandri, G. Pezzotti, Wear, **262**, 5229 (2007)
4. J. Dusza', P. Sajgalik, M. Steen, E. Semerad, Mater. Sci. Eng., A **291**, 250 (2000)

Chapter 15
Static Fatigue

Abstract Static fatigue considers fracture occurring at a stress level less than that it is required for tensile fracture. Static fatigue is also termed delayed fracture and often the influence of hydrogen on the premature failure of components is observed; classical example for delayed fracture in steel occurs by hydrogen. Static fatigue failure depends on the presence of water vapor in the air also. In other words static fatigue is caused by the combined action of tensile stress and moisture on a component in particularly ceramics. Static fatigue can be induced with constant stress acting over time and in such tests the goal is to determine the time-to-failure of a specimen. In such tests a series of specimens at decreasing constant loadings results in higher lifetimes allowing the determination of the time to failure of a component. In ceramics, static fatigue is one of the important factors for life determination. All substances considered in this book are exposed to static fatigue.

Keywords Static fatigue · Fracture · Water vapor · Alumina · Y-TZP · SiC · B_4C · Si_3N_4

15.1 1 Introduction

Static fatigue commonly considers fracture occurring at a stress level less than that it is required for tensile fracture. It is associated with adverse effect of environment on material. Static fatigue is also termed delayed fracture and often the influence of hydrogen on the premature failure of components is emphasized. The term delayed fracture is associated with the long period of time that a crack takes to grow large enough to cause spontaneous failure. It is a form of material embrittlement that may occur in various materials (not only ferrous) and different environments. Thus, static fatigue failure depends on the presence of water vapor in the air also. In other words static fatigue is caused by the combined action of tensile stress and moisture on a component in particularly ceramics. One should note that there is no need to produce an artificial crack in the material; it is sufficient to have natural inherent flaws in the material to induce static fatigue on load application, which is very sensitive to the surface residual stresses. From the what is said above, static fatigue can be induced with constant stress acting over time and in such tests the goal is to determine the

J. Pelleg, *Cyclic Deformation in Oxides, Carbides and Nitrides*, Structural Integrity 22,
https://doi.org/10.1007/978-3-030-86118-6_15

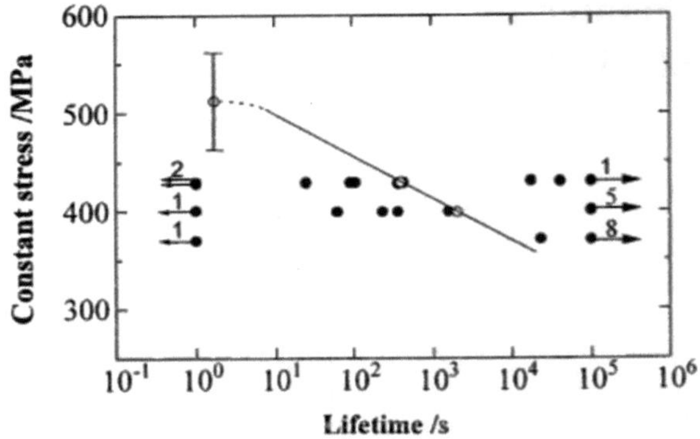

Fig. 15.1 Stress-lifetime curve for alumina in the static fatigue tests. Jin et al. [1]. Free access

time-to-failure of a specimen or a structural component. In such tests a series of specimens at decreasing constant loadings results in higher lifetimes allowing the determination of the time to failure of a component. In ceramics, static fatigue is one of the important factors for life determination.

15.2 Static Fatigue in Alumina

Polycrystalline alumina of 99.5% purity and mean grain size of 2 μm is considered in this section in terms of static fatigue. The specimens for the fatigue tests were smooth. Specimens that failed before the maximum testing stress and those that survived after the running time are included in the figure for the analysis of the lifetime and therefore the scattered distribution is representative of the defects and the size of the cracks. In Fig. 15.1 static fatigue curves of the stress-lifetime conducted at three constant stress levels of 430, 400 and 370 MPa are shown. The empty circles, (○) indicate the median value of the static fatigue lifetime for each estimated stress level tested in air. The residual strength distribution after the static fatigue tests using the stress levels of 400 and 370 MPa was evaluated as 365 MPa. This is illustrated in Fig. 15.2. The complete stress-lifetime curves including cyclic faigue curve also is seen in Fig. 15.3.

In summary, static test in pure smooth-surface alumina under several stress levels, was discussed. The statistical treatment includes imperfect data that failed below the maximum applied static stress.

15.3 Static Fatigue in Yttria-Stabilized Tetragonal Zirconia (Y-TZP)

One of the applications of Y-TZP is for dental use. Since this material undergoes transformation (tetragonal to monoclinic), understanding its effect on crack growth is

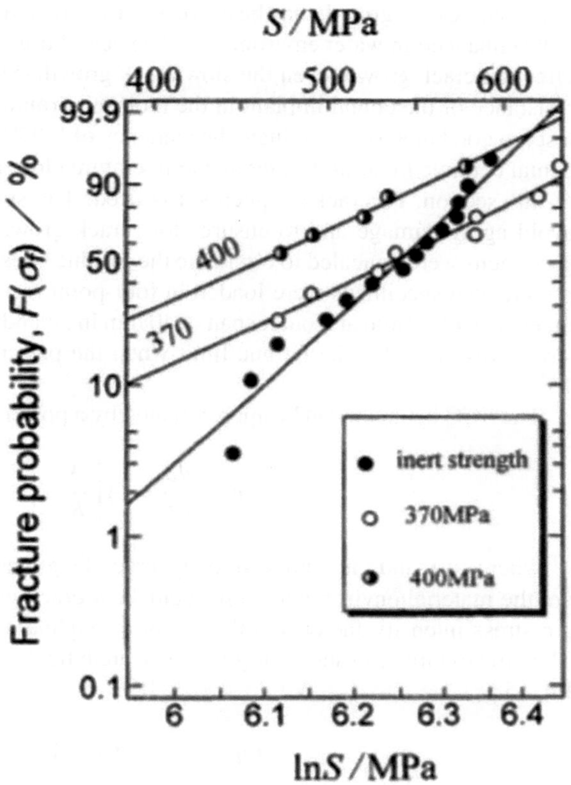

Fig. 15.2 Weibull plot for the original residual strength distributions of specimens surviving after the running time in the static fatigue tests. Jin et al. [1]. Free access

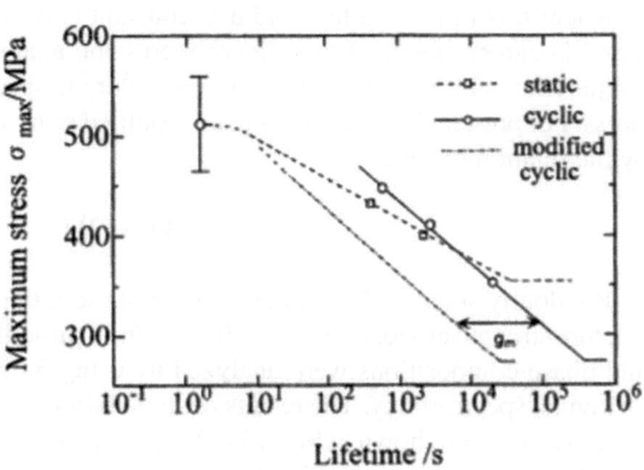

Fig. 15.3 Comparison od stress-lifetime curves obtained in the static and cyclic fatigue tests. Jin et al. [1]. Free access

important. Crack growth may be the reason to static fatigue failure which is considred to be enhanced in water environment. The dental use is in oral environment, and the effect on crack growth even the slow crack growth (SCG) is critical to ensure good resistance of the dental implant in the moist environment application. Static fatigue test is a good method to evaluate the tendency of Y-TZP for SCG. Comparison of this dental ceramic in air and water at a temperature close to RT, say 37 °C, is presented in this section. Precracked specimen is used. The specimens were pre-cracked to avoid aging damage and to ensure slow crack growth dominating. After that, the specimens were annealed to eliminate the residue stress by machine processing. The pre-cracked specimens were loaded in four-point rectangular beam flexure with an inner span of 20 mm and outer span of 40 mm in air and in water at 37 °C. The applied stress was taken for the fatigue limit when the pre-cracked specimens survived to 10^5 s.

The SCG behavior can be approximated by a power law relation given as Eq. 15.1

$$v = \frac{da}{dt} = A\left[\frac{K_I}{K_{Ic}}\right]^n \tag{15.1}$$

where v, a, and t are crack velocity, crack length and time, respectively. A and n are the material/environment dependent slow crack growth parameters, K_I and K_{Ic} are stress intensity factor and the fracture toughness of the material, respectively. The time to failure in static fatigue can be analytically derived (see Bouten [2]) from Eq. 15.1.

$$\log t_f = n \log \sigma + \log D \tag{15.2}$$

where t_f is time to failure and σ is constant applied stress. The SCG parameters n and D can be obtained by a linear regression analysis with experimental data in conjunction with Eq. 15.2 when time to failure is specified as a function of applied stress. For pre-cracked bars n has to be modified as n' which is related to n of Eq. 15.2 by the relation given as

$$n = \left(4n' - 2\right)/3 \tag{15.3}$$

Randomly selected 5–6 specimens for static fatigue loads to evaluate the mean fracture strength at a loading rate of 0.5 mm/min. In addition, the phase transformation and phase compositions were analyzed by using X-ray diffraction analysis (XRD) or Raman spectroscopy. The results of static fatigue as plots of applied stress against time to failure are shown in Fig. 15.4. Evidence for t → m (tetragonal → monoclinic) transformation is seen in Fig. 15.5 where x-ray diffraction is shown in addition to the microstructure. XRD analyses of the SCG region on the fracture surfaces showed that a significant amount of m-ZrO_2 could be detected and confirmed that t → m phase transformation did occur during SCG. The amount of the monoclinic phase during SCG in water at 37 °C increased compared to that in air, which is an indication of the tendency of the tetragonal phase transformation to the monoclinic

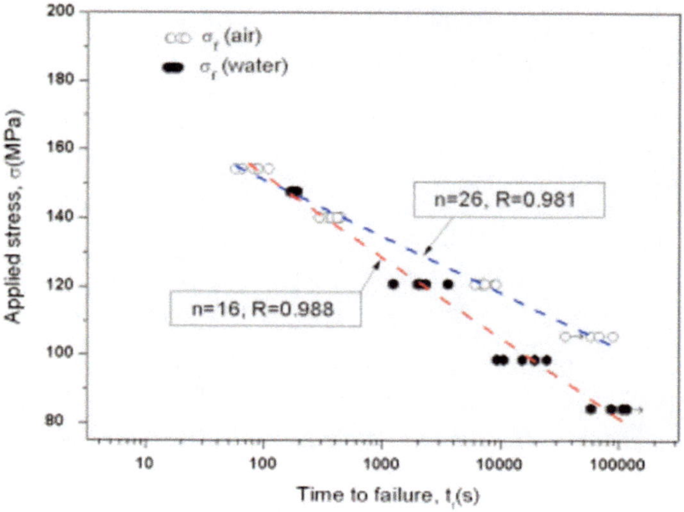

Fig. 15.4 Static fatigue testing for pre-cracked specimens in air and in water at 37 °C. The fracture strengths in respective circumstance were included for comparison. Gui and Xie [3]. With kind permission of Elsevier

phase in water environment. Raman measurements provide direct evidence for the t → m phase transformation that resulted from the interaction of stress and moisture at the crack tip. Raman spectra were taken at different distances at the crack tips of the initial cracks. The results are shown in Fig. 15.5b. There is an increase of transformation zones around the crack tip by propagating in water at 37 °C, which suggests that the SCG behavior in water was a result of cracking through a zone of monoclinic material ahead of the crack tip, formed by water/stress induced t → m phase transformation. It is assumed that the crack growth velocity is controlled by the t → m phase transformation, which can be expressed by

$$\nu \propto \exp\left(-\frac{\Delta G^*}{RT}\right) \tag{15.4}$$

Here, ΔG* is the effective activation energy barrier for the t → m phase transformation, which includes the influence of both stress and water. The Y_2O_3 retains a portion of the metastable tetragonal zirconia which has a higher Gibbs free energy than the monoclinic phase. In air, the energy reduces the free energy, ΔG_w^*. Further decrease in free energy occurs by the high stress. The stress assisted energy ΔG_k^* is the contribution of the external load. In water ΔG_k^* is expected to be larger because the crack was propagating in water, and it can be expressed as ΔG_a^* which is high and therefore the transformation is difficult, unlike in water which

$$\Delta G^* = \Delta G_w^* - \Delta G_k^* \tag{15.5}$$

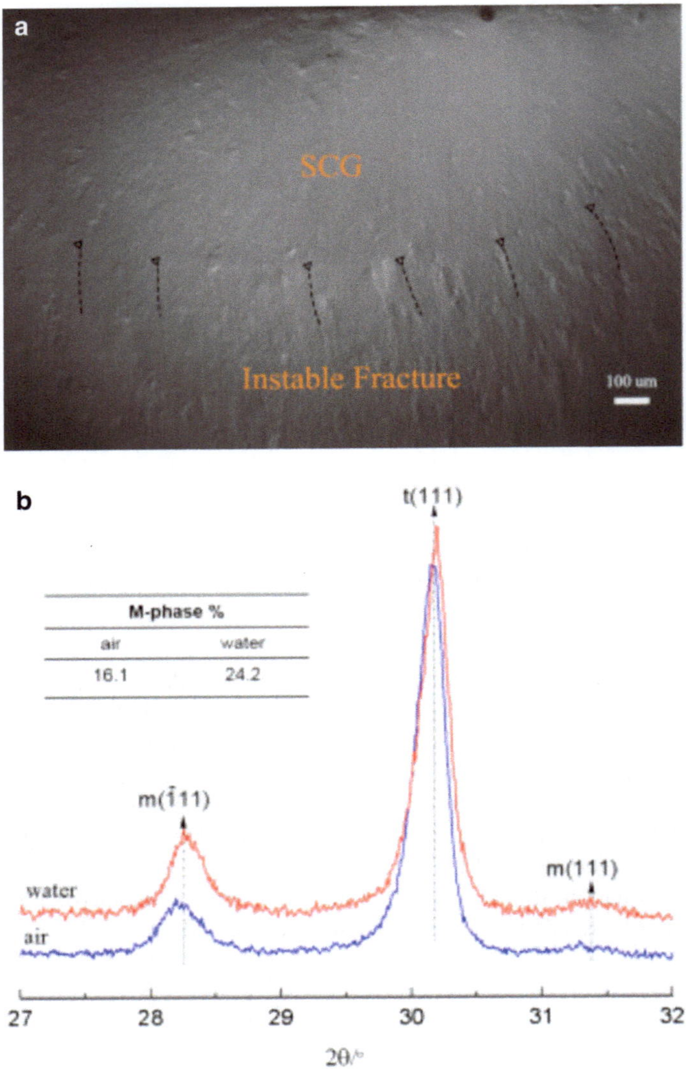

Fig. 15.5 Fracture surface of pre-cracked specimens that failed at about 10^5 s in the static fatigue testing. Triangles marked front of smooth area. Black lines marked direction change of the striations (**a**); XRD analysis of SCG region on the fracture surfaces (**b**). Gui and Xie [3]. With kind permission of Elsevier,

Substituting Eqs. 15.4 and 15.5 one obtains the crack growth rate as

$$\nu \propto \exp\left(\frac{\Delta G_w^* - \Delta G_k^*}{RT}\right) \tag{15.6}$$

In summary, static fatigue performed on dental Y-TZP precracked specimens indicates that they are more susceptible to SCG in water than in air and takes place with larger growth velocity. By Raman analysis quantitative assesments of phase transformatio of t → m, it was shown that transformation zones are existing at the crack tip and they are much larger than in air. It has been shown that cracking occurs through a zone of monoclinic phase ahead of the crack tip. The crack growth rate is proportional to the probability of successful completion of water/stress induced t → m phase transformation. Raman spectroscopic assesment provides an understanding of the SCG of Y-TZP during static fatigue in different environments.

15.4 Static Fatigue in SiC

Slow crack growth (SCG) of hot pressed SiC at high temperature is considered below, deformed under constant stress, i.e., under static fatigue conditions. The flexural test in air was performed at 1300 °C at constant stress, namely, at static fatigue conditions. The interest in high temperature deformation is because SiC is a high temperature ceramic and a candidate for structural applications in heat engines. One of the major limitations in high temperature applications is delayed failure, where inherent flaws can occur until a critical size crack growth attaining catastrophic failure. Consequently, it is important to understand the SCG behavior and evaluate the lifetime within a reasonable prediction to ensure performance under a specicic loading conditions.

The major room-temperature physical and mechanical properties of the silicon carbide are shown in Table 15.1.

Table 15.1 Young's modulus, density, and fracture toughness of NC 203 silicon carbide at room temperature. Choi et al. [4]. With kind permission of Springer Nature. NC stands for Norton Company

Young's modulus[a] (GPa)	Density[b] ($g\,cm^{-3}$)	Fracture toughness[c] K_{Ic} ($MPa\,m^{1/2}$)
443 ± 2	3.31 ± 0.01	3.99 ± 0.20

[a]By the impulse excitation technique, ASTM C 1259 (with 31 flexural specimens).
[b]By the specimen mass/volume method (with 31 specimens).
[c]By the single-edge pre-cracked beam method (with five specimens).

For static fatigue (constant stress) testing, the time to failure, t_f at constant stress is given by

$$t_f = G\sigma^{-n} \tag{15.7}$$

G is a parameter associated with n, the inert strength, the fracture toughness, and the crack geometry factor. G and n are SCG parameters and can be obtained by linear regression from the slope and intercept, respectively, from the curve of the time to failure versus the applied stress. This curve based on relation 15.7 is seen in Fig. 15.6. The constant stress results are summarized in Table 15.2. The Weibull parameters m_{sf}, the intercept and the Weibull median time to failure are included in the Table. The SCG parameter was found to be $n = 31.8 \pm 1.5$ based on Eq. 15.7.

The Weibull time-to-failure distributions obtained at each applied stress are presented in Fig. 15.6. The Weibull moduli m_{sf} was estimated to be 0.70, 0.60, 0.45 and 0.34 at applied stresses of 395 MPa, 360 MPa, 330 MPa and 300 MPa, respectively, resulting in an average value of $m_{sf} = 0.52 \pm 0.16$. The Weibull modulus, m_{sf},

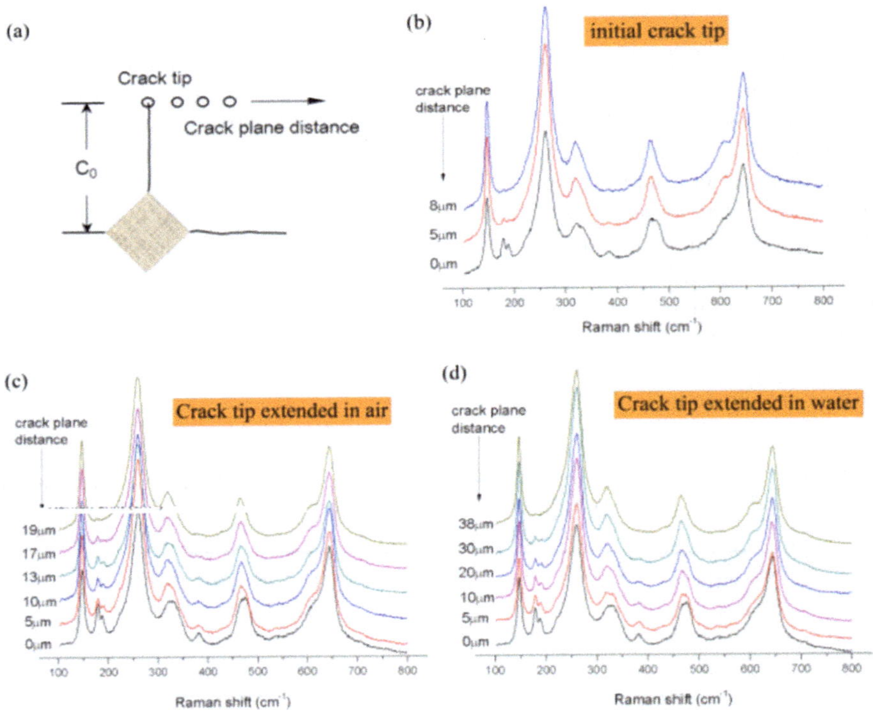

Fig. 15.6 Places analyzed by Raman spectroscopy around crack tip (**a**), Raman spectra obtained at the initial crack tip (**b**), obtained at the crack tip propagating in air (**c**), obtained at the crack tip propagating in water at 37 °C (**d**). Gui and Xie [3]. With kind permission of Elsevier,

Table 15.2 A summary of constant-stress testing results of NC 203 silicon carbide in flexure at 1300 °C in air. Choi et al. [4]. With kind permission of Springer Nature

Applied stress, σ (MPa)	Number of specimens	Mean time to failure*, t_f (s)	Weibull parameters		
			Weibull modulus, m_{sf}	Intercept (ln(s))	Weibull median time to failure (s)
395	20	16.4 (21.8)	0.70	1.769	7.5
360	20	187.1 (420.3)	0.60	2.640	45.6
330	20	5020.9 (17632.3)	0.45	3.035	356.9
300	20	87228.6 (369414.5)	0.34	2.571	677.0

*The values in parentheses represent ± one standard deviation.

is related to the Weibull modulus, m_i, of inert strength by

$$n_{sf} = \frac{m_i}{n-2} \tag{15.8}$$

With parameters $n \approx 30$ and $m_i \approx 14$ (from Table 15.2 of the work of Choi et al.) a value of $m_{sf} = 0.5$ is obtained which is in good agreement with the average value of 0.52 determined from Fig. 15.7. Fracture surfaces shown in Fig. 15.8 reveal slow crack growth which is the governing mechanism for failure in the constant-stress (static fatige) testing. The fracture surfaces and their corresponding tensile surfaces are subjected to low and high applied stresses as seen in Fig. 15.7.

In summary, failure of SiC at 1300 °C in air is thought to be the result of slow crack growth when tested at constant stress (static fatigue). An average value of the Weibull modulus evaluated is $m_{sf} = 0.52 \pm 0.16$.

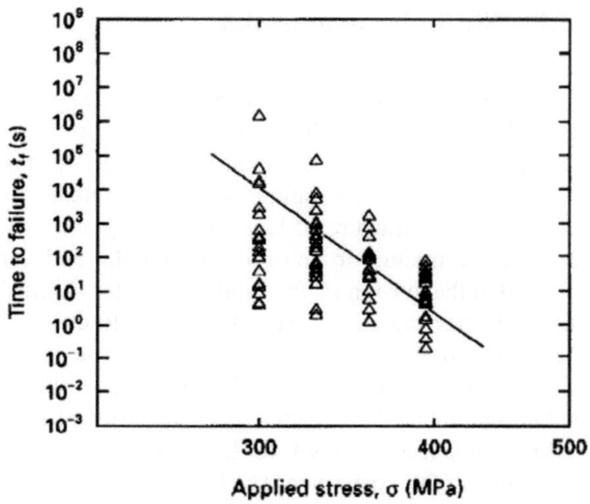

Fig. 15.7 Results of constant stress ("static fatigue" or "stress rupture") testing of NC 203 silicon carbide in flexure at 1300 °C in air. The solid line represents a prediction from the arithmetic mean data obtained from the constant-stress-rate testing. Choi et al. [4]. With kind permission of Springer Nature

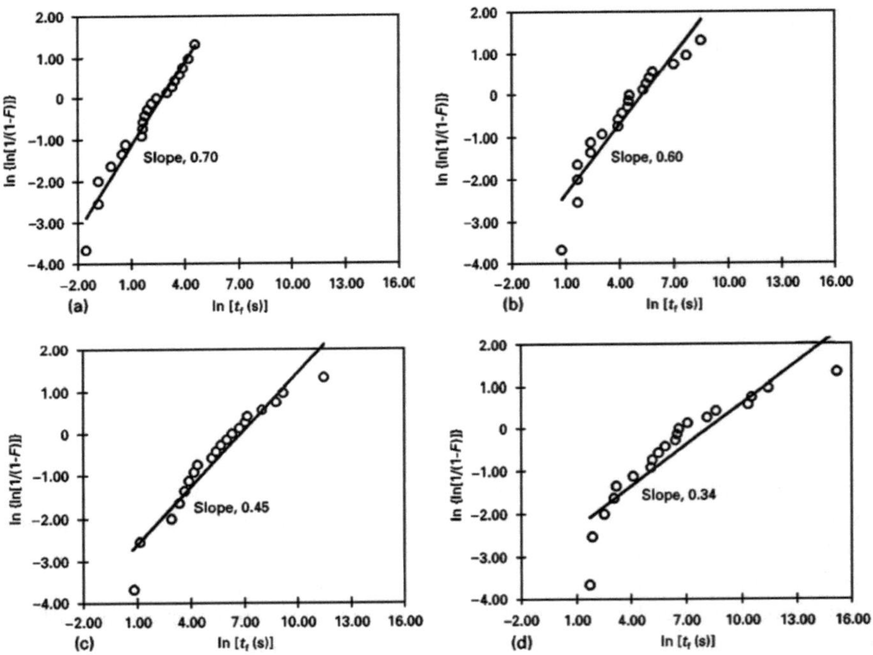

Fig. 15.8 Weibull time-to-failure distributions for different applied stresses obtained for NC 203 silicon carbide at 1300 °C in air: **a** 395 MPa; **b** 360 MPa; **c** 330 MPa; **d** 300 MPa. Choi et al. [4]. With kind permission of Springer Nature

15.5 Static Fatigue in B_4C

Very limited information exists on static fatigue—if at all- on pure non-compositional B_4C. However, in an earlier communication on static and cyclic fatigue of ceramic materials, of the Army Research Center Report, mention is being made on static report in B_4C. The static fatigue limit determined in three-point flexure is 43000 psi. Examination or fracture surface by scanning and transmission electron microscopy revealed the fracture to be of a transgranular and intragraular nature. The fracture initiated at the tension surface and propagated immediately to failure. There was no change in fracture appearance. The microstructure of hot pressed B_4C is shown in Fig. 15.9 and the properties in Table 15.3. The static fatigue test was conducted at atmospheric conditions and a plot of stress vs. life is shown in Fig. 15.10, which includes alumina also. Fracture occurred after 61 min.

In summary, static fatigue for B_4C was determined by applying a dead-weight load to the center of the specimen. The fracture surface is of transgranuar and intergranular nature. The fracture initiates at the tension surface, and once the crack is formed it propgates instantaneously.

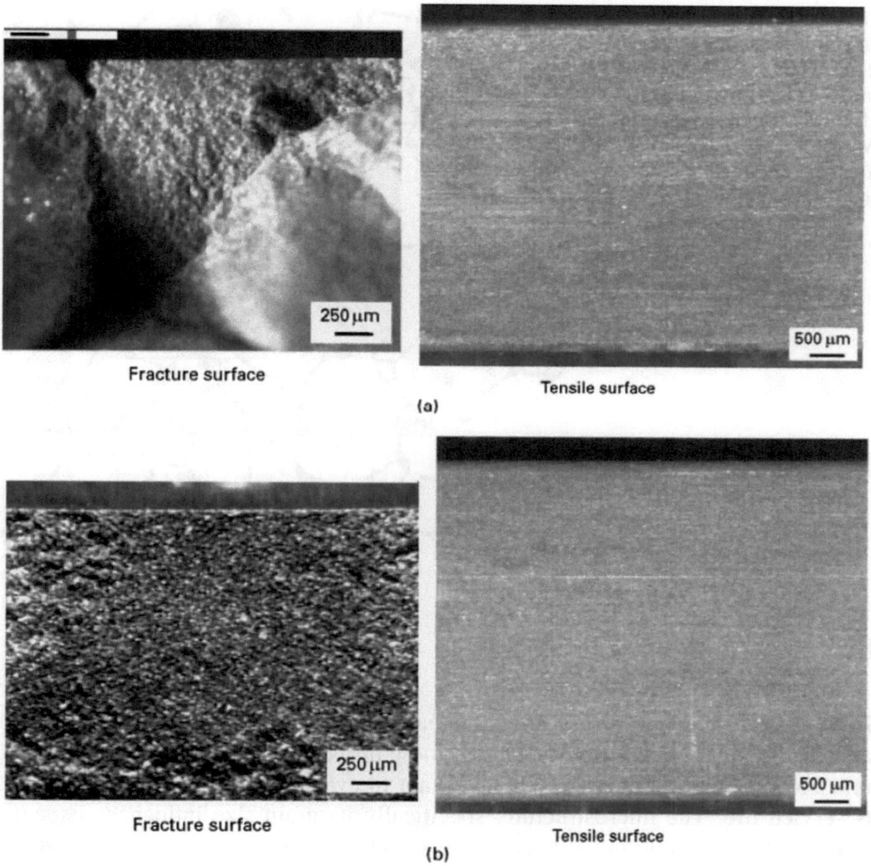

Fig. 15.9 Typical fracture surfaces and tensile surfaces of NC 203 silicon carbide specimens subjected to constant stress ("static fatigue") testing in flexure at 1300 °C in air: **a** $\sigma = 395$ MPa, $t_f = 1.5$ s; **b** $\sigma = 300$ MPa, $t_r = 460$ h. Choi et al. [4]. With kind permission of Springer Nature

Table 15.3 Physical Properties. S. J. Acquaviva and R. Chait, Army Mater. Mechanics Res, Center Watertown, Ma 02,172. Approved for public release, distribution unlimited

Material	Flexural Strength	Modulus	Density	Grain Size
B_4C	51,000 psi	60×10^6 psi	2.43 g/cc	10μ

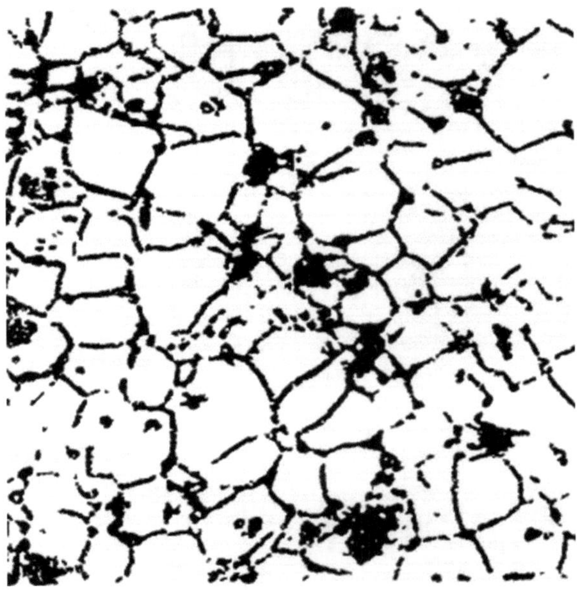

Fig. 15.10 Photomicrograph of material tested. 500x. S. J. Acquaviva and R. Chait, Army Mater. Mechanics Res, Center Watertown, Ma 02,172. Approved for public release, distribution unlimited

15.6 Static Fatigue in Si_3N_4

Si_3N_4 is a brittle material possesing high hardness, wear resistance, corrosion resistance and good thermal stability. Due ro these properties Si_3N_4 is a potential structural component, and as such it is important to evaluate its service reliability and its service life. The microstructure, specifically its grain size influences its static fatigue behavior. Sintering aids are usually added for the preparation of the structural components. The behavior of these material depends on the presence of cracks and their growth. Thus, it is of interest to characterize the crack growth behavior of this brittle material. A power law usually is used to characterize in static loading the crack growth dependenc on the stress intensity factor, K, according to Eq. (15.9) A and n are constants, n being the stress corrosion exponent. For a type I crack on the surface of a material the stress intensity factor is expressed by (15.10).

$$V = A(K)^n \tag{15.9}$$

$$K_I = Y\sigma_a a^{\frac{1}{2}} \tag{15.10}$$

Here Y is a geometric factor, σ is the applied stress and a is the crack length. For stating loading the lifetime, t_f can be calculated by integrating Eq. 15.9 to obtain

$$f = B\sigma_a^{n-2}\sigma_a^{-n}\left[1 - (\sigma_a/\sigma_c)^{n-2}\right] \tag{15.11}$$

where σ_a is the inherent strength, namely the strength in the absence of a subcritical crack extension and B is given by

$$B = 2/\left[AY^2(n-2)K_I\sigma^{n-2}\right] \tag{15.12}$$

B is a constant which depends on the material and environment but is independent on the applied load. The value of σ_a/σ_c is much smaller than 1 and n for ceramic is generally higher. Thus, Eq. 15.10 can be simlified to

$$t_f = B\sigma_n^{n-2}\sigma_a^{-n} \tag{15.13}$$

Expressing Eq. 15.13 in a log–log form one can write

$$\log\sigma_a = (-1/n)\log t_f + (1/n)\log B + [(n-2)/n]\log\sigma_c \tag{15.14}$$

It is possible to fit from Eq. 15.14 the relation between the applied load (σ_a) and the lifetime (t_f) to a straight line on a log–log representation whose slope is (1/n) and thus the stress corrosion exponent can be obtained. Following specimen preparation (grinding and polishing), precracks were introduced by indentation in the centre of the sintered samples before the fatigue tests. Experiments were performed with three-point bending in an in-house designed testing system. All the experiments were carried out at room temperature in three different media, namely H_2O, air, and kerosene. At each environment specimens were tested at static load.

Two SEM specimens showing the microstructures are illustrated in Fig. 15.11, indicating that the main phase is columnar β-Si_3N_4. ZrO_2 in sample 2 (designated as N2) is distributed on the grain boundaries as small particles. The mechanical properties of the two samples are shown in Table 15.4. The static fatigue plots at the environments indicated are shown in Fig. 15.12 according to Eq. 15.14 for the two samples. From the slope of the lines in Fig. 15.12 n can be evaluated. The static fatigue paramentes of the two materials in the three media (environments) are listed in Table 15.5. One can see from Table 15.5 that the values, n, lnA, and lnB in the two materials are different from one another. The variation of the crack growth rate of the two materials shown in Fig. 15.11 in all ebvironments at the same stress intensity factor is the greatest in water, followed by air and then kerosene. This illustration shows that the tested materials all appear to be sensitive to the environment. The reason for the greatest crack growth in water is associated with the fact that water is known to chemisorb onto strain bonds in SiO_2 and Al_2O_3, which seems to occur in Si_3N_4 based materials. However, the viscosity of kerosene inhibits this chemisorption reaction, and thus decreases the crack growth rate as observed in specimen N1. This does not occur in N2 and the lines are very close to one another, where the n values are higher than that of N1. This indicates that the N2 samples have a higher crack propagation resistance and are less sensitive to the environmental medium (Fig. 15.13). The higher crack growth rate of the N1 sample is illustrated in a plot of (Recall that logV–log K relation is given in Eq. 15.9 and is related to crack growth) Fig. 15.12. The n values in ceramics are related to the intensity factor according to

Material	Flexural Strength	Modulus	Density	Grain Size
B_4C	51,000 psi	60×10^6 psi	2.43 g/cc	10μ

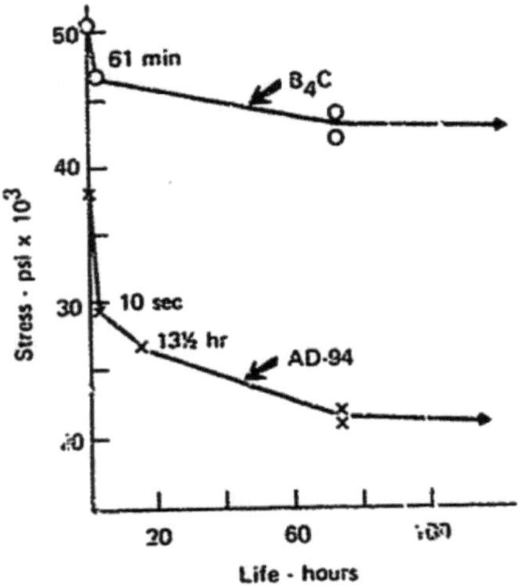

Fig. 15.11 Static fatigue of ceramics. S. J. Acquaviva and R. Chait, Army Mater. Mechanics Res, Center Watertown, Ma 02,172. Approved for public release, distribution unlimited. AD-94 is a commercially available alumina

Table 15.4 Mechanical and microstructural properties of the materials. Hongjie et al. [5]. With kind permission of Springer Nature

Material	σ_b^a(MPa)	K_{IC}^b(MPa m$^{1/2}$)^b	Aspect ratio	Morphology
N1	700	6.25	7	Columnar
N2	579	7.94	5	Columnar and equaxial grain

[a] Determined by three-point bending method
[b] Determined by indentation method (196N)

Fig. 15.12 SEM micrographs of samples **a** N1 and **b** N2. Hongjie et al. [5]. With kind permission of Springer Nature

Table 15.5 The static fatigue parameters for samples N1 and N2 obtained in the three investigated media. Hongjie et al. [5]. With kind permission of Springer Nature

	n		Ln *B*		Ln *A*	
	N1	N2	N1	N2	N1	N2
Air	647.8	1196.46	− 416.78	− 1718.19	− 223.62	− 843.46
Water	425.33	596.88	− 324.51	− 285.9	− 457.78	− 136.64
Kerosene	84.25	228.13	− 41.19	− 147.89	− 114.40	− 141.71

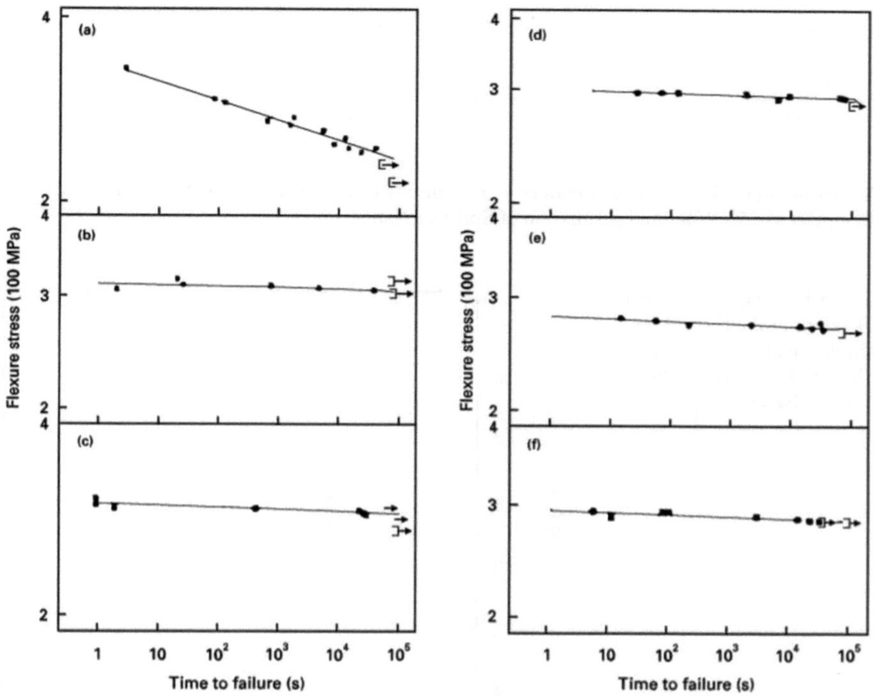

Fig. 15.13 The static fatigue data for sample N1 **a** in kerosene, **b** in water and **c** in air and also sample N2 in **d** kerosene, **e** water and **f** air. W. Hongjie, W. Yonglan, J. Zhihao and Z. Huijiu, J. Matr. Sci., 32, 5775 (1997). With kind permission of Springer Nature

$$n = cK_{Ic} \tag{15.9}$$

where c is a constant. Thus, Eq. 15.13 can be rewritten as

$$t_f = B\sigma_a^{-cK_{Ic}} \tag{15.10}$$

Thus, it can be concluded that the higher the fracture toughness the higher the n value and the longer the lifetime at the same stress level. The n value of N2 sample is

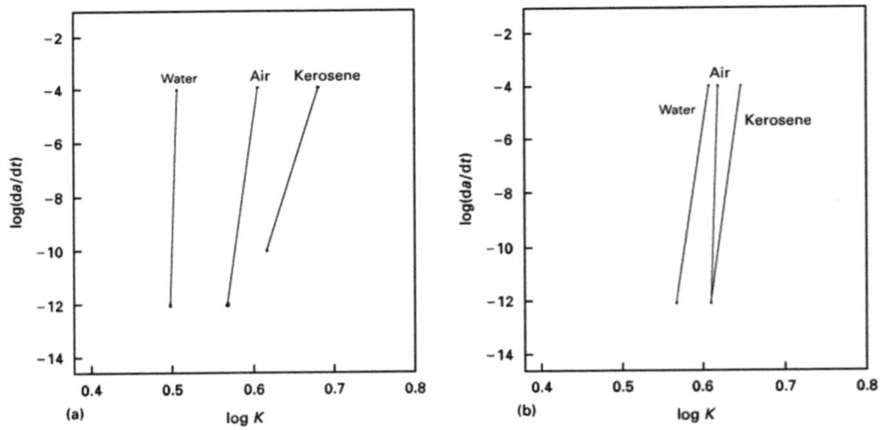

Fig. 15.14 logV- log K curves obtained in the various media for **a** sample N1 and **b** sample N2. Hongjie et al. [5]. With kind permission of Springer Nature

Fig. 15.15 The static fatigue behaviour of the N1 and N2 samples in air. Hongjie et al. [5]. With kind permission of Springer Nature

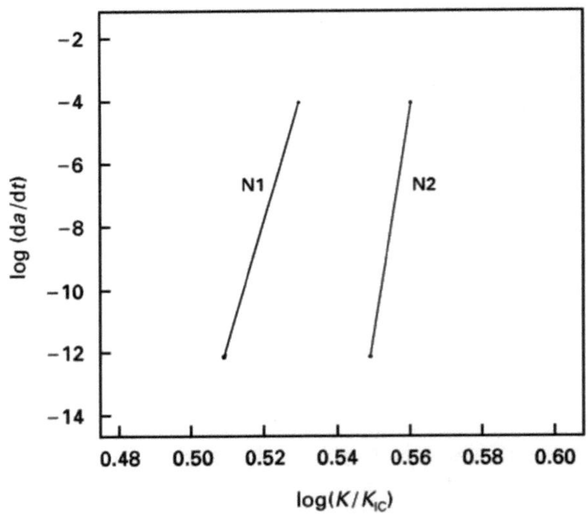

higher and the crack propagation resistance is also higher because its higher fracture toughness (Fig. 15.14).

In summary, the sintered samples considered in this section contained a glassy phase at the grain boundary. Their static fatigue behavior in various environmental media indicates that the rate of crack growth is greatest in water and the least in kerosene. The crack propagation resistance of the N2 sample is higher that that of N1 because its specific microstructure and is also less sensitive to environment tha N1. The crack propagation resistance is related to the microstructure through n which is related to the fracture toughness through Eq. 15.9. Thus, the higher the fracture toughness the higher is the n value and the longer is the lifetime at the same stress level (Fig. 15.15).

References

1. G. Jin, A. Honjoh, H. Awaji, J. Cer. Soc. Japan **108**, 614 (2000)
2. P.C.P. Bouten, Philos. Mag. A **82**, 2059 (2002)
3. J. Gui, Z. Xie, Mater. Sci. Eng. A **676**, 531 (2016)
4. S.R. Choi, J.A. Salem, N.N. Nemet, J. Matr. Sci. **33**, 1325 (1998)
5. W. Hongjie, W. Yonglan, J. Zhihao, Z. Huijiu, J. Matr. Sci. **32**, 5775 (1997)

Chapter 16
Thermal Fatigue

Abstract Thermal fatigue is a fatigue failure mechanism that is induced by cyclic stresses from repetitive fluctuations in the temperature. The degree of damage depends on the magnitude and frequency of the temperature changes. Macroscopic cracks can be induced by cyclic stresses or strains due to temperature changes. Thermal fatigue cannot be prevented, but there are ways to prevent failure due to thermal fatigue. Avoiding temperature fluctuations in particular during shutdown and start-up of a component can help to prevent thermal fatigue.

Keywords Thermal fatigue · Temperature fluctuations · Alumina · YSZ · Boron carbide

16.1 Introduction

Thermal fatigue is a fatigue failure mechanism that is induced by cyclic stresses from repetitive fluctuations in the temperature. The degree of damage depends on the magnitude and frequency of the temperature changes. Temperature variations can occur in service of a component. Macroscopic cracks can be induced by cyclic stresses or strains due to temperature changes. The damage may appear from one or more cracks at the surface of a component in service which eventually can grow to macroscopic cracks, propagate to failure unless remedied. All material without exception can be affected to various degrees by thermal fatigue. However, thermal fatigue cannot be prevented, but there are ways to prevent failure due to thermal fatigue. The most effective way to prevent failure in operating equipment due to thermal fatigue, is to minimize as far as possible thermal stresses and thermal cycling already in the design stage. Avoiding temperature fluctuations in particular during shutdown and start-up can help to prevent thermal fatigue. Further, reducing thermal gradients can help if not completely eliminate thermal failure. Difference in thermal expansion coefficient variation in multi-component equipment should be minimized by selecting parts with minimum difference in thermal expansion coefficients between them. Thermal fatigue occurs not only in multi-component equipment, but in other systems as well. For example, in castings a part experiences high temperature, while other parts of the casting remain at lower temperature during cooling down inducing

J. Pelleg, *Cyclic Deformation in Oxides, Carbides and Nitrides*, Structural Integrity 22,
https://doi.org/10.1007/978-3-030-86118-6_16

thermal variation which eventually can lead to thermal fatigue. Stress raisers should be minimized as far as possible. As mentioned, thermal fatigue cannot always be prevented, but it is essential to inspect the tendency to failure. Visual inspections are important but not as effective as ultrasonic technique which can detect internal cracks. In the next sections thermal fatigue will be considered in several components, relevant to this book.

16.2 Thermal Fatigue in Alumina

Application of ceramics such as alumina for high temperature structural application is mainly due to their high melting point and good mechanical strength at high temperature. Their application, however, is limited due to catastrophic failure when subjected to sudden temperature change (thermal shock) or repetitive cycling of temperatures (thermal fatigue). Regarding thermal fatigue, in the wake of cyclic temperature changes micro-crack with initial crack size a_0 ($a_0 < a_c$) growth incrementally in each cycle till it reaches the critical crack sizem a_c at which the sample fails catastrophically. For experimental details and methods consult Panda et al. Thickness and grain size have an influence on the thermal fatigue. The failure temperature of alumina of different thicknesses is shown in Table 16.1. Vickers indentations were made by applying 5, 10 and 20 kg loads and the average crack length measured. ΔT_c was evaluated after the indented samples were subjected to the thermal shock tests at 600 °C for 4 h and the results are summarized in Table 16.2 and a graph of ΔT_c versus average crack length in Fig. 16.1. The temperature distribution measured in alumina for three trials is presented in Fig. 16.2. From the failure temperatures of alumina of different thickness, the temperature values at different radial distances were extrapolated and presented in Fig. 16.3.

In summary, the thermal stress is maximum at periphery of the top surface, where the failure is initiated. The critical temperature difference, ΔT_c depends on the thickness of the sample and is higher for thicker sample (because higher temperature is required to develop the same stress level in a thicker sample). Further, ΔT_c decreases with increase in the initial crack length. And also the number of thermal fatigue cycles decreases with increase in temperature difference, ΔT_c of the cycles.

Table 16.1 Failure temperature and ΔT_C of alumina of different thickness. P. K. Panda, T. S. Kannan, J. Dubois, C. Olagnon and G. Fantozzi, J. Eur. Cer. Soc., 22, 2187 (2002). With kind permission of Elsevier

2	342±33	317±33
3	390±39	365±39
4	557±75	532±75
5	647±100	622±100
6	700±60	675±60

Table 16.2 ΔT_C obtained for samples with different indented crack lengths. P. K. Panda, T. S. Kannan, J. Dubois, C. Olagnon and G. Fantozzi, J. Eur. Cer. Soc., 22, 2187 (2002). With kind permission of Elsevier

No.	Load (kg)	Average crack length (μm)	Failure temperature (°C)	ΔT_C (°C)
1	20	481	342	317
2	20	555	405	380
3	20	546	367	342
4	10	343	348	323
5	10	395	268	343
6	5	178	412	387
7	5	149	381	356

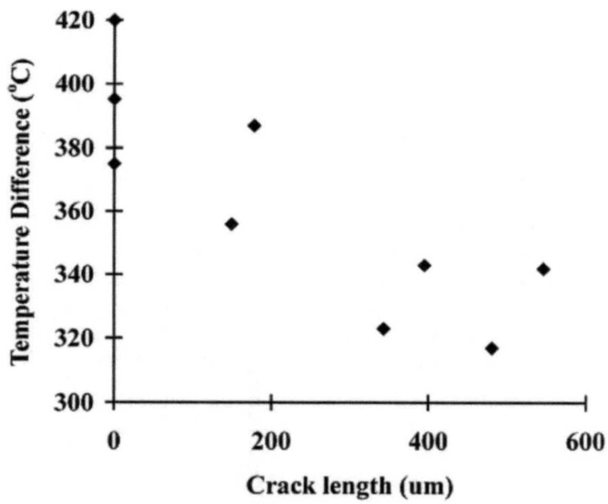

Fig. 16.1 Critical temperature difference versus crack length. P. K. Panda, T. S. Kannan, J. Dubois, C. Olagnon and G. Fantozzi, J. Eur. Cer. Soc., 22, 2187 (2002). With kind permission of Elsevier

16.3 Thermal Fatigue in YSZ

Yttria stabilized zirconia (YSZ) is mainly used as thermal barrier coating (TBC) consisting of a topcoat and bond coat NiCoCrAlY. It is used to protect NiCoCrAlY super alloys for high temperature applications due to its corrosion resistant properties and resistance to thermal fatigue. The topcoat and bond coat were deposited on the substrate using plasma spraying techique, followed by thermal fatigue treatment. The treatment consists of heating up to 900 °C for 5 h and cooling down to 25 °C for 15 min. Such a process is called one cycle. The microstructers of the specimens used for thermal fatigue tests at 900 °C are compared with the as received specimen with those subjected to several cycles by thermal fatigue in Fig. 16.4. Cracks are observed after 5 cycles (Fig. 16.4c). In Fig. 16.4d widened cracks in two areas are seen after 42 cycles. SEM images are compared in Fig. 16.5 of as received specimen (after

Fig. 16.2 Temperature distribution measured in an alumina disc (5 mm thick) during thermal shock test. P. K. Panda, T. S. Kannan, J. Dubois, C. Olagnon and G. Fantozzi, J. Eur. Cer. Soc., 22, 2187 (2002). With kind permission of Elsevier

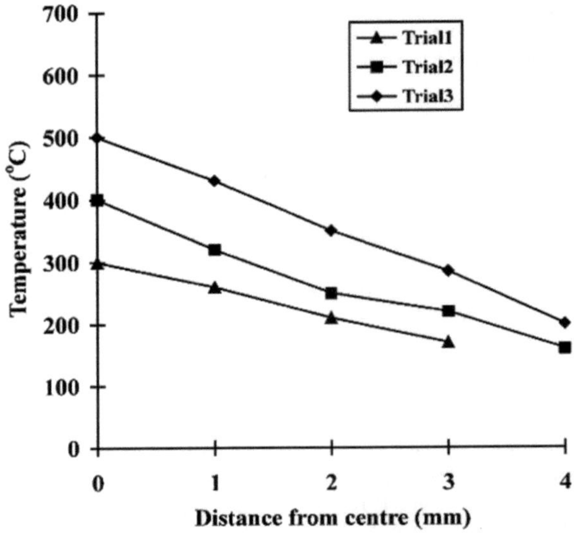

Fig. 16.3 Extrapolated temperatures for alumina of different thickness. P. K. Panda, T. S. Kannan, J. Dubois, C. Olagnon and G. Fantozzi, J. Eur. Cer. Soc., 22, 2187 (2002). With kind permission of Elsevier

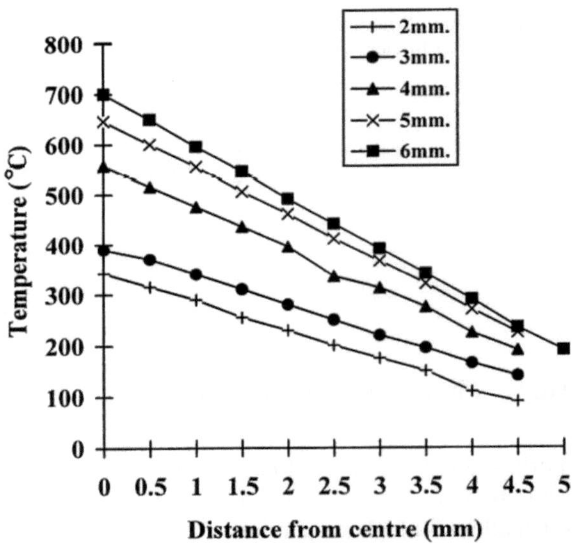

900 °C heating) with specimen subjected to 42 cycles thermal fatigue. A depletion layer in the topcoat is 10 μm (the topcoat is YSZ). This layer recives the highest heat load. Electron microscope analyses performed after heat treatment shows the interface between the bond coat and substrates. It can be seen that the image shows some parts which are oxidized, primarily at the interface between the bond coats to the substrate. Light and dark phase dominates the overall image as seen in Fig. 16.8.

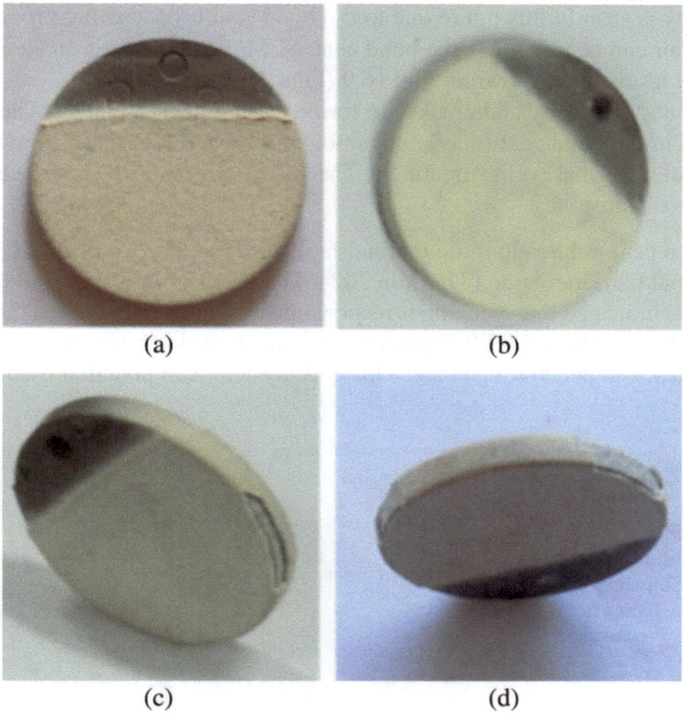

Fig. 16.4 Macrostructure analysis on specimen as received **a** and specimen after thermal fatigue at 900 °C, **b** 2 cycles, **c** 6 cycles, **d** 42 cycles. H. Purwaningsih, L. Noerochim, R. Fajarain, J. A. Hakim and Sulistijono, IPTEK, J. for Techn. Sci., 21, No 4 (2010). Free to share

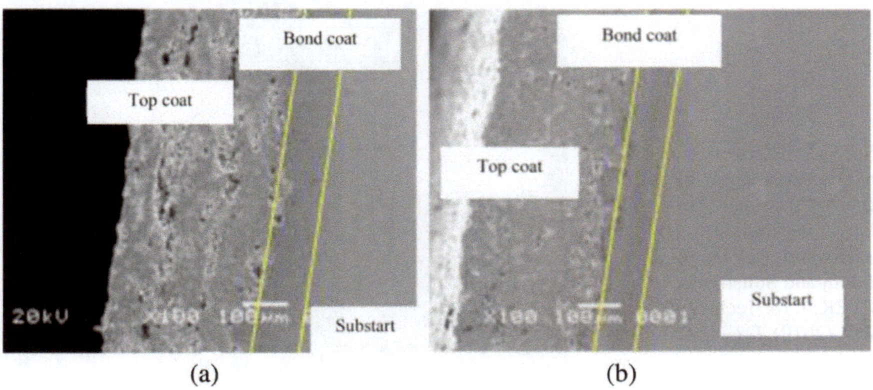

Fig. 16.5 Electron microscope analysis on specimen **a** as-received; **b** after thermal fatigue 42 cycles. H. Purwaningsih, L. Noerochim, R. Fajarain, J. A. Hakim and Sulistijono, IPTEK, J. for Techn. Sci., 21, No 4 (2010). Free to share

No porosity is seen in this figure due to the bond coat layer coating process where very little air can be trapped in the bond coat. SEM/ EDX analysis on some regions (outline by a square) is shown in Fig. 16.9 (Figs. 16.6 and 16.7).

In summary, porosity and cracks are formed at the interface layer. The porosity is induced in the bond coat (due to unhomogenity) when it is spark sprayed on the substrate after thermal cycling treatment. Ther is no porosity on the interface between the bond coat/substrate.

In a more recent publication, the thermal fatigue (TF) behavior of plasma sprayed YSZ thermal barrier coating (TBC) on 2024 Al alloy at 600 °C was evaluated. The results have indicated that cracking is responsible for the thermal fatigue failure. The cracks formed on the YSZ-TBC system are vertical cracks, horizontal crack, propagating crack and penetrating crack. This TBC system is in potential use in internal

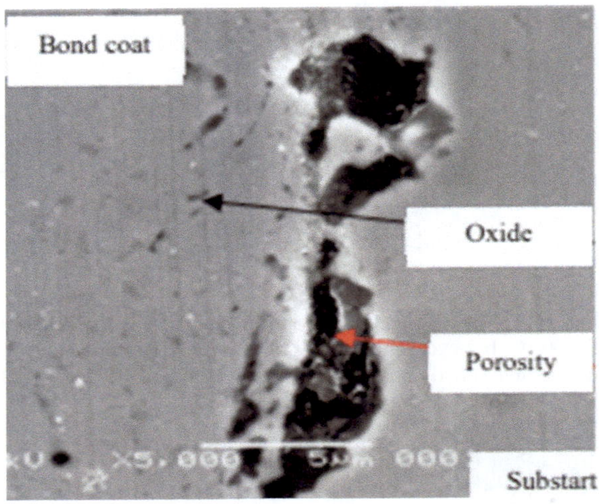

Fig. 16.6 Electron microscope analysis on as received specimen. H. Purwaningsih, L. Noerochim, R. Fajarain, J. A. Hakim and Sulistijono, IPTEK, J. for Techn. Sci., 21, No 4 (2010). Free to share

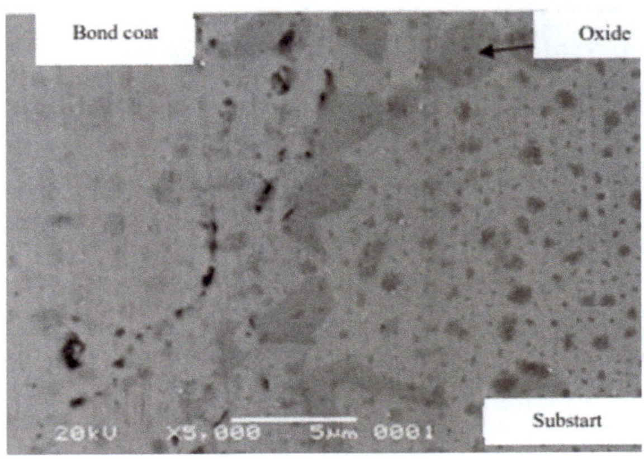

Fig. 16.7 Electron microscope analysis on as received specimen. H. Purwaningsih, L. Noerochim, R. Fajarain, J. A. Hakim and Sulistijono, IPTEK, J. for Techn. Sci., 21, No 4 (2010). Free to share

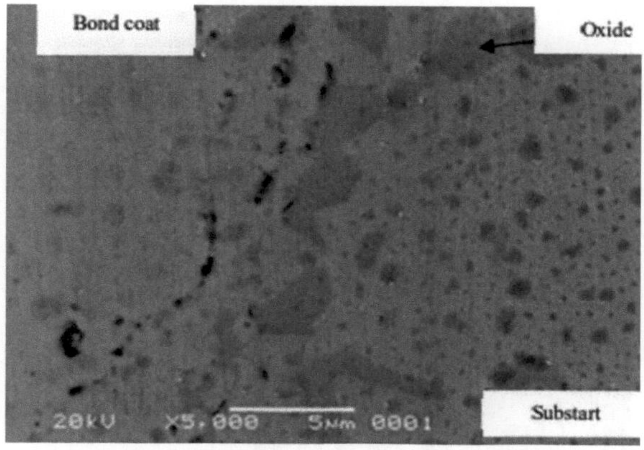

Fig. 16.8 Electron microscope analysis on specimen after thermal fatigue at 900 °C—42 cycles. It is shown dark area and bright area indicating the difference on morphology. H. Purwaningsih, L. Noerochim, R. Fajarain, J. A. Hakim and Sulistijono, IPTEK, J. for Techn. Sci., 21, No 4 (2010). Free to share

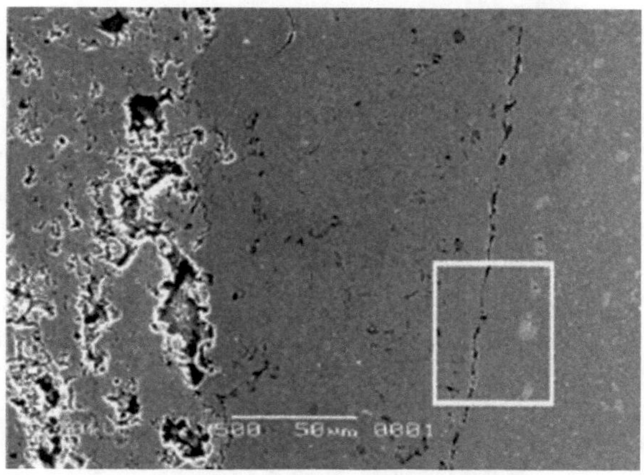

Fig. 16.9 SEM/EDX analysis on thermal fatigue specimen, this analysis was done to observe element diffusion along the interface. H. Purwaningsih, L. Noerochim, R. Fajarain, J. A. Hakim and Sulistijono, IPTEK, J. for Techn. Sci., 21, No 4 (2010). Free to share

combustion engine components and gas turbines, and to achieve high thermal efficiency in advanced diesel engines it is essential to improve in-service performance. TBC-often consisting of monolithic ceramics—is of interest in these applications. The YSZ-TBC system on Al alloy is very important in the auromotive industry and its high temperature application is of interest. The Al alloys substrate was bond coated by NiCoCrAlY followed by another coat, the topcoat which is a TSZ deposited by air plama spray (APS). The average of three specimens were used for the analysis. The TBC was cracked under thermal fatigue penetrating from the coated surface to substrate (penetration cracking) along horizontal and vertical directions. The initiation stage of the crack is seen in Fig. 16.10a, which cannot be seen by open eye, and the fractures surface morphology is shown in Fig. 16.10b. Fractured surface is seen in Fig. 16.11 by field emission scanning electron microscope (FESEM). The fractured topcoat (spherical particle numbered 1) and the fractured surface (red ciecle) due to penetrating cracking is seen in this figure. The micro columnar grains seen in this location were solidified from the melted fraction of YSZ top coat powder.

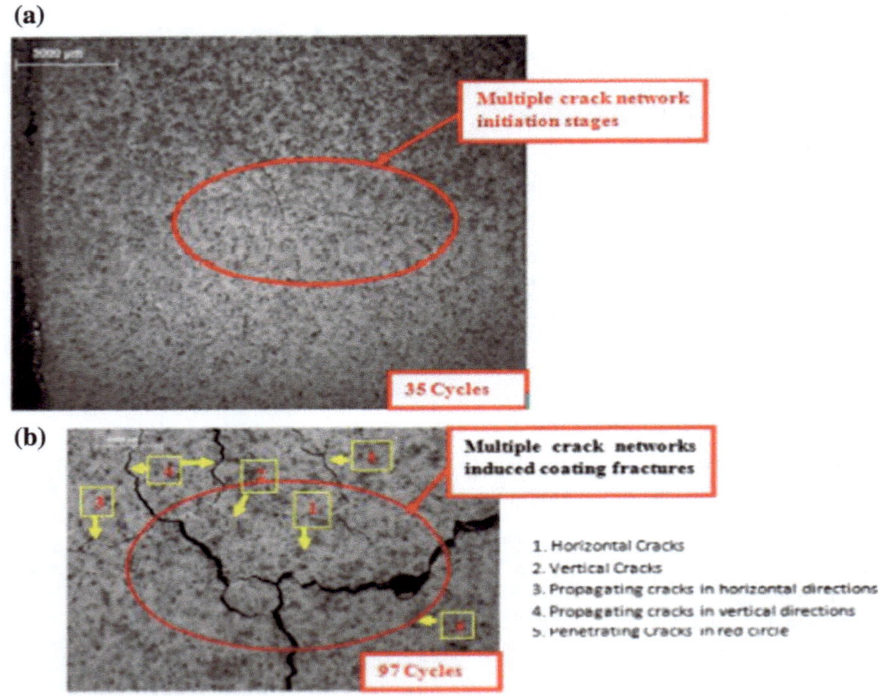

Fig. 16.10 **a** and **b** Optical micrograph of fractured plasma sprayed TBC specimens with thickness of 250 μm after TF testing. D. Kumar and K. N. Pandey, International J. Mechanical and Production Engineering, 2, 2320 (2014). Open access

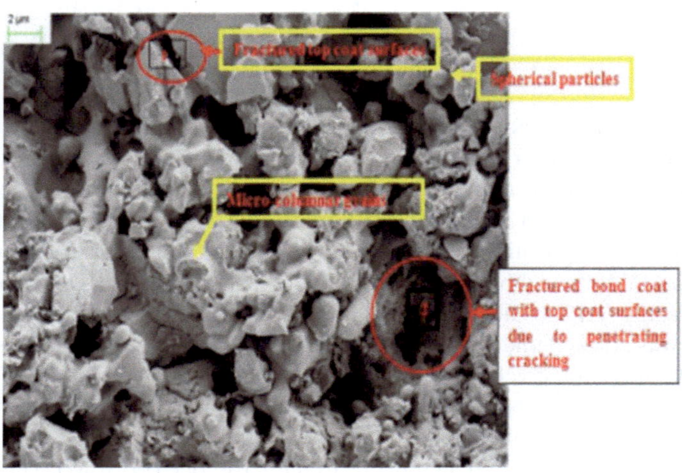

Fig. 16.11 FESEM micrograph of the fractured surface of as sprayed YSZ TBC. D. Kumar and K. N. Pandey, International J. Mechanical and Production Engineering, 2, 2320 (2014). Open access

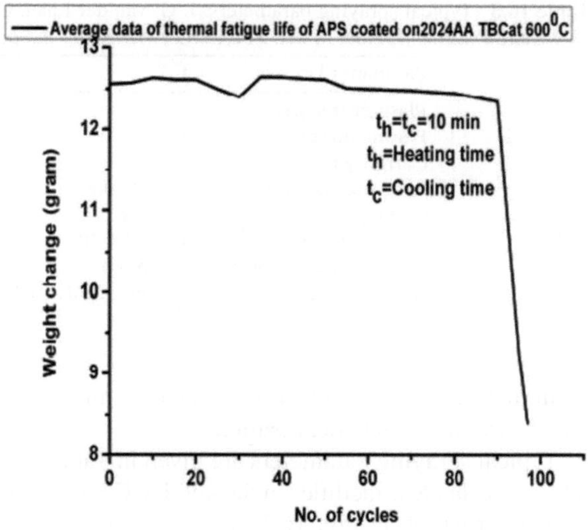

Fig. 16.12 Weight change as a function of cycle number YSZ TBCs during thermal fatigue testing. D. Kumar and K. N. Pandey, International J. Mechanical and Production Engineering, 2, 2320 (2014). Open access

The failure started from the edges of the samples and then propagated to the surface areas as inferred from Fig. 16.10. The start of the failure from the edges (and corners) can be associated with the heating and cooling inducing thermal stresses between different layers during the thermal fatigue loading conditions. Weight changes occur in the TBC specimens, first increasing due to substrate free surface oxidation and then reducing suddenly after 30 cycles due to multiple crack formation. After the sudden weight loss in 30 cycles, a weight gain is observed resulting from the substrate and bond oxidation. This is illustrated in Fig. 16.12. Gradual weight loss can be observed after 52 cycles indicating thermal degradation of the coated speciens. After 97 cycles a sudden weight loss is an indication of coating damage as a consequence of cracking.

In summary, thermal fatigue of plasma sprayed TBC induced cracking which initiates at edges or corners. The failure of YSZ TBC system is due to multiple cracking of various types such as veritical, horizontal, propagating and penetrating. The penetrating vertical cracks started to penetrate from topcoat to the substrate surface at 600 °C.

16.4 Thermal Fatigue in Boron Carbide (B_4C)

Pasma-sprayed B_4C coatings was deposited on 316L stainless steel (SS) substrate in the thickness range of 0.3–2.0 mm. The use of this coating is for first-wall protection in a Next Step Device (nuclear system) is due to its ability to protect the structure against excessive thermal fatigue from disruptions and for intrinsic resistance to thermal shock as well. Moreover, thermal fatigue of the heated surface (subsurface)

Table 16.3 Typical spraying parameters. J. G. van der Laan, G. Schnedecker, E. V. van Osch, R. Duwe and J. Linke, J. Nuc. Mater. 211,135(1994). With kind permission of Elsevier.

Parameter lab.	CEA	KFA
Plasmogenic gas	Ar–H_2 4:1; 30:12	Ar–H_2 2:1
Plasma power	50 kW	45 kW
Carrier gas	Ar	Ar
Spraying distance	100–120 mm	200–300 mm
Cooling	liquid Ar	none
Spray atmosphere	Ar ($[O_2] < 40$ ppm)	Ar
Top-coating thickness	0.5–1.5 mm	0.1–0.3 mm
Intermediate layer	≈ 0.5 mm 316L/B_4C	

seems to be dominant, and due to the layered microstructure of the coating the erosion rate clearly has a statistical nature.

Typical spraying parameters are given in Table 16.3. In this Table CEA and KFA refer to the nuclear facilities in Saclay (i,.e CEA) and Jūlich (KFA), respectively. The SEM micrograph in Fig. 16.13 illustrates the cross section of the plasma sprayed B_4C top coating on the SS 316L substrate. The samples from CEA had a graded coating sprayed on the substrate as an intermedeiate or adaption layer. The samples of KFA were coated with B_4C without an intermediate layer. The coating-substrate combination and the designation of the samples is shown in Table 16.4. Experiments have been performed (both the CEA and KFA coatings) with Judith electron beam facility (at KFA Jūlich). Typical surface effects are sintering, melting, and chipping (chip thickness is ~0.05 mm) as shown in Fig. 16.14. The calculated maximum surface temperature rise of B_4C coatings versus energy density for surfaces and volumetric energy deposition is illustrated in Fig. 16.15. The surface layer can heat up to the melting point and it depends on the duration of the energy deposition. The calculated surface melting thresholds are given in Fig. 16.16 as a function of the duration of

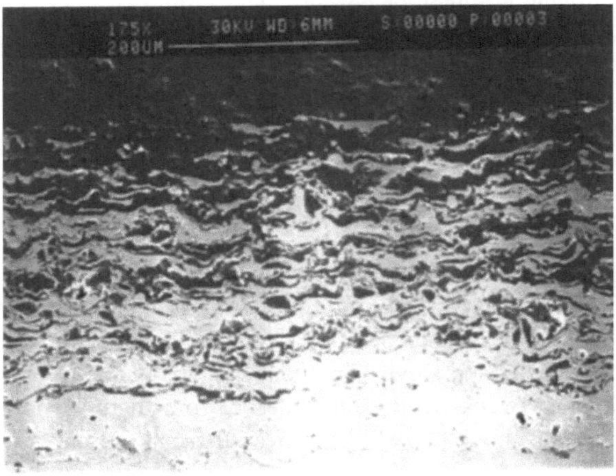

Fig. 16.13 SEM micrograph of cross section of plasma-sprayed boron carbide coating (dark, top) on type 316L stainless steel substrate. The structure of the graded 'adaptation' layer, a mixture of boron carbide and stainless steel, is clearly visible. J. G. van der Laan, G. Schnedecker, E. V. van Osch, R. Duwe and J. Linke, J. Nuc. Mater. 211, 135 (1994). With kind permission of Elsevier

Table 16.4 Coating-substrate combinations for plasma-sprayed B4C in high heat flux tests. J. G. van der Laan, G. Schnedecker, E. V. van Osch, R. Duwe and J. Linke, J. Nuc. Mater. 211,135(1994). With kind permission of Elsevier

Designation	Substrate/ thickness (mm)	Bond-coat/ thickness (mm)	Top-coat/ thickness (mm)	Tested
A1	316L/5.0	-	0.26	JEBIS, Laser
A2	316L/5.0	0.5–0.6	0.4	JEBIS
A3	316L/5.0	0.5–0.6	0.8	JEBIS
B1	316L/5.0	-	0.24	JEBIS
C1	316L/3.0	0.4	1.2	JUDITH, FE200, Laser
C2	316L/3.0	0.5	1.5	JUDITH, FE200, Laser
C3	316L/3.0	0.5	1.2	JUDITH, FE200, Laser
C4	316L/2.0	0.5	1.2	JUDITH, FE200, Laser
D1	St.37/10	-	0.3	JUDITH, FE200, Laser

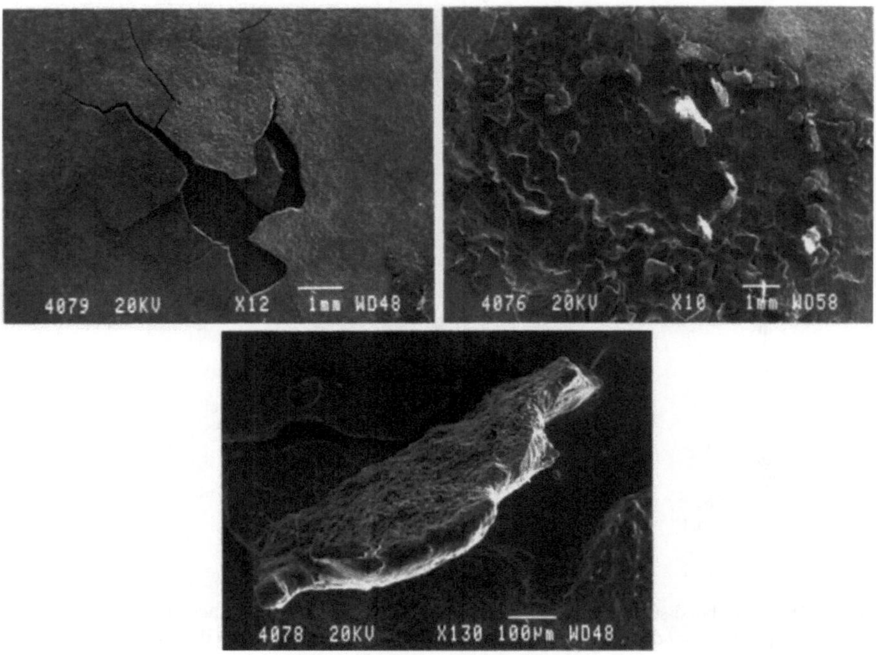

Fig. 16.14 SEM micrographs of coating surface after testing in JUDITH at E" ≈ 2.4 MJ/m^2 in 10 ms on ≈ 10 mm^2 area: **a** specimen Cl after 5 shots; **b** specimen C3 after 100 shots and **c** detail of 'chip' from **b**. J. G. van der Laan, G. Schnedecker, E. V. van Osch, R. Duwe and J. Linke, J. Nuc. Mater. 211,135(1994). With kind permission of Elsevier

energy deposition in a double logarithmic plot. For the surface absorption case the curves depend on the inverse of the square root of pulse duration. For increasing depth of energy deposition, Z_e the surface temperature rises more slowly. It should be noted that in case of volumetric energy deposition the temperature may rise faster at some distance from the surface and may actually reach the melting point before surface

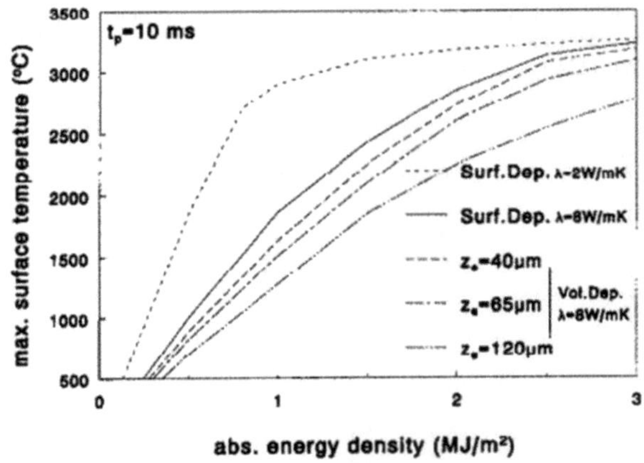

Fig. 16.15 Calculated maximum surface temperature rise of B_4C coating versus energy density for surface and volumetric (z, = 40, 65, 120 nm) energy depositions in 10 ms. J. G. van der Laan, G. Schnedecker, E. V. van Osch, R. Duwe and J. Linke, J. Nuc. Mater. 211,135(1994). With kind permission of Elsevier

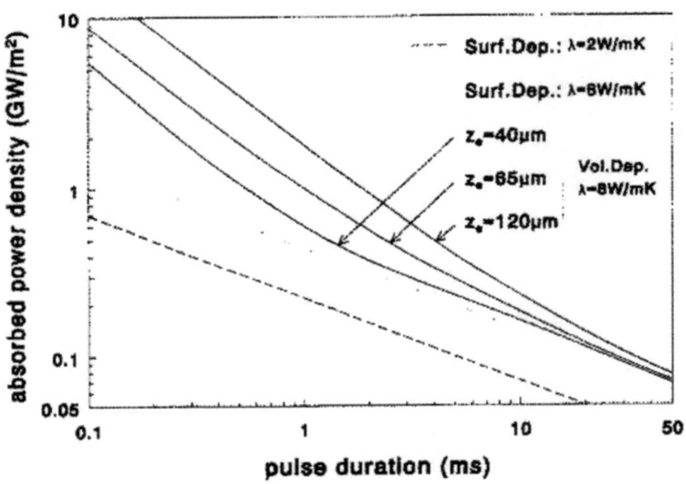

Fig. 16.16 Calculated surface melting threshold of B_4C coatings for surface and volumetric (z, = 40, 65, 120 μm) energy depositions as a function of the duration of energy deposition. J. G. van der Laan, G. Schnedecker, E. V. van Osch, R. Duwe and J. Linke, J. Nuc. Mater. 211,135 (1994). With kind permission of Elsevier

melting occurs. At longer pulse duration the divergence decreases and convergence of calculated threshols occurs as seen in Fig. 16.16.

Experimental and numerical results are summarized in Fig. 16.17. The difference between electron beam and laser test results can be mainly attributed to the effects of volumetric energy deposition by electrons. The erosion rates in the electron beam tests do not correlate directly with the calculated melting and evaporation behaviour due to the volumetric energy deposition. Moreover, thermal fatigue of the heated

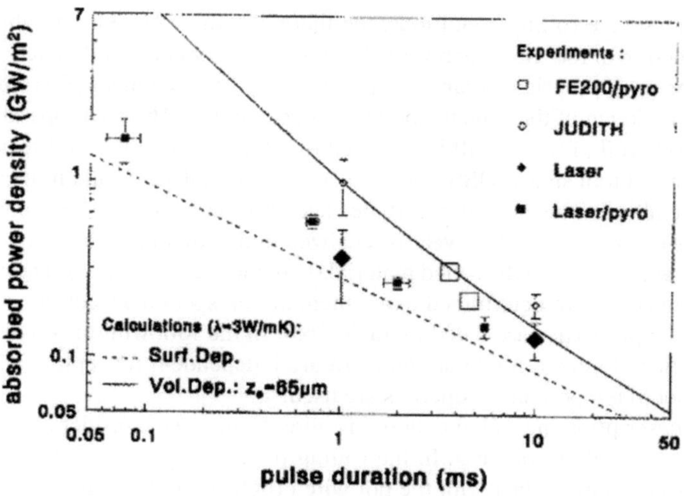

Fig. 16.17 Experimental and numerical results for the melting threshold of coatings in case of laser and 100 keV electron beam tests. Experimental values obtained include results from post-test analysis as well as from fast pyrometry. Numerical values have been derived from the calculations as shown in Fig. 16.16. J. G. van der Laan, G. Schnedecker, E. V. van Osch, R. Duwe and J. Linke, J. Nuc. Mater. 211,135(1994). With kind permission of Elsevier

(sub)-surface seems to be dominant, and due to the layered microstructure of the coating the erosion rate clearly has a statistical nature.

In summary, boron carbide coatings by plasma spraying with thicknesses up to about 1.5 mm on type 316L substrate were manufactured and have shown a reasonable performance in simulated disruption heat load experiments. It seems also that some surface melting can be tolerated without coating failure. No gross failure or burn-through of the coatings occurs even at higher loads of repeated shots. The results suggest that thick plasma sprayed B_4C coating have a potential to survive repeated thermal shocks from disruption heat loads. The applicability of coatings for first-wall protection in a Next Step Device depends partially on its ability to protect the structure against excessive thermal fatigue from disruptions as well as its intrinsic resistance to thermal shocks.

16.5 Thermal Fatigue in Si_3N_4

Si_3N_4 rolls are routinly used in rolling mills replacing cemented carbides which were used successfully as material for the rolling tools. The lifetime of cemented carbides rolls is restricted by wear and at very high operating temperatures cemented carbides rolls suffer thermal fatigue cracking. For the processing, some type of super alloys the rolling temperature can reach up to 1100 °C and stresses up to 750 MPa.

Under these severe rolling conditions, the thermal fatigue cracking might increase with a reduction of tool life. Therefore, there is a need for better roll materials. Si_3N_4 seems to be an appropriate ceramic as predicted by finite element (FE) calculation. A schematic sketch of the rolls in operation is seen in Fig. 16.18a in operation and a silicon nitride roll in (b). A mill is designed to roll a high number of different steels like high-speed tool steels (HSS), hot and cold-work tool steels and more and more nickel-base alloys. The cemented carbides need regrinding due to severe wear after 500 tonnes of rolled wire. The wear is localized in the middle of the roll groove and causes surface damage in the rolled material (here inconel 625 wires). The properties of silicon nitride were determined experimentally on specimens cut out of ceramic rolls, and the properties are listed in Table 16.5. In the following it is assumed that the properties of silicon nitide are temperature independent (except the CTE), and therefore room temperature properties are used.

In this description the rolled material is wire. Hence, we consider as an example this case in the rolling process. In the rolling process a given position on the roll groove surface is in contact with the hot wire (1060 °C) only on a short part of the circumference and for a short time and for the rest the revolution at this position is cooled with water. In the first revolution the groove surface reaches a high temperature (900 °C) in a very short time called the contact time and when the contact between rolls and wire ends and cooling begins the surface temperature sharply decreases. At the end of the first cooling cycle the surface temperature is about 70 °C. Then the described surface element comes in contact with the hot wire again in the next cycle and the process is repeated. The compressive strength of ceramic materials is many times higher than its tensile strength. Only tensile stresses developed are responsible for failure of the roller. The very severe cyclic heating and cooling of the surface elements in the roll groove (between 130 and 930 °C) causes predominantly large compressive stresses at the surface. Some reasonable tensile stresses only occur in the

Fig. 16.18 Schematic sketch of a pair of rolls in operation **a** and a picture of a variant of the analysed rolls **b**. The rolls (diameter 225 mm) were manufactured from silicon nitride. R. Danzer and M. Lengauer, Engineering Failure Analysis 17, 596 (2010). With kind permission of Elsevier

Table 16.5 Material properties of the Si_3N_4 rolls at room temperature. R. Danzer and M. Lengauer, Engineering Failure Analysis 17, 596 (2010). With kind permission of Elsevier

Material properties	HC Starck N3214 measured	FCT FSN 10 measured	Values used for calculations
Density ρ (kg/m³)[a]	3.224	3.256	3.24
Hardness (Vickers, 5 kg) (GPa)[a]	15.1	15.1	–
Characteristic bending strength σ_{b0} (MPa)[a]	945	943	–
90%-confidence interval of σ_{b0} (MPa)[a]	925–965	921–966	–
Weibull modulus m[a]	16	14	–
90%-confidence interval for m[a]	12–19	11–17	–
Fracture toughness, IF-method[a]	7.1	6.4	6.5
K_{Ic} (MPa $\sqrt{m}$) SENB-S-method[a]	5.4	6.4	
Young's modulus E (GPa)[a]	293	311	305
Poisson's ratio μ[b]	0.28	0.28	0.28
Thermal conductivity λ (W/mK)[b]	25	30	25
Specific heat capacity c_p (J/kgK)[b]	1100	1100	1100
Thermal expansion coefficient α (K^{-1})[b]	3.4×10^{-6}	3.2×10^{-6}	3.3×10^{-6}

a

OWN measurements.

b

Data from producers data sheet.

interior of the rolls. Thus, it can be stated that for silicon nitride rolls the contribution of the temperature variation to the overall stress is considered harmless. Compare the situation in other ceramic material, say cemented carbide, which was used as the roller material. Thermal stresses scale with the product Young's modulus and CTE which is 4 times larger for silicon carbide than for silicon nitride and its strength is about half of that of silicon nitride. It means that in the carbide rolls the very large compressive stresses which occur at the heated surface area in the roll groove, may cause thermal fatigue. In fact, thermal fatige cracks were observed in the roll groove of thermally loaded cemented carbide.

Due to the deformation of the wire very high contact stresses occur, which is analysed in more detail in the FE-model. Results of the model calculations are shown in Fig. 16.19.

In summary significant tensile stresses occur due to contact pressure and friction. The behaviour of a circumferential surface crack at the highest loaded position in the roll groove is analyzed. For any time, the stress intensity factor can be determined. The result is shown in 16.20. In each of the stress intensity factor there is a maximum position at different points along the crack path and which depends on time. The maximum loading has to be described by the time envelope. It is interesting to note

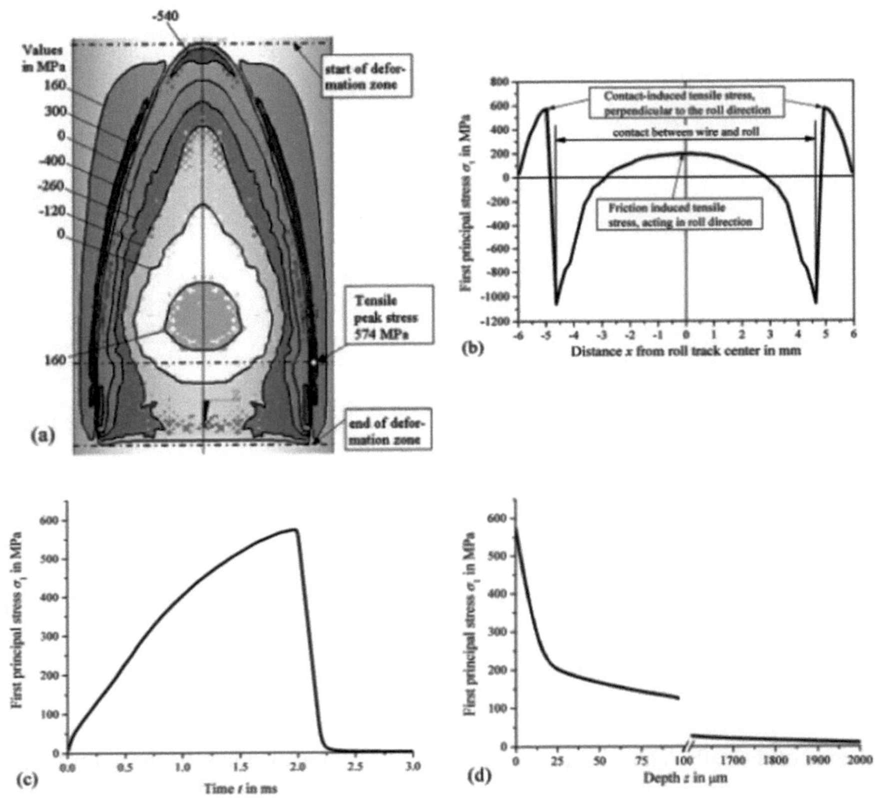

Fig. 16.19 First principal stresses in the rolls: **a** distribution of the stress on the surface of the roll groove on height of the tensile peak stress, **b** stresses along the dashed line shown in (a), **c** development of the first principal stress over the time at the position of the peak stress and **d** slope of the first principal stress perpendicular to the roll groove surface starting at the position of the peak stress. R. Danzer and M. Lengauer, Engineering Failure Analysis 17, 596 (2010). With kind permission of Elsevier

that each of the temporal stress intensity factors own a maximum and that in the analysed cases the maximum of the envelope is at a crack depth of about 350 μm. Also shown in Fig. 16.20 is the fracture toughness of the investigated silicon nitride ceramic. In 16.20b four regions are indicated describing the behavior of cracks which depends on their length. In region I, the stress intensity factor is smaller than the

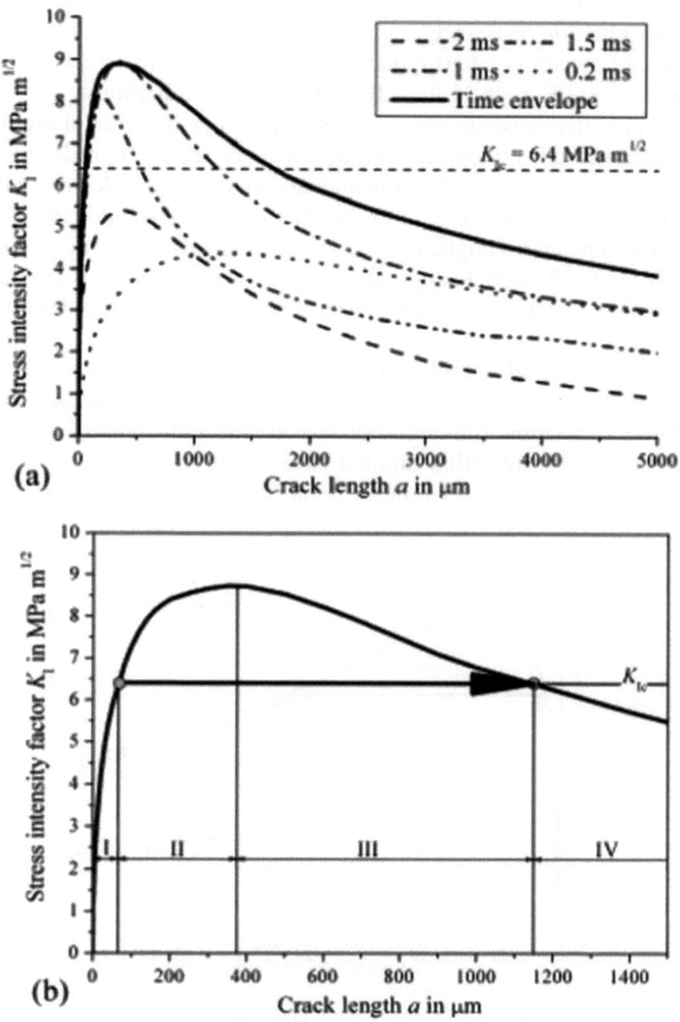

Fig. 16.20 Stress intensity factors of a circumferential crack in the roll groove in dependence of crack depth *a*. **a** The maximum loading along the crack path during the contact is given by the time–envelope curve. Also indicated are the fracture toughness of the silicon nitride rolls. **b** Detail from (a). R. Danzer and M. Lengauer, Engineering Failure Analysis 17, 596 (2010). With kind permission of Elsevier

fracture toughness at any time and the crack will not grow by brittle fracture. In region II the stress intensity factor increases with time and at a determined time it exceeds the fracture toughness, and the crack will spontaneously grow. If the stress intensity factor further increases with time, the crack stably extens up to the end of this region. Cracks in region IV do not extend by spontaneously crack growth. For a high enough crack depth—the stress intensity factor decreases with increasing crack depth. The stress intensity factor depends on the rolling conditions. It increases with increasing deformation resistance.

The course of the stress intensity factor of a circumferential crack (at peak stress position) is shown in Fig. 16.21 for rolling different wire qualities.

The crack growth rate versus the stress intensity range is shown in Fig. 16.22. For full experimental details on thermal fatigue effects on the Si_3N_4 roller performance and on other related aspects, all mathematical relations, and derived calculations under the specific performance of the roller, the interested should consult the work of Danzer and Lengauer. Fatigue crack growth mechanism in ceramics is a major issue of their work. FE model supported by experimental results is the basis of the expected and observed roller mill performance for hot working of high strength metal wires.

In summary, silicon nitride rolls can be used at the highest loaded positions in a wire rolling mill to process soft and hard metals and alloys. When rolling superalloys, the temperature in forming rolls made of silicon nitride fluctuates between RT and more than 900 °C. The resulting thermal stresses are predominantly compressive and therefore harmless. When rolling superalloys severe tensile stresses occur in

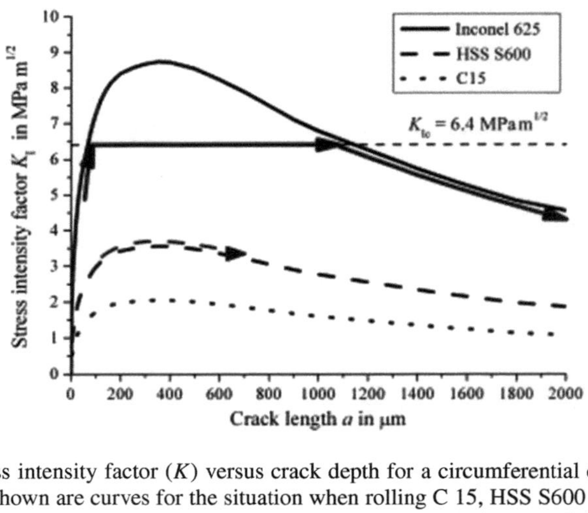

Fig. 16.21 Stress intensity factor (K) versus crack depth for a circumferential crack at the peak stress position. Shown are curves for the situation when rolling C 15, HSS S600 and Inconel 625, respectively. At peak stress position the minimum stress in a revolution is zero and the stress intensity factor equals to the range of the stress intensity factor ($K = \Delta K$). The crack extension by fatigue and spontaneously crack growth is indicated by arrows. R. Danzer and M. Lengauer, Engineering Failure Analysis 17, 596 (2010). With kind permission of Elsevier

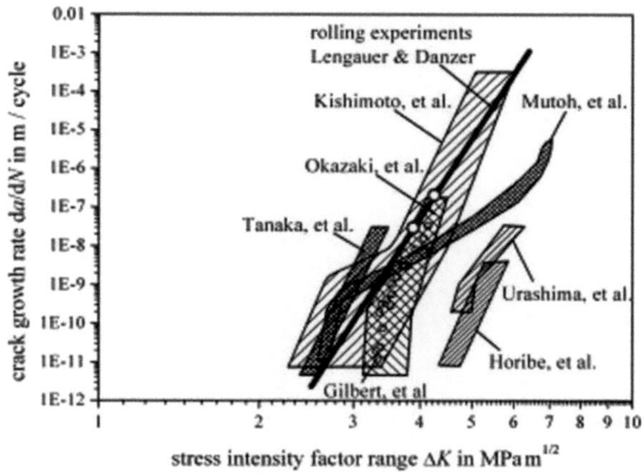

Fig. 16.22 Fatigue crack growth rate versus range of stress intensity factor for silicon nitride materials. The experiments are in the tensile dwell regime, i.e., $R = 0$–0.1. The thick straight line corresponds to the fatigue crack growth determined in rolling experiments. For a stress intensity factor of about 3 MPa$\sqrt{m}$ the crack growth rate is almost similar to the wear rate. Therefore, for smaller cracks fatigue is assumed to be insignificant. R. Danzer and M. Lengauer, Engineering Failure Analysis 17, 596 (2010). With kind permission of Elsevier

the roll groove of the rolls due to the contact pressure and by friction. The stress intensity factor of surface cracks has a maximum at a crack depth of about 350 μm. On application of high enough load deep cracks are found in the roll groove of silicon nitride rolls after rolling. The lifetime of the Si_3N_4 rolls suffering from wear is about twice of that of cemented carbide rolls. Improvements are possible by a roll design which reduces tensile stresses in the rolls, or by use of tougher and more fatigue resistant materials. The resistance to thermal fatigue cracking is increased with an increased tool life when cemented carbide is replaced by Si_3N_4. Silicon nitride guiding rolls are already routinely used for all processed materials.

References

1. K. Panda, T.S. Kannan, J. Dubois, C. Olagnon, G. Fantozzi, J. Eur. Cer. Soc. **22**, 2187 (2002)
2. H. Purwaningsih, L. Noerochim, R. Fajarain, J. A. Hakim and Sulistijono, IPTEK, J. Techn. Sci. **21**(4) (2010)
3. D. Kumar, K.N. Pandey, Int. J. Mech. Prod. Eng. **2**, 2320 (2014)
4. J.G. van der Laan, G. Schnedecker, E.V. van Osch, R. Duwe, J. Linke, J. Nuc. Mater. **211**, 135 (1994)
5. R. Danzer, M. Lengauer, Eng. Fail. Anal. **17**, 596 (2010)

References

Index

J. Pelleg, *Cyclic Deformation in Oxides, Carbides and Nitrides*, Structural Integrity 22,
https://doi.org/10.1007/978-3-030-86118-6

MIX
Papier aus verantwortungsvollen Quellen
Paper from responsible sources
FSC® C105338

If you have any concerns about our products,
you can contact us on
ProductSafety@springernature.com

In case Publisher is established outside the EU,
the EU authorized representative is:
Springer Nature Customer Service Center GmbH
Europaplatz 3, 69115 Heidelberg, Germany

Printed by Libri Plureos GmbH
in Hamburg, Germany